Lecture Notes in Physics

Edited by H. Araki, Kyoto, J. Ehlers, München, K. Hepp, Zürich
R. Kippenhahn, München, D. Ruelle, Bures-sur-Yvette
H. A. Weidenmüller, Heidelberg, J. Wess, Karlsruhe and J. Zittartz, Köln
Managing Editor: W. Beiglböck

371

K.W. Morton (Ed.)

Twelfth International Conference on Numerical Methods in Fluid Dynamics

Proceedings of the Conference
Held at the University of Oxford, England
on 9–13 July 1990

Springer-Verlag

Berlin Heidelberg New York London
Paris Tokyo Hong Kong Barcelona

Editor

K. W. Morton
ICFD, Numerical Analysis Group
Oxford University Computing Laboratory
11 Keble Road, Oxford OX1 3QD, England

ISBN 3-540-53619-1 Springer-Verlag Berlin Heidelberg New York
ISBN 0-387-53619-1 Springer-Verlag New York Berlin Heidelberg

This work is subject to copyright. All rights are reserved, whether the whole or part of the material is concerned, specifically the rights of translation, reprinting, re-use of illustrations, recitation, broadcasting, reproduction on microfilms or in other ways, and storage in data banks. Duplication of this publication or parts thereof is only permitted under the provisions of the German Copyright Law of September 9, 1965, in its current version, and a copyright fee must always be paid. Violations fall under the prosecution act of the German Copyright Law.

© Springer-Verlag Berlin Heidelberg 1990
Printed in Germany

Printing: Druckhaus Beltz, Hemsbach/Bergstr.
Bookbinding: J. Schäffer GmbH & Co. KG., Grünstadt
2153/3140-543210 – Printed on acid-free paper

EDITOR'S PREFACE

The present volume of Lecture Notes in Physics constitutes the Proceedings of the Twelfth International Conference on Numerical Methods in Fluid Dynamics, held at the University of Oxford, England on 9-13 July 1990. It has been arranged to reflect the layout of the conference, which gave particular emphasis to the six themes:

> Algorithm Development   Parallel Computing
> Hypersonic Flows   Transition and Turbulence
> Environmental Flows   Propulsion Systems.

Seven papers based on invited lectures appear first, in the order that they were given. Then all the contributed papers, which were presented in fourteen lecture sessions and twelve poster sessions, have been grouped under six topic headings. These are in the order that the topics first appeared on the conference programme and, under each heading, papers based on lectures appear first followed by those presented in poster/discussion sessions. An Author Index enables particular papers to be found quickly.

The 141 contributed papers that were presented at the conference were selected, from 224 abstracts submitted from all over the world, by four paper selection committees. As Conference Chairman I am particularly indebted to these committees and their chairmen: M Holt (covering USA), V Rusanov (USSR, Eastern Europe), R Temam (Western Europe, Canada, Israel) and K Oshima (South Asia, Pacific Rim).

The conference was attended by 261 people from 21 countries. Its success owes much to the efforts of the International Organising Committee, my associates M J Baines, T V Jones and L C Woods on the Local Organising Committee, and above all to Mrs Bette Byrne, the Conference Secretary, and our colleagues in the Oxford University Computing Laboratory. Thanks are also due to Bette Byrne for preparing the Proceedings and to Professor Dr W Beiglböck and the editorial staff of Springer-Verlag for their assistance in this task.

K W Morton
Oxford, September 1990

Acknowledgements

The Organising Committee wish to express their gratitude to the following organisations for their generous financial support for the Conference and this volume:

British Aerospace
BP Research
ICI Chemicals & Polymers
Oxford University Press
Rolls-Royce
Sun Microsystems
US Office of Naval Research
US Air Force Office of Scientific Research
The Royal Society, London

K W Morton
Oxford, September 1990

CONTENTS

INVITED LECTURES

B E LAUNDER: Turbulence Modelling for the Nineties: Second-Moment Closure ... and Beyond? 1

B L ROZHDESTVENSKY, M I STOYNOV: Simulation of Turbulent Flows by Numerical Integration of Navier-Stokes Equations 19

E KRAUSE: The Solution to the Problem of Vortex Breakdown 35

J-P VEUILLOT, L CAMBIER: Computation Techniques for the Simulation of Turbomachinery Compressible Flows 51

A N STANIFORTH, J CÔTÉ: Semi-Lagrangian Integration Schemes and their Application to Environmental Flows 63

T D TAYLOR: Computer Trends for CFD 80

T AKI: Computation of Shock-Shock Interference Around a Cylindrical Leading Edge at Hypersonic Speeds 90

CONTRIBUTED PAPER SESSIONS

Transition, Turbulence and Solution of Navier-Stokes Equations

T H PULLIAM, J A VASTANO: Chaotic Flow Over an Airfoil 106

K P DIMITRIADIS, M A LESCHZINER: Modelling Shock/Turbulent-Boundary-Layer Interaction with a Cell-Vertex Scheme and Second-Moment Closure 111

T DUBOIS, F JAUBERTEAU, R TEMAM: The Nonlinear Galerkin Method for the Two and Three Dimensional Navier-Stokes Equations 116

G ERLEBACHER, M Y HUSSAINI, C G SPEZIALE, T A ZANG: On the Large-Eddy Simulation of Compressible Isotropic Turbulence 121

Y NAKAMURA, J P WANG, M YASUHARA: A Calculation of the Compressible Navier-Stokes Equations in Generalized Coordinates by the Spectral Method 127

D KWAK, S E ROGERS, C KIRIS, S YOON: Numerical Methods for Viscous Incompressible Flows 132

G KURUVILA, M D SALAS: Three-Dimensional Simulation of Vortex
    Breakdown.. 137

S C R DENNIS, J D HUDSON: Accurate Finite-Difference Methods for
    Solving Navier-Stokes Problems Using Green's Identities................ 142

M SANCHEZ, N K MITRA, M FIEBIG: Influence of Longitudinal Vortices
    on the Structure of the Three Dimensional Wake of a Partially
    Enclosed Body 147

F J BRANDSMA, J G M KUERTEN: The ISNaS Compressible Navier-Stokes
    Solver; First Results for Single Airfoils................................. 152

T B GATSKI, E J KERSCHEN: Leading-Edge Effects on Boundary-Layer
    Receptivity.. 157

R ARINA, C CANUTO: The Viscous-Inviscid Coupling via a Self-Adaptive
    Domain Decomposition Technique 162

N-H CHO, C A J FLETCHER, K SRINIVAS: Efficient Computation of Wing
    Body Flows 167

H I ANDERSSON, J T BILLDAL, P ELIASSON, A RIZZI: Staggered and
    Non-Staggered Finite-Volume Methods for Nonsteady Viscous
    Flows: A Comparative Study 172

O M BELOTSERKOVSKII: Direct Numerical Modelling of Turbulence:
    Coherent Structures and Transition to Chaos 177

A KARAGEORGHIS, T N PHILLIPS: Conforming Chebyshev Spectral
    Collocation Methods for the Solution of the Incompressible
    Navier-Stokes Equations in Complex Geometries....................... 179

I N SIMAKIN, S E GRUBIN: Direct Simulation of Transition and Turbulent
    Boundary Layer 181

H TOKUNAGA, K ICHINOSE, N SATOFUKA: Numerical Simulation of
    Transition to Turbulence Using Higher Order Method of Lines 183

M FERRY, J PIQUET: A New Fully-Coupled Method for the Solution of
    Navier-Stokes Equations.. 186

R RADESPIEL, N KROLL: A Multigrid Scheme with Semicoarsening for
    Accurate Computations of Viscous Flows 188

G DANABASOGLU, S BIRINGEN, C L STREETT: Numerical Simulation of Spatially-Evolving Instability in Three-Dimensional Plane Channel Flow 190

D PRUETT, G ERLEBACHER, L NG: Secondary Instability and Transition in Compressible Boundary Layers 192

K B WINTERS, E A D'ASARO, J J RILEY: Three-Dimensional Internal Wave Breaking at a Critical Level 195

F F GRINSTEIN, J P BORIS: Numerical Simulation of Transitional Bluff-Body Near-Wake Shear Flows 197

V THEOFILIS, P W DUCK, D I A POLL: On the Stability of the Infinite Swept Attachment Line Boundary Layer 199

T KAJISHIMA, Y MIYAKE, T NISHIMOTO: Large Eddy Simulation of Turbulent Flow in a Duct of Square Cross Section 202

R I LEIGHTON, T F SWEAN JR: Direct Numerical Simulation of Turbulent Flow in an Open Channel 205

H L CHEN, K OSHIMA: Direct Numerical Simulation of a Complicated Transition-Breakdown of a Circular Pipe Flow through Axisymmetric Sudden Expansions 207

Y KATO, T SATO, T SAWADA, T TANAHASHI: Numerical Simulation of Unsteady Incompressible Viscous Flows by Improved GSMAC-FEM 209

P LE QUÉRÉ, A DULIEU, T PHUOC LOC: Investigation of 2-D and 3-D Instability of the Driven Cavity Flow 211

B TROFF, T H LÊ, T PHUOC LOC: Numerical Simulation of Three-Dimensional Unsteady Incompressible Viscous Flow for Separated Configurations 213

G A OSSWALD, K N GHIA, U GHIA: Flow Over Arbitrary Axisymmetric Multi-Body Configurations, Using Direct Navier-Stokes Simulations 215

Algorithm Development and Parallel Computing

A A SAMARSKY, A P FAVORSKY, V F TISHKIN, V F VASILEVSKY, K V VYAZNIKOV: Construction High Order Monotonous Difference Schemes on Nonregular Grids 219

H-C KU, A P ROSENBERG, T D TAYLOR: High-Order Time Integration Scheme and Preconditioned Residual Method for Solution of Incompressible Flow in Complex Geometries by the Pseudospectral Element Method 223

D J MAVRIPLIS: Turbulent Flow Calculations Using Unstructured and Adaptive Meshes 228

V VENKATAKRISHNAN, J H SALTZ, D J MAVRIPLIS: Parallel Preconditioned Iterative Methods for the Compressible Navier-Stokes Equations 233

P VAN RANSBEECK, C LACOR, C HIRSCH: A Multidimensional Cell-Centered Upwind Algorithm Based on a Diagonalization of the Euler Equations 238

P I CRUMPTON, J A MACKENZIE, K W MORTON, M A RUDGYARD, G J SHAW: Cell Vertex Multigrid Methods for the Compressible Navier-Stokes Equations 243

J Y YANG, C-A HSU, T H LEE: A Numerical Study of Third-Order Nonoscillatory Schemes for the Euler Equations 248

M H CARPENTER: A High-Order Compact Numerical Algorithm for Supersonic Flows 254

H C YEE, P K SWEBY, D F GRIFFITHS: A Study of Spurious Asymptotic Numerical Solutions of Nonlinear Differential Equations by the Nonlinear Dynamics Approach 259

M MEINKE, D HÄNEL: Simulation of Unsteady Flows 268

P L ROE, H DECONINCK, R J STRUIJS: Recent Progress in Multi-Dimensional Upwinding 273

J LAMINIE, F PASCAL, R TEMAM: Nonlinear Galerkin Method with Finite Element Approximation 278

H WU, K OSHIMA: Three-Dimensional Highly Accurate MmB Schemes for Viscous, Compressible Flow Problems 283

P O'BRIEN, M G HALL: A Comparison of the Effects of Grid Distortion on Finite-Volume Methods for Solving the Euler Equations 287

J K DUKOWICZ, B J A MELTZ: Vorticity Errors in Multidimensional Lagrangian Codes 291

J P BORIS, E S ORAN, F F GRINSTEIN, E F BROWN, C LI, R KOLBE, R O WHALEY: Three Dimensional Large Eddy Simulations with Realistic Boundary Conditions Performed on a Connection Machine..... 297

D TROMEUR-DERVOUT, L TA PHUOC, L MANE: A 3D Navier-Stokes Solver on Distributed Memory Multiprocessor.......................... 303

R W LELAND, J S ROLLETT: Parallel Extrapolation Methods for Computational Fluid Dynamics.. 308

L B WIGTON, R C SWANSON: Variable Coefficient Implicit Residual Smoothing.. 313

E S ORAN, J P BORIS, D A JONES: Reactive-Flow Computations on a Connection Machine... 318

S YOON: Navier-Stokes Solvers Using Lower-Upper Symmetric Gauss-Seidel Algorithm.. 323

J CASPER: An Extension of Essentially Non-Oscillatory Shock-Capturing Schemes to Multi-Dimensional Systems of Conservation Laws........... 325

S-H CHANG: On a Modified ENO Scheme and Its Application to Conservation Laws with Stiff Source Terms.............................. 327

N CLARKE, D M INGRAM, D M CAUSON, R SAUNDERS: Convenient Entropy-Satisfying TVD Schemes with Applications..................... 329

E DICK: Second Order Defect-Correction Multigrid Formulation of the Polynomial Flux-Difference Splitting Method for Steady Euler Equations... 333

B ENGQUIST, B SJÖGREEN: Projection Shock Capturing Algorithms...... 335

P Å WEINERFELT: Multigrid Technique Applied to the Euler Equations on a Multi-Block Mesh.. 337

A J FORESTIER: Treatment of Inert and Reactive Flows by a TVD Formulation in 2D Space in Unstructured Geometry..................... 339

K NAKAHASHI, K EGAMI: Unstructured Upwind Approach for Complex Flow Computations... 342

L STOLCIS, L J JOHNSTON: Implementation of a High Resolution TVD Scheme for Compressible Inviscid Flow on Unstructured Grids.......... 344

S T ZALESAK: Generalized Finite Volume (GFV) Schemes on Unstructured
    Meshes: Cell Averages, Point Values, and Finite Elements............... 346

J LAVERY: Solution of Steady-State Conservation Laws by l_1
    Minimization.. 348

C-C ROSSOW: Flux Balance Splitting — A New Approach for a Cell Vertex
    Upwind Scheme... 350

R STRUIJS, H DECONINCK, P DE PALMA: Multidimensional Upwind
    Schemes Using Fluctuation Splitting and Different Wave Models for the
    Euler Equations... 352

R M S M SCHULKES: Interactions of a Viscous Fluid with an Elastic
    Solid: Eigenmode Analysis... 355

R C HALL, D J DOORLY: Parallel Solution Procedures for Transonic
    Flow Computations 357

L A BLINOVA, Yu E EGOROV, A E KUZNETSOV, M A ROTINIJAN,
    M Kh STRELETS, M L SHUR: Compressibility Scaling Method
    for Multidimensional Arbitrary Mach Number Navier-Stokes
    Calculations of Reactive Flows.................................... 359

C K LOMBARD, S K HONG, J BARDINA, J OLIGER, S SUHR,
    J CAPORALETTI, W H CODDING, D WANG: Simple, Efficient
    Parallel Computing with Asynchronous Implicit CFD Algorithms on
    Multiple Grid Domain Decompositions............................... 361

D W FOUTCH, D R McCARTHY: Validation of an Upwind Euler Solver 363

J SHI, D HITCHINGS: Finite Element Analysis of Airfoil Immersed
    in a Sinusoidal Gust.. 365

S A E G FALLE: Flame Capturing 367

C BASDEVANT, M HOLSCHNEIDER, J LIANDRAT, V PERRIER,
    Ph TCHAMITCHIAN: Numerical Resolution of the Burgers Equation
    Using the Wavelet Transform....................................... 369

B LI, M G SMITH, U T EHRENMARK, P S WILLIAMS: Application of
    Boundary Element Method to a Shallow Water Model 371

M O BRISTEAU, R GLOWINSKI, J PÉRIAUX, M RAVACHOL, Y XIANG:
    Stabilization of the Navier-Stokes Solutions via Control Techniques...... 373

N DÉBIT, Y MADAY: Approximation of the Stokes Problem with a Coupled Spectral and Finite Element Method.................................... 375

P M GOORJIAN, S OBAYASHI, G P GURUSWAMY: Streamwise Upwind Algorithm Development for the Navier-Stokes Equations................ 377

A DADONE, B GROSSMAN: A Domain of Dependence Upwind Scheme for the Euler Equations.. 379

B FAVINI, M DI GIACINTO: Numerical Approximation by Godunov-Type Schemes of Shocks and Other Waves.................................... 381

F E CANNIZZARO, A ELMILIGUI, N D MELSON, E VON LAVANTE: A Multiblock Multigrid Three-Dimensional Euler Equation Solver....... 385

F GRASSO, F BASSI, A HARTEN: ENO Schemes for Viscous High Speed Flows.. 387

Transonic, Supersonic and Hypersonic Flows

M HAFEZ, J AHMAD: Design of Airfoils in Transonic Flows................ 389

P M HARTWICH: Split Coefficient Matrix (SCM) Method with Floating Shock Fitting for Transonic Airfoils.................................... 394

P K KHOSLA, T E LIANG, S G RUBIN: Supersonic Viscous Flow Calculations for Axisymmetric Configurations.......................... 400

Y TAMURA, K FUJII: Visualization Method for Computational Fluid Dynamics with Emphasis on the Comparison with Experiments......... 406

C MARMIGNON, H HOLLANDERS, F COQUEL: Navier-Stokes Calculations of Hypersonic Flow Configurations with Large Separation by an Implicit Non-Centered Method 411

S BORRELLI, M PANDOLFI: An Upwind Formulation for the Numerical Prediction of Non-Equilibrium Hypersonic Flows....................... 416

F G BLOTTNER: Navier-Stokes Results for the Hypersonic Blunt Body Problem with Equilibrium Air Properties and with Shock Fitting........ 421

P LARDY, H DECONINCK: A Polynomial Flux Vector Splitting Applied to Viscous Hypersonic Flow Computation............................. 426

M PFITZNER: Simulations of Inviscid Equilibrium and Nonequilibrium Hypersonic Flows... 432

N SATOFUKA, K MORINISHI, Y SAKAGUCHI: Numerical Solutions of the Boltzmann Equation for Hypersonic Flows 437

N KROLL, C ROSSOW: A High Resolution Cell Vertex TVD Scheme for the Solution of the Two and Three Dimensional Euler Equations 442

C-H BRUNEAU, J LAMINIE: Computation of Hypersonic Flows Around Blunt Bodies and Wings 447

K D JONES, F C DOUGHERTY, H SOBIECZKY: Hypersonic Flows About Waveriders with Sharp Leading Edges 449

B MÜLLER: Upwind Relaxation Method for Hypersonic Flow Simulation 451

N QIN, B E RICHARDS: PNS Solution Using Sparse Quasi-Newton Method for Fast Convergence 453

M L SAWLEY, S WÜTHRICH: Numerical Simulation of Non-Equilibrium Hypersonic Flow Using the Second-Order Boundary Layer Equations 455

J B VOS, C M BERGMAN: Chemical Equilibrium and Non-Equilibrium Inviscid Flow Computations Using a Centered Scheme 457

S ASO, M HAYASHI: Numerical Simulations of Separated Flows Around Bluff Bodies and Wings by a Discrete Vortex Method Combined with Panel Method 459

M G MACARAEG, Q I DAUDPOTA: Numerical Studies of Transverse Curvature Effects on Compressible Flow Stability 461

L I TURCHAK: Conservative Schemes for Supersonic Flows with Internal Shocks 463

Propulsion Systems

T L JIANG: The Flow Computation of a Liquid Rocket Engine Combustor of Complex Geometry 464

K NAITOH, K KUWAHARA, E KRAUSE: Computation of the Transition to Turbulence and Flame Propagation in a Piston Engine 469

S VENKATESWARAN, C L MERKLE: Coupled Navier-Stokes Maxwell Solutions for Microwave Propulsion 475

J FOŘT, M HUNĚK, K KOZEL, M VAVŘINCOVÁ: Comparison of Several Numerical Methods for Computation of Transonic Flows Through a 2D Cascade 480

S YAMAMOTO, H DAIGUJI: Numerical Simulation of Unsteady Turbulent Flow through Transonic and Supersonic Cascades 485

G FERNANDEZ, K Z TANG, H A DWYER: A Comparative Study of the Low Mach Number and Navier-Stokes Equations with Time-Dependent Chemical Reactions 490

S LECHELER, H-H FRUEHAUF: An Accurate and Fast 3-D Euler-Solver for Turbomachinery Flow Calculation 495

Z BAR-DEROMA, M WOLFSHTEIN: Computation of the Flow in Elliptical Ducts 497

Environmental Flows

M H MAWSON, M J P CULLEN: A Fully Implicit Method for Solving the 3-Dimensional Quasi-Equilibrium Equations on the Sphere 499

S H MIGORSKI, R F SCHAEFER: The Existence Aspects of Dupuit and Boussinesq Filtration Models Using Finite Element Method 504

G L BROWNING, W R HOLLAND, S J WORLEY, H-O KREISS: An Accurate Hyperbolic System for Approximately Hydrostatic and Incompressible Oceanographic Flows 509

A F GHONIEM, H R BAUM, R G REHM: Vortex Simulation of Particulate Plume Dispersal and Settling 514

K D NGUYEN: A 3D Numerical Diagnostic Simulation for the General Circulation in the Gulf of Lion (France) 516

Mesh Generation and Adaption

R TILCH: Unstructured Grids, Adaptive Remeshing and Mesh Generation for Navier-Stokes 519

M J BOCKELIE, P R EISEMAN, R E SMITH: A General Purpose Time Accurate Adaptive Grid Method 524

S CHIBA, K KUWAHARA: A Finite Difference Formulation for Free Surface Flow Problems Using a Free Surface Conforming Grid System 529

N MAMAN, B LARROUTUROU: On the Use of Dynamical Finite-Element Mesh Adaptation for 2D Simulation of Unsteady Flame Propagation.. 534

B PALMERIO, A DERVIEUX: On Weak and Strong Coupling Between Mesh Adaptors and Flow Solvers 540

H TAKEDA, C-Z HSU: A Finite-Difference Method for Incompressible Flows Using a Multi-Block Technique.................................... 545

S L KABALKIN: Effective Method for Calculation of Gas Flow About Blunt Bodies in a New Approach to Constructing Three-Dimensional Grids.... 550

J J QUIRK: An Adaptive, Embedded Mesh Refinement Algorithm for the Euler Equations.. 552

K SAWADA: Numerical Simulation of Flows over Complete Aircraft Using Block Structured Grid Systems.. 554

R LÖHNER, J D BAUM: Further Algorithmic Improvements of Adaptive H-Refinement for 3-D Transient Problems............................... 556

AUTHOR INDEX.. 559

Turbulence Modelling for the Nineties: Second Moment Closure ...... and Beyond?

Brian E. Launder
UMIST, Manchester, UK

1 Introduction

If sheer numbers of publications are any guide, no-one would guess, from reading the literature of the Eighties on numerical fluid mechanics, that the vast majority of flows of engineering or environmental interest are turbulent. Technical journals and conference proceedings concerned with numerical methods in fluid mechanics abound with papers on laminar - even inviscid - flow problems yet offer only an occasional genuflexion in the direction of turbulent flow. This paradoxical situation partly reflects the fact that the Reynolds-averaged Navier Stokes equations are seen as too complicated to form a practicable starting point for most flows of engineering interest and partly to a natural inclination among the academic community to tackle problems that are soluble rather than ones that, for all their importance, may not be.

The last few years have seen a remarkable advance in the power of CFD methods, however, springing both from improved algorithms (e.g. multi-grid schemes) and improved computer hardware. The next few years will surely be as exciting as any as algorithms and software emerge that will efficiently exploit the potential of parallel computers. These developments will have an especially large impact on turbulent flow for they bring a vastly greater number of industrially interesting problems within reach. Research sponsors will thus increasingly aim to support work in turbulent flows and this, in time, is likely to mean that a substantially greater proportion of the CFD community than at present will be swimming in the turbulent flow pool. A major question is thus how the newcomers will get in: down the steps at the shallow end or by making a big splash at the deep end? To clarify my meaning, a relatively painless way to advance from laminar-flow computations to turbulent is to represent the Reynolds stresses via an eddy viscosity. This may be initially obtained by an algebraic model (a *mean-field* closure) before introducing transport equations first for the turbulence energy (k) and then for a second scalar parameter of the turbulence such as the rate of dissipation of k, denoted ε . This step-by-step route would have much to commend it were it not for the fact that the turbulent stress tensor is not well described by an eddy viscosity model. True, for simple two-dimensional shear flows close to local equilibrium such approaches *do* give a reasonably accurate estimate of the turbulent shear stress; but these are not the flow problems requiring resolution by CFD in the Nineties. One needs to handle three-dimensional flows with recirculation and strong streamline curvature.

If one had to pick an animal whose character most resembled turbulence my choice would be an irritable, unpredictable old goat, liable to react violently when prodded. Modelling turbulence by an eddy viscosity scheme is roughly like replacing it by an inoffensive, docile sheep. For the Nineties one needs a model that retains something of the old goat in turbulence. One class of closure that does that - indeed, the simplest class - is second-moment closure; it is the application and refinement of this type of model that forms the focus of this paper. The sections that follow will attempt to provide convincing evidence that models of this type do indeed capture the quirky behaviour that turbulence so often exhibits when removed from an equilibrium state of simple shear.

Even if he or she accepts the arguments provided, however, the numerical analyst un-used to models of this type may conclude that they are too complicated in concept and that the greater stiffness they bring to the mean-flow transport equations may render the equation set unworkably difficult to solve. It is true that the rather loose coupling between the mean-strain and stress fields does require modifications to the algorithmic practices used to handle turbulent flows with an eddy viscosity model. Remedial strategies are well known however (e.g. ref [1-4]) ; moreover, the paper by Dimitriadis and Leschziner [5] at this conference provides a detailed consideration of the application of second-moment closure to a separated transonic flow. Numerical aspects will thus not be considered further in the present paper. A further limitation that space constraints demand is that discussion will be limited to incompressible flows and, save for a few general remarks, to the 'dynamic' second moments i.e. the Reynolds stresses. Less narrowly drawn or differently focused recent articles on second-moment closure have been provided in ref [6-8]. Suffice it to say here that the principles and practices involved in modelling the Reynolds stresses described below carry over intact to the treatment of turbulent heat or mass fluxes.

Section 2 below introduces the Reynolds stress transport equations, considers briefly their response characteristics and presents the very simple closure — here termed the *Basic Model* — that has been used in the great majority of the engineering applications of second-moment closures reported to date. Sections 3 and 4 consider more elaborate approaches, the former strictly within the framework of a single-time-scale second moment closure, the latter with more varied approaches. Finally Section 5 draws a few overall conclusions.

2 The Reynolds Stress Equations and the Basic Model

An exact equation for the transport of the kinematic Reynolds stresses $\overline{u_i u_j}$ (which render the averaged Navier Stokes equations unclosed) is obtained by multiplying the instantaneous x_i momentum equation by the fluctuating velocity u_j, adding it to the mirror equation with indices i and j interchanged and averaging. The resultant equation is shown in Fig 1 which gives both the physical groupings forming the right hand side of the equation and the symbolic forms by which it will be convenient to denote them. Here upper-case U's denote mean velocities, β_i is the product of the gravitational acceleration vector and the volumetric expansion coefficient, Ω_k is the angular velocity of the coordinate system and p and θ denote fluctuating pressures and temperatures.

$$\underbrace{\frac{\partial \overline{u_i u_j}}{\partial t} + U_k \frac{\partial \overline{u_i u_j}}{\partial x_k}}_{C_{ij}} = \underbrace{-\left(\overline{u_i u_k}\frac{\partial U_j}{\partial x_k} + \overline{u_j u_k}\frac{\partial U_i}{\partial x_k}\right)}_{P_{ij}}$$

$$\underbrace{-2\Omega_k\left(\overline{u_j u_m}\varepsilon_{ikm} + \overline{u_i u_m}\varepsilon_{jkm}\right)}_{F_{ij}}$$

$$\underbrace{-\left(\overline{u_j \theta}\beta_i + \overline{u_i \theta}\beta_j\right)}_{G_{ij}}$$

$$\underbrace{+\overline{\frac{p}{\rho}\left(\frac{\partial u_i}{\partial x_j} + \frac{\partial u_j}{\partial x_i}\right)}}_{\phi_{ij}} \underbrace{- 2\nu \overline{\frac{\partial u_i}{\partial x_k}\frac{\partial u_j}{\partial x_k}}}_{-\varepsilon_{ij}}$$

$$\underbrace{-\frac{\partial}{\partial x_k}\left(\overline{u_k u_i u_j} + \overline{\frac{p}{\rho}u_i}\delta_{jk} + \overline{\frac{p}{\rho}u_j}\delta_{ik}\right)}_{d_{ij}}$$

Fig 1. Exact Reynolds stress transport equations for weakly non-uniform density flow at high Reynolds number

The most important advantage of second-moment closure over any simpler approach is that the generation terms, whether they be associated with mean strain (P_{ij}), coordinate rotation (Coriolis) effects (F_{ij}) or body forces (G_{ij}) comprise only second-moment and mean-flow quantities: they can thus be handled *exactly* without modelling approximations. The vastly greater potential that this brings to the model can be illustrated by a single example. Weak streamline curvature has long been known [9] to exert a disproportionately large effect on turbulence. Why? The answer - or much of it - is to be found in the shear production tensor. Take axes x_1 and x_2 aligned with the flow direction and velocity gradient in an initially parallel near-wall shear flow; the imposition of a section of concavely curved wall introduces a secondary strain $\partial U_2/\partial x_1$ of positive sign. The production rate of the turbulent shear stress $\overline{u_1 u_2}$ (which is the stress component primarily affecting the mean flow) is:

$$P_{12} = -\overline{u_2^2}\frac{\partial U_1}{\partial x_2} - \overline{u_1^2}\frac{\partial U_2}{\partial x_1}$$

As the wall is approached it is well known that streamwise fluctuations become increasingly larger than those normal to the wall (the ratio being about 4:1 in the near-wall region where the velocity increases as the logarithm of the distance from the wall); thus the coefficient multiplying the secondary strain is markedly larger than that for the primary strain. There is a secondary effect: the primary shear does not contribute to the production of $\overline{u_2^2}$ but from the secondary shear there arises:

$$P_{22} = -2\overline{u_1 u_2}\frac{\partial U_2}{\partial x_1}$$

The shear stress $\overline{u_1 u_2}$ will be negative in the indicated coordinates (i.e. $\overline{u_1 u_2}$ takes the same sign as P_{12}) so the secondary strain tends to raise the level of $\overline{u_2^2}$ which in turn will augment P_{12}. Through this intercoupling the shear-stress production is an order of magnitude more sensitive to the secondary than the primary strain.

Numerous other paradoxical features of turbulent flow fall into place by considering production agencies in the second-moment equations. For example, why, in a heated turbulent boundary layer, twice as much heat flows in the stream direction by turbulent mixing as *across* the boundary layer despite temperature gradients in the latter direction being typically two orders of magnitude larger than in the former. Or yet again, why temperature fluctuations in a rising vertical plume (i.e. where the vertical motion is induced by the plume's density being less than the surroundings) are twice as large as in an equivalent heated jet driven by forced convection. These results can be seen to spring directly from the form of the generation rate of heat flux $\overline{u_i \theta}$, $P_{i\theta}$:

$$P_{i\theta} \equiv -\overline{u_i u_j}\frac{\partial \Theta}{\partial x_j} - \overline{u_j \theta}\frac{\partial U_i}{\partial x_j} - \beta_i \overline{\theta^2} \tag{1}$$

The first two terms of eq (1) ensure that, in a shear flow, heat fluxes are generated in directions other than that of the mean temperature gradient; indeed, the second can only be important in the mean-flow direction. The third term in (1), arising from gravitational effects, is what provokes increased temperature fluctuation levels in the rising plume

$$\varepsilon_{ij} = \frac{2}{3}\delta_{ij}\varepsilon \quad ; \quad \frac{\partial \varepsilon}{\partial t} + U_k \frac{\partial \varepsilon}{\partial x_k} = d_\varepsilon + \frac{1}{2}c_{\varepsilon 1}(P_{kk} + G_{kk})\frac{\varepsilon}{k} - c_{\varepsilon 2}\frac{\varepsilon^2}{k}$$

$$\phi_{ij} = \phi_{ij1} + \phi_{ij2} + \phi_{ij3} + (\phi_{ij}^w)$$

$$\phi_{ij1} = -c_1 \frac{\varepsilon}{k}\left[\overline{u_i u_j} - \frac{1}{3}\delta_{ij}\overline{u_k u_k}\right]$$

$$\phi_{ij2} = -c_2\left[P_{ij} - C_{ij} + F_{ij} - \frac{1}{3}\delta_{ij}(P_{kk} - C_{kk})\right]$$

$$\phi_{ij3} = -c_3\left[G_{ij} - \frac{1}{3}\delta_{ij}G_{kk}\right]$$

$$d_{ij} = \frac{\partial}{\partial x_k}\left[c_s \frac{k}{\varepsilon}\overline{u_k u_\ell}\frac{\partial \overline{u_i u_j}}{\partial x_\ell}\right] \quad ; \quad d_\varepsilon = \frac{\partial}{\partial x_k}\left[c_\varepsilon \frac{k}{\varepsilon}\overline{u_k u_\ell}\frac{\partial \varepsilon}{\partial x_\ell}\right]$$

Wall Flows Only

$$\phi_{ij}^w = \left\{ c_1' \frac{\varepsilon}{k}\left[\overline{u_k u_m}n_k n_m \delta_{ij} - \frac{3}{2}\overline{u_i u_k}n_k n_j - \frac{3}{2}\overline{u_k u_j}n_k n_i\right]\right.$$
$$+ \, c_2'\left[\phi_{km2}n_k n_m \delta_{ij} - \frac{3}{2}\phi_{ik2}n_k n_j - \frac{3}{2}\phi_{kj2}n_k n_i\right]$$
$$+ \left. c_3'\left[\phi_{km3}n_k n_m \delta_{ij} - \frac{3}{2}\phi_{ik3}n_k n_j - \frac{3}{2}\phi_{kj3}n_k n_i\right]\right\} \frac{k^{3/2}}{c_\ell \varepsilon x_n}$$

c_1	c_2	c_3	c_s	c_ε	$c_{\varepsilon 1}$	$c_{\varepsilon 2}$	c_1'	c_2'	c_3'	c_ℓ
1.8	0.6	0.5	0.22	0.18	1.44	1.92	0.5	0.3	0	2.5

Fig 2. The Basic Second-Moment Closure for High-Re Flows

[10]. This brief excursion into heat-transport processes has perhaps served to show that second-moment closure has as important a role to play there as in the prediction of the Reynolds stress field to which attention is now limited.

While the generation terms appearing above the line in Fig 1 can be handled directly, the other processes require approximation in terms of known or knowable quantities. It is usually supposed that the fine-scale interactions responsible for viscous dissipation are isotropic thus allowing ε_{ij} to be represented as indicated in Fig 2, ε being the energy dissipation rate of turbulence energy. In the simplest and most widely used form of second-moment closure, ε itself is obtained from essentially the same empirical transport equation adopted in the k-ε eddy viscosity model as shown in Figure 2.

Concerning the pressure-strain process, ϕ_{ij}, it is readily shown that local pressure fluctuations are directly affected by both fluctuating and mean-velocity gradients and, indeed, by fluctuating force fields too. Consequently most workers nowadays include separate models of these different influences. The simple model presented in Fig 2 has, in all but one respect, been in use for the past fifteen years. It incorporates Rotta's original return-to-isotropy model (ϕ_{ij1}) together with the isotropization of production concept [12, 13] applied to mean-strain and buoyant influences. The mean-strain process (ϕ_{ij2}) is modified from the original proposal, however, by the inclusion of the convective transport tensor as well as P_{ij}. The reason for doing so is that the tensor P_{ij} is not independent of the observer's frame of reference. The problem is not eliminated by regarding the Coriolis terms associated with a rotating reference frame as part of the production. Addition of the tensor C_{ij} in the manner indicated in Fig 2 does, however, produce a model of ϕ_{ij2} that renders the resultant behaviour independent of the observer — as, of course, it should be [14]. The addition is usually of little importance except in swirling flows where its influence can be of decisive benefit.

In near-wall flows an addition to ϕ_{ij} is required supposedly to account for the reflection of pressure fluctuations from the rigid boundary that impedes the transfer of fluctuating energy from the streamwise direction to that normal to the wall (the 'echo' effect). The indicated correction, ϕ_{ij}^w (in which n_i denotes the unit vector normal to the wall and x_n the normal distance to the wall), is specifically designed for use with the free-flow version of ϕ_{ij} given in Fig 2, [15]. Although wall-reflection is certainly an important contributor, it seems that the model of ϕ_{ij}^w is also trying to account for the effects of the very steep inhomogeneities that arise near a no-slip boundary. Because the effects of inhomogeneities die out more rapidly as the distance from the wall increases than do those of wall reflection (and indeed arise from different parameters) the modelled form of ϕ_{ij}^w is inevitably a compromise. In internal flows, for example, it is sometimes found that better agreement is achieved if the factor $k^{3/2}/c_\ell \varepsilon x_n$, the ratio of the turbulent length scale to normal distance to the wall, is raised to the power two rather than unity [16].

In most common flows the modelling of stress (or ε) diffusion processes is not vitally important and, in the interests of simplicity and stability, the simple generalized gradient diffusion hypothesis [17] indicated in Fig 2 is usually adopted.

The closure outlined above has, over the past 15 years, been applied to at least 50 different turbulent shear flows with some twenty of these relating to complex elliptic or 3-dimensional flows. The early comparisons, limited to simple shear flows, displayed only

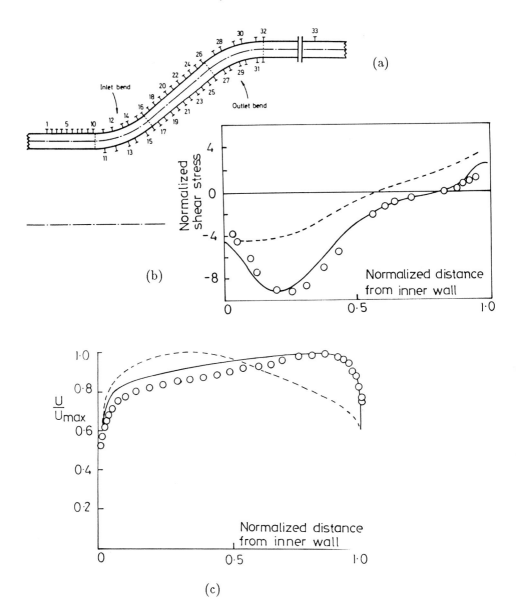

Fig 3. Boundary-layer development in annular diffuser, after Jones and Manners [18]

(a) Diffuser configuration
(b) Shear stress profiles at Station 16
(c) Mean velocity profiles at Station 32
Symbols: experiments [19]; Lines: computations [18]; broken: $k - \varepsilon$ EVM; continuous: Basic Model

insubstantial advantages over the k-ε eddy viscosity model (EVM). As attention shifted to progressively more complex flows involving streamline curvature and other complex strain or force fields, however, decisive benefits started to appear from second-moment modelling. An impression of these benefits is provided by the two illustrative test cases considered in Figures 3 and 4. The first relates to the faired annular diffuser, Fig 3a, computed by Jones and Manners [18]. Such diffusing passages carry air from the gas-turbine compressor outlet to the combustion chamber inlet. An important feature to be able to predict is how the velocity at exit differs from a symmetric pattern. The flow field is in fact dominated by the flow's passage around the inlet bend. In line with our earlier discussion, the strong streamline curvature strongly damps the shear stress near the convex outer wall and augments it on the concave inner one leading to the highly asymmetric profile in Fig 3b. While the second-moment closure captures this asymmetry with good accuracy, the k-ε EVM gives a shear-stress distribution similar to that in a straight annulus. Due to the adverse pressure gradient through the diffuser the large differences between the measured stresses on the inner and outer walls just downstream of the bend show little signs of diminishing as the flow develops through the diffuser. As a result, the measured mean velocity profile at exit from the diffuser shows a marked maximum towards the inside of the passage whereas the EVM predictions show a maximum on the outside. One might well conclude that here the EVM predictions are worse than no predictions at all. The results from the second-moment closure in contrast give a satisfactory imitation of the actual flow distribution.

The second example, taken from [20] relates to the rotating plane channel flow shown in Fig 4a. (In the figure Cartesian coordinates are adopted with $x \equiv x_1$ denoting the flow direction, etc). In this case, because the Coriolis forces make no contribution to the turbulence energy equation, the k-ε EVM model would predict the same symmetric mean flow distribution found in a non-rotating channel flow. The real flow is very sensitive to rotation, however, as is clear from the mean velocity profile in Fig 4b for an inverse Rossby number $\Omega D/\overline{U}$ of 0.21. The sensitivity arises from the Coriolis generation terms F_{ij} which introduce sources of $+4\overline{u_1 u_2}\Omega$ and $-4\overline{u_1 u_2}\Omega$ into the equations for the normal stresses in the directions of the mean flow and normal to the duct walls respectively. There is also a source $-2(\overline{u_1^2} - \overline{u_2^2})\Omega$ in the shear stress equation. Because $\overline{u_1 u_2}$ is of different signs on each side of the channel the Coriolis terms tend to augment $\overline{u_2^2}$ and $|\overline{u_1 u_2}|$ on the pressure side and diminish them on the suction side. Fig 4c compares the predicted profile of $\sqrt{\overline{u_2^2}}$ for a rotating and non-rotating case with the large-eddy simulations of Kim [21]; it is clear that the second-moment closure does broadly capture the changes produced in the turbulent stress field by the rotation and the resultant modifications to the mean velocity field, Fig 4b.

3 Some Recent Developments in Second-Moment Closure

Despite the continuing success being achieved with the Basic Model described in Section 2 it should be regarded as the base-camp rather than the final summit of second-moment closure. Its weaknesses are not hard to uncover; in particular (as we shall examine more closely a little later) it is particularly unreliable in predicting free turbulent shear flows.

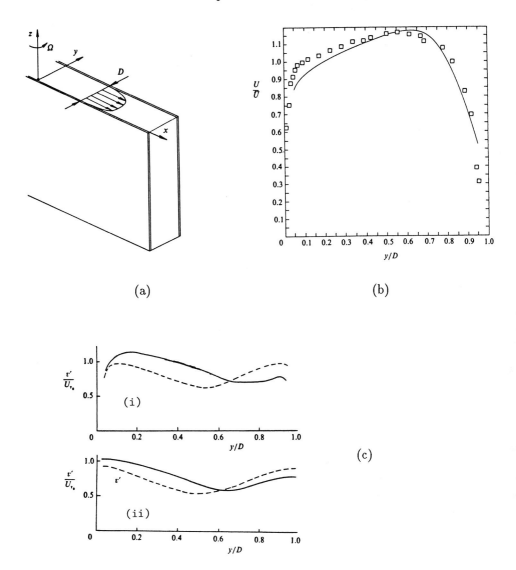

Fig 4. Flow in Plane Channel Rotating in Orthogonal Mode [20]

a) Flow configuration
b) Mean velocity profile for $\Omega D/\overline{U} = 0.21$
Symbols: Experiment; Line: Basic Model
c) Rms velocity fluctuations normal to wall: (i) LES [21]; (ii) Basic Model; Broken line: $\Omega = 0$; Continuous line: $\Omega D/\overline{U} = 0.07$

This relative frailty should scarcely surprise us, however, as the models of both ϕ_{ij} and ε are based mainly on intuition - and are as simple in form as one can imagine.

Here we outline a more analytical and capacious approach to closure. Its origins perhaps spring from the work of Schumann [22] and continue through that of Lumley and colleagues [23-25], Reynolds [26], Lecointe et al [27], Speziale [28] and the writer and his associates [14, 29-31]. Schumann's [22] contribution was to suggest that models for any process should be devised so that they were incapable of generating impossible values for the turbulence variables, for example negative normal stresses or cross-correlations violating Scharz's inequality. This requirement that the model should be *"realizable"* has fairly close parallels with the practice of devising discretization schemes for convective transport that are accurate but bounded. In practice realizability in turbulence modelling is achieved by arranging that the model should display *compliance with extreme states of turbulence, CEST*. As an illustration, immediately before the stress field reaches an unrealizable state (a negative value of $\overline{u_2^2}$, say) it must have been at an *extreme but just possible* state (a value of $\overline{u_2^2}$ equal to zero). Thus if the model is designed so that as it reaches the limiting state, the models of the different processes exactly comply with any requirements that the *exact* forms may impose, the stress-transport equation will never lead to unrealizable values of the stresses being returned.

Perhaps the most important of the "extreme states" to satisfy is two-component turbulence, i.e. where turbulent fluctuations lie entirely in a plane. This is the limit to which wall turbulence reduces as the wall is approached. Considering the budget of the Reynolds normal stress perpendicular to this plane (x_2 say), we readily conclude that at the wall $\phi_{22} = 0$. If we compare this requirement with the limiting value given by the Basic Model of ϕ_{ij1} in Fig 2, it is readily verified that that model causes ϕ_{221} to take its *maximum* value in this limit.

The usual route to removing this problem is for c_1 to become a function of some parameter that vanishes in the limit of two-component turbulence. It is here appropriate to introduce the two non-zero invariants of the Reynolds stress tensor, here denoted A_2 and A_3:

$$A_2 \equiv a_{ij}a_{ji} \; ; \; A_3 \equiv a_{ij}a_{jk}a_{ki}$$

where

$$a_{ij} \equiv (\overline{u_i u_j} - \frac{1}{3}\delta_{ij}\overline{u_k u_k})/k$$

Lumley [23] was the first to note that in the two-component limit A_2 always exceeds A_3 by 8/9. Thus we may define a *"flatness"* factor

$$A \equiv 1 - \frac{9}{8}(A_2 - A_3)$$

that vanishes in that limiting case. By letting c_1 in Figure 2 depend on A raised to a positive power one can thus ensure appropriate vanishing of ϕ_{ij1}, [24, 30]. Reynolds [26] shows that alternatively the inclusion of non-linear elements also allows one to make the model consistent with two-component turbulence. The form employed extensively at UMIST in free-shear flow calculations given in Fig 5, adopts both A and non-linear terms.

$$\phi_{ij1} = -c_1\varepsilon\left(a_{ij} + c'_1\left(a_{ik}a_{kj} - \frac{1}{3}\delta_{ij}A_2\right)\right) - \varepsilon a_{ij}$$

$$\phi_{ij2} = -0.6\left[P_{ij} + \frac{1}{2}F_{ij} - \frac{1}{3}\delta_{ij}P_{kk}\right] + 0.3\varepsilon a_{ij}(P_{kk}/\varepsilon)$$

$$- 0.2\left\{\frac{\overline{u_k u_j}\,\overline{u_\ell u_i}}{k}\left[\frac{\partial U_k}{\partial x_\ell} + \frac{\partial U_\ell}{\partial x_k}\right] - \frac{\overline{u_\ell u_k}}{k}\left[\overline{u_i u_k}\left(\frac{\partial U_j}{\partial x_\ell} + \varepsilon_{jm\ell}\Omega_m\right)\right.\right.$$

$$\left.\left. + \overline{u_j u_k}\left(\frac{\partial U_i}{\partial x_\ell} + \varepsilon_{im\ell}\Omega_m\right)\right]\right\}$$

$$- r\left[A_2(P_{ij} + F_{ij} - D_{ij}) + 3a_{mi}a_{nj}(P_{mn} + F_{mn} - D_{mn})\right]$$

$$D_{ij} \equiv - \left\{\overline{u_i u_k}\frac{\partial U_k}{\partial x_j} + \overline{u_j u_k}\frac{\partial U_k}{\partial x_i}\right\}$$

$$\frac{D\varepsilon}{Dt} = d_\varepsilon + c_{\varepsilon 1}\frac{\varepsilon}{k}(P_{kk} + G_{kk}) - c_{\varepsilon 2}\frac{\varepsilon^2}{k}$$

c_1	c'_1	r	$c_{\varepsilon 1}$	$c_{\varepsilon 2}$
$3.1(A_2 A)^{1/2}$	1.2	0.6	1.0	$1.92/(1 + 0.7 A_2^{1/2} A)$

Fig 5. A New Model for Determining ϕ_{ij} and ε in Free Shear Flows

Flow	Basic (IP) Model	Recommended exptl. values	New Model
Plane plume	0.078	0.120	0.118
Round plume	0.088	0.112	0.122
Plane jet	0.100	0.110	0.110
Round jet	0.105	0.094	0.101

Table 1. Spreading rates for self-preserving free shear flows

Formal approximations for the mean-strain, Coriolis and buoyant parts of ϕ_{ij} may be developed by replacing the fluctuating pressure by the volume and surface integrals associated with the Poisson equation for p. In free flows the surface integrals are negligible and so, assuming a spatially homogeneous flow, we may, for example, write the mean-strain part ϕ_{ij2} as [11]

$$\phi_{ij2} = \frac{\partial U_\ell}{\partial x_m} \left(a^{mi}_{\ell j} + a^{mj}_{\ell i} \right)$$

where

$$a^{mi}_{\ell j} \equiv -\frac{1}{2\pi} \int \frac{\partial^2 \overline{u_m u_i}}{\partial r_\ell \partial r_j} \frac{\partial vol}{|\mathbf{r}|}$$

The tensor $a^{mi}_{\ell j}$ is commonly approximated in terms of a polynominal series in the anisotropic stress, a_{ij}, in which all or most of the unknown coefficients are determined by imposing the symmetry, contraction and, perhaps, other properties of the original integral. If only linear terms are retained the two-component limit cannot be satisfied [23]. At quadratic level it *can* be, however, and the resultant form is uniquely determined with no freely assignable coefficients [25, 30].

$$\phi_{ij2} = - 0.6[P_{ij} - \frac{1}{3}\delta_{ij}P_{kk}] + 0.3\varepsilon a_{ij}(P_{kk}/\varepsilon)$$
$$- 0.2 \left\{ \frac{\overline{u_k u_j}\,\overline{u_\ell u_i}}{k} \left[\frac{\partial U_k}{\partial x_\ell} + \frac{\partial U_\ell}{\partial x_k} \right] - \frac{\overline{u_\ell u_k}}{k} \left[\overline{u_i u_k}\frac{\partial U_j}{\partial x_\ell} + \overline{u_j u_k}\frac{\partial U_i}{\partial x_\ell} \right] \right\} \quad (2)$$

An interesting feature of eq (2) is that its leading term - indeed the *only* linear term - is just the Basic Model, including the value of the coefficient. Unfortunately the form shown is not satisfactory since, in a simple shear, $U_1(x_2)$, it gives $\overline{u_2^2} > \overline{u_3^2}$ contrary to experiment. If cubic terms are added two undetermined coefficients appear. In most work at UMIST one of these is set to zero (partly for expedience as the use of a non-zero value brings in such a large number of additional terms); this truncated cubic version appears in Fig 5. Even more elaborate forms have been proposed e.g [27, 32], though, so far as the writer is aware, none has been tested in inhomogeneous flows. The corresponding treatment of the Coriolis terms is also given in Fig 5 (Tselepidakis, personal communication) though for reasons of space the buoyant contribution, due to T. Craft (see [31]) is omitted.

It is found that the new model of ϕ_{ij} does much better in predicting the effects of changes in the relative strain rate on the stress field than does the Basic Model. While both models return roughly the correct level of $\overline{u_1 u_2}/k$ in a simple shear ($U_1(x_2)$) in local equilibrium, i.e. $1/2\,P_{kk}/\varepsilon = 1$, when this ratio is increased to 1.5 the Basic Model predicts a 10% increase in $\overline{u_1 u_2}/k$ whereas experiments and the new model in Fig 5 suggest a 10% decrease. It is paradoxical that by giving attention to satisfying the two-component limit which pertains at a rigid wall the model should lead to much improved predictions of free-shear-flow phenomena. Even a goat is occasionally friendly!

We have seen above that features of turbulence structure, in the form of the stress-anisotropy invariants, have begun to appear in the modelling of ϕ_{ij}. The same trends are evident in the transport equation for ε. While the use of A_2 as an element in the ε equation dates back to the mid '70's [33] the benefits from this inclusion have only become incontrovertible since the arrival of improved models for ϕ_{ij}. At present only

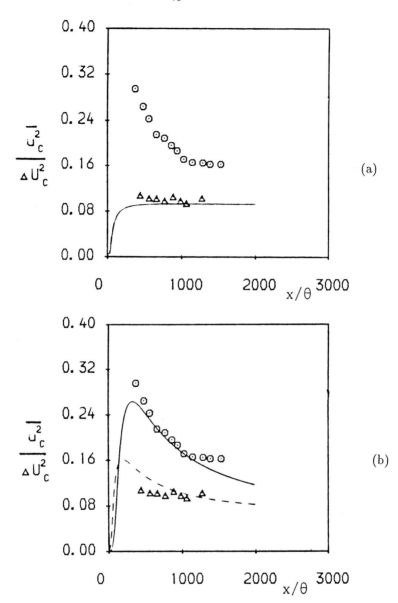

Fig 6. Development of Plane Wakes with Different Initial Conditions

Symbols: Experiments [34] ⊙ airfoil; △ solid strip
Lines: Computations [35] (a) Basic Model (b) New Model

fairly simple dependences on the invariants have been proposed all, perforce, relying on experimental calibration. The adaptation to $c_{\varepsilon 1}$ and $c_{\varepsilon 2}$ shown in Fig 5 is typical: notice that the direct importance of mean strain in creating ε is less than in the Basic Model, a modification that weakens the linkage between energy generation and dissipation rates and that, consequently, leads to predicted flow developments where greater departures from local equilibrium occur.

The above changes in the way ϕ_{ij} and ε are determined have led to substantial improvements in the prediction of free shear flows. Table 1, for example, compares observed rates of spread of four equilibrium self-preserving shear flows with the behaviour predicted by the Basic and New Models. Not only are rates of spread more accurately predicted with the latter, they show better internal consistency also; for example, the rates of spread of both the round jet and the round plume are now predicted modestly too high - an indication that something is not quite right in axisymmetric shear flows.

An even sharper illustration of the greater "freedom" that the New Model permits is provided in Fig 6. It shows the evolution of the centre-line streamwise normal stress plotted as a function of downstream distance (normalized by the wake momentum thickness) for two plane wakes of the same Reynolds number produced by different bodies, a stubby airfoil and a solid strip. It is clear from the experimental data [34] that the associated differences in initial conditions produce large changes in the development of the fluctuating velocity field. In modelling these two flows the initial mean velocity and turbulent length scale profiles were taken as identical but, for the strip-generated wake, the turbulent kinetic energy was assumed to be five times as large as for the airfoil. With the Basic Model the effects of these differences rapidly wash out and the predicted development of the two wakes is essentially identical. The New Model however captures the sensitivity of the far field to the initial conditions rather well. The differences in this case between the predicted developments with the two models are due largely to the different form of ε equation. Of course, many more such evolving flows need to be tested before any conclusive view can be reached about the adequacy of the model shown in Fig 5. It is encouraging, however, that the directions for improvements in second-moment closure advocated over more than a decade are now being implemented by several groups with tangible benefits.

At present no fully satisfactory proposals have been published for extending these new-style second-moment closures up to the wall itself. Shih and Lumley [36] have addressed the easier problem of the near-wall fully turbulent region under equilibrium conditions while two preliminary statements have appeared on our work at UMIST [37, 38]. The latter study emphatically brought out that adopting a model which was *compatible* with the two-component limit by no means ensured that the computed Reynolds stress field actually *did* revert to two-component turbulence, $a_{22} = -2/3$ at the wall ($x_2 = 0$). Indeed the variation of $\overline{u_2^2}$ across the viscous sublayer and buffer regions has proved to be difficult to predict accurately. It is important that it should be, however, for only then can one expect to compute reliably the effects of Prandtl (Schmidt) number on heat (mass) transport in other than equilibrium flows parallel to the wall.

4 More Extensive Refinements

Section 3 limited attention to developments within what might be called "classical" second-moment closure, namely models based on transport equations for the Reynolds stresses and a scalar quantity, usually (but not necessarily) ε. The latter supplies, *inter alia*, a characteristic time scale (k/ε) for the rate of progress of different turbulent processes. Several workers have remarked on the apparent imbalance between the attention accorded the stresses and the time scales. Turbulence is composed of a wide range of eddy sizes which respond at different rates to external disturbances; it might be argued that to adopt just a single turbulent time scale is inadequate - at any rate if one is solving six equations for the stress field. Certainly, to add a further scale equation would, in these circumstances, have little effect on computer resources.

Two-time-scale models have, in fact, been proposed by Schiestel [39] and Wu et al [40] while Hanjalic et al [41] have developed a somewhat more elaborate scheme in which the turbulent energy spectrum is divided into two slices, a "production" and a "transfer" region with separate energy and scale equations being solved for each. All these proposals, however, were developed within the framework of an eddy viscosity stress-strain connection. Whatever the merits of two- or multi-scale approaches, the intrinsic weakness of the eddy viscosity concept means that, as presented, such models will have very limited utility. However, in the writer's view the inclusion of a two-time-scale approach within a second-moment closure would be a worthwhile and relatively simple extension that could bring benefits that were substantial in relation to cost.

The multi-scale approach of [41] is, conceptually at any rate, easily extended to provide rate equations for the stresses and it has been used in that form by Schiestel [42] and others for computing homogeneous turbulent flows. Since, however, in that formulation, transport equations are solved for the Reynolds stresses in both the "production" and "transfer" ranges it is hardly a minor elaboration in computational terms (up to fourteen variables to be obtained from transport equations — double that of a single-scale second-moment closure). There is no consensus in the turbulence modelling community that, if one adds a second second-rank tensor to one's list of dependent variables, that tensor should be for a spectral slice of the Reynolds stresses. Certainly other proposals have been floated. The direct numerical simulation work at Stanford's Center for Turbulence Research seems to suggest that the return of the stress field to isotropy, driven by pressure interactions, depends a lot more on the anisotropy of the dissipation tensor than on a_{ij} and, in consonance with that view, that ε_{ij} is far less isotropic than is generally supposed. So, there may be a case for solving equations for ε_{ij}, a direction that has previously been suggested by a number of workers [43, 44]. A somewhat similar elaboration would be a system of transport equations for a tensorial length scale ℓ_{ij}, a route advocated in [45, 46].

It should be stressed that none of the schemes involving transport equations for two second-rank tensors (one being the Reynolds stresses) has yet acquired any significant track record. It is possible their day will come one day. At the moment, however, with the emergence of better second-moment closures (perhaps augmented by a time-scale transport equation) it isn't clear that there will be a significant engineering need for any more elaborate approach short of large-eddy simulation.

5 Concluding Remarks

The paper has attempted to convey an impression of the capabilities of second-moment closure for mimicking the behaviour of the turbulent stress field in complex flows. Despite its rudimentary form, the Basic Model now being incorporated in several commercial CFD packages easily outperforms any eddy viscosity treatment. More elaborate approaches in which the components of the model are constrained to comply with extreme states of turbulence and where "structural" dependence is incorporated by the use of the stress invariants look certain to return realistic answers over a still wider range of flows; already several apparently irreconcilable flows have been predicted with a single model. There is also an indication that these realizable models can lead, in elliptic flows, to a reduction in computing time by reducing the number of iterations needed to achieve convergence [31].

For the future, the addition of a transport equation for a turbulent time scale appears to offer a way of extending further the capabilities of second-moment closure - perhaps especially for rapidly evolving flows - for little additional cost. Other proposed refinements involving solution of a second second-rank tensor bring such an increase in computational load in their wake that it is not by any means clear that they will be useful for engineering calculations.

The paper has been produced in LaTeX format by Mrs J. Buckley with appreciated help from Gary Page.

References

1. Pope, S.B. and Whitelaw, J.H., *J. Fluid Mech.*, 73, 9-32, 1976
2. Huang, P.G. and Leschziner, M.A., *Proc. 5th Symp. Turb. Shear Flows*, 20.7-20.12, Cornell University, Ithaca, 1985
3. Iacovides, H. and Launder, B.E., *Proc. 4th Int. Conf. on Numerical Methods in Laminar and Turbulent Flows*, 1023-1044, Pineridge Press, Swansea, 1985
4. Leschziner, M.A. *Int. J. Heat and Fluid Flow*, 10 (3), 186-202, 1989
5. Dimitriadis, K. and Leschziner, M.A. 'Modelling shock/turbulent boundary layer interaction with a cell-vertex scheme and second-moment closure', *Proc. 12th ICNMFD*, Oxford, 1990
6. Launder, B.E., *Int. J. Heat Fluid Flow*, 10, 282-300, 1989
7. Launder, B.E., *ASME J. Heat Trans.*, 110, 1112-1128, 1988
8. Launder, B.E., *Advances in Turbulence-2*, (Ed. H. Fernholz and H.E. Fiedler) 338-358, Springer, 1989
9. Bradshaw, P., 'Effects of streamline curvature on turbulent flow', AGARDograph 169, 1973
10. Haroutunian, V. and Launder, B.E., in Stably Stratified Flow and Dense Gas Dispersion, (Ed. J.S. Puttock), 409-430, Oxford, 1988
11. Rotta, J., *Z. Phys.*, 129, 547-573, 1951
12. Naot, D., Shavit, A. and Wolfshtein, M., *Israel J. Tech.*, 8, 259-269, 1970
13. Launder, B.E., *J. Fluid Mech.*, 67, 569-581, 1975

14. Fu, S., Launder, B.E. and Leschziner, M.A., *Proc. 6th Symp. on Turbulent Shear Flows*, Paper 17.6, Toulouse, 1987
15. Gibson, M.M. and Launder, B.E., *J. Fluid Mech.*, 86, 491-511, 1978
16. Demuren, O. and Rodi, W., *J. Fluid Mech.*, 140, 189, 1984
17. Daly, B.J. and Harlow, F.H., *Phys. Fluids*, 13, 2634, 1970
18. Jones, W.P. and Manners, A., *Turbulent Shear Flows-6*, (ed. J.C. André et al), 18-31, Springer, Heidelberg, 1989
19. Stevens, S.J. and Fry, P., *J. Aircraft*, 10, 73, 1973
20. Launder, B.E., Tselepidakis, D.P. and Younis, B.A., *J. Fluid Mech.*, 183, 63, 1987
21. Kim, J., *Proc. 4th Symp. on Turbulent Shear Flows*, Karlsruhe, 6.14, 1983
22. Schumann, U., *Phys. Fluids*, 20, 721-725, 1977
23. Lumley, J.L., *Advances in Appl. Mech.*, 18, 123-175, 1978
24. Lumley, J.L. and Newman, G.R., *J. Fluid Mech.*, 82, 161-178, 1977
25. Shih, T.-H. and Lumley, J.L., Report FDA-85-3, Sibley School of Mech. Aero. Engng, Cornell University, 1985
26. Reynolds, W.C., in *"Turbulence models and their applications"* (by B.E. Launder, W.C. Reynolds and W. Rodi), 149-294, Eyrolles, 1984
27. Lecointe, Y., Piquet, J. and Visonneau, M., *Proc. 5th Symp. on Turbulent Shear Flows*, 12.7-12.12, Cornell, 1985
28. Speziale, C.G., Gatski, T.B. and Mhuiris, N.M.G., *Proc. 7th Symp. Turb. Shear Flows*, Paper 27-3, Stanford University, 1989
29. Craft, T.J. and Launder, B.E., *Proc. 7th Symp. Turbulent Shear Flows*, Paper 17-1, Stanford University, California, 1989
30. Fu, S., Launder, B.E. and Tselepidakis, D.P., UMIST Mech. Engng. Dept. Report TFD/87/5, 1987
31. Cresswell, R., Haroutunian, V., Ince, N.Z., Launder, B.E. and Szczepura, R.T., Paper 12.4, *Proc. 7th Symp. on Turbulent Shear Flows*, Stanford, 1989
32. Reynolds, W.C., Notes for Class ME261B-Turbulence, Stanford University, 1989
33. Lumley, J.L. and Khajeh-Nouri, B.J., Advances in Geophysics (*Proc. 2nd IUGG-IUTAM Symp. on Atmospheric Diffusion in Environmental Pollution*), 18A, 169, 1974
34. Wygnanski, I., Champagne, F. and Marasli, B., *J. Fluid Mech.*, 168, 31, 1986
35. El Baz, A.R., Launder, B.E. and Nemouchi, Z., Open Forum, *7th Symp. on Turbulent Shear Flows*, Stanford, 1989
36. Shih, T.-H., and Lumley, J.L., *Phys. Fluids.*, 29, 971-975, 1986
37. Launder, B.E. and Tselepidakis, D.P., *Proc. Zoran Zaric Memorial Conf. on Near-Wall Turbulence*, (Ed. S.J. Kline & N.H. Afgan), 819-833, Hemisphere, 1990
38. Launder, B.E. and Tselepidakis, D.P., *Proc. 7th Symp. Turbulent Shear Flows*, Open Forum, Paper T-7, Stanford, 1989
39. Schiestel, R., Thèse Docteur es Sciences, Université de Nancy, 1974
40. Wu, C.-T., Ferziger, J. and Chapman, D.R., *Proc. 5th Symp. on Turbulent Shear Flows*, 17.13-17.19, Cornell, 1985
41. Hanjalic, K., Launder, B.E. and Schiestel, R., *Turbulent Shear Flows-2*, (ed. L.J.S. Bradbury et al), 36-49, Springer, Heidelberg, 1980
42. Schiestel, R., *Phys. Fluids*, 30, 722-731, 1987

43. Lin, A. and Wolfshtein, M., *Turbulent Shear Flows-1*, (ed. F. Durst et al), 327-343, Springer, Heidelberg, 1979
44. Morse, A.P., PhD Thesis, Faculty of Engineering, University of London, 1980
45. Donaldson, C. du P. and Sandri, G., 'On the inclusion of eddy structure in second-order closure models of turbulent flow', AGARD Rep., 1982
46. Kolovandin, B.A. and Vatutin, I.A., *Int. J. Heat Mass Transfer*, <u>15</u>, 2371, 1972

SIMULATION OF TURBULENT FLOWS BY NUMERICAL INTEGRATION OF NAVIER-STOKES EQUATIONS

B. L. Rozhdestvensky and M. I. Stoynov
Keldysh Institute of Applied Mathematics,
Academy of Sciences of the USSR, Moscow, USSR

Introduction

In this paper the numerical methods intended for simulation of turbulent fluid flows by direct numerical integration of incompressible Navier–Stokes (NS) equations are discussed. This approach has been successful for turbulent incompressible viscid fluid flows in channels and pipes with a variety of boundary and external conditions. Two- and three-dimensional turbulent flows in a plane channel with fixed (or moving) walls, under the action of constant (or variable in time) space-averaged pressure gradient (or flux) were computed.

Turbulent Poiseuille flows, Couette flows, and pulsating flows periodic in homogeneous directions were obtained. Some turbulent Poiseuille flows in circular and annular pipes were also obtained. At transition Reynolds numbers the 3D-simulations reproduce the basic characteristics of turbulent flows with rather good accuracy, but 2D-simulations cannot correctly reproduce them.

The finest resolution in numerical simulation of turbulent flows in a plane channel was achieved by J. Kim et al [1]. The authors used the grid of about $4 \cdot 10^6$ points for description of the velocity field. They stated that this grid resolution was adequate to turbulence simulation. The database obtained in this simulation is of great value for the studies of the near wall turbulence structures. The other big computation with resolution 128×128×128 is reported in [2]. These computations are discussed here in some details.

On the other hand, dozens of numerical simulations of turbulent 3D-flows with much coarser resolution were performed over the past decade. A few dozens of simulations of 3D- turbulent Poiseuille, Couette and pulsating Poiseuille flows in a plane channel were performed [4-10]. An aim of these computations was to take into account the interactions between the basic large eddies which have a dominated portion of the energy of fluctuations. These 3D– computations, even with about 2700 grid points, reproduce rather good basic integral characteristics of turbulent flows for transition Reynolds numbers.

The 3D–simulation of even simplest turbulent flows is rather expensive and it needs not only the best computers, but the best numerical methods and codes too. Hence, it is very important to have methods for checking stability and accuracy of the numerical technique. Such a low–cost method for checking the simulations of linear stages of flow evolution, based on the work [3], is presented and discussed.

Some other questions on mathematical and numerical formulation of the problem are also discussed.

1. Mathematical formulation of the problem

The incompressible viscous, periodic in x and y directions fluid flows in a plane channel $K=\{ x=(x,y,z): (x,y) \in R^2, z \in [-1,+1] \}$ are described by the solutions of the NS eqns:

$$V_t + (V \cdot \nabla) \cdot V + \nabla p - \nu \Delta V = 0; \qquad \nabla \cdot V = 0, \qquad (1)$$

that satisfy the no-slip conditions on the walls:

$$V(x,y,z=\pm 1,t) = 0 \qquad (2)$$

and the periodicity conditions:

$$V(x+X,y,z,t) = V(x,y+Y,z,t) = V(x,y,z,t), \qquad (3)$$

as well as one of the next two external conditions:

a) $<\nabla p> = -2\nu i,$ b) $<V> = 2/3 i$ (4a,b)

Here the brackets mean the averaging of pressure gradient ∇p and velocity V over the whole channel K; $i=(1,0,0)$; $V=(u,v,w)$, the fluid density is equal to the unity and ν is the kinematic viscosity. The condition (4a) fixes the mean pressure gradient, while (4b) means that the mass flux is constant. To eqns (1)–(4) the initial condition $V_0(x,0) = V_0(x)$ ($\nabla V_0=0$) must be added.

To any flow in channel K two Reynolds numbers R_P and R_Q are ascribed:

$$R_P(t) = |<\nabla p(x,t)>|/2\nu^2 \, ; \qquad R_Q(t) = 3|<V(x,t)>|/2\nu$$

(see [4]). For the flows that meet condition (4a) $R_P(t)=1/\nu$, and if (4b) is valid, then $R_Q(t)=1/\nu$. For fully developed turbulent flows $R_P(t) > R_Q(t)$.

Note that generally it is much cheaper to obtain in numerical simulation a turbulent flow using the external condition (4b) rather than (4a).

2. Numerical methods: spatial discretization

The conditions of periodicity (3) may be easily taken into account in numerical model, if the approximate velocity and pressure gradient fields are represented as the trigonometrical polynomials in the streamwise (x) and spanwise (y) directions. In all computations of turbulent flows in channels and pipes, performed in the eighties, this advantage of Fourier series was used.

For discretization in normal (z) direction several methods were used: the Galerkin method with a variety of the base function systems, the collocation methods and the usual grid methods. To be efficient all these methods have to describe the velocity field near the walls much more accurate than in the middle of the channel.

Spectral methods – the Fourier series in x and y variables, the polynomial expansions in z as well as the Chebyshev tau method – were employed for spatial discretization of NS eqns and the boundary conditions [11,1,2,4,7,10]. In these papers the approximate solution may be represented in the form:

$$\mathbf{V}(\mathbf{x},t) = \sum_{m,n} \mathbf{V}_{mn}(z,t) \cdot \exp(im\alpha_o x + in\beta_o y); \quad |m| \leq M, \quad |n| \leq N \tag{5}$$

$$p(\mathbf{x},t) = p_x(t)x + p_y(t)y + \sum_{m,n} p_{mn}(z,t) \cdot \exp(im\alpha_o x + in\beta_o y),$$

where $\mathbf{V}_{mn}$ and p_{mn} are polynomials in z of the degree at most L and $\alpha_o = 2\pi/X$, $\beta_o = 2\pi/Y$.

In our new numerical technique [10] the representation (5) of numerical solution is also used but we apply the collocation method in which the collocation points $z_1,...,z_{L-1}$ are the zeros of the Jacobi polynomial $P_{L-1}^{(1,1)}(z)$. To obtain the numerical scheme the representation (5) for approximate solution is substituted into eqn (1). Then the left hand side of momentum eqn (1) is projected onto the system of harmonics $\psi_{mn}(x,y) = \exp(im\alpha_o x + in\beta_o y)$ and equated to zero at the collocation points $z_1,...,z_{L-1}$. These relations form the system of ODE. To preserve the conservation properties of the nonlinear term $\mathbf{F} = (\mathbf{V} \cdot \nabla)\mathbf{V}$ we use the next truncation:

$$F^{tr} = (V \times \text{rot} V)^{MN} - 1/2 \, \nabla \, [(V^2)^{MN}]. \tag{6}$$

Here truncation $[f]^{MN}$ means the function g which has the Fourier coefficients $g_{mn} = f_{mn}$ if $|m| \leq M$ and $|n| \leq N$, and $g_{mn} = 0$ otherwise.

We require the condition $\nabla \cdot V = 0$ to be fulfilled for each pair (m,n) at $L+1$ points $z_0 = -1, z_1, \ldots, z_{L-1}, z_L = 1$, i.e. require the approximate velocity field (5) to be solenoidal at any $t \geq 0$ and $x \in K$. These linear relations together with no-slip conditions (2) and one external condition (4a) or (4b) complete a definition of the ODE system solutions.

The described method of spatial discretization has a good accuracy like other spectral methods. Its main advantage is that the system of ODE obtained has the discrete conservation laws for mass, impulse and energy, which are the finite-dimensional analogs and approximations of the physical conservation laws.

3. Numerical methods: the time discretization

Different time advancement schemes were used for numerical simulations of turbulent flows in a plane channel [11,1,2,4,8,10–12]. Their main idea is to employ the second order accuracy of explicit time advancement for the nonlinear terms (Adams–Bashforth scheme) and the explicit or implicit (Crank–Nicolson) schemes for the viscous terms.

We have developed [10] the fully implicit second order time advancement scheme:

$$(V_{mnl}^{k+1} - V_{mnl}^{k})/\tau = -(\nabla p^{k+0.5})_{mnl} + \nu \, (\Delta V^{k+0.5})_{mnl} + (F^{tr})_{mnl}^{k+0.5} \tag{7}$$

$$l = 1, \ldots L-1; \quad |m| \leq M; \quad |n| \leq N,$$

$$\nabla \cdot V^{k+1} \equiv 0 \tag{8}$$

Here $V^k = V(x, \tau k)$, $V^{k+0.5} = 0.5(V^k + V^{k+1})$, $(F^{tr})^{k+0.5} = [F(V^{k+0.5})]^{tr}$.

To this scheme the no-slip and external conditions are added. Note, that the identity (8) is equivalent to $L+1$ linear relations for each pair (m,n).

This time discretization uses the Crank–Nicolson type scheme for the pressure gradient and the viscous terms and the Kasahara type scheme for the nonlinear terms. The implicit scheme (7),(8) has again the discrete analogs of the mass, momentum and energy conservation laws. To solve these nonlinear eqns the method of successive approximations is applied. For the most part of our computations two or three iterations were sufficient to achieve acceptable accuracy. This scheme has rather high limit for time step $\tau \simeq 0.2$ at moderate numbers M, N $\simeq$ 20, L $\simeq$ 60 and $\alpha_0, \beta_0 \sim 1$.

4. Checking of numerical techniques

It is very difficult to determine, even roughly, the optimal parameters of the mathematical model (the periods X,Y and the full integration time T) and of the numerical algorithms (α_0, β_0; M,N,L;τ and some other parameters) for a realistic simulation of turbulent flow in a channel (pipe) at the given Reynolds number R_Q (or R_P).

It is clear that all these parameters must be corrected in the process of numerical simulation by means of comparisons between the numerical and experimental results. Such an approach to checking problem is very expensive.

On the other hand, the mathematical investigation of the influence of these parameters on the results of numerical simulation is also possible. Generally such an investigation is very complicated due to the nonlinearity of the NS and numerical equations and it may be even more expensive than the numerical simulation itself.

However, there exists a possibility to obtain easily some results on this matter, which are very informative although not final.

Let us consider the evolution of infinitesimal disturbances against a background of a stationary solution of NS eqns, depending only on coordinate z. The evolution of infinitesimal disturbances of the Poiseuille, Couette and some other stationary flows has been investigated reasonably well in the linear theory of hydrodynamic stability (LTHS), but the results obtained are not sufficiently comprehensive for serious tests of numerical methods.

In LTHS all disturbances may be presented in the form (5) and almost all of them are the linear combinations of the solutions of linearized NS eqns, which have the form:

$$\mathbf{V}_j(\mathbf{x}, t) = \mathbf{V}_j(z) \cdot \exp(\lambda_j t + \alpha x + i\beta y) \qquad (9)$$

The exponents $\{\lambda_j(R, \alpha, \beta)\}_{j=1}^{\infty}$, ($\mathrm{Re}\lambda_j \geq \mathrm{Re}\lambda_{j+1}$) are the ordered eigenvalues of the known boundary value problem for the Orr–Sommerfeld equation.

Analogous investigation of evolution of infinitesimal disturbances to a stationary solution of numerical scheme is also possible for all schemes whose linearization with respect to stationary solution $V_0(z)$ allows the decomposition of discrete variables. These requirements are satisfied in all numerical schemes from the papers cited here.

For such numerical methods the evolution of arbitrary infinitesimal disturbance may be described by the linear combination of the next solutions of linearized numerical scheme

$$V_j(x,k\tau)=V_j^k(x)=q_j^k \cdot \hat{V}_j(z) \cdot \exp(i\alpha x + i\beta y), \quad \alpha=m\alpha_o, \quad \beta=n\beta_o. \qquad (10)$$

The coefficients $q_j=q_j(R,\alpha,\beta,\tau,L)$, $(j=1,2,...,J)$, and the number J essentially depend also on the type of numerical scheme; they may be ordered by their seniority: $|q_j| \geq |q_{j+1}|$. Small disturbances of the form (10) may be called the numerical Tollmien–Schlichting (NTS) waves.

The numerical method approximates the evolution of infinitesimal disturbances if for any fixed j_o

$$\hat{\lambda}_{j_o} - \lambda_{j_o} \to 0 \;; \qquad \hat{V}_{j_o}(z) / V_{j_o}(z) \to \text{Const}, \qquad (11)$$

hen $L \to \infty$, $\tau \to 0$. Here $\hat{\lambda}_j = 1/\tau \cdot \ln(q_j)$.

Thus, the investigation of the spectrum $\{q_j\}$ or $\{\hat{\lambda}_j\}$ at large L and small τ (or $\tau=0$) allows one to obtain the spectrum $\{\lambda_j(R,\alpha,\beta)\}$ of the Orr–Sommerfeld eqn in any details and to define the parameters L,M,N,τ, which provide the required accuracy in numerical simulation of evolution of small disturbances of the stationary solution. It is clear that these requirements to the numerical method are the necessary conditions for correct simulation of turbulent flows.

The problem of computation of spectrum $\{q_j\}$ and eigenfunctions $\{\hat{V}_j(z)\}$ is reduced to the full problem on the eigenvalues and eigenfunctions of the complex $J \times J$ matrix, which depends on numerical methods. Due to the developed methods and the standard matrix solver packages the computation of spectrum $\{q_j\}$ is rather inexpensive. The determination of spectrums $\{\hat{\lambda}_j(R,\alpha,\beta,L,\tau)\}$ and their comparisons with spectrums $\{\lambda_j(R,\alpha,\beta)\}$ as well as obtaining of the conclusions on the accuracy, stability and efficiency of numerical method we call the spectral analysis of numerical method.

Our investigation of numerical schemes from [1,2,10] gives an example of very useful application of spectral analysis of numerical methods. Here are some of the results obtained.

1. $J=L-3$ for the scheme (7),(8) ; $J=2(L-2)$ for scheme [1] and $J=4(L-2)$ for scheme [2].

2. The quality of spatial discretization in these three schemes is approximately the same. For $R = 2 \cdot 10^3 - 2 \cdot 10^4$, $L=128$ it appears that at almost any $\alpha, \beta \in [0.5, 50.]$ only about $1/3 \cdot L$ eigenvalues $\hat{\lambda}_j$ approximate the eigenvalues λ of the Orr–Sommerfeld equation with "100 percent accuracy", i.e. $|\text{Re}(\hat{\lambda}_j - \lambda_j)/\text{Re}\lambda_j| \leq 1$.

It means also that these schemes (at $\tau = 0$) approximate the evolution of less than $1/3 \cdot L$ TS waves. The rest (nearly $2/3 \cdot L$ NTS waves in the scheme [10]; $5/3 \cdot L$ in [1] and $11/3 \cdot L$ in [2]) are "the numerical rubbish". Such quality of spatial resolution is rather usual for spectral methods. It may be noted that all these three schemes have no spurious NTS waves.

3. The spatial and time resolution becomes much worser if the time step τ is large. Thus, for the scheme [1] at $R_Q = 4200$, $\tau = 0.008781$ (the time step used in [1]), $\alpha \geq 40$ and certain β the exponent $\hat{\lambda}_1(R, \alpha, \beta, \tau, L)$ has a positive real part. For example: $\text{Re}\hat{\lambda}_1 = +0.579$ at $\alpha = 48$, $\beta = 0$ while $\text{Re}\lambda_1 = -0.526$.

It means that for $\alpha \geq 40$ this scheme has the unstable NTS waves, which increase in time very quickly (in linear approach), and that all NTS waves at $\alpha \geq 40$ should be considered as numerical rubbish. The stability of NTS waves is restored when $\tau \leq 0.007$.

At the same time the large time step leads to a very bad approximation of the exponent λ_j for $\alpha > 24$ even for small j. The satisfactory approximation of TS waves for all $\alpha \leq 48$ is restored at time step $\tau \leq 0.0013$.

The analogous, but much coarser violations of the numerical scheme stability by too large time step τ were allowed also in work [2].

As to the scheme (7),(8) from [10], it is stable (in linear approach) at any time step τ, according to the relation:

$$\hat{\lambda}_j(\tau) = 1/\tau \cdot \ln[(1 + 0.5\hat{\lambda}_j(0) \cdot \tau)/(1 - 0.5\hat{\lambda}_j(0) \cdot \tau)], \tag{12}$$

if the spatial discretization provides the correct approximation of the NS eqns. However, at large τ some difficulties may arise with the convergence of iterative process, if it is applied to solving the implicit numerical equations.

Thus, computations [1,2] were performed with too large time steps. This may be the cause of incorrect description of the waves with large wavenumbers m, n. The spectral analysis of numerical scheme may check this defect easily and inexpensively. Of course, the convective instability of numerical simulation in essentially nonlinear stages of computation does not follow from its linear instability.

More detailed description of the spectral analysis of numerical techniques are presented in Appendix.

5. Conclusion

The problem on simulation of turbulent flows in channel and pipes by direct numerical integration of NS eqns is very difficult. The periodic in x and y solutions depend on the Reynolds number R and the periods X and Y. The numerical solutions depend also on the numbers M, N, L and time step τ. Increasing M, N, L and reducing τ we improve the accuracy of numerical solution of the posed mathematical problem.

On the other hand, increasing periods X, Y we vary the mathematical problem itself and bring it nearer to the perfect problem on the turbulent flows in the infinite (in two directions) plane channel. Simultaneously the numbers M, N have to be increased proportional to X and Y; maybe the integration time T must be increased too. We may see that even for a single Reynolds number R the numerous investigations of numerical solutions are desirable.

By now, however, no one has investigated the dependence of 3D turbulence simulation on the periods X, Y when X,Y $\to \infty$.

We have performed such investigations for 2D simulation of turbulent flow at R_p=5000 and X $\to \infty$. It was found that at X $\simeq 64\pi$ the main integral characteristics of 2D-turbulent flow did not change essentially with further increase of X.

The choice of the proper time step τ and the integration time T are also of great importance. Usually the time step choice is checked by reproducing numerically the growth (damping) rate of some eigenfunctions of the Orr-Sommerfeld eqn at rather small wave numbers α, $\beta \simeq 1$. It is quite insufficient, especially for the cases of large numbers N, M, L, to be sure that the numerical scheme is stable even in a linear approach. In our opinion, the correct choice of τ should be checked by full spectral analysis of numerical algorithm.

The discussed approach to computer simulation of turbulent flows is very promising, but the problems to be solved are too extensive and rather expensive. Therefore, the most important factor in solving many principal problems on turbulence is, in our opinion, the international cooperation and coordination of efforts in these investigations.

Appendix
1°. Spectral analysis of numerical algorithm [1]

In numerical algorithm [1] NS equations are reduced to yield the equations for the normal velocity component $w(\mathbf{x},t)$, the normal vorticity component $g(\mathbf{x},t)=u_y-v_x$ and $f(\mathbf{x},t)=u_x+v_y$ as follows:

a) $\Delta w_t = 1/R \cdot \Delta\Delta w + h_w$; b) $g_t = 1/R \cdot \Delta g + h_g$; c) $f + w_z = 0$; (A1)

d) $w = f = g = 0$, at $z = \pm 1$,

where $h_w = -\frac{\partial}{\partial z}[\frac{\partial}{\partial x}F^u + \frac{\partial}{\partial y}F^v] + [\frac{\partial^2}{\partial x^2} + \frac{\partial^2}{\partial y^2}]F^w$, $h_g = \frac{\partial}{\partial y}F^u - \frac{\partial}{\partial x}F^v$,

$\mathbf{F}=(F^u, F^v, F^w)=[\mathbf{v}, \text{rot}\,\mathbf{v}]$ are the convective terms of the NS eqns.

The time advancement is carried out by semi-implicit scheme: Crank–Nicolson for the viscous terms and Adams–Bashforth for the nonlinear terms. Equations (A1) are approximated by the system

$\Delta[(w^{n+1}-w^n)/\tau] = 1/(2R) \cdot \Delta\Delta[(w^{n+1}+w^n)] + 3/2 \cdot h_w^n - 1/2 \cdot h_w^{n-1}$ (a)

$\Delta[(g^{n+1}-g^n)/\tau] = 1/(2R) \cdot \Delta[(g^{n+1}+g^n)] + 3/2 \cdot h_g^n - 1/2 \cdot h_g^{n-1}$ (b) (A2)

$f^{n+1} + w_z^{n+1} = 0$, (c)

$w^{n+1} = f^{n+1} = g^{n+1} = 0$, at $z = \pm 1$. (d)

The periodic in x and y directions solutions of (A2) are sought in the form:

$$\{w, f, g\} = \sum_{p=0}^{P} \sum_{m=-M}^{M} \sum_{n=-N}^{N} \{w_{mnp}(t), f_{mnp}(t), g_{mnp}(t)\} \cdot e^{im\alpha_0 x + in\beta_0 y} \cdot T_p(z).$$

Equations (A2) are solved by the Chebyshev–tau method for each wave number ($\alpha = \alpha_0 m$, $\beta = \beta_0 n$) after they have been Fourier transformed in the streamwise and spanwise directions. To eliminate spurious instability of numerical algorithm the fourth-order equation (A2a) is reduced to the system of two second-order equations in the following way (here and below the indexes m and n are omitted):

$(\psi_p^{n+1} - \psi_p^n)/\tau = 1/(2R) \cdot \sum_{p'=0}^{P} A_{pp'}(\psi_{p'}^{n+1} + \psi_{p'}^n) + [3/2 \cdot h_w^n - 1/2 \cdot h_w^n]_p$ (a)

$\psi_p^{n+1} = \sum_{p'=0}^{P} A_{pp'} w_{p'}^{n+1}$ (b) (A3)

$$\sum_{p=0}^{P} w_p^{n+1} = \sum_{p=0}^{P} (-1)^p \cdot w_p^{n+1} = 0; \tag{c}$$

$$\sum_{p=0}^{P} p^2 \cdot w_p^{n+1} = \sum_{p=0}^{P} (-1)^p \cdot p^2 \cdot w_p^{n+1} = 0; \tag{d}$$

equation (A2b) is replaced by the Chebyshev–tau approximation

$$(g_p^{n+1} - g_p^n)/\tau = 1/(2R) \cdot \sum_{p'=0}^{P} A_{pp'} \cdot [(g_{p'}^{n+1} + g_{p'}^n)] + [3/2 \cdot h_g^n - 1/2 \cdot h_g^{n-1}]_p \tag{A4}$$

$$\sum_{p=0}^{P} g_p^{n+1} = \sum_{p=0}^{P} (-1)^p \cdot g_p^{n+1} = 0,$$

where $p = 0, 1, \ldots, P-2$, w_p^{n+1}, g_p^{n+1} are the coefficients in the Chebyshev expansion of $w_{mn}^{n+1}(z,t)$ and $g_{mn}^{n+1}(z,t)$; matrix $A_{pp'} = D2_{pp'} - (\alpha^2 + \beta^2) \cdot \delta_{pp'}$ approximates the Laplace operator Δ.

The stationary solution of equations (A2)–(A4) coincides with stationary solution $V_0(z) = (U_0(z), 0, 0) = (1 - z^2, 0, 0)$ of the NS eqns (A1). The time–evolution of infinitesimal disturbances of this solution are described by system (A3)–(A4), linearized with respect to $V_0(z)$. This system is obtained from equations (A3),(A4) by replacing h_w, h_g by linear operators $l_w = U_0'' \cdot w_x - U_0 \cdot \Delta w_x$, $l_g = -U_0' \cdot w_y - U_0 \cdot g_x$ ($h_{(w,g)} = l_{(w,g)} + O(v^2)$). The solutions of this linear system can be represented in the form

$$\{w_p^n, \psi_p^n, g_p^n\} = q^n \cdot \{w_p, \psi_p, g_p\}. \tag{A5}$$

The complex constants $\{q\}$ are determined from discrete analogs of spectral problems in the theory of hydrodynamics stability. So, from equations (A3) a discrete analog of Orr–Sommerfeld equations is derived. Substituting (A5) into (A3a) we obtain

$$(q-1) \cdot \psi_p / \tau = (q+1)/(2R) \cdot \sum_{p'=0}^{P} A_{pp'} \cdot \psi_{p'} + \left(\frac{3}{2} - \frac{1}{2q}\right) \cdot \sum_{p'=0}^{P} L_{pp'} \cdot w_{p'}, \tag{A6}$$

where, $p = 0, 1, \ldots, P-2$, and $L_{pp'}$ are defined by the relations $[l_w]_p = \sum L_{pp'} \cdot w_{p'}$.

The spectral problem (A6) can readily be reduced to the standard matrix eigenvalue problem. From equations (A3b) and (A3c) the relationship between w and ψ is derived as

$$w_p = \sum_{p'=0}^{P-2} \hat{A}^{-1}_{pp'} \psi_{p'}, \qquad (A7)$$

where

$\hat{A}_{pp'} = A_{pp'}$, $p=0,1,\ldots,P-2$, $\hat{A}_{P-1,p'}=1$, $\hat{A}_{P,p'}=(-1)^{p'}$, $p'=0,1,\ldots,P$.

Boundary conditions for ψ are determined by substitution of (A7) in the equations (A3d). The result is

$$\sum_{p=0}^{P-2} r_p \cdot \psi_p = \sum_{p=0}^{P-2} s_p \cdot \psi_p = 0, \qquad (A8)$$

where $r_p = \sum_{p'=0}^{P-2} p'^2 \cdot \hat{A}^{-1}_{p'p}$, $s_p = \sum_{p'=0}^{P-2} (-1)^{p'} \cdot p'^2 \cdot \hat{A}^{-1}_{p'p}$.

Boundary conditions (A8) imply that only (P+1)−2 out of the P+1 coefficient ψ_p are independent, and thus any two of them (for example, ψ_0 and ψ_1) can be expressed in terms of remaining coefficients. Thus equations (A6) can be rewritten in the form

$$q^2 \cdot \psi_{p+1} = q \cdot \sum_{p'=1}^{P-1} B_{pp'} \psi_{p'+1} + \sum_{p'=1}^{P-3} C_{pp'} \psi_{p'+1}, \quad p=1,2,\ldots,P-1, \qquad (A9)$$

where B and C are complex (P−1)x(P−1) matrices depending on R, α, β, τ, and P.

These equations may be reduced to standard eigenvalues problem

$$qx = Tx, \qquad (A10)$$

where T is the (2P−2)x(2p−2) matrix $\begin{bmatrix} B & C \\ I & 0 \end{bmatrix}$, I is the unit matrix; components of the vector x are determined as follows: $x_i = q \cdot \psi_{i+1}$, $x_{i+P-1} = \psi_{i+1}$, $i=1,2,\ldots,P-1$. Spectrum of the matrix T is easily computed by usual eigenvalues package.

It is convenient to introduce the numerical algorithm eigenvalues

$$\hat{\lambda}_j(R,\alpha,\beta,P,\tau) = 1/\tau \cdot \ln(q_j). \qquad (A11)$$

The approximate solutions of eqn (A3a) can be represented in terms of $(w_p^j(x,t), \psi_p^j(x,t)) = (w_p^j(z), \psi_p^j(z)) \cdot \exp(\hat{\lambda}_j \cdot t + i\alpha x + i\beta y)$, $j=1,\ldots,2P-2$. Thus, the comparison of the sets $\{\hat{\lambda}_j(R,\alpha,\beta,\tau,P)\}_{j=1}^{2P-2}$ and $\{\lambda_j(R,\alpha,\beta)\}_{j=1}^{\infty}$ enables one to evaluate the accuracy and stability of the numerical techniques in linear approach.

It follows from eqns (A6),(A7) and (A9) that in the spectrum of the problem (10) there are two degenerated eigenvalues: q=0 and q=-1 (the eigenvalues corresponding to each of them are $\psi_p=\delta_{p,P-1}$ and $\psi_p=\delta_{p,P}$). These q do not depend on R,α,β,τ and P. They are generated by numerical algorithm and are not true eigenvalues. As is follows from eqn (A7) the coefficients w_p corresponding to q=0 and q=-1 are zeros. Hence, the presence of this false eigenvalues is not a defect of the numerical algorithm [1], and below it is not taken into account.

Tables 1,2 and 4 present results of computations of $\{\hat{\lambda}\}$ and $\{\lambda\}$ for values of parameters R, α, β, τ and P corresponding to those of the run [1]: R=4200, α∈[0.5,48], β=0, τ≤0.008791, P=128 (Reynolds number $R=3U_m h/2\nu$, whe h is the channel half-width, U_m is the bulk mean velocity).

We observe from Table 1 that there is rapid convergence $\hat{\lambda}_j$ with decreasing τ. For τ≤0.001 these $\hat{\lambda}_j$ practically coincide with accurate values of λ_j (see Table 4). Thus the accuracy of spatial approximation of numerical algorithm [1] is quite well for all α∈[0.5,48].

At the same time the large time step τ=0.008791 (this time was used in the run [1]) leads to very bad approximation of $\hat{\lambda}$ in a wide range of α ∈ [24,48]. Disagreement between $\{\hat{\lambda}\}$ and $\{\lambda\}$ grows with increasing α. For α≥40 and many β the exponent λ_1 has a positive real part (see Table 2). We say that for α≥40 the scheme is unstable in linear approach and that all numerical TS waves with α≥40 should be considered as "the numerical rubbish". These facts indicate that time step in the run [1] is too large and correctness of the description of high wavenumber eddies (with α≥24) is very doubtful.

2.° Spectral analysis of numerical algorithm [10]

The linearization of the scheme [10] is obtained by replacing the nonlinear operator F^{tr} in the eqns (7) and (8) by linear operator $L[v]=[V_0,\text{rotv}]+ [\text{rotv},V_0]$. Here v are the infinitesimal disturbances of the base flow $V_0(z)=(1-z^2,0,0)$. It is convenient to rewrite the linearized equation (7) in the form:

$$(v_{mn}^{k+1}-v_{mn}^k)/\tau=-(\nabla p^{k+0.5})_{mn}+\nu(\Delta v^{k+0.5})_{mn}+ L_{mn}^{k+0.5}+ r_{mn}^{k+1}\cdot\omega_0(z) + s^{k+1}\cdot\omega_L(z) \quad (A12)$$

where $v_{mn}^k(z)$ are polynomials of the degree at most L; $L^{k+0.5}= L[v^{k+0.5}]$; r^k

and $\mathbf{s}^k$ are vectors of the discrepancy for the linearized eqn (7) at the $z=\pm 1$; $\omega_0(z)=\prod_{k=1}^{L-1}(z-z_k)/(z_0-z_k)$, $\omega_L(z)=\prod_{k=0}^{L-1}(z-z_k)/(z_L-z_k)$, $\{z_k\}_{k=0}^{L}$ are collocation points.

Solutions of eqn (A12) may be described by linear combination of the next solution:

$$(\mathbf{v}^k_{mn,j}, \mathbf{r}^k_{mn,j}, \mathbf{s}^k_{mn,j}) = q_j^k \cdot (\mathbf{v}_{mn,j}, \mathbf{r}_{mn,j}, \mathbf{s}_{mn,j}). \tag{A13}$$

The complex coefficients q_j, functions $\mathbf{v}_{mn,j}(z)$, constants $\mathbf{r}_{mn,j}$ and $\mathbf{s}_{mn,j}$ are determined from the system, which is an analog of the Orr–Sommerfeld equation. To obtain this system the representation (A13) is substituted into eqn (A12). Then applying operator $[\mathbf{k}\cdot\mathrm{rot}^2(\cdot)]_{mn} = [\partial/\partial z \cdot \mathrm{div}(\cdot)]_{mn} - \mathbf{k}\cdot[\Delta(\cdot)]_{mn}$ to this equation and using the obvious relations

$[\mathbf{k}\cdot\mathrm{rot}^2(\mathbf{v})]_{mn} = -(\Delta w)_{mn}$,
$[\mathbf{k}\cdot\mathrm{rot}^2(\mathbf{r}\omega_0)]_{mn} = [im\alpha_0 \cdot r^u_{mn} + in\beta_0 \cdot r^v_{mn}]\cdot\omega_0'(z) + [(m\alpha_0)^2 + (n\beta_0)^2 \cdot]\cdot r^w_{mn}\cdot\omega_0(z)$,
$[\mathbf{k}\cdot\mathrm{rot}^2(\nabla p)]_{mn} = 0$.

we obtain the next analog of the Orr–Sommerfeld equation:

$$\bar{\lambda}\cdot[(\Delta w)_{mn} + \tilde{r}_{mn}\cdot\omega_0' + \tilde{s}_{mn}\cdot\omega_L' + ((m\alpha_0)^2 + (n\beta_0)^2)\cdot(r^w_{mn}\cdot\omega_0 + s^w_{mn}\cdot\omega_L)] = $$
$$= \nu\cdot(\Delta^2 w)_{mn} - (\mathbf{k}\cdot\mathrm{rot}^2[\mathbf{L}])_{mn} \tag{A14}$$

where

$\tilde{r}_{mn} = (im\alpha_0\cdot r^u_{mn} + in\beta_0\cdot r^v_{mn})\cdot(q/\bar{\lambda})$, $\tilde{s}_{mn} = (im\alpha_0\cdot s^u_{mn} + in\beta_0\cdot s^v_{mn})\cdot(q/\bar{\lambda})$

$$\bar{\lambda} = (2/\tau)\cdot(q-1)/(q+1), \tag{A15}$$

To this equation four boundary conditions are added

$$w_{mn} = w'_{mn} = 0, \text{ at } z = \pm 1 \tag{A16}$$

For each pair (m,n) the $w_{mn}(z)$ are the polynomials of the degree at most L. Therefore, the eqn (A14) is equivalent to L+1 linear relations. These relations may be obtained by equating the left and right hand sides of (A14) at the collocation points $\{z_l\}_{l=0}^{L}$. Hence, the system (A14),(A16) can be reduced to standard matrix eigenvalue problem

$$\bar{\lambda}\cdot A\mathbf{x} = B\mathbf{x}, \tag{A17}$$

where A and B are complex $(L-3)\times(L-3)$ matrices, $x_l = w_{mn}(z_{l+1})$, $l=1,2,\ldots,L-3$.

From (A15) we have the relations:

$$\hat{\lambda}_j = 1/\tau \cdot \ln[(1+0.5\bar{\lambda}_j \cdot \tau)/(1-0.5\bar{\lambda}_j \cdot \tau)], \qquad (A18)$$

where $\hat{\lambda}(R,\alpha,\beta,L,\tau) = 1/\tau \cdot \ln(q)$.

Table 3 gives the eigenvalues of the scheme [10]. We observe from this Table, that the scheme [10] is stable at any time step τ and has good accuracy at enough large $\tau=0.0088$. This scheme is more accurate rather than scheme [1] (see Table 1 and 3).

Tables.

Table 1. Spectral characteristics of the scheme [2]. $R=4200, \beta=0, P=128$.

α	τ	$\hat{\lambda}_1$	$\hat{\lambda}_2$	$\hat{\lambda}_3$
0.5	0.0088	−0.028 −0.094i	−0.035 −0.209i	−0.039 −0.461i
	0.005	−0.028 −0.094i	−0.035 −0.209i	−0.039 −0.461i
24	0.0088	−0.146 −24.33i	−0.265 −24.24i	−0.376 −24.14i
	0.005	−0.191 −24.02i	−0.306 −23.93i	−0.413 −23.83i
	0.001	−0.199 −23.88i	−0.312 −23.79i	−0.417 −23.70i
	0.0005	−0.199 −23.88i	−0.312 −23.79i	−0.417 −23.69i
48	0.0088	0.579 −51.82i	0.379 −51.66i	−0.015 −51.34i
	0.005	−0.471 −49.08i	−0.636 −48.93i	−0.799 −48.77i
	0.001	−0.626 −47.90i	−0.781 −47.75i	−0.933 −47.60i

Table 2. Spectral characteristics of the scheme [2].
R=4200, β=0, P=128, τ=0.0088

α	$\hat{\lambda}_1$	$\hat{\lambda}_2$	$\hat{\lambda}_3$
36	−0.063 −37.48i	−0.218 −37.35i	−0.368 −37.21i
38	−0.012 −39.77i	−0.173 −39.63i	−0.330 −39.50i
40	0.056 −42.10i	−0.112 −41.96i	−0.277 −41.82i
42	0.145 −44.47i	−0.031 −44.32i	−0.203 −44.17i
46	0.401 −49.33i	0.210 −49.17i	0.022 −49.02i
48	0.579 −51.82i	0.379 −51.66i	0.181 −51.50i

Tables 3. Spectral characteristics of the scheme [10].
R=4200, β=0, L=128.

α	τ	$\hat{\lambda}_1$	$\hat{\lambda}_2$	$\hat{\lambda}_3$
0.5	0.0088	−0.028 −0.094i	−0.035 −0.209i	−0.039 −0.461i
	0.005	−0.028 −0.094i	−0.035 −0.209i	−0.039 −0.461i
24	0.0088	−0.197 −24.79i	−0.308 −23.70i	−0.413 −23.61i
	0.005	−0.198 −23.85i	−0.311 −23.76i	−0.416 −23.66i
	0.001	−0.199 −23.87i	−0.312 −23.79i	−0.417 −23.69i
	0.0005	−0.199 −23.88i	−0.312 −23.79i	−0.417 −23.69i
48	0.0088	−0.599 −47.20i	−0.747 −47.06i	−0.893 −46.92i
	0.005	−0.617 −47.66i	−0.769 −47.51i	−0.920 −47.37i
	0.001	−0.626 −47.88i	−0.780 −47.73i	−0.932 −47.58i

Table 4. The eigenvalues of the Orr-Sommerfeld eqn.
R=4200, β=0.

α	λ_1	λ_2	λ_3
0.5	−0.028 −0.094i	−0.035 −0.209i	−0.039 −0.461i
24	−0.199 −23.88i	−0.312 −23.79i	−0.417 −23.70i
48	−0.626 −47.89i	−0.781 −47.74i	−0.933 −47.60i

References:

1. Kim,J. Moin,J. Moser,R. (1987) Turbulence statistics in fully developed channel flow at low Reynolds number//J. Fluid Mech. V. 177. P. 133–166.

2. Gilbert,N. (1988) Numerical Simulation of Transition from Laminar to Turbulent Channel Flow // Dissertation, DFVLR, Gottingen, 1988.

3. Rozhdestvensky,B.L.(1973). On the applicability of difference methods for Navier–Stokes equations at high Reynolds number// Dokl. Acad. Nauk USSR, V.211. N 2. P. 308–311 (in Russian).

4. Rozhdestvensky,B.L. Simakin,I.N. (1984) Secondary flows in a plane channel: their relationship and comparison with turbulent flows // J. Fluid Mech. V.147 P. 261–289.

5. Rozhdestvensky,B.L. Simakin,I.N. Stoynov,M.I. (1989) Numerical simulation of turbulent Couette flow in a plane channel//Prikl. Mech. i Technic. Phys. N 2(174), P.60–68 (in Russian).

6. Rozhdestvensky,B.L. Simakin,I.N. Stoynov,M.I. (1986) Study of harmonic disturbances influence on turbulent flows in a plane channel. Preprint Inst. Appl. Math. of the USSR Academy Sciences N 147(in Russian

7. Priymak,V.G. Rozhdestvensky,B.L. (1987) Secondary flows of incompressible fluid in a pipe and their statistical properties. //Dokl. Acad. Nauk USSR 1987. V.297. N 6. P. 1326– 1330 (in Russian).

8. Ponomarev,S.G. Priymak,V.G. Rozhdestvensky,B.L. (1988) Statistically stationary solutions of the Navier – Stokes equations in an annuli. Hydrodynamical characteristics and space – time structures. // J. Vychisl. Mat. i Mat. Phys.V.28. N 9. P.1354 –1366 (in Russian).

9. Ponomarev,S.G. Priymak,V.G. Rozhdestvensky,B.L. (1989) Numerical simulation of heat transfer of viscous incompressible fluid in a plane channel.//Dokl. Acad. Nauk USSR, V.306. N 3. P. 570–574 (in Russian).

10. Rozhdestvensky,B.L. Stoynov,M.I. (1987) The algorithms of integration of the Navier–Stokes equations, which have the analogs to conservation lows.// Preprint Inst. Appl. Mathem. of the USSR Academy Sciences, N 119 (in Russian).

11. Orszag,S.A. Kells,L.C. (1980) Transition to turbulence in plane Poiseuille and plane Couette flows.// J. Fluid Mech. V.96. P. 159–205.

12. Gilbert,N. Kleiser,L. (i987) Low resolution simulation of transitional and turbulent channel flows // Proc. Int. Conf. on Fluid Mech.,Peking Univ. Press,Beijing,China, P. 67–72.

The Solution to the Problem of Vortex Breakdown[*]

Egon Krause
Aerodynamisches Institut, RWTH Aachen
Templergraben 55, D - 5100 Aachen

Abstract

A numerical solution of the Navier-Stokes equations is discussed for time-dependent, three-dimensional, incompressible flow. The solution is based on the artificial compressibility concept and a dual-time stepping procedure. An implicit Roe-type flux-difference splitting is used for the discretization of the convective and central differencing for the viscous terms. The difference equations are written for a Cartesian grid. A new approximation for the downstream boundary conditions is derived. The solution was employed to describe the process of vortex breakdown in isolated slender vortices. It is shown, that stable breakdown can be achieved by iterating the side boundary conditons through a new integral technique. Comparison with experimental flow visualizations shows good qualitative agreement between the vortex structures observed in experiments and simulations.

Introduction

Slender vortices are characterized by a small radial extension of their core in comparison to their length in the axial direction. This is a consequence of the small radial velocity component, which in magnitude is only a fraction of the axial and the circumferential component. If the axial velocity component is reduced, the pressure increases along the axis of the vortex, and upon further reduction, the flow begins to stagnate, and also the circumferential velocity component approaches zero in the vicinity of the stagnation point. The vortex bursts or breaks down, as this phenomenon is referred to, meaning that the origional flow structure in the core of the vortex is suddenly destroyed. Several modes of breakdown were found experimentally, for example, the bubble- and the spiral-type breakdown. Early investigations, as described in the review articles [1] and [2], tried to establish breakdown criteria and explained the breakdown mechanism through approximate theories. One of the major conclusions arrived at in these investigations was, that the flow near the free stagnation point, the breakdown point, is generally no longer axially symmetric and, of equal importance, no longer steady. In order to investigate this problem a numerical solution of the Navier-Stokes equations for time-dependent, incompressible, three-dimensional flow was constructed. The solution is based on the principle of artificial compressibility, extended to time-dependent flow behaviour by the concept of dual-time stepping. Higher-order upwinding is used for the discretization of the terms describing the convective acceleration and the pressure gradient, and central differencing for the terms describing the shear stresses. The solution is established on a Cartesian grid with a matrix splitting of the flux Jacobians and an implicit line-Gauss-Seidel relaxation for each physical time step. The details are described in [3].

---

[*] Dedicated to Prof. K.W. Morton on his 60th birthday

It is known from experimatal and numerical investigations [4], that the boundary conditions imposed are decisive for the breakdown process. For example, the location of the breakdown point may strongly depend on the side boundary conditions, chosen for large radial distances. When breakdown is initiated, it can happen, that the breakdown point moves up- or downstream until it hits the inflow or outflow cross-section. This motion of the bursted vortex structure is usually referred to as unstable breakdown. At least for technical applications it is of great importance to find side boundary conditions for stable breakdown, i. e. a time-independent or almost time-independent location of the breakdown point. Since side boundary conditions compatible with this requirement are in general not known, the question was posed, whether they can be determined. It will be shown, that this is possible through an iteration procedure, which is based on an integral formulation for the pressure far away from the axis of the vortex. This formulation is obtained by integrating the radial momentum equation in the radial direction. Possible upstream motion of the bursted part of the vortex mentioned earlier can then be suppressed by assuming that the local acceleration has to vanish asymptotically upstream and by evaluating all other terms in the pressure integral from the data obtained in the latest step of integration for the Navier-Stokes equations. With a new pressure distribution so obtained for large radial distances, new side boundary conditions for the velocity components can be determined, for example, by integrating the Euler equations for the far field.

The other boundary conditions were prescribed in the following way: For the inflow cross-section of the domain of integration, the distributions of the axial and the circumferential velocity components were assumed to be given. A compatible radial velocity component was obtained by integrating the continuity equation. The static pressure distribution in the inflow cross-section was determined from the radial momentum equation.

A new approximate formulation will be described for the boundary conditions in the outflow cross-section of the domain of integration. Since the outfow velocity is not known in general, it was expanded in a Taylor series in the direction of the main flow. The first and second derivatives in that direction can be expressed through crosswise derivatives of the velocity components and two vorticity components. The latter can be determined in a time expansion, so that the outflow boundary conditions are approximated exact up to second order, provided the axial velocity component remains positive. The initial conditions were assumed to be identical with the inflow boundary conditions.

In the following the flow in slender vortices will be discussed first. The influence of the pressure on the development of the flow in the core of the vortex will be explained in a simplified analysis. Then the method of integration will briefly be reviewed. Thereafter the inflow and the initial conditions will be discussed. Special attention is given to the determination of the side and the outflow boundary conditions. They are discussed in separate chapters. Finally some results of flow calculations will be given. The results obtained so far demonstrate, that boundary conditions for stable breakdown can be determined and that the main features of the bursted vortex structures can be simulated through numerical integration of the Navier-Stokes equations.

Flow in Slender Vortices

The flow in slender vortices is in general described by three velocity components: The component in the circumferential direction

$$w = w(x, r, \Theta, t)$$

describing the rotation of the flow around the axis of the vortex, the component in the radial direction

$$v = v(x, r, \Theta, t)$$

describing the widening or the narrowing of the vortex core, and the component in the axial direction

$$u = u(x, r, \Theta, t)$$

describing the axial motion of the flow. The circumferential and the axial velocity components cause a spiralling of the flow, as indicated in Fig. 1 below.

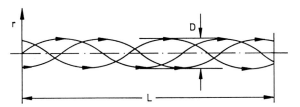

Fig. 1 Schematic of flow in slender vortices

A vortex is called slender, if the diameter of its core D, to be defined later, is small in comparison to the length L, over which the spiralling is observed, i. e. $D/L \ll 1$.

The radial profiles of the velocity components may have different shapes. The circumferential component increases almost linearly with the radial coordinate in the immediate vicinity of the axis, reaches a maximum in a non-linear fashion at some distance away from the axis, and falls off again with $1/r$ or other powers of r. The position of the maximum may be used to define the diameter of the core D (See Fig. 2 below).

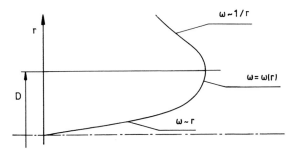

Fig. 2 Typical radial profile of the circumferential velocity component

The radial profile of the axial velocity component can be uniform, jet-like, wake-like, or combinations thereof. The typical shapes are depicted in Fig. 3.

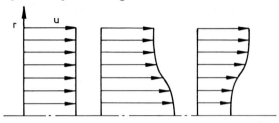

Fig. 3 Typical radial profiles of the axial velocity component (uniform, jet-like, and wake-like)

The velocity profiles are strongly influenced by the pressure distribution in the vortex. This will be shown in the following simplified analysis. Consider a steady, axially symmetric, inviscid, incompressible flow: For $D/L \ll 1$, it follows from the radial momentum equation (see Ref. [4]), that the governing terms are

$$\frac{\rho w^2}{r} = \frac{\partial p}{\partial r}.$$

Differentiation with respect to x yields

$$2\frac{\rho w}{r}\frac{\partial w}{\partial x} = \frac{\partial^2 p}{\partial x \partial r}.$$

The momentum equation for the circumferential direction

$$u\frac{\partial w}{\partial x} + v\frac{\partial w}{\partial r} + \frac{vw}{r} = 0$$

can be written in the form

$$\frac{\partial w}{\partial x} = -\frac{v}{ur}\frac{\partial}{\partial r}(rw)$$

so that

$$\frac{\partial^2 p}{\partial x \partial r} = -2\frac{\rho w}{r^2}\frac{v}{u}\frac{\partial}{\partial r}(rw).$$

Finally, the integration of the last expression in the radial direction gives the following relation for the axial pressure gradient

$$\frac{\partial p}{\partial x}(x,r) = \frac{\partial p}{\partial x}(x, r \to \infty) + 2\rho \int_r^\infty \frac{w}{r'^2}\left(\frac{v}{u}\right)\frac{\partial}{\partial r'}(r'w)dr',$$

from which the following preliminary conclusions can be drawn:

1. If $\frac{\partial p}{\partial x}(x, r \to 0) > 0$, then the axial flow in the vortex core is decelerated. The pressure increases in the direction of the main flow. If rigid body rotation is assumed for the flow in the vicinity of the axis, namely

$$\rho\frac{\omega^2 r^2}{2} = p(r) - p(0),$$

it can be seen, that if $p(0)$ increases and if $p(r)$ remains unchanged, then the circumferential velocity component decreases also.

2. It is known from experiment, that w approaches zero suddenly. Therefore the notion "breakdown" was introduced.

3. Even for $\frac{\partial p}{\partial x}(x, r \to \infty) = 0$, breakdown can occur. The necessary condition for this to happen is, that $w(x,r)$ has such a radial profile that $\frac{\partial p}{\partial x}(x, r \to 0) > 0$.

4. The above relations make clear, that breakdown can be influenced by choosing $\frac{\partial p}{\partial x}(x, r \to \infty)$ appropriately.

5. In the frame of this simplified analysis the location of the breakdown point x_{BD} can be determined. The total differential

$$dp = \frac{\partial p}{\partial x} dx + \frac{\partial p}{\partial r} dr$$

gives with

$$\frac{\partial p}{\partial x}(x, 0) = \frac{\partial p}{\partial x}(x, r \to \infty) + I$$

$$I = 2\rho \int_0^\infty \frac{w}{r^2} \left(\frac{v}{u}\right) \frac{\partial}{\partial r}(rw) dr$$

$$\frac{\partial p}{\partial r} = \frac{\rho w^2}{r} = 0 \quad for \quad r = 0$$

$$dx = \frac{dp}{\left(\frac{\partial p}{\partial x}(x, r \to \infty) + I\right)}$$

the location of the breakdown point

$$x_{BD} = \int_{p(0,0)}^{p_0(0,0)} \frac{dp}{\left(\frac{\partial p}{\partial x}(x, r \to \infty) + I\right)}$$

6. Finally, it can be shown, that the axial pressure gradient $\frac{\partial p}{\partial x}(x, r \to \infty)$ can be expressed through an integral relation, even when the complete conservation equations are used.

Remarks on the Integration of the Conservation Equations
Inflow- and Initial Conditions

Experiments have shown, that for large Reynolds numbers the flow in the bursted part of the vortex is no longer axially symmetric nor steady. It was therefore decided to integrate the Navier-Stokes equations for time-dependent, three-dimensional, incompressible flows. Although cylindrical coordinates could be used, Cartesian coordinates were chosen in order to avoid the difficulties and inaccuracies associated with the singularity on the axis.

The conservation equations

$$\nabla \cdot \vec{V} = 0$$

$$\rho\left(\frac{\partial \vec{V}}{\partial t} + (\vec{V}\cdot\nabla)\vec{V}\right) = -\nabla p + \eta \nabla^2 \vec{V}$$

were written in the usual dimenssionless vector-matrix form:

$$\overline{R}\frac{\partial Q}{\partial t} + \frac{\partial F}{\partial x} + \frac{\partial G}{\partial y} + \frac{\partial H}{\partial z} = \frac{1}{Re}\left(\frac{\partial R}{\partial x} + \frac{\partial S}{\partial y} + \frac{\partial T}{\partial z}\right)$$

The meaning of the various terms is clear and will not be discussed here. It is only mentioned, that the matrix $\overline{R}$ is defined as

$$\overline{R} = \begin{pmatrix} 0 & & & \\ & 1 & & \\ & & 1 & \\ & & & 1 \end{pmatrix}$$

signalizing the singular behaviour of the solution for vanishing density changes. This difficulty can be bypassed by adding an artificial time step $\hat{R}\frac{\partial Q}{\partial t}$, in which the matrix $\hat{R}$ is defined as

$$\hat{R} = \begin{pmatrix} \frac{1}{\beta^2} & & & \\ & 1 & & \\ & & 1 & \\ & & & 1 \end{pmatrix}$$

In contrast to the matrix $\overline{R}$, the matrix $\hat{R}$ in the artificial time step is not singular. The new element in $\hat{R}$ is given by the inverse of the square of the artificial speed of sound β. The equations can now readily be integrated in an iteration procedure, in which the artificial time step is recomputed for each physical time step, until the former vanishes within prescribed error bounds. The details of the discretization process and the solution of the difference equations is given in Ref. [3] by M. Breuer and D. Hänel.

Since the solution is constructed for time-dependent flows, in addition to in-, outflow and side boundary conditions, initial conditions must be prescribed. For reasons to be clear later, the inflow and initial conditions will be discussed here first. Special attention is then paid to the formulation of the outflow and side boundary conditions, which will be discussed separately in the following sections.

In order to initiate the computation, radial profiles of the axial and the circumferential velocity components must be specified in the inflow cross-section of the domain of integration:

$$u(x = 0, r, \Theta, t) = f_1(r, \Theta)$$

$$w(x = 0, r, \Theta, t) = f_2(r, \Theta)$$

With the assumption that the inflow conditions do not depend on time, a compatible profile for the radial velocity component can be computed from the continuity equation, and a compatible pressure distribution from the radial momentum equation. An example for inflow conditions used in recent computations may be found in [3]. Investigations showing the influence of different radial profiles of the inflow conditions on the solution have not been carried out so far in a systematic manner. This is still left for future investigations.

The initial conditions are chosen in such a way that a suitable streamwise variation of the pressure is prescribed. The inflow profiles for the circumferential velocity component can then also be used as initial condition for the entire flow field. Naturally, other initial conditions can be prescribed. In fact, it should be very interesting to study their influence on the development of the solution in time. But if other conditions are imposed, it must be guaranteed, that they are compatible with the inflow conditions specified. Next, a new approximate formulation for the outflow conditions will be discussed.

Outflow Boundary Conditions

The mathematically necessary condition for solving the conservation equations for mass and momentum, namely $\vec{V} = F(x_{out}, r, \Theta, t)$, is generally not available. For that reason physically meaningful assumptions have to be introduced, through which outflow conditions can be determined in an approximation. The question then arises, in what fashion physically meaningful assumptions can be found. Most certainly, the outflow conditions should represent a good approximate solution of the Navier-Stokes equations. One such approximation consists in dropping the term $\frac{1}{Re}\vec{V}_{xx}$ and constructing the outflow conditions from the remaining part of the conservation equations.

Another approximation can be constructed, when there is no backflow in the outflow cross-section. In order to demonstrate the behaviour of the flow there, the velocity vector $\vec{V}$ is expanded in a Taylor series in the x-direction, which is assumed to be the direction of the main flow. If the subscript i denotes the time level t and N the x-coordinate of the outflow cross-section, the expansion for $\vec{V}$ can be written as

$$\vec{V}_{i,N} = \vec{V}_{i,N-1} + \left(\frac{\partial \vec{V}}{\partial x}\right)_{i,N-1} \Delta x + \frac{1}{2}\left(\frac{\partial^2 \vec{V}}{\partial x^2}\right)_{i,N-1} \Delta x^2 + 0(\Delta x^3)$$

the derivatives $\left(\frac{\partial \vec{V}}{\partial x}\right)_{i,N-1}$ and $\left(\frac{\partial^2 \vec{V}}{\partial x^2}\right)_{i,N-1}$ can only be determined numerically, if the velocity vector $\vec{V}$ in the outflow cross-section is known. This difficulty can be circumnavigated by introducing the vorticity vector

$$\vec{\omega} = \xi\vec{i} + \eta\vec{j} + \zeta\vec{k} \quad ,$$

and making use of the continuity equation. The first derivatives in the x-direction can then be written as

$$\frac{\partial u}{\partial x} = -\left(\frac{\partial v}{\partial y} + \frac{\partial w}{\partial z}\right)$$

$$\frac{\partial v}{\partial x} = \frac{\partial u}{\partial y} + 2\zeta$$

$$\frac{\partial w}{\partial x} = \frac{\partial u}{\partial z} - 2\eta$$

Differentiation and subsequent substitution of the above relations gives the following relations for the second derivatives in the x-direction:

$$\frac{\partial^2 u}{\partial x^2} = -\left(\frac{\partial^2 u}{\partial y^2} + \frac{\partial^2 u}{\partial z^2}\right) + 2\left(\frac{\partial \eta}{\partial z} - \frac{\partial \zeta}{\partial y}\right)$$

$$\frac{\partial^2 v}{\partial x^2} = -\left(\frac{\partial^2 v}{\partial y^2} + \frac{\partial^2 w}{\partial y \partial z}\right) + 2\frac{\partial \zeta}{\partial x}$$

$$\frac{\partial^2 w}{\partial x^2} = -\left(\frac{\partial^2 v}{\partial z \partial y} + \frac{\partial^2 w}{\partial z^2}\right) - 2\frac{\partial \eta}{\partial x}$$

With the definition of the following operator matrices and vectors $\vec{B}$ and $\vec{V}_R$

$$D_1 = \begin{pmatrix} 0 & -\frac{\partial}{\partial y} & -\frac{\partial}{\partial z} \\ \frac{\partial}{\partial y} & 0 & 0 \\ \frac{\partial}{\partial z} & 0 & 0 \end{pmatrix} \quad ; \quad D_2 = \begin{pmatrix} \frac{\partial^2}{\partial y^2} + \frac{\partial^2}{\partial z^2} & 0 & 0 \\ 0 & \frac{\partial^2}{\partial y^2} & \frac{\partial^2}{\partial y \partial z} \\ 0 & \frac{\partial^2}{\partial y \partial z} & \frac{\partial^2}{\partial z^2} \end{pmatrix}$$

$$D_3 = \begin{pmatrix} 0 & -\frac{\partial}{\partial y} & -\frac{\partial}{\partial z} \\ 0 & \frac{\partial}{\partial x} & 0 \\ 0 & 0 & -\frac{\partial}{\partial x} \end{pmatrix} \quad ; \quad \vec{B} = \begin{pmatrix} 0 \\ \zeta \\ -\eta \end{pmatrix} \quad ; \quad \vec{V}_R = \begin{pmatrix} 0 \\ w \\ -v \end{pmatrix}$$

the Taylor series expansion for $\vec{V}$ can be written as

$$\vec{V}_{i,N} = \vec{V}_{i,N-1} + (D_1\vec{V})_{i,N-1}\Delta x + \frac{1}{2}\left(D_2\vec{V}\right)_{i,N-1}\Delta x^2$$
$$+ 2\vec{B}_{i,N-1}\Delta x + (D_3\vec{B})_{i,N-1}\Delta x^2 + 0(\Delta x^3).$$

The components of the vector $\vec{B}$, the reduced vorticity vector, can be expanded in another Taylor series in t, namely

$$\vec{B}_{i,N-1} = \vec{B}_{i-1,N-1} + \left(\frac{\partial \vec{B}}{\partial t}\right)_{i-1,N-1}\Delta t + 0(\Delta t^2).$$

Since all flow quantities are known at time level $i-1$, the vector $\vec{B}$ can be determined from the velocity vector $\vec{V}_{i-1,N-1}$ and its derivatives. The time derivative of the reduced vorticity vector $\left(\frac{\partial \vec{B}}{\partial t}\right)_{i-1,N-1}$ can be expressed through the reduced vorticity transport equation

$$\left(\frac{\partial \vec{B}}{\partial t}\right)_{i-1,N-1} = \left[-(\vec{V}\cdot\nabla)\vec{B} + (\vec{\omega}\cdot\nabla)\vec{V}_R + \nu\nabla^2\vec{B}\right]_{i-1,N-1}.$$

With the above relation the expression for $\vec{V}_{i,N}$ can be written as

$$\vec{V}_{i,N} = \vec{V}_{i,N-1} + (D_1\vec{V})_{i,N-1}\Delta x + \frac{1}{2}\left(D_2\vec{V}\right)_{i,N-1}\Delta x^2$$
$$+ 2\vec{B}_{i-1,N-1}\Delta x + (D_3\vec{B})_{i-1,N-1}\Delta x^2$$
$$+ 0(\Delta t \Delta x) + 0(\Delta t \Delta x^2) + 0(\Delta t^2) + 0(\Delta x^3).$$

For explicit integration $\Delta t \sim \Delta x^2$, and, therefore, the terms containing the time derivatives can be neglected, and the last expression can further be simplified to

$$\vec{V}_{i,N} = \vec{V}_{i,N-1} + (D_1\vec{V})_{i,N-1}\Delta x + 2\vec{B}_{i-1,N-1}\Delta x + 0(\Delta x^2) + 0(\Delta t^2).$$

The accuracy of this approximation can be tested with comparison computations. It is noted, that this consideration can also be extended to compressible flows.

The boundary conditon for the pressure in the outflow cross-section can be determined from the momentum equation or from the Poisson equation for the pressure. The side boundary conditions will be considered next.

Choice of Side Boundary Conditions and Stabilization of the Solution

For an isolated slender vortex the side boundary conditions can be constructed from the assumption of inviscid flow, prevailing for large radial distances. Breakdown can be enforced by prescribing a pressure distribution $p(x, r \to \infty)$ with $\frac{\partial p}{\partial x}(x, r \to \infty) > 0$. However, a pressure distribution $p(x, r \to \infty)$ with $\frac{\partial p}{\partial x}(x, r \to \infty) > 0$, in general generates a time-dependent vorticity distribution in the flow field, and, as a consequence the breakdown point may move up- or downstream. If the breakdown point moves all the way up to the inflow cross-section, the solution can no longer converge. If the breakdown point is to be held at one and the same position, the side boundary conditions must be changed. Since side boundary conditions compatible with this requirement are not known at the start of the computation of the flow field, the possibility of finding a search algorithm was investigated. Preliminary considerations showed, that this can be done in several ways. One such method, which was tested in the meantime by M. Breuer of the Aerodynamisches Institut, will briefly be described in the following. It consists out of several steps. They are listed below.

1. The integration of the Navier-Stokes equations is initiated with a pressure distribution $p(x, r \to \infty)$ with $\frac{\partial p}{\partial x}(x, r \to \infty) > 0$.

2. If breakdown of the vortex structure is observed, and if the breakdown point moves upstream, the integration is halted, and a new pressure distribution is determined from the previously obtained results and additional assumptions to be discussed next.

3. The pressure distribution $p(x, r \to \infty)$ is expressed as an integral relation, that can be obtained from the momentum equations. For the sake of simplicity, the integral relation will be derived here for cylindrical coordinates. The derivation can readily be extended to other coordinates. The complete momentum equation for the radial direction can be written in the following way:

$$\frac{\partial p}{\partial r} = -\rho \frac{\partial v}{\partial t} - L_1(\vec{V})$$

The quantity $L_1(\vec{V})$ represents the sum of all other terms in the radial momentum equation. Integration in the radial direction yields

$$p(x, r \to \infty) = p(x, 0) - \rho \int_0^\infty \frac{\partial v}{\partial t} dr - \int_0^\infty L_1(\vec{V}) dr.$$

The pressure on the axis $p(x, 0)$ is determined by integrating the axial momentum equation:

$$p(x, 0) = p(0, 0) - \rho \int_0^x \frac{\partial u}{\partial t} dx' - \int_0^x L_2(\vec{V}) dx'$$

The term $L_2(\vec{V})$ represents the rest of the axial momentum equation. Combining the last two relations gives

$$p(x, r \to \infty) = p(0, 0) - \rho \int_0^x \frac{\partial u}{\partial t} dx' - \rho \int_0^\infty \frac{\partial v}{\partial t} dr - \int_0^\infty L_1(\vec{V}) dr - \int_0^x L_2(\vec{V}) dx'.$$

The pressure for large radial distances $p(x, r \to \infty)$ can now be recomputed from this relation. In particular, it offers the possibility to introduce variations in the local accelerations $\frac{\partial u}{\partial t}$ and $\frac{\partial v}{\partial t}$, and thereby construct an interation procedure for the side boundary conditions. This can be done in the following way:

In order to prevent the bursted part of the vortex from travelling upstream, it is assumed, that the local accelerations $\frac{\partial u}{\partial t}$ and $\frac{\partial v}{\partial t}$ vanish asymptotically, far enough upstream of the breakdown point. This is schematically depicted in Fig. 4.

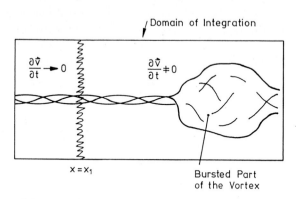

Fig. 4 Schematic diagram of assumptions for stabilizing the flow

If in the relation for $p(x, r \to \infty)$ the local accelerations are assumed to vanish for $x < x_1$, then a new pressure distribution $p(x, r \to \infty)$ can be obtained for each physical time step in the integration of the Navier-Stokes equations with

$$\int_0^x \frac{\partial u}{\partial t} dx' = \int_0^\infty \frac{\partial v}{\partial t} dr = 0 \quad for \quad x \leq x_1$$

For $x > x_1$ the local accelerations $\frac{\partial u}{\partial t}$ and $\frac{\partial v}{\partial t}$ are taken from the latest step of integration, as are the other terms in the integrals.

4. The side boundary conditions for the velocity components can then be obtained by integrating the Euler equations, into which the newly computed pressure is inserted. This iteration procedure was successfully employed by M. Breuer, and some of the results obtained with his search algorithm, which differs slightly from the technique described above, will be discussed in the next section. In the computations carried out so far, only a fraction of the pressure difference $\Delta p(x, r \to \infty)$ computed between two consecutive time steps was imposed in the new boundary condition. This was done since it was not known a priori, for what changes in pressure the solution would converge. Although a number of computations was successfully completed in the manner described, the limits of the technique are not known yet.

Discussion of Results

In this final section some of the computed results will be compared with experimentally determined flow patterns, obtained by flow visualization techniques. The first few figures show the flow structure in the bursted part of an isolated vortex. Fig. 5 shows a picture of the bubble-type breakdown made visuable with fluorescent dye in a test stand for vortex breakdown by R. Teichmann and W. Limberg of the Aerodynamisches Institut.

Fig. 5 Bubble-type breakdown. Picture taken by R. Teichmann and W. Limberg in a test stand of the Aerodynamisches Institut

The dye is injected into the main flow through a thin injection pipe, positioned on the axis of the vortex. The filament of dye, which is spread out over the bubble, as can be seen in the left and middle part of the figure, is finally entrained in the spiralling wake of the bubble. Figs. 6 and 7 show some results of flow computations, carried out by S. Menne, Ref. [5]. Fig. 6 shows traces of particles, released in the inflow cross-section, and Fig. 7 the corresponding vortex lines. Although quatitative comparison is not possible, since the boundary conditions of the experiment could not be determined, the similarity between the experimentally observed and the computed patterns is striking.

Fig. 6 Particle traces for bubble-type breakdown, computed by S. Menne [5]

Fig. 7 Vortex lines of bubble-type breakdown, computed by S. Menne [5]

The characteristic behaviour of unstable breakdown is depicted in the following three figures. Fig. 8 shows a sequence of pictures of bubble-type breakdown as reported by M. Escudier in Ref [6].

Fig. 8 Unstable bubble-type breakdown. Experimental observations. Pictures taken by M. Escudier; a: $t = 12s$, b: $t = 20s$, c: $t = 52s$. Ref. [6]

The flow was made visuable with the light-sheet technique. Shown is the flow structure in the bubble and its wake for three consecutive times. The upstream motion of the bursted part of the vortex, explained earlier, is clearly visuable in the sequence of pictures, each of which showing a cut through the vortex structure in the plane of the light sheet. Comparison with particle traces as computed by S. Menne, Ref. [5], reveals very similar traces in vertical and horizontal planes, as can be seen in Fig. 9, taken from [5].

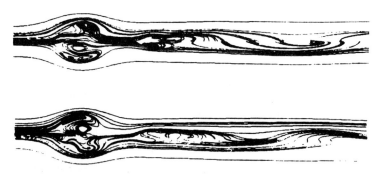

Fig. 9 Particle traces in a vertical(a) and a horizontal(b) plane, computed by S. Menne for unstable bubble-type breakdown, reported in Ref. [5]. Compare with picture in Fig.8

The upstream motion of the bubble in unstable breakdown as observed in the experiments of M. Escudier (Ref. [6]) is also caught in the computations reported by M. Breuer and D. Hänel in [3]. This can be seen in Fig. 10, where the side view of streamline patterns of bubble-type breakdown is shown for $Re = 2000$ for three dimensionless times.

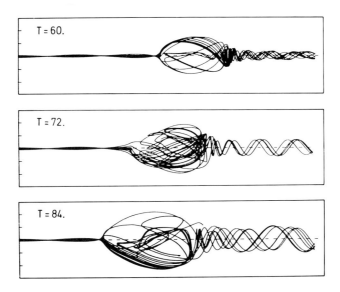

Fig. 10 Unstable bubble-type breakdown. Computed upstream motion of the bubble. $Re = 2000$. Data are from Ref. [3]. Compare with the picture in Fig. 8

The critical position of the bubble during its motion upstream in unstable breakdown was computed in Ref.[3]. This can be seen in Fig. 11. Therein the streakline pattern is depicted for a Reynolds number of $Re = 200$ and a dimensionless time $T = 99$. The bubble is now very close to the inflow cross-section, and the computation cannot be carried much further.

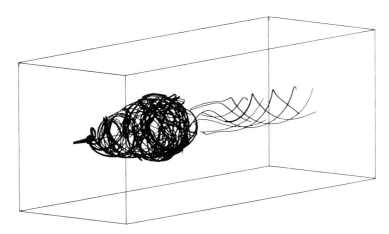

Fig. 11 Unstable bubble-type breakdown. Computed upstream motion of the bubble. $Re = 200$. Data are from Ref. [3]

The solution can be stabilized, if the search algorithm for compatible side boundary conditions is incorporated in the integration of the Navier-Stokes equations. This can be seen in the following three figures. Fig. 12 shows the streakline pattern computed for the flow with the same inflow conditions as that of Fig 11 for a dimensionless time of $T = 270$.

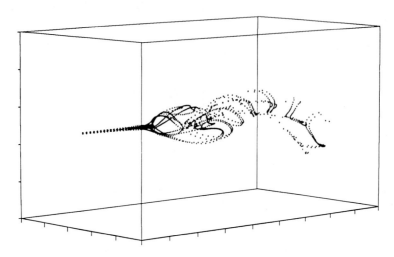

Fig. 12 Stable breakdown. Transition from bubble-type to spiral-type breakdown. $Re = 200$. Computation of Ref. [3]

It is seen, that the bubble-type structure changes to the spiral-type structure, which is even more evident in Figs. 13 and 14. The corresponding dimensionless times are $T = 894$ and $T = 1803$. It is noted, that the breakdown point remains in almost the same position. These results demonstrate, that side boundary conditions compatible for stable breakdown can be found with the method described. It is unknown yet, what is causing the transition from bubble-type to spiral-type breakdown. The clarification of this point needs additional investigation.

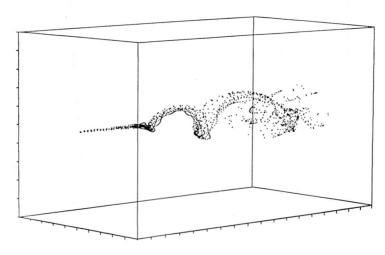

Fig. 13 Stable spiral-type breakdown. $Re = 200$. Computation of Ref. [3]

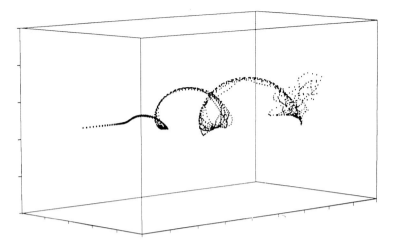

Fig. 14 Stable spiral-type breakdown. $Re = 200$. Computation of Ref. [3]

Finally, it is mentioned, that very similar spiral-type breakdown was observed in experimental investigations of the flow in diffusors of Francis turbines. Fig. 15 shows a picture of such a flow, taken by T. Jacob of the Institut de Machines Hydrauliques et Mechanic de Fluides of the Ecole Polytechnique Federal de Lausanne. The core of the vortex-like flow in the diffusor is visuable, since the pressure in the core is lower than the vapor pressure, and, as a consequence, cavitation bubbles are formed.

Fig. 15 Stable spiral-type breakdown. Experiment by T. Jacob [7].

They remain at the site of the lowest pressure, the core, and enable the observation of its distortion due to vortex bursting without any additional effort.

Conclusion

It was shown that breakdown of slender vortices can be simulated with a numerical solution of the Navier-Stokes equations for time-dependent, three-dimensional, incompressible flows. The vortex structures, which are being generated in the course of the integration, depend strongly on the side boundary conditions. If for large radial distances the pressure increases in the direction of the main flow, the bursted part of the vortex tends to move upstream. The outflow and side boundary conditions were investigated in detail. In this analysis a search algorithm was constructed, through which side boundary conditions compatible with stable breakdown can be found. Sample computations showed the upstream motion of unstable breakdown and the transiton from bubble-type to spiral-type breakdown when side boundary conditions compatible with stable breakdown were implemented. Further investigations are necessary in order to clarify the transition from one mode to another.

References

[1] Hall, M.G.: Vortex breakdown, Ann. Rev. Fluid Mech. 4, 195-217, 1972.

[2] Leibovich, S.: The structure of Vortex breakdown, Ann. Rev. Fluid Mech. 10, 221-246, 1978.

[3] Breuer, M. ; Hänel, D.: Solution of the 3-D, incompressible Navier-Stokes equations for the simulation of Vortex breakdown. Proceedings of the Eight GAMM-Conference on Numerical Methods in Fluid Mechanics, vol. 29, 42-51, 1990.

[4] Krause, E.: A contribution to thew problem of vortex breakdown. AGARD CP No. 342, Aerodynamics of vortical type flows in three dimensions. Paper 26, 1983.

[5] Menne, S.: Simulation of vortex breakdown in tubes, AIAA-Paper 88-3575, 1st National Fluid Dynamics Congress, Cincinatti, Ohio,, July 25-28, 1988.

[6] Escudier, M.: Vortex breakdown: Observation and Explanations, Progress in Aerospace Sciences, vol. 25, No. 2, pp. 189-229, 1988.

[7] Jacob, T.: Private Communication, 1990.

COMPUTATION TECHNIQUES FOR THE SIMULATION OF TURBOMACHINERY COMPRESSIBLE FLOWS

by J.P. Veuillot and L. Cambier
Office National d'Etudes et de Recherches Aérospatiales (ONERA)
BP 72 - 92322 CHATILLON (France)

1. Introduction

The turbojets which ensure the propulsion of transport or combat aircrafts are certainly among the most complex and elaborated manufactured systems. The performances (thrust level, propulsion and thermal efficiencies,...) of such propulsive systems contribute in a very large way to the commercial success of the aircraft and therefore justify the gigantic amount of manpower spent on developing a new engine.

A turbojet is mainly composed of a gas generator which delivers air flow with high total temperature and total pressure and of a propulsive nozzle through which the total enthalpy of the inlet flow is transformed into kinetic energy, hence providing an outlet flow with a high ejection velocity V_e. Thus the thrust F of the engine is given by : $F = \dot{m}(V_e - V_0)$, where $\dot{m}$ is the mass flow rate and V_0 is the aircraft velocity.

Considering the gas generator, it consists of three main parts: the compressor, the combustion chamber and the turbine. Through the compressor, the mechanical work is converted into total enthalpy with an efficiency η_{comp}; through the combustion chamber, the total enthalpy increases again due to the heat provided to the gas, and through the turbine, part of the total enthalpy variation is converted, with an efficiency η_{turb}, into mechanical work which is transmitted to the compressor. The thermal efficiency η of such a process can be defined as the ratio of the total enthalpy which can be recovered at the turbine exit (for instance, by assuming an isentropic expansion of the gas down to the free stream static pressure) over the amount of heat provided to the gas flow through the combustion chamber. In Fig. 1, this efficiency is plotted as a function of the temperature T_B at the compressor exit and is compared to the efficiency η_C of a Carnot cycle (which only depends on the temperature T_C at the turbine inlet) and to the efficiency η_S of an ideal turbojet (i.e. a turbojet in which the compressor and the turbine are supposed to work isentropically: $\eta_{comp} = \eta_{turb} = 1$). The efficiency deficit ($\eta_C - \eta$) can be splitted into two parts: ($\eta_C - \eta_S$) and ($\eta_S - \eta$). The first contribution ($\eta_C - \eta_S$) results from the fact that the flow heating is not performed at the maximum temperature T_C, thus it can be called thermodynamical "losses". It is obvious that the closer the temperature T_B approaches the temperature T_C, the closer these "losses" tend to zero. The second contribution ($\eta_S - \eta$) is due to the aerodynamical losses, which are maximum when the blading is highly loaded (i.e. when the temperature T_B is maximum). These two opposite effects lead to a maximum for the thermal efficiency which mainly depends on the efficiencies of the compressor (η_{comp}) and of the turbine (η_{turb}).

Therefore, from an industrial viewpoint, the challenge is to design a turbojet by minimizing (at least for the cruise conditions) the aerodynamical losses. The origin of these losses is sketched in Fig. 2. This goal requires tools able to analyse and evaluate the aerodynamical performances of the compressor and turbine bladings. The flow analysis can be experimentally done by testing a large number of configurations but clearly this purely experimental process is very expensive and requires a very long time, of the order of several years. Today, due to the development of numerical methods and computing facilities, the trend is to reduce the cost and the time spent by using numerical tools in order to select a few number of configurations that have to be validated by experimental testing. Of course, this combined approach requires the study of numerical methods and of their implementation in codes which have to be robust (for industrial use), efficient (in CPU time and in memory core) and accurate especially for the simulation of complex flows as shown in Fig. 2. The aerodynamical losses are due to the rotational features of the flow (shocks, vortex sheets, viscous layers,...); consequently, the numerical simulation must be performed through the Euler model (with boundary layer calculation, uncoupled or coupled) or through the Reynolds averaged Navier-Stokes model. Therefore this paper is focused on the state-of-the-art of the computation techniques for the simulation of turbomachinery compressible flows by the numerical solution of the Euler and Navier-Stokes equations.

2. Euler and Navier-Stokes equations for turbomachinery flow calculations

This section is devoted to features which are typical of the formulation of the Euler and/or Navier-Stokes equations for turbomachinery flow calculations: the different ways for writing the governing equations in a rotating system, the quasi-3D approximation which allows to save CPU time when comparing to the full 3D Navier-Stokes model and finally the turbulence modelling which can be considered as the weak side in the presented numerical methods.

2.1. Equations for a rotating system

Let us denote $\overline{V}$ the absolute velocity and $\overline{W}$ the relative velocity, these two velocities are related by: $\overline{V} = \overline{W} + \overline{\Omega}.\overline{r}$, where $\overline{\Omega}$ is the antisymmetrical rotation tensor and where $\overline{r}$ is the radius vector. For conciseness, we consider only the Euler equations. For the Navier-Stokes equations, the stress tensor expressing the diffusion due to the molecular viscosity is not affected by the rotation (which is a solid body motion), but the Reynolds tensor expressing the turbulent diffusion may be largely affected by the rotation (see section 2.3.2.). In order to formulate the governing equations in the rotating frame of reference, two main points have to be taken into consideration. The first one concerns the sense of the time derivative. It is clear that the mesh used in the discretization process rotates with the blading, so we have to consider the time derivative $\delta\phi/\delta t$ of any function $\phi(M,t)$, defined with a point M fixed in the rotating wheel; this time derivative is related to the classical time derivative $\partial\phi/\partial t$ (i.e. defined with a point M fixed in the absolute frame of reference), by: $\delta\phi/\delta t = \partial\phi/\partial t + \nabla\phi.\overline{\Omega}.\overline{r}$. The second point is relative to the vectorial momentum equation which have to be projected in a coordinate system $(x_i, \overline{e}_i)$. In the case of a rotating coordinate system which is most frequently used, it is very convenient to define the relative time derivative of a vector (for instance the momentum vector $\overline{Q} = \rho\overline{V}$) in which the basis vectors are supposed not to be time dependent. This relative time derivative $(\delta\overline{Q}/\delta t)_R$ is related to $\delta\overline{Q}/\delta t$ by: $\delta\overline{Q}/\delta t = (\delta\overline{Q}/\delta t)_R + \overline{\Omega}.\overline{r}$. If the flow is steady in the relative frame of reference, only the relative time derivative tends to zero.

In a first approach, the governing equations can be written in the rotating system by choosing the relative velocity $\overline{W}$ in order to define the conservative variables and by considering a relative coordinate system:

$$\delta\rho/\delta t + div\,(\rho\overline{W}) = 0$$
$$(1) \quad (\delta\rho\overline{W}/\delta t)_R + div\,(\rho\overline{W}\,\overline{W} + p\,\overline{I}) = -\rho\nabla\Phi_C - 2\rho\overline{\Omega}.\overline{W}$$
$$\delta\rho E_R/\delta t + div\,(\rho H_R\overline{W}) = -\rho\nabla\Phi_C.\overline{W},$$

where E_R (resp. H_R) is the relative total energy (resp. enthalpy), and $\Phi_C = -\Omega^2 r^2/2$ is the centrifugal potential function. After some manipulations, the energy equation can be classically rewritten as follows: $\delta\rho(E_R + \Phi_C)/\delta t + div\,(\rho J\,\overline{W}) = 0$, where $J = H_R + \Phi_C$ is the rothalpy. The set of equations (1), which is not strictly in conservative form, is used most of the time in numerical codes mainly because when starting from a code written in the absolute frame of reference, it only requires the adding of source terms. But the main drawback of system (1) is that the absolute uniform flow is generally not an exact solution of the discretized equations (the discretized form of $div\,(\rho\overline{W}\,\overline{W})$ does not balance the source term $-\rho\nabla\Phi_C$).

In a second approach, in order to eliminate this drawback, one can define the conservative variables with the absolute velocity $\overline{V}$. In this case, the governing equations are the following:

$$\delta\rho/\delta t + div\,(\rho\overline{W}) = 0$$
$$(2) \quad (\delta\rho\overline{V}/\delta t)_R + div\,(\rho\overline{V}\,\overline{W} + p\,\overline{I}) = -\rho\overline{\Omega}.\overline{W}$$

$\delta \rho E / \delta t + div\, (\rho E\, \overline{W} + p\overline{I}.\overline{V}) = 0$,

The implementation of such a system in a code requires not only the adding of a source term, but also the modification of the fluxes.

2.2. Quasi-3D formulation

In this section we briefly describe the quasi-3D (or 2.5D) Navier-Stokes model which is very typical of some turbomachinery flow calculations. The numerical solution of the full 3D Navier-Stokes equations currently entails calculation costs that are hardly compatible at all with industrial constraints. To remedy this state of affairs, and while waiting for significant progress in the performance of computers and algorithms, one classical approach consists of adopting the hypothesis that the actual 3D flow can be broken into a meridian flow (or through flow) exhibiting axial symmetry and obtained by averaging according to the azimuthal variable θ, and into a family of blade-to-blade flows included between two infinitely close stream surfaces of the meridian flow. This quasi-3D modelling is two-dimensional in the sense that it involves only partial derivatives with respect to two space variables, but integrates three-dimensional effects due to the geometry of the stream surface chosen, to the stream tube thickness and to the rotation speed of the blade row. This approach can be considered as a simplification of the well known C.H. Wu's decomposition, proposed in 1952, of the through flow in "S1" and "S2" stream surfaces. The methodology for the formulation of the blade-to-blade flow on a given stream surface of the through flow is described hereafter. In a first step, a curvilinear coordinate system (α, β, θ) adapted to the geometry of the considered stream surface is defined as follows: $\alpha = cst.$ defines an axisymmetric stream surface of the through flow; $\beta = cst.$ defines an axisymmetric surface generated by a curve orthogonal to the curves $\alpha = cst.$ in a meridional plane; $\theta = cst.$ defines a meridional plane.

An orthogonal and normalized reference frame can be associated with this coordinate system by: $\overline{e}_\alpha = \partial \overline{M} / h_\alpha \partial \alpha$, $\overline{e}_\beta = \partial \overline{M} / h_\beta \partial \beta$ and $\overline{e}_\theta = \partial \overline{M} / r \partial \theta$. It must be noticed that the metric coefficients h_α, h_β and r are independent of θ. After the system of four equations (continuity, energy and momentum in the $\overline{e}_\beta$ and $\overline{e}_\theta$ directions on a given $\alpha = \alpha_0$ stream surface) has been written, further conditions must be found to eliminate the derivatives with respect to α. The Navier-Stokes equations are then written in the (α, β, θ) coordinates, let alone the projection of the momentum equation on the $\overline{e}_\alpha$ direction which is discarded. For an inviscid flow, the only additional assumption is: $V_\alpha = 0$, which is consistent with the previous approach. As far as the Navier-Stokes equations are concerned, we must impose two more conditions corresponding to zero fluxes in the $\overline{e}_\alpha$ direction ($\tau_{\alpha\beta} = \tau_{\alpha\theta} = 0$, but $\tau_{\alpha\alpha} \neq 0$). In practice, on the stream surface $\alpha = \alpha_0$, we choose the coordinates (m, θ), in which m is the normalized curvilinear coordinate in the meridional direction instead of the coordinates (β, θ) so that $dm = h_\beta d\beta$. Then h_α can be considered as a dimensionless stream tube thickness which is generally called b: $b(m) = h_\alpha(\alpha_0, \beta)$. The quasi-3D Navier-Stokes equations to be solved can be found in [1].

2.3. Turbulence modelling

2.3.1. Modelling concepts

There exist two broad classes of turbulence models.
The first and simpler class consists of the eddy viscosity models. In these models, the Reynolds stress tensor $\overline{\tau}_R$ and the turbulent enthalpy diffusion flux $\overline{q}_t$ are related to the gradients of the mean flow by the following expressions:

$$\overline{\tau}_R = -2/3\,(\rho k + \mu_t\, div\overline{V})\overline{I} + \mu_t\,[\nabla \overline{V} + (\nabla \overline{V})^T]\ \text{ and }\ \overline{q}_t = -\frac{\mu_t C_p}{Pr_t}\nabla T$$

In that case, turbulence modelling comes down to evaluating three scalar quantities: the turbulence kinetic energy k, the eddy viscosity coefficient μ_t (calculated from a length scale and a time scale characteristic of the turbulence) and the turbulent Prandtl number Pr_t (generally constant and equal to 0.9).

The eddy viscosity model class is divided into two sub-classes: the algebraic models and the transport equation models. In the algebraic models, the length and time scales are expressed in terms of the mean flow quantities. Some examples of algebraic models are the Michel et al. model [2],the Cebeci-Smith model [3] and the Baldwin-Lomax model [4]. In the transport equation models, supplementary transport equations are used to realize closure. The more common transport equation models are two-equation models giving both length and time scales. Examples of these models are given by the Jones-Launder (k,ε) model [5], the Wilcox-Rubesin (k,ω^2) model [6] and the Coakley (q,ω) model [7]. The second broad class of turbulence models removes the eddy viscosity hypothesis and is based on the Reynolds-stress equations. The full Reynolds stress models (for example, see [6]) that solve transport equations for each of the stresses belong to that class. However, full Reynolds stress models are very complex, and some approximate models have been developed; for instance, the ASM models only use two transport equations, which are complemented by algebraic relations involving the Reynolds stresses and the quantities given by the transport equations.

2.3.2. Turbulence modelling in turbomachinery (today)

A large majority of today's computations for turbulent compressible turbomachinery flows is carried out using simple algebraic turbulence models (e.g. [8-16]). Among these, the most commonly used is the Baldwin-Lomax model [4]. Fairly often, the boundary layer is actually resolved down to laminar sublayer scale, which requires at high Reynolds numbers a very small wall mesh size. In order to reduce the computational cost, some authors still use wall functions, especially for unsteady simulations. A few compressible turbulent flow calculations have been done with two-transport equation models (e.g. [17-20]). Among those calculations, approximately half are done with wall functions and half without wall functions. Generally, curvature and rotation effects are not included in the models. To our knowledge, only one author, C. Hah [21], used a non-eddy viscosity model for compressible turbomachinery flow calculations. The calculations were done with an ASM model taking into account centrifugal effects, and associated with wall functions. To end with this quick review of turbulence modelling used in turbomachinery calculations, we have to note that very often nothing is done concerning transition. In the rare calculations where transition is taken into account, this is done in a very crude way (for instance, by setting to zero the turbulent quantities in some a-priori defined region).

2.3.3. Turbulence modelling in turbomachinery (near future)

In our opinion, the focus in turbulence modelling for compressible turbomachinery flows is going to shift in a near future from algebraic turbulence models to two-transport equation models. As a matter of fact, the two-transport equation models can provide a better representation of the physics ; they are intended to be more appropriate to situations (such as shock wave/boundary layer interactions) where the mean motion and turbulence are not in equilibrium. Furthermore, two-transport equation models are well suited for the introduction of special physical phenomena. In particular, it is desirable for turbomachinery flows to take into account the following phenomena: freestream turbulence, transition phenomena, streamline curvature effects and rotation effects; and it is impossible to do all that with a simple algebraic turbulence model. The advantages obtained from two-transport equation models should overcome the drawbacks, and in particular the increasing complexity and cost (both in terms of CPU time and memory core). It is also known that computations using two-transport equation models are prone to instability and that there are some initialization problems and time step limitations associated with these models.

3. Grid generation for turbomachinery flow calculations

3.1. Generalities

Grid generation is a primary part of obtaining reliable numerical solutions to flows inside turbomachinery. To compute such flows, one has to deal with complex geometries. Complexity arises first from the variety of geometries to consider : axial or centrifugal machines, supersonic compressors or highly cambered turbines, ... Besides, taking into account additional elements such as splitters or part-span dampers is no easy matter. Tip clearance representation also leads to difficult mesh choices. The desirable mesh qualities for turbomachinery flow calculations are the following: good regularity, good orthogonality, easy introduction of periodicity condition and high density for Navier-Stokes calculations. A look at the papers dealing with turbomachinery flow calculations shows that there is a large variety of mesh choices, even in the simplest case of structured grids for a 2D cascade. We are going to discuss here the 2D choices, as 3D meshes are generally built as a stacking of 2D meshes (cf. Fig. 3 [22]).

3.2. Discussion about mesh types

If we first consider a single (or mono-domain) grid, three types of grids can be used for meshing a 2D blade-to-blade channel. Let us now describe the advantages and the drawbacks attached to each grid type. (Unfortunately, both are always present.) A single H-grid (Fig. 4 [19]) is easy to generate, has good far field properties and leads to an easy application of the periodicity condition. However, an H-grid gives a poor resolution of leading and trailing edges. Moreover, an H-grid tends to skew significantly particularly in the case of supersonic compressors or highly cambered turbine blades. A single C-grid (Fig. 5 [23]) provides good resolution in the leading edge region and in the wake, and allows a reduction of the number of mesh points (in particular, for Navier-Stokes calculations and in comparison with an H-grid). Nevertheless, a C-grid has the disadvantages of becoming skewed at the inflow and periodic boundaries, and leading to a more difficult treatment of the inflow and periodicity conditions. Finally, a single O-grid (Fig. 6 [20]) gives good resolution in both leading and trailing edge regions, and allows a saving of mesh points in comparison with an H-grid or a C-grid. However, using an O-grid around a blade also has some drawbacks. First of all, due to the spatial periodicity, a single O-grid inevitably becomes skewed at the inflow, outflow and periodic boundaries. Then, the inflow, outflow and periodicity condition treatments are more difficult than with an H-grid. Finally, for Navier-Stokes calculations, the application of the turbulence model in an O-mesh can also be more difficult, and some authors think that there might be some flow unsteadiness in the trailing edge region. All the previous considerations lead to the conclusion that no single grid system offers all the desirable properties for meshing a 2D cascade channel. Nevertheless, there are two ways to try to satisfy these properties: either to consider structured meshes associated with a multi-domain methodology, or to use unstructured meshes.

3.3. Multi-domain grid systems

A multi-domain grid system allows a combination of the advantages of different mesh types. As an example, Fig. 7 shows a decomposition of the computational mesh into an O-mesh around the blade and two H-meshes respectively located upstream and downstream from the O-mesh. This splitting leads to an accurate description of the round leading and trailing edges while presenting good far field properties and also good regularity and orthogonality qualities. For complex 3D geometries including, for instance splitter blades or tip clearance, a multi-domain grid system can be a quasi-necessity if structured meshes are used. However, the drawback of a multi-domain approach is an increase in complexity, both considering the sub-domain decomposition choice and the information passage between sub-domains. This last problem can be particularly important when implicit numerical schemes are considered.

3.4. Unstructured meshes

The use of unstructured meshes for compressible turbomachinery flow calculations is recent and not yet common. Figure 8 shows the mesh used by Mavriplis [15] in a turbine blade-to-blade channel. The main advantage of using an unstructured mesh is the easy description of complex geometries. Two important advantages are the ability to perform adaptive refinement (with local mesh enrichment) and the reduction of the overall grid size for a given physical problem. The main drawback is the important increase in algorithm complexity. Other drawbacks are often related to the wall region resolution for Navier-Stokes equations and to difficulties in the implementation of a turbulence model. Moreover, application to 3D flows is not well-developed at the present time. Note that some authors patch structured and unstructured meshes. Figure 9 shows the grid used by Nakahashi et al. [16] to compute the viscous flow in a 2D turbine cascade, and composed of a structured C-mesh near the blade and a non-structured mesh in the remaining part of the blade-to-blade channel.

4. Numerical methods

Except for the above mentioned features concerning the formulation of the governing equations, the numerical methods used for the turbomachinery flow calculations do not differ from those used for other applications. Therefore there is a very large variety of numerical methods, which cannot be exhaustively listed in this limited paper, beginning from the very simple MacCormack explicit scheme up to the implicit upwind third order scheme used by M.M. Rai et al. [11]. Most of the approaches are unsteady or pseudo-unsteady (assuming, for instance, that the total enthalpy or the rothalpy is constant) approaches, but steady formulations of the governing equations are considered by J.G. Moore [24] or by C. Hah [17]. The space discretization is mainly performed through the Finite-Difference-Method (FDM) or through the Finite-Volume-Method (FVM), but the Finite-Element-Method (FEM) is also used, for instance, by D.J. Mavriplis [15], by D.G. Holmes et al. [25] or by K. Nakahashi et al. [16]. For the FDM and for the FVM, the centered schemes with artificial dissipation are the most often implemented, but some authors (M.M. Rai et al. [11], C.J. Knight et al. [19]) use upwind schemes. Generally, and especially for steady flow calculations, the convergence of the time marching process is accelerated by special techniques: the classical local time stepping, the implicit methods (Briley-McDonald, Beam-Warming, implicit residual smoothing) and multigrid techniques (Ni, Jameson, Couaillier).

From an industrial viewpoint, one can consider that the codes for the steady quasi-3D Euler, steady quasi-3D Navier-Stokes and steady 3D Euler codes are commonly used for industrial flow analysis, the unsteady 3D Euler and steady 3D Navier-Stokes methods are still being developed and will be used in the near future. But the following step, using the unsteady 3D Navier-Stokes equations, is for the far future.

5. Examples of Navier-Stokes calculations

Due to the limited length of the paper, the numerical applications presented here by increasing approximation level are restricted to some significant results obtained by solution of the Reynolds averaged Navier-Stokes equations.

5.1. 2D steady turbine flow

In the work [17] by C. Hah, the steady Navier-Stokes equations with a low Reynolds number version of the (k,ε) turbulence model are solved by means of an pressure relaxation technique. Both 2D and 3D results are presented in the paper [17] but only 2D results are reported here. The free stream turbulence intensity is taken into account in the calculation. The computation is performed on a H-O composite grid, the O-grid being wrapped around the blade to represent the high gradients near the surface. This fine 110x240 mesh allows a good prediction of thermal effects as it is proved by Fig. 10 which shows comparison with experimental data of the calculated blade surface heat transfer. For this type of turbine flow, the author claims that it is necessary to have about one hundred nodes in the blade passage to have a grid-independent solution. The flow pattern with a detached shock system at the trailing edge is illustrated by the iso-Mach lines given in Fig. 11.

5.2. Quasi-3D steady compressor flow

C. Vuillez and J.P. Veuillot [1] have computed the flow in several sections of a highly twisted fan blading by solving the quasi-3D Navier-Stokes equations (assuming a constant rothalpy) with an algebraic turbulence model. The solver uses an explicit time-marching centered finite-volume scheme with local time stepping and multigrid convergence acceleration. The results presented here have been obtained with a C-mesh around the blade and with an upstream H-mesh in order to have good farfield discretization. The number of mesh points is about 20,000. Figures 12 and 13 represent the calculated iso-Mach lines in two sections near the foot and the tip of the blade.

5.3. Quasi-3D unsteady rotor/stator interaction

In the paper [26] by P.C.E. Jorgenson and R.V. Chima, the unsteady quasi-3D thin layer Navier-Stokes equations are integrated in time with an explicit four stage Runge-Kutta scheme. The Baldwin-Lomax two-layer eddy viscosity model is used. The discretization of the stator and rotor blade-to-blade channels is done in C-type grids generated by the well-known GRAPE code. The solution on the interface between stator and rotor is determined by interpolating the flux variables through a grid overlapping. We present the results obtained for a multipassage configuration with a 2:3 blade count. Each stator blade-to-blade channel has 115x31 grid points. Each rotor blade-to-blade channel has 197x41 grid points. Figures 14 and 15 represent the calculated iso-Mach lines at two time levels corresponding to a half rotor pitch rotation difference and illustrate the interaction of the stator wakes with the rotor leading edges.

5.4. 3D steady turbine flow

The paper [14] by L. Cambier and B. Escande deals with the analysis of secondary flow phenomena in a linear turbine cascade flow by solution of the 3D Navier-Stokes equations. The turbulence model is a 3D extension of a mixing length model taking into account the presence of several walls through the generalized wall distance proposed by Buleev. The numerical method is similar to the one used by C. Vuillez and J.P. Veuillot for the application presented in 5.2.. The meshing of the domain (limited by a lateral wall and by the symmetry plane) is done by a stacking of 33 H-O-H composite 2D grids. The total number of mesh points is about 315,000 and allows a good representation of the secondary flow and especially of the horseshoe vortex. The results presented here correspond to the iso-total pressure lines (Fig. 16) and to the velocity vectors (Fig. 17) in a plane located downstream from the trailing edge. The flow pattern with a total pressure loss core in the wake and with a large vortex is typical for that configuration.

5.5. 3D steady compressor flow with tip clearance

W.N. Dawes [27] presents an attempt to resolve the tip clearance flow in a transonic compressor rotor by using the Navier-Stokes equations with the Baldwin-Lomax turbulence model. The numerical method is based on a finite volume time-marching implicit algorithm very similar to a two-step Runge-Kutta scheme with implicit residual smoothing. A multigrid acceleration technique is implemented. The domain is discretized by a 83x33x33 H-type mesh, the 33x33 mesh of a cross section is shown in Fig. 18. The blade tip is treated by simply reducing the blade thickness smoothly to zero and the clearance region is represented by only four cells. The iso-Mach lines in a cross section of the blading are presented in Fig. 19.

5.6. 3D unsteady rotor/stator interaction

In the work [11] by N.K. Madavan, M.M. Rai and S. Gavali, the 3D unsteady thin-layer Navier-Stokes equations are solved using an upwind implicit finite difference algorithm. This algorithm is third order accurate in space and second order accurate in time. The turbulence model is a modified version of the Baldwin-Lomax model. The rotor/stator configuration is discretized using a multi-domain grid consisting of five sub-domains. O-grids are used around the blades and in the rotor tip clearance region, and overlap with external H-grids. The clearance effects are represented by five cells. The total number of grid points used is 203,055 for a coarse grid calculation and 409,970 points for a fine grid calculation (CPU time for that very complex application is not given). Results from the fine grid calculation are presented here. Figures 20 and 21 depict time-averaged limiting streamlines respectively on the stator hub surface and on the stator suction surface. In particular Fig. 20 shows a saddle point of separation upstream from the leading edge and corresponding to the formation of a horseshoe vortex.

References

[1] C. VUILLEZ and J.P. VEUILLOT - Quasi-3D Viscous Flow Computations in Subsonic and Transonic Turbomachinery Bladings - AIAA Paper 90-2126, July 1990.

[2] R. MICHEL, C. QUEMARD et R. DURANT - Application d' un schéma de longueur de mélange à l' étude des couches limites turbulentes d' équilibre - ONERA NT No 154, 1969.

[3] T. CEBECI and A.M.O. SMITH - Analysis of Turbulent Boundary Layers, Academic Press, 1974.
[4] B.S. BALDWIN and H. LOMAX - Thin Layer Approximation and Algebraic Model for Separated Turbulent Flows - AIAA Paper 78-257, Jan. 1978.
[5] W.P. JONES and B.E. LAUNDER - The Prediction of Laminarization with a Two-Equation Model of Turbulence - International Journal of Heat and Mass Transfer, Vol. 15, 1972.
[6] D.C. WILCOX and M.W. RUBESIN - Progress in Turbulence Modeling for Complex Flow Fields Including Effects of Compressibility - NASA Technical Paper 1517, 1980.
[7] T.J. COAKLEY - Turbulence Modeling Methods for the Compressible Navier-Stokes Equations - AIAA Paper 83-1693, July 1983.
[8] R.V. CHIMA and J.W. YOKOTA - Numerical Analysis of Three-Dimensional Viscous Internal Flows - AIAA Journal, Vol. 28, No 5, May 1990.
[9] R.L. DAVIS, R.-H. NI and J.E. CARTER - Cascade Viscous Flow Analysis Using the Navier-Stokes Equations - AIAA Paper 86-0033, Jan. 1986.
[10] W.N. DAWES - Development of a 3D Navier-Stokes Solver for Application to all Types of Turbomachinery - ASME Paper 88-GT-70, June 1988.
[11] N.K. MADAVAN, M.M. RAI and S. GAVALI - Grid Refinement Studies of Turbine Rotor-Stator Interaction - AIAA Paper 89-0325, Jan. 1989.
[12] J.N. SCOTT and W.L. HANKEY,Jr. - Navier-Stokes Solutions of Unsteady Flow in a Compressor Rotor - Journal of Turbomachinery, Vol. 108, Oct. 1986.
[13] S.V. SUBRAMANIAN and R. BOZZOLA - Numerical Simulation of Three-Dimensional Flow Fields in Turbomachinery Blade Rows Using the Compressible Navier-Stokes Equations - AIAA Paper 87-1314, June 1987.
[14] L. CAMBIER and B. ESCANDE - Calcul de l' écoulement tridimensionnel turbulent dans un aubage rectiligne de turbine - 74th AGARD/PEP Meeting, Luxemburg, August 1989.
[15] D.J. MAVRIPLIS - Euler and Navier-Stokes Computations for Two-Dimensional Geometries Using Unstructured Meshes - NASA Contractor Report 181977 and ICASE Report No. 90-3, January 1990.
[16] K. NAKAHASHI, O. NOZAKI, K. KIKUCHI and A. TAMURA - Navier-Stokes Computations of Two- and Three-Dimensional Cascade Flow Fields - AIAA Paper 87-1315, June 1987.
[17] C. HAH - Numerical Study of Three-Dimensional Flow and Heat Transfer Near the Endwall of a Turbine Blade Row - AIAA Paper 89-1689, June 1989.
[18] J.-S. LIU, P.M. SOCKOL and J.M. PRAHL - Navier-Stokes Cascade Analysis With a Stiff $k-\varepsilon$ Turbulence Solver - AIAA Paper 88-0594, Jan. 1988.
[19] C.J. KNIGHT and D. CHOI - Development of a Viscous Cascade Code Based on Scalar Implicit Factorization - AIAA Paper 87-2150, June 1987.
[20] O. KEY KWON - Navier-Stokes Solution for Steady Two-Dimensional Transonic Cascade Flows - ASME Paper 87-GT-54, June 1987.
[21] C. HAH - Navier-Stokes Calculation of Three-Dimensional Compressible Flow Across a Cascade of Airfoils with an Implicit Relaxation Method - AIAA Paper 86-1689, Jan. 1986.
[22] K.F. WEBER, D.W. THOE and R.A. DELANEY - Analysis of Three-Dimensional Turbomachinery Flows on C-type Grids Using an Implicit Euler Solver - ASME Paper 89-GT-85, June 1989.
[23] Y.K. CHOO, W.-Y. SOH and S. YOON - Application of a Lower-Upper Implicit Scheme and an Interactive Grid Generation for Turbomachinery Flow Field Simulations - ASME Paper 89-GT-20, June 1989.
[24] J.G. MOORE - An Elliptic Calculation Procedure for 3D Viscous Flow - AGARD LS-140, June 1985.
[25] D.G. HOLMES, S.H. LAMSON and S.D. CONNELL - Quasi-3D Solutions for Transonic, Inviscid Flows by Adaptive Triangulation - ASME Paper 88-GT-83,1988.
[26] P.C.E. JORGENSON and R.V. CHIMA - Explicit Runge-Kutta Method for Unsteady Rotor-Stator Interaction - AIAA Journal, Vol. 27, No 6, June 1989.
[27] W.N. DAWES - Analysis of 3D Viscous Flows in Transonic Compressors - VKI Lecture Series 1988-03, Feb. 1988.

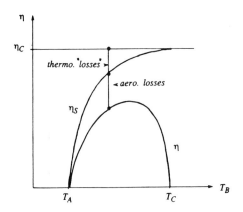

Fig.1 - Thermal efficiency of a turbojet

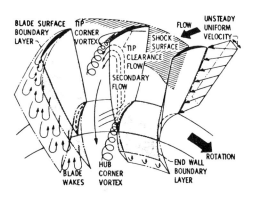

Fig.2 - Flow phenomena in a rotor [NASA]

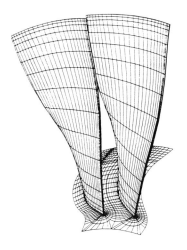

Fig.3 - 3D grid for a fan rotor [22]

Fig.4 - H-grid for a rotor blade [19]

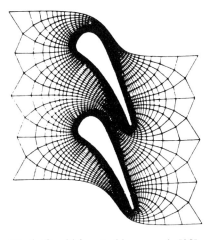

Fig.6 - O-grid for a turbine cascade [20]

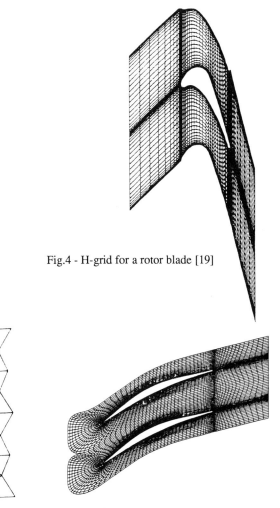

Fig.5 - C-grid for a compressor cascade [23]

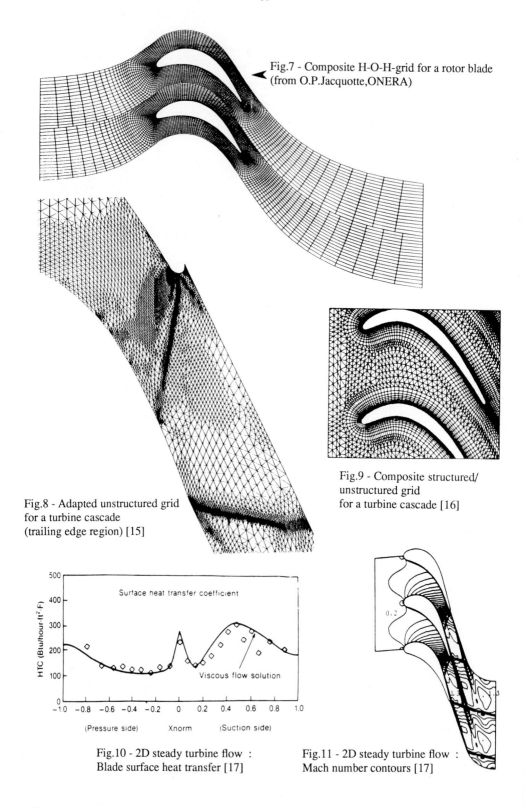

Fig.7 - Composite H-O-H-grid for a rotor blade (from O.P.Jacquotte,ONERA)

Fig.8 - Adapted unstructured grid for a turbine cascade (trailing edge region) [15]

Fig.9 - Composite structured/unstructured grid for a turbine cascade [16]

Fig.10 - 2D steady turbine flow : Blade surface heat transfer [17]

Fig.11 - 2D steady turbine flow : Mach number contours [17]

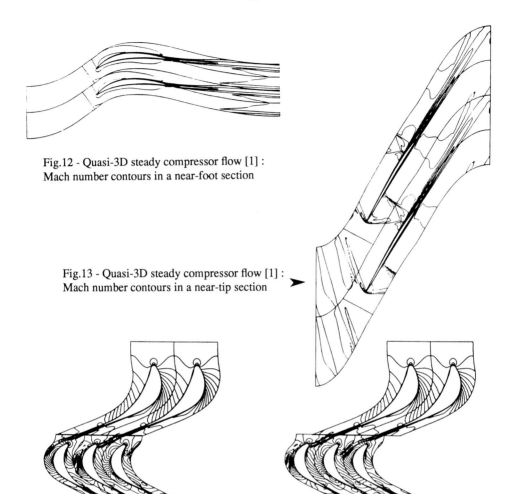

Fig.12 - Quasi-3D steady compressor flow [1] :
Mach number contours in a near-foot section

Fig.13 - Quasi-3D steady compressor flow [1] :
Mach number contours in a near-tip section

Fig.14 - Quasi-3D unsteady rotor/stator interaction : Fig.15 - Quasi-3D unsteady rotor/stator interaction :
Mach number contours [26] Mach number contours (1/2 rotor pitch rotation) [26]

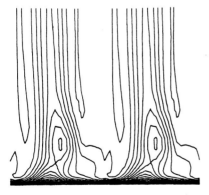
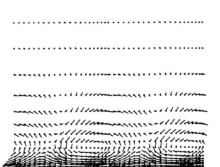

Fig.16 - 3D steady turbine flow [14] : Total pressure contours downstream from the trailing edge

Fig.17 - 3D steady turbine flow [14] : Velocity vectors downstream from the trailing edge

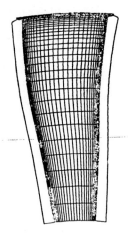 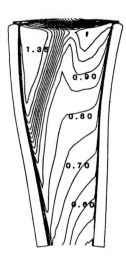

Fig.18 - 3D steady compressor flow with tip clearance [27] : Mesh of a cross section

Fig.19 - 3D steady compressor flow with tip clearance [27] : Mach number contours in a cross section

 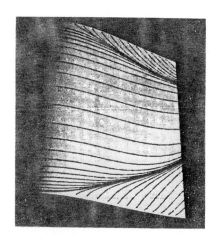

Fig.20 - 3D unsteady rotor/stator interaction : Time-averaged limiting streamlines on the stator hub [11]

Fig.21 - 3D unsteady rotor/stator interaction : Time-averaged limiting streamlines on the stator suction side [11]

SEMI-LAGRANGIAN INTEGRATION SCHEMES AND THEIR APPLICATION TO ENVIRONMENTAL FLOWS

Andrew Staniforth and Jean Côté
Recherche en prévision numérique, Environment Canada
2121 Route Trans-canadienne, porte 508
Dorval, Québec
CANADA, H9P 1J3

INTRODUCTION

Accurate and *timely* forecasts of weather elements are of great importance to both the economy and public safety. Weather forecasters rely on guidance provided by Numerical Weather Prediction (NWP), a computer-intensive chain of operations beginning with the collection of data from around the world and culminating in the production of weather charts and computer-worded messages. At the heart of the system are the numerical models used to assimilate the data and to forecast future states of the atmosphere. The accuracy of the forecasts depends among other things on model resolution. Increased resolution, given the *real-time* constraints, can only be achieved by judiciously combining the most efficient numerical methods on the most powerful computers with the most appropriate programming techniques.

A longstanding problem in the integration of numerical weather prediction models is that the maximum permissible timestep has been governed by considerations of stability rather than accuracy. For the integration to be stable, the timestep has to be so small that the time truncation errors are very much smaller than the spatial truncation errors, and it is therefore necessary to perform many more timesteps than would otherwise be the case. The choice of time integration scheme is therefore of crucial importance when designing an efficient weather forecast model, and this is also true when designing Environmental Emergency Response (EER) models. Early NWP models used an explicit leapfrog scheme, whose timestep is limited by the propagation speed of gravitational oscillations. By treating the linear terms responsible for these oscillations in an implicit manner, it is possible to lengthen the timestep by about a factor of six, at little additional cost and without degrading the accuracy of the solution [Kwizak and Robert (1971)]. Such a scheme is termed *semi-implicit* in the meteorological literature. Nevertheless, the maximum stable timestep still remains much smaller than seems necessary from considerations of accuracy alone.

Discretization schemes based on a semi-Lagrangian treatment of advection [Krishnamurti (1962)] have elicited considerable interest in the past several years for the efficient integration of weather-forecast models, since they offer the promise of allowing larger timesteps (with no loss of accuracy) than Eulerian-based advection schemes (whose timestep length is overly-limited by considerations of stability). To achieve this end it is essential to associate a semi-Lagrangian treatment of advection with a sufficiently-stable treatment of the terms responsible for the propagation of gravitational oscillations. By associating a *semi-Lagrangian* treatment of advection with a *semi-implicit* treatment of gravitational oscillations, Robert (1981) demonstrated a further increase of a factor of six in the maximum stable timestep, at some additional cost. This idea was demonstrated in the context of a *three*-time-level shallow-water finite-difference model in Cartesian geometry, and resulted in the time truncation errors finally being of the same order as the spatial ones. Since Robert's seminal paper, the method has been extended in several important ways. These are discussed below, after a description of the fundamentals of semi-Lagrangian advection.

SEMI-LAGRANGIAN ADVECTION

In an *Eulerian* advection scheme an observer watches the world evolve around him at a fixed geographical point. Such schemes work well on regular Cartesian meshes (facilitating vectorisation and parallelisation of the resulting code), but often lead to overly-restrictive timesteps due to considerations of computational stability. In a *Lagrangian* advection scheme an observer watches the world evolve around him as he travels with a fluid particle. Such schemes can often use much larger timesteps than Eulerian ones, but have the disadvantage that an initially regularly-spaced set of particles will generally evolve to a highly-irregularly-spaced set at later times, and important features of the flow may consequently not be well represented. The idea behind *semi-Lagrangian* advection schemes is to try to get the best of both worlds: the regular resolution of Eulerian schemes and the enhanced stability of Lagrangian ones. This is achieved by using a different set of particles at each timestep, the set of particles being chosen such that they arrive exactly at the points of a regular Cartesian mesh at the end of the timestep.

Passive advection in 1-d

To present the basic idea behind the semi-Lagrangian method in its simplest context, we apply it to the 1-d advection equation

$$\frac{dF}{dt} \equiv \frac{\partial F}{\partial t} + \frac{dx}{dt}\frac{\partial F}{\partial x} = 0 , \tag{1}$$

where

$$\frac{dx}{dt} = U(x,t) , \tag{2}$$

and U(x,t) is a given function. Eq.(1) states that the scalar F is constant along a fluid path (or trajectory or characteristic). In Fig. 1, the *exact* trajectory in the (x-t) plane of the fluid particle that arrives at meshpoint x_m at time $t_n+\Delta t$ is denoted by the solid curve AC, and an *approximate* straight-line trajectory by the dashed line A'C. Let us assume that we know F(x,t) at all meshpoints x_m at times $t_n-\Delta t$ and t_n, and that we wish to obtain values at the same meshpoints at time $t_n+\Delta t$. The essence of semi-Lagrangian advection is to approximately integrate (1) along the approximated fluid trajectory A'C. Thus

$$\frac{F(x_m, t_n+\Delta t) - F(x_m-2\alpha_m, t_n-\Delta t)}{2\Delta t} = 0 , \tag{3}$$

where α_m is the distance BD the particle travels in x in time Δt, when following the approximated trajectory A'C. Thus if we know α_m, then the value of F at the arrival point x_m at time $t_n+\Delta t$ is just its value at the upstream point $x_m-2\alpha_m$ at time $t_n-\Delta t$. However we have not as yet determined α_m: even if we had, we only know F at meshpoints, and generally it still remains to evaluate F somewhere between meshpoints.

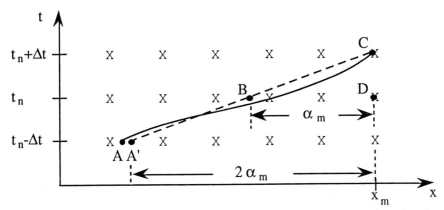

Fig. 1: Schematic for actual (solid curve) and approximated (dashed line) trajectories that arrive at meshpoint x_m at time $t_n+\Delta t$. Here α_m is the distance the particle is displaced in x in time Δt.

To determine α_m, we note that U evaluated at the point B of Fig. 1 is just the inverse of the slope of the straight line A'C, and this gives us the following $O(\Delta t^2)$ approximation to (2)

$$\alpha_m = \Delta t\, U(x_m-\alpha_m, t_n) . \tag{4}$$

Eq.(4) may be iteratively solved for the displacement α_m, for example by

$$\alpha_m^{(k+1)} = \Delta t\, U(x_m-\alpha_m^{(k)}, t_n) , \tag{5}$$

with some initial guess for $\alpha_m^{(0)}$, provided U can be evaluated between meshpoints. To evaluate F and U between meshpoints, spatial interpolation is used. The semi-Lagrangian algorithm for passive advection in 1-d in summary is thus:

(i) Solve (5) iteratively for the displacements α_m for all meshpoints x_m, using some initial guess (usually its value at the previous timestep), and an interpolation formula.
(ii) Evaluate F at upstream points $x_m-2\alpha_m$ at time $t_n-\Delta t$ using an interpolation formula.
(iii) Evaluate F at arrival points x_m at time $t_n+\Delta t$ using (3).

We defer to a little later the discussion of interpolation details, and first generalize the above algorithm to forced advection in several space dimensions.

Forced advection in multi-dimensions

Consider the forced advection problem

$$\frac{dF}{dt} + G(\mathbf{x},t) = R(\mathbf{x},t) , \tag{6}$$

where

$$\frac{dF}{dt} = \frac{\partial F}{\partial t} + V(x,t) \cdot \nabla F, \qquad (7)$$

$$\frac{dx}{dt} = V(x,t), \qquad (8)$$

∇ is the gradient operator, x is the position vector (in 1,2 or 3-d), and G and R are forcing terms. A semi-Lagrangian approximation to (6) and (8) is then:

$$\frac{F^+ - F^-}{2\Delta t} + \frac{1}{2}[G^+ + G^-] = R^0, \qquad (9)$$

$$\alpha = \Delta t \, V(x-\alpha, t), \qquad (10)$$

where the superscripts "+", "0" and "-" respectively denote evaluation at the arrival point $(x, t+\Delta t)$, the midpoint of the trajectory $(x-\alpha, t)$ and the departure point $(x-2\alpha, t-\Delta t)$. Here, x is an arbitrary point of a regular (1,2 or 3-d) mesh. The above is a centered $O(\Delta t^2)$ approximation, where we have evaluated G as the time average of its values at the endpoints of the trajectory, and R at the midpoint of the trajectory. The trajectories are calculated by iteratively solving (10) for the vector displacements α in an analogous manner to the 1-d case for passive advection [eq.(5)]. If G is known (we assume that R is known since it involves evaluation at time t), then the algorithm proceeds in an analogous manner to the 1-d passive advection one. If G is not known at time $t+\Delta t$ (for instance if it involves another dependent variable in a set of coupled equations), then this leads to a coupling to other equations (more on this later).

Interpolation

A priori any interpolation could be used to evaluate F and U (or V) between meshpoints in the above algorithm. In practice the choice of interpolation formula has an important impact on the accuracy and efficiency of the method. Various interpolations have been tried including: linear; quadratic Lagrange; cubic Lagrange; and cubic spline. For step (ii) of the algorithm, it is found [see e.g. McDonald (1984) and Pudykiewicz & Staniforth (1984) for analysis] that cubic interpolation is a good compromise between accuracy and computational cost. Cubic interpolation gives 4th-order spatial accuracy with very little damping (it is very scale selective, affecting primarily the smallest scales), whereas linear interpolation has unacceptably large damping (it is also scale selective, but has a much less sharp response). Cubic spline interpolation has the useful property that it conserves mass [Bermejo (1990)].

For step (i), the order of the interpolation is much less important. Theoretically McDonald (1984) has shown that one should use an interpolation of order one less than for step (ii), e.g. quadratic interpolation of U when using cubic interpolation of F. In practice however, and in the context of coupled systems of equations in several spatial dimensions, it is found that there is no advantage in using anything but linear interpolation for the computation of the displacements, which is very economical. It is also found that there is no advantage in using more than two iterations for solving the displacement equation [step (i].

Stability and accuracy (and connection with other advection methods)

Analyses of the stability properties of the semi-Lagrangian advection scheme [e.g. Bates & McDonald (1982), Pudykiewicz & Staniforth (1984)] show that the maximum timestep isn't limited by the maximum wind speed, as is the case for Eulerian advection schemes, and consequently it is possible to stably integrate with Courant numbers (C =U Δt/Δx) that far exceed unity. To illustrate this point we reproduce (with permission) the results of Bermejo (1990) for the slotted cylinder test [introduced by Zalesak (1979)]. In Fig. 2a we show the slotted cylinder at initial time, and in Fig. 2b the corresponding result after *six* revolutions of solid-body rotation at uniform angular velocity about the domain center. The experiment was conducted using a cubic-spline interpolator at a Courant number of 4.2. This is recognized as being a challenging test, and the result is remarkably good. In particular the results illustrate the scheme's ability to handle sharp discontinuities without disastrous consequences (even though it was not designed specifically to do so) and the absence of noticeable dispersion problems (which are typically present for Eulerian advection schemes).

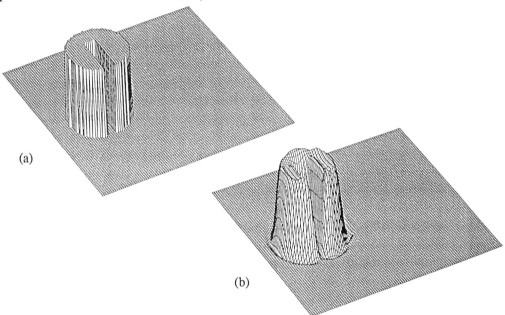

Fig. 2: The 'slotted' cylinder, (a)- at initial time, and (b)- after 6 revolutions.

In general we find that semi-Lagrangian advection compares favorably with Eulerian advection with respect to accuracy, but it has the added advantage that this accuracy can be achieved at less computational cost, since it can be integrated stably with timesteps that far exceed those of the limiting timestep of Eulerian schemes.

Semi-Lagrangian advection is intimately connected with several other advection methods that have appeared in the literature over the years, including particle-in-cell [e.g. Raviart (1985)] and characteristic Galerkin [e.g. Morton (1985)] methods. Indeed for uniform advection in 1-d, the simplest semi-Lagrangian advection scheme (using linear interpolation, and not recommended) is equivalent to both classical upwinding and to the simplest characteristic Galerkin method : and semi-Lagrangian advection using cubic-spline interpolation is equivalent to a recommended higher-order characteristic Galerkin method described in Morton (1985). Further, under more general conditions (including non-uniform advection in 2 and 3-d), Bermejo (1990) has shown that semi-Lagrangian advection using cubic-spline interpolation can be viewed as being a particle-in-cell finite-element method.

It is also equivalent for uniform 1-d advection to several other methods, but what distinguishes it from other methods is that it generalizes differently to non-uniform advection in multi-dimensions. The principal difference is the use of (10) for the trajectory calculations. Of particular importance is that *the trajectory computation is $O(\Delta t^2)$ accurate*. It is possible to use a simpler, and cheaper, $O(\Delta t)$ accurate method to approximate the displacement equation (8), as in e.g. Bates & McDonald (1982), but this can dramatically deteriorate the accuracy of the scheme, as demonstrated by Staniforth & Pudykiewicz (1985), and is consequently not recommended.

Two-time-level advection schemes (and a pollutant-transport application)

Present semi-Lagrangian schemes are based on discretization over either *two* or *three* time levels, and thus far we have restricted our attention to three-time-level schemes. The principal advantage of *two-time-level* schemes over three-time-level ones is that they are potentially *twice* as fast. This is because three-time-level schemes require timesteps half the size of two-time-level ones for the same level of time truncation error [Temperton & Staniforth (1987)]. It is however important to maintain second-order accuracy in time in order to reap the full benefits of a two-time-level scheme (since enhanced stability with large timesteps is of no benefit if it is achieved at the expense of diminished accuracy). Early two-time-level schemes for NWP models unfortunately suffered from this deficiency [e.g. Bates & McDonald (1982), Bates (1984), McDonald (1986)]. The crucial issue is how to efficiently determine the trajectories to at least second-order accuracy in time [Staniforth & Pudykiewicz (1985)].

This problem arises in the context of self-advection of momentum. To see this we reexamine the above algorithm for 1-d advection. Provided U is known at time t_n, *independently of F at the same time*, then it is possible to evaluate the trajectory, and then leapfrog the value of F from time $t_n-\Delta t$ to $t_n+\Delta t$, without knowing any value of F at time t_n. Proceeding in this way, $F(t_n+3\Delta t)$ is then obtained using values of $F(t_n+\Delta t)$ and $U(t_n+2\Delta t)$. Thus we have two decoupled independent integrations, one using values of F at *even* timesteps and U at *odd* timesteps, the other using values of F at *odd* timesteps and U at *even* timesteps. Either of these two independent solutions is sufficient, thus halving the computational cost, and we obtain a two-time-level scheme (for the advected quantity F) by merely relabelling time levels $t_n-\Delta t$, t_n and $t_n+\Delta t$ respectively as t_n, $t_n+\Delta t/2$ and $t_n+\Delta t$ (see Fig. 1). Note that values of U (assumed known) only appear at time level $t_n+\Delta t/2$, and they are *solely* used to estimate the trajectories.

This is the essence of the algorithm described and analyzed in Pudykiewicz & Staniforth (1984). It led to the development of a three-dimensional pollutant transport model [Pudykiewicz et al (1985)], where a family of chemical species are advected and diffused in the atmosphere using winds and diffusivities: these are either provided by a NWP model (for real-time prediction) or from analyzed data (for post-event simulations). This model has evolved into Canada's Environmental Emergency Response Model, and is designed to provide real-time guidance in the event of an environmental accident. The model employs a variable-resolution Cartesian mesh, with a uniform-resolution sub-domain that is focused over the disaster area. It has been used to successfully simulate the dispersion of nuclear debris from the Chernobyl reactor accident [Pudykiewicz (1989)].

Returning to the problem of self-advection of momentum, the above argument breaks down in the special case where F = U in (1), i.e. when the transported quantity U is advected by itself, as is the case for the momentum equations of fluid-dynamic problems in general, and NWP models in particular. This problem was addressed simultaneously and independently by Temperton & Staniforth (1987) and McDonald & Bates (1987), opening the way towards stable *and* accurate two-time-level schemes. The key idea here is to time-extrapolate the winds [with an $O(\Delta t^2)$-accurate extrapolator] to time-level $t+\Delta t/2$ using the known winds at time levels t and $t-\Delta t$: these winds are then used to obtain sufficiently-accurate [$O(\Delta t^2)$] estimates of the trajectories, which in turn are used to advance the dependent variables from time level t to $t+\Delta t$. Thus the two-time-level algorithm to solve (6)-(8), analogous to the three-time-level one given by (9)-(10), is

$$\frac{F^+ - F^0}{\Delta t} + \frac{1}{2}[G^+ + G^0] = R^{1/2}, \qquad (11)$$

where

$$\alpha = \Delta t\, V^*(x-\alpha/2, t+\Delta t/2), \qquad (12)$$
$$V^*(x, t+\Delta t/2) = (3/2)V(x,t) - (1/2)V(x, t-\Delta t) + O(\Delta t^2), \qquad (13)$$

the superscripts "+", "1/2" and "0" now respectively denote evaluation at the arrival point $(x, t+\Delta t)$, the midpoint of the trajectory $(x-\alpha/2, t+\Delta t/2)$ and the departure point $(x-\alpha, t)$, and α is still the distance the fluid particle is displaced in time Δt.

In the above formulation the evaluation of $R^{(1/2)}$ involves extrapolated quantities and therefore could potentially lead to instability. Temperton & Staniforth (1987) didn't find this to be a problem when some weak nonlinear metric effects were evaluated in this way, but it is preferable to evaluate all non-advective terms (i.e. G in the above) as time averages along the trajectory. [Subsequently Côté (1988) showed how to avoid evaluating the above-mentioned metric terms in terms of extrapolated quantities.] When all non-advective terms are evaluated implicitly as time averages along the trajectory, as we recommend, then the extrapolated quantities are used *solely* for the purpose of obtaining a sufficiently-accurate estimate of the trajectories.

APPLICATION TO COUPLED SETS OF EQUATIONS

To illustrate how semi-Lagrangian advection can be advantageously used to solve coupled systems of equations, we apply it to the discretization of the shallow-water equations

$$\frac{dU}{dt} + \phi_x - fV = 0, \qquad (14)$$
$$\frac{dV}{dt} + \phi_y + fU = 0, \qquad (15)$$
$$\frac{d\ln\phi}{dt} + U_x + V_y = 0, \qquad (16)$$

where U and V are the wind components, ϕ (=gz) is the geopotential height (i.e height multiplied by g) of the free surface of the fluid above a flat bottom, and f is the Coriolis parameter.

These equations are often used in NWP to test new numerical methods, since they are a 2-d prototype of the 3-d equations that govern atmospheric motions (they can be derived from them under certain simplifying assumptions). They share several important properties with their progenitor. A linearization of the equations reveals that there are two basic kinds of motion associated with these equations, slow-moving Rossby modes (which most affect the weather, and which move to leading order at the local wind speed) and small-amplitude fast-moving gravitational oscillations (which cannot in any case be adequately represented at initial time due to the paucity of the observational network). From a numerical standpoint this has the important implication that the timestep of an explicit Eulerian scheme (e.g. leapfrog) is limited by the speed of the fastest-moving gravity mode. Since for atmospheric motions this speed is six times faster than those associated with the Rossby modes that govern the weather, this leads to timesteps that are six times shorter than those associated with an explicit treatment of advection. A time-implicit treatment of the pressure-gradient term of the vector momentum equation [2nd terms of (14) and (15)] and horizontal divergence of the continuity equation [2nd and 3rd terms of (16)], introduced in Kwizak and Robert (1971) and termed the *semi-implicit* scheme, allows stable integrations with no loss of accuracy using timesteps that are six times longer than that of the leapfrog scheme. The price to be paid for this increase in timestep length is the need to solve an elliptic-boundary-value problem once per timestep: nevertheless this improves efficiency by approximately a factor of five. Analysis shows that the maximum-possible timestep length is then limited by the Eulerian treatment of advection, and this motivated Robert (1981) to associate a semi-Lagrangian treatment of advection with a semi-implicit treatment of the terms responsible for gravitational oscillations, and permits five-times-longer timesteps in the context of a *three-time-level* scheme.

We discretize (14)-(16) using the *two*-time-level semi-implicit semi-Lagrangian scheme of Temperton & Staniforth (1987), which permits a further doubling of efficiency. Thus

$$\frac{U^+ - U^0}{\Delta t} + \frac{\phi_x^+ + \phi_x^0}{2} - \frac{1}{2}[(f V)^+ + (f V)^0] = 0, \qquad (17)$$

$$\frac{V^+ - V^0}{\Delta t} + \frac{\phi_y^+ + \phi_y^0}{2} + \frac{1}{2}[(f U)^+ + (f U)^0] = 0, \qquad (18)$$

$$\frac{\ln \phi^+ - \ln \phi^0}{\Delta t} + \frac{1}{2}[(U_x + V_y)^+ + (U_x + V_y)^0] = 0, \qquad (19)$$

where we have discretized (14)-(16) using (11) with R set to zero. Here advection terms are treated as time-differences along the trajectories and all other terms are treated as time-averages along the trajectories, leading to an $O(\Delta t^2)$-accurate scheme. Where traditional (three-time-level) semi-implicit time discretizations have an explicit time-treatment of the Coriolis terms we have a time-implicit treatment, since we want an $O(\Delta t^2)$-accurate scheme and explicitly evaluating these terms at time t would in any case be unstable. The trajectories are computed using (12)-(13). For the 1-d shallow-water equations it can be shown that there are three characteristic velocites in the coupled set, one being the local wind speed and associated with the slow Rossby modes that govern weather motions, the other two being associated with the propagation of gravitational oscillations. Thus the coupling of a semi-Lagrangian treatment of advection with a semi-implicit treatment of gravitational oscillations corresponds to integrating along the most important characteristic direction of the problem: this is somewhat similar in spirit to a suggestion given on p. 860 of Morton (1985).

Eqs. (17)-(18) can be manipulated to give

$$U^+ = -\frac{\Delta t}{2}\left[a\,\phi_x^+ + b\,\phi_y^+\right] + \text{known}, \tag{20}$$

$$V^+ = -\frac{\Delta t}{2}\left[a\,\phi_y^+ - b\,\phi_x^+\right] + \text{known}, \tag{21}$$

where $a = [1+(f\Delta t/2)^2]^{-1}$ and $b = (f\Delta t/2)\,a$. Taking the divergence of (20)-(21) and eliminating this in (19) then leads to the elliptic-boundary-value problem

$$\left[(a\,\phi_x)_x + (a\,\phi_y)_y + (b\,\phi_y)_x - (b\,\phi_x)_y - 4\frac{\ln\phi}{\Delta t^2}\right]\Big|_{(x,t+\Delta t)} = \text{known}. \tag{22}$$

We can now summarize the above as the following algorithm:

(i) Extrapolate **V** using (13) and solve (12) iteratively for the displacements α_m for all meshpoints x_m, using values at the previous timestep as initial guess, and an interpolation formula. Note that it is only necessary to perform this computation once per timestep, since the same trajectory is used for all three advected quantities.

(ii) Compute upstream (superscript 0) quantities in (17)-(19) by first computing derivative terms (e.g. U_x) and *then* evaluating quantities upstream (these two operations are *not* commutative!). Here it is more efficient to collect together all terms to be evaluated upstream in a given equation before interpolating (the distributive law applies here).

(iii) Solve the elliptic-boundary-value problem (22) for $\phi(x,t+\Delta t)$.

(iv) Back substitute $\phi(x,t+\Delta t)$ into (20)-(21) to obtain $U(x,t+\Delta t)$ and $V(x,t+\Delta t)$.

The above elliptic-boundary-value problem is weakly non-linear and is solved iteratively using ϕ at the previous timestep as a first guess. It is only marginally more expensive to solve than the Helmholtz problem associated with traditional three-time-level semi-implicit discretizations.

RECENT ADVANCES

When Robert (1981) proposed associating a semi-Lagrangian treatment of advection with a semi-implicit treatment of gravitational oscillations, it was thought that this approach was restricted to three-time-level schemes in Cartesian geometry using a finite-difference discretization. This has happily proved not to be the case, and in this section we discuss some important extensions of the approach. Although important, the extension to *two-time-level schemes* has already been discussed in some detail, and will therefore only be briefly discussed in this section in the context of other extensions.

Finite-element discretizations and variable-resolution

Pudykiewicz & Staniforth (1984) coupled semi-Lagrangian advection with a uniform-resolution finite-element discretization of the diffusion terms in the solution of the 2-d advection-diffusion equation, and this was extended to the 3-d case in Pudykiewicz et al (1985). Staniforth &

Temperton (1986) extended the methodology in the context of a coupled system of equations (the shallow-water equations) in two ways. Firstly they showed that in this context the semi-Lagrangian method could be coupled to a spatial discretization scheme other than a finite-difference one, viz. a *finite-element* discretization, and secondly that it could also be applied on a *variable-resolution* Cartesian mesh. A set of comparative tests demonstrated that with a six-times-longer timestep it is as accurate as its analogous semi-implicit Eulerian version [Staniforth & Mitchell (1978)] when run with its maximum-possible timestep (which in turn uses a six-times-longer timestep than an Eulerian leapfrog scheme).

Temperton & Staniforth (1987) then demonstrated a further doubling of efficiency by replacing the three-time-level scheme of the Staniforth & Temperton (1986) model by a two-time-level one. Both these models use a differentiated (vorticity-divergence) form of the governing equations. This has the advantage of easily allowing variable resolution, but has the disadvantage of incurring additional interpolations and the need to solve two Poisson problems, resulting in an approximately 20% overhead when compared to the ideal. This overhead can be eliminated by the use of the *pseudo-staggered* scheme proposed in Côté et al (1990).

Non-interpolating schemes

The interpolation in a semi-Lagrangian scheme, as mentioned previously, leads to some damping of the smallest scales. While this damping is very scale selective, it has been argued that it would unacceptably degrade accuracy for very long simulations (e.g. many decades in the context of a climate model). To address this problem Ritchie (1986) proposed a non-interpolating version of semi-Lagrangian advection. The basic idea here is to decompose the trajectory vector into the sum of two vectors, one which goes to the nearest meshpoint, the other being the residual. Advection along the first trajectory is done via a semi-Lagrangian technique that displaces a field from one meshpoint to another (and therefore requires no interpolation), while the advection along the second vector is done via an undamped three-time-level Eulerian approach such that the residual Courant number is always less than one. Thus the attractive stability properties of interpolating semi-Lagrangian advection are maintained but without the consequent damping. There are however a couple of disadvantages of this approach for problems where the small damping of the interpolating scheme is acceptable: this is generally the case for NWP applications. Firstly, the non-interpolating method will have the dispersive properties of its Eulerian component, which are not generally as good as those of interpolating semi-Lagrangian advection schemes. Secondly, being based on a three-time-level scheme it is potentially twice as expensive as a two-time-level interpolating scheme.

Shape-preserving interpolation

Although most authors have adopted polynomial schemes for the interpolatory steps of semi-Lagrangian schemes, other interpolators are also possible. Williamson & Rasch (1989) and Rasch & Williamson (1990) have examined several different possible interpolators, designed to better preserve the shape of advected fields and to maintain monotinicity. They performed experiments in both Cartesian and spherical geometry, and concluded that this is a viable approach. The difficulty with this approach is to decide precisely the required attributes of the interpolator, and how to tailor it to respect them, since there is no universal best choice.

Spherical geometry

The convergence of the meridians at the poles of an Eulerian finite-difference model in spherical geometry leads to unacceptably-small timesteps being required in order to maintain computational stability. The usual approach to this problem is to somehow filter the dependent variables in the vicinity of the poles. While this procedure does relax the stability constraint, it unfortunately deteriorates accuracy. Ritchie (1987) demonstrated that it is possible to passively advect a scalar over the pole using semi-Lagrangian advection with timesteps far exceeding the limiting timestep of Eulerian advection schemes. This paved the way to applications in global *spherical* geometry. The first such application was to couple semi-Lagrangian advection with a *spectral* representation (i.e. expansion in terms of spherical harmonics) of the dependent variables to solve the shallow-water equations over the sphere [Ritchie (1988)]. A new problem arose here associated with the stable advection of a *vector* quantity (momentum). The solution proposed by Ritchie (1988) is to introduce a tangent plane to avoid a weak instability due to a metric term. An alternative solution, proposed by Côté (1988), is to use a Lagrange multiplier method. Both methods give good results which are almost indistinguishable in practice. Ritchie (1988) successfully integrated his model with a timestep six-times longer than that of the limiting timestep of the corresponding Eulerian semi-implicit spectral model (which in turn uses a six-times-longer timestep than that of an Eulerian leapfrog model). Côté & Staniforth (1988) then doubled the efficiency of the Ritchie (1988) model, by replacing its three-time-level scheme by a two-time-level one analogous to that of Temperton & Staniforth (1987) for Cartesian geometry.

The spectral method (i.e. expansion of the dependent variables in terms of spherical harmonics) has been the method of choice during the past decade for the horizontal discretization of global NWP models. The spectral method ultimately becomes very expensive at high-enough resolution, due to the $O(N^3)$ cost of computing the Legendre transforms, where N is the number of degrees of freedom around a latitude circle. Finite-difference and finite-element methods on the other hand have a potential $O(N^2)$ cost. This, and the success of the semi-Lagrangian method in addressing the pole problem, suggests that it would be highly advantageous to use a semi-Lagrangian treatment of advection in a finite-difference or finite-element global model for medium-range forecasting. A first tentative step in this direction was taken by McDonald and Bates (1989), who introduced semi-Lagrangian advection into a two-time-level global semi-implicit shallow-water model. Although their scheme was stable with time steps that exceeded the limiting time step of an Eulerian treatment of advection, the enhanced stability was unfortunately achieved at the expense of accuracy. The degradation of accuracy is attributable to the use of a time-decentered integration scheme. The solution to the problem is to adopt a more time-centered scheme [Bates et al (1990)] very similar to that employed in Ritchie (1988) and Côté & Staniforth (1988), and results in significant improvements in accuracy. Nevertheless, Bates et al (1990) found it necessary to use divergence damping (with what appears to be a rather large coefficient) in order to integrate to 5 days, suggesting that there still remain some accuracy and/or stability problems. Côté & Staniforth (1990) replaced the spectral discretization in the Côté & Staniforth (1988) model by a pseudo-staggered finite-element one [analogous to that described in Côté et al (1990)], to obtain a two-time-level semi-implicit semi-Lagrangian global model of the shallow-water primitive equations. Its performance at comparable resolution matched that of their corresponding 1988 model based on a spectral discretization, and this performance was achieved without recourse to any divergence damping.

3-d NWP applications

Thus far we have only discussed 2-d applications for NWP. To be useful the method must be applicable in 3-d. A first step in this direction was taken in Bates & McDonald (1982) who employed a semi-Lagrangian treatment of *horizontal* advection in a 3-d (baroclinic primitive equations) model with a split-explicit time scheme. This scheme is only $O(\Delta t)$ accurate and although stable with long timesteps, the increase in timestep is limited very much by accuracy considerations. The 3-d model formulated in McDonald (1986) is also $O(\Delta t)$ accurate and is limited in the same way, but has the advantage that the revised time scheme makes it more efficient. McDonald & Bates (1987) subsequently modified the trajectory calculations of the McDonald (1986) model to make them $O(\Delta t^2)$ accurate, which improves accuracy and allows longer timesteps. Nevertheless the resulting scheme still has some $O(\Delta t)$ truncation errors. Robert et al (1985) introduced a three-time-level $O(\Delta t^2)$-accurate 3-d model with a semi-Lagrangian treatment of *horizontal* advection. The limitation of this model, and the other 3-d models mentioned above, is that the timestep is limited by the stability of the explicit treatment of *vertical* advection: or put another way, vertical resolution is limited when using a large timestep. To remove this limitation, Tanguay et al (1989) proposed a three-time-level model that uses semi-Lagrangian advection in all *three* space dimensions: this finite-element model uses a timestep which is three-times longer than that of the corresponding Eulerian version [Staniforth & Daley (1979)]. It is currently used by the Canadian Meteorological Center to operationally produce weather forecasts to 48 h twice daily. Ritchie (1990) has recently introduced semi-Lagrangian advection into a three-time-level 3-d global spectral model in two different ways. The first way uses an interpolating semi-Lagrangian scheme in all three dimensions, as in Tanguay et al (1989), whereas the second uses an interpolating semi-Lagrangian scheme for horizontal (2-d) advection and a non-interpolating scheme for vertical advection.

SOME RESULTS

2-d shallow-water models in spherical geometry

To illustrate, under highly-controlled conditions, the efficiency gains possible when using semi-Lagrangian advection we show some 2-d results to appear in Côté & Staniforth (1990). The initial geopotential height ϕ for all experiments is shown in Fig. 3a, and the high-resolution control 5-day forecast in Fig. 3b. The control forecast was run using a 3-time-level semi-implicit Eulerian spectral model at a resolution of T213 (i.e the spherical harmonic expansions are triangularly truncated at wave number 213) with a timestep of 6 mins. In Figs. 3c and 3d respectively we show the 5-day forecasts of two two-time-level semi-implicit semi-Lagrangian models [those of Côté & Staniforth (1988) and Côté & Staniforth (1990)], both run at comparable, but lower, resolution with 2-hour timesteps. These two models differ in their spatial discretization. The first employs a *spectral* discretization, and was run at T126 resolution using a 384 x 190 Gaussian computational grid (which eliminates quadratic aliasing), whereas the second employs a *finite-element* discretization using a 375 x 187 uniform lat-lon grid.

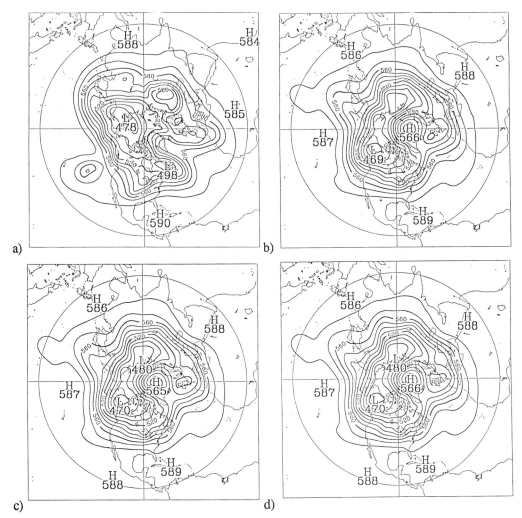

Fig. 3: Geopotential height in dam (contour interval : 10 dam). (a) At initial time. (b) High-resolution 5-day Eulerian spectral T213 control forecast, $\Delta t=6$ min. (c) 5-day semi-Lagrangian spectral T126 (computational grid: 384 x 190) forecast, $\Delta t=2$ h. (d) 5-day semi-Lagrangian finite-element (computational grid: 375 x 187) forecast, $\Delta t=2$ h.

Qualitatively the two 5-day semi-Lagrangian forecasts agree very well with the control one. In order to investigate the growth of time truncation error as a function of timestep length, a series of forecasts was run for each model using different values of Δt. The results at 5 days are displayed in Fig. 4 where $\Delta \tau$ is the interval over which time derivatives are calculated. Thus $\Delta \tau = \Delta t$ for the two semi-Lagrangian integrations, whereas $\Delta \tau = 2\Delta t$ for the T126 Eulerian integration.

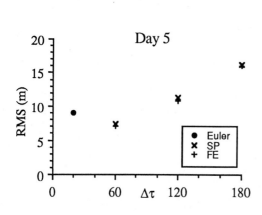
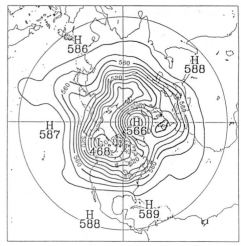

Fig.4 R.m.s. height differences (m) at day 5 with respect to the Eulerian spectral T213 control, as a function of Δτ (min). Eulerian spectral T126 (computational grid: 384 x 190) - •; semi-Lagrangian spectral T126 (computational grid: 384 x 190) - ×; semi-Lagrangian finite-element (computational grid: 375 x 187) - +.

Fig.5 Geopotential height in dam (contour interval : 10 dam). 5-day Eulerian spectral T126 (computational grid: 384 x 190), Δt=10 min.

The solid circle shows the r.m.s. difference from the high-resolution (T213) control Eulerian forecast when using the same Eulerian scheme (shown in Fig. 5) but at the resolution (T126) of the Côté & Staniforth (1988) integration. Since the time step of the T126 Eulerian integration is necessarily small (10 mins, for stability reasons) this provides an approximate measure of the spatial truncation error associated with the resolution of our two semi-Lagrangian integrations. We see that the two semi-Lagrangian integrations are of almost identical quality for fixed Δt.

The choice of an acceptable level of accuracy (and consequently a scheme and time step) is somewhat arbitrary. We arbitrarily use the criterion of 3 m /day (i.e. 15 m after 5 days) on the basis that such errors are insignificant when compared with typical 5-day forecast errors. Thus we see that for the two semi-Lagrangian schemes we can use time steps almost as long as 3 h. This is approximately 18 times longer than the limiting time step of a semi-implicit Eulerian spectral model at equivalent resolution, and more than 100 times larger than that of an explicit leapfrog spectral model.

3-d baroclinic primitive equations model in Cartesian geometry

To illustrate a 3-d application we reproduce some results from Tanguay et al (1989). We compare two 48 h forecasts performed using a 125 x 101 x 19 variable mesh. The 125 x 101 horizontal mesh (Fig. 6) is on a polar-stereographic projection (true at 60 deg North) and has a 84 x 58 uniform-resolution subdomain (of meshlength 100 km) covering N America: for clarity, only every other line of the mesh is shown. This is the configuration used operationally. Both

3-time-level finite-element models solve the (3-d) hydrostatic primitive equations, and both have the same semi-implicit treatment of gravitational oscillations. The only difference between the models is that the first (see Fig. 7a for its 48 h forecast of mean-sea-level pressure) has an Eulerian treatment of advection with a 400 sec timestep (close to the limit), whereas the second (see Fig. 7b for its forecast of the same field) has a (3-d) semi-Lagrangian scheme with a three-times-longer timestep. We see that the differences are acceptably small. Of interest is how well the semi-Lagrangian version has preserved the strong gradient around the low over the Bering sea. This is particularly noteworthy since this low came from the mid-Pacific and deepened by 31 mb during the first 36 h of the forecast to become a cold coastal low.

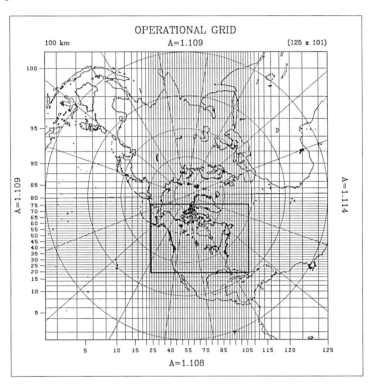

Fig. 6: A 125 x 101 variable-resolution horizontal grid on a polar-stereographic projection, having uniform 100 km resolution over the 84 x 58 window. For clarity, only every other line of the mesh is shown.

CONCLUSIONS

Recent developments in applying semi-Lagrangian methods to 2-d and 3-d atmospheric flows in both Cartesian and spherical geometries have been reviewed. The models described are generally found to be at least an order of magnitude more efficient than corresponding explicit leapfrog-based models run at the same resolution. The efficiency gains are presently more spectacular for 2-d models than 3-d ones, since they have been further developed, but considerable gains have already been realised for 3-d models with the hope of more to come.

Applications include: an operational 3-d short-range (<48 h) weather forecast model; and an operational 3-d emergency-response model for use during environmental accidents such as the release of toxic gases. The challenge now is to develop $O(\Delta t^2)$-accurate two-time-level 3-d models, and obtain the further doubling of efficiency (demonstrated in 2-d) that accrues from the use of a *two-time-level* scheme (work in progress). It is possible that the discussed methodology may be advantageously applied to other sub-disciplines of fluid dynamics.

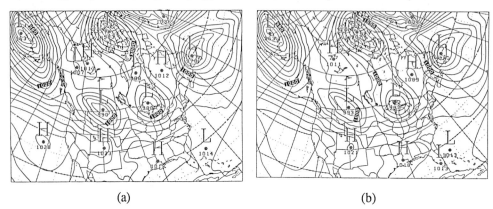

(a) (b)

Fig. 7 (a) 48 h forecast (over uniform-resolution window) of the MSL pressure from the Eulerian finite-element model with a timestep of 400 s valid 0000 UTC 23 December 1978. Contour interval is 4 mb. (b) As in (a), but for the three-dimensional semi-Lagrangian model with a timestep of 1200 s.

ACKNOWLEDGEMENTS

The authors gratefully acknowledge the assistance kindly provided by Sylvie Gravel and Michel Roch in the preparation of this paper.

REFERENCES

Bates, J.R., and A.McDonald, Multiply-upstream, semi-Lagrangian advective schemes: analysis and application to a multi-level primitive equation model, *Mon.Wea.Rev.* 112, 2033-2047 (1982).

Bates, J.R., An efficient semi-Lagrangian and alternating-direction implicit method for integrating the shallow-water equations, *Mon.Wea.Rev.* 112, 2033-2047 (1984).

Bates, J.R., F.H.M. Semazzi and R.W. Higgins, Integration of the shallow-water equations on the sphere using a vector semi-Lagrangian scheme with a multigrid solver, *Mon.Wea.Rev.* 118, in press (1990).

Bermejo, R., On the equivalence of semi-Lagrangian and particle-in-cell finite-element methods, *Mon.Wea.Rev.* 118, 979-987 (1990).

Côté, J., A Lagrange multiplier approach for the metric terms of semi-Lagrangian models on the sphere, *Q.J.Roy.Met.Soc.* 114, 1347-1352 (1988).

Côté,J., and A. Staniforth, A two-time-level semi-Lagrangian semi-implicit scheme for spectral models, *Mon.Wea.Rev.* 116, 2003-2012 (1988).

Côté,J., S. Gravel and A. Staniforth, Improving variable-resolution finite-element semi-Lagrangian integration schemes by pseudo-staggering, *Mon.Wea.Rev.* 118, in press (1990).

Côté, J., and A. Staniforth, An accurate and efficient finite-element global model of the shallow-water primitive equations, *Mon.Wea.Rev.* 118, in press (1990).

Krishnamurti, T.N., Numerical integration of primitive equations by a quasi-Lagrangian advective scheme, *J.Appl.Met.*, 1, 508-521 (1962).

Kwizak, M., and A.J. Robert, A semi-implicit scheme for gridpoint atmospheric models of the primitive equations, *Mon.Wea.Rev.* 99, 32-36 (1971).

McDonald, A., Accuracy of multiply-upstream, semi-Lagrangian advective schemes, *Mon.Wea.Rev.* 112, 1267-1275 (1984).

McDonald, A., A semi-Lagrangian and semi-implicit two-time-level integration scheme, *Mon.Wea.Rev.* 114, 824-830 (1986).

McDonald, A., and J.R. Bates, Improving the estimate of the departure point position in a two time-level semi-Lagrangian and semi-implicit model, *Mon.Wea.Rev.* 115, 737-739 (1987).

McDonald, A., and J.R. Bates, Semi-Lagrangian integration of a gridpoint shallow-water model on the sphere, *Mon.Wea.Rev.* 117, 130-137 (1989).

Morton, K.W., Generalised Galerkin methods for hyperbolic problems, *Comp.Meth.Appl.Mech.Eng.* 52, 847-871 (1985).

Pudykiewicz, J., and A. Staniforth, Some properties and comparative performance of the semi-Lagrangian method of Robert in the solution of the advection-diffusion equation, *Atmos.Ocean* 22, 283-308 (1984).

Pudykiewicz, J., R. Benoit and A. Staniforth, Preliminary results from a partial LRTAP model based on an existing meteorological forecast model, *Atmos.Ocean* 23, 267-303 (1985).

Pudykiewicz, J., Simulation of the Chernobyl dispersion with a 3-D hemispheric tracer model, *Tellus* 41B, 391-412 (1989).

Rasch, P., and D. Williamson, On shape-preserving interpolation and semi-Lagrangian transport, *SIAM J.Sci.Stat.Comput.*, in press (1990).

Ritchie, Eliminating the interpolation associated with the semi-Lagrangian scheme, *Mon.Wea.Rev.* 114, 135-146 (1986).

Ritchie, H., Semi-Lagrangian advection on a Gaussian grid, *Mon.Wea.Rev.* 115, 608-619 (1987).

Ritchie, H., Application of the semi-Lagrangian method to a spectral model of the shallow-water equations, *Mon.Wea.Rev.* 116, 1587-1598 (1988).

Ritchie, H., Application of the semi-Lagrangian method to a multi-level spectral primitive equations model, submitted to *Q.J.Roy.Met.Soc.* (1990).

Robert, A., A stable numerical integration scheme for the primitive meteorological equations, *Atmos.Ocean* 19, 35-46 (1981).

Robert, A., T.L. Yee and H. Ritchie, A semi-Lagrangian and semi-implicit numerical integration scheme for multilevel atmospheric models, *Mon.Wea.Rev.* 113, 388-394 (1985).

Staniforth, A., and R. Daley, A baroclinic finite-element model for regional forecasting with the primitive equations, *Mon.Wea.Rev.* 107, 107-121 (1979).

Staniforth, A., and H. Mitchell, A variable-resolution finite-element technique for regional forecasting with the primitive equations, *Mon.Wea.Rev.* 106, 439-447 (1978).

Staniforth, A., and J. Pudykiewicz, Reply to comments on and addenda to "Some properties and comparative performance of the semi-Lagrangian method of Robert in the solution of the advection-diffusion equation.", *Atmos.Ocean* 23, 195-200 (1985).

Staniforth, A., and C. Temperton, Semi-implicit semi-Lagrangian integration schemes for a barotropic finite-element regional model, *Mon.Wea.Rev.* 114, 2078-2090 (1986).

Tanguay, M., A. Simard and A. Staniforth, A three-dimensional semi-Lagrangian scheme for the Canadian regional finite-element forecast model, *Mon.Wea.Rev.* 117, 1861-1871 (1989).

Temperton, C., and A. Staniforth, An efficient two-time-level semi-Lagrangian semi-implicit integration scheme, *Q.J.Roy.Met.Soc.* 113, 1025-1039 (1987).

Williamson, D., and P. Rasch, Two-dimensional semi-Lagrangian transport with shape-preserving interpolation, *Mon.Wea.Rev.* 117, 102-129 (1989).

Zalesak, S.T., Fully multi-dimensional flux-corrected transport, *J.Comput.Phys.* 31, 335-362 (1979).

COMPUTER TRENDS FOR CFD

Dr. Thomas D. Taylor
Center for Naval Analyses
Alexandria, Virginia

Background

- SINCE THE BEGINNING OF ELECTRONIC COMPUTING THE COMPUTER DESIGNER HAS PRODUCED MACHINES THAT HE LIKED AND THEN RENTED THEM TO CFD PEOPLE BECAUSE THEY WERE TOO EXPENSIVE FOR THE SMALL USER TO BUY:
    - IBM
    - CDC
    - CRAY

- SOLID STATE TECHNOLOGY BROUGHT CHEAPER MACHINES WITH MORE PROCESSORS AND THE DESIGNERS THEN BEGAN TO PRODUCE MACHINES THAT CFD USERS COULD BEGIN TO PURCHASE
    - CONNECTION MACHINE
    - N-CUBE
    - HYPERCUBE
    - ALLIANT
    - ETC.

The advent of new computer chip technology allowed these transitions to occur. It meant that for CFD the computer began to be under the control of the user and computer costs became more affordable for the researcher. This was a big change, however the user still had little influence on computer design.

The Classic CFD Problem

- THE COMPUTER CENTER APPROACH

WE DO NOT ACCEPT COMPLAINTS FROM USERS

RATES/TURNAROUND	
PLATINUM PROJECT ($10,000/HR)	ASAP
GOLD PROGRAM ($5,000/HR)	1-2 DAYS
CFD DEADBEAT ($1,000)	WHEN WE GET TO IT
NO REFUNDS	

The classic computer center approach often is an "insult" to a person performing CFD work. It usually serves the accounting and software branches of the organization as a first priority. The costs are frequently not competitive with outside vendors and any attempt to point this out is frequently met with hostility. However, this is changing with the advent of more cost effective computing.

Single-User Supercomputers Price/Performance

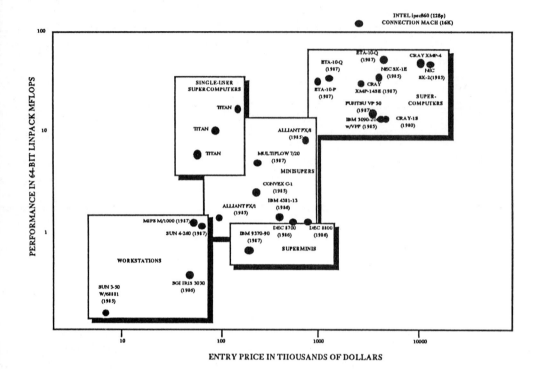

Price Performance (Dollars/MFLOP)

```
CONNECTION MACHINE (64K).................................300
INTEL iPSC/860 (128) ........................................450
CRAY YMP/8 ................................................ 7,000
IBM 3090 - 600E/SF.......................................21,000
VAX 9000/440 ............................................... 7,800
```

    The price performance of the new parallel computing machines are shown here. Note there is more than a factor of ten improvement. This figure however does not guarantee that for a given problem you will always benefit by running on the parallel computer. It is a good possibility, however, if the problem is structured correctly on the computer.

Some Typical CFD Performance Levels
(sec/grid point - time step)

NCUBE (1024 HYPERCUBE) $\sim 0.2 \times 10^{-4}$

CONNECTION MACHINE (16K) $\sim \left\{ \begin{array}{l} 0.1 \times 10^{-4} \text{ (2D)} \\ 0.04 \times 10^{-4} \text{ (3D)} \end{array} \right.$

CRAY X-MP/18 $\sim 0.15 \times 10^{-4}$ (2D+3D)

ALLIANT FX/8 $\sim 0.25 \times 10^{-3}$ (3D)

The CFD performance levels shown are for finite difference calculations. The principal point is that the connection machine out performed the CRAY X-MP by a factor of four on a 3-D problem. These results are from a paper by Oran, Boris and Brown [1].

Why Do We (The User) Care

- SOLUTION OF PRACTICAL 3-D FLUIDS PROBLEMS FOR REAL SYSTEMS STILL IS BEYOND OUR REACH IN MANY AREAS
    - COMPUTER TIME
    - COST

Practical systems CFD simulation problems still place large demands on computers and are very expensive due to need for large memory and CPU time. The CFD community, of course, will always have a problem bigger than the computer for the foreseeable future. Some examples of the needs are shown next.

Example 1
(assumes memory available)

- CALCULATION OF SHIP FLOW FIELD AND WAKE
    - MULTIPLE LENGTH SCALES (cm to km)
    - DETAILED RESOLUTION CAN REQUIRE 10^6 TO 10^9 GRID POINTS
    - SMALL TIME STEP REQUIREMENTS CAN RESULT IN 10^4 OR MORE TIME STEPS
    - COMPUTE TIME FOR CURRENT MACHINES 10^5 TO 10^8 SEC OR 1 DAY TO 10^3 DAYS

The calculation of large scale incompressible flows is a difficult CFD problem. From the times shown it is clear that these types of calculations push the limits of current computers. The next example is for compressible flows.

Example 2
(assuming memory available)

- AIRCRAFT FLOW SIMULATIONS WITH VISCOUS EFFECTS
    - RESOLUTION OF BOUNDARY LAYER AND INVISID FIELD REQUIRES
      10^6 to 5×10^6 Points
    - TIME STEP TO CONVERGE SOLUTION
      10^3 to 4×10^3
    - COMPUTE TIME
      10^4 to 20×10^4 sec or 2 hrs to 2 days

The compressible flow aircraft CFD problem is not quite as demanding as the ship wake problem. One can still require days of computing however.

What is Maximum Performance We Can Expect From Computers

- A RESULT PER GRID POINT/TIME STEP EVERY FLOATING POINT CLOCK CYCLE FOR EACH PROCESSOR

$$\sim 100 \times 10^{-9}$$

- FOR N PROCESSORS PRODUCING RESULTS SIMULTANEOUSLY

$$\sim \frac{100 \times 10^{-9}}{N}$$

If one assumes that the average floating point clock cycle is in the range of 100 nanoseconds then the best output from a single computer that one can expect per clock cycle is that shown. If one increases the number of processors, then the output can be expected to increase proportionally.

Today vs Tomorrow in CFD Computing

- TODAY RESULT TIME $\sim 10^{-5}$ sec/grid point/timestep

- TOMORROW RESULT TIME $\sim \dfrac{10^{-7}}{N}$ sec/grid point/timestep

The typical time to obtain result on current day computers is 10^{-5} seconds per grid point per time step. If, however, one looks to the future then one can possibly obtain a speed up of two or three orders of magnitude. This is a big advance.

What is Needed

- A BETTER MATCHING OF COMPUTER DESIGNS TO ALGORITHMS
- POSSIBILITIES
    - NEAR TERM
        -- USE PARALLEL ARCHITECTURE WITH LOW COST DISTRIBUTED PROCESSORS THAT CAN OPERATE INDEPENDENTLY (MIMD)
        -- DIVIDE UP MACHINE INTO ASSEMBLY LINES THAT USE M PROCESSORS TO STEP BY STEP ASSEMBLE ALGORITHM RESULT
        -- MACHINE THAT WILL GENERATE A RESULT FROM EACH SET OF M PROCESSORS ROUGHLY EVERY FLOATING POINT OR MESSAGE PASSING CLOCK CYCLE

In order to obtain the maximum CFD output from the digital computer one has to be able to efficiently match algorithms to computers. This has not been the approach in the past because computer designers never talked to CFD people and visa versa. This is changing, however, because the CFD community is beginning to understand what they need from computers and they are in a position to begin building their own. The optimum architecture would be a machine that could partition itself into cells that represented CFD algorithm equivalents. An example of this is shown next.

Example

@ $\frac{\partial T}{\partial t} + \frac{\partial T}{\partial x} = \alpha \frac{\partial^2 T}{\partial x^2}$ I.C. $T = T(x)$,   B.C. $T = T_0(t,0)$, $T = T_2(t,1)$

@

@ $T_I^{n+1} = a\, T_{I+1}^n + b\, T_I^n + c\, T_{I-1}^n$

@ Data Flow

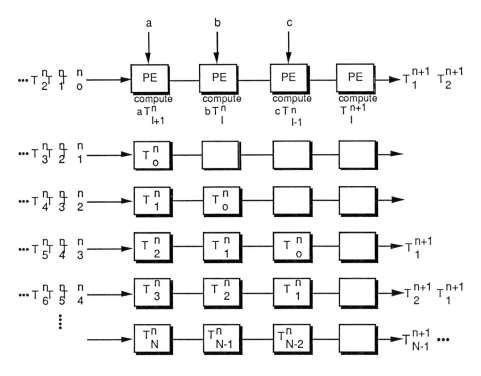

A computer made up of the nodes shown below has the possibility of accomplishing this task. This node has been developed recently by INTEL Corporation and they volunteered this information to the author.

An Example Architecture Node

- INTEL iPSC/860

iPSC/860 Compute Node

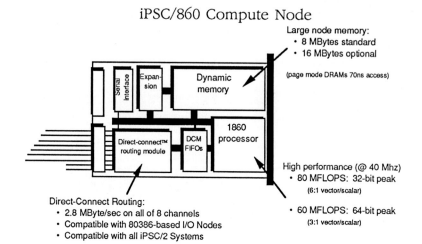

Beyond the INTEL iPSC/860 technology will be the ability to map algorithms directly onto silicon.

Implications for the Future

- NEAR OPTIMUM MATCHING OF CFD ALGORITHMS WITH COMPUTER HARDWARE WILL BE POSSIBLE

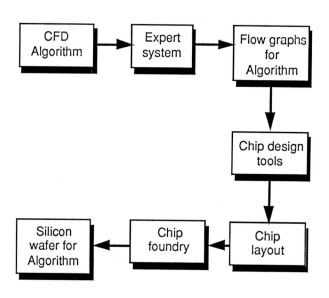

Implications for the Future (cont)

- SILICON WAFER ALGORITHM EQUIVALENTS INSTEAD OF SOFTWARE WILL BE BUILDING BLOCKS OF CFD IN THE FUTURE
- THIS PROCESS COULD IMPROVE THE SPEED OF PRODUCING RESULTS PER GRID POINT PER TIME STEP BY ORDERS OF MAGNITUDE

1   AIAA-70-0335, 28th Aerospace Sciences Meeting, Reno, Nevada, *Fluid Dynamic Computations on a Connection Machine,* by E.S. Oran, J.P. Boris and E.F. Brown, 8-11 Jan 1990

Computation of shock-shock interference around a cylindrical leading edge
at hypersonic speeds

T. Aki

National Aerospace Laboratory
Tokyo 182, Japan

1. Introduction

Recent demands for hypersonic flight which has come the realization among several countries within the aerospace community are stimulating to advance CFD technology further as can be seen from the fact that hypersonic flows have been emphasized for twice consecutively in this series of conference. Currently proposed hypersonic flights demand a variety of vehicle's configurations other than that of the existing reentry vehicle, and cause flow fields different from each vehicle. The flow problems associated with the hypersonic flight, however, will be continued essentially unchanged. Therefore, reconsideration on problem areas and problems associtated with the hypersonic flows would be fruitless. But, needless to say, problem areas concerning the hypersonic flows cover a wide variety of disciplines extending over science and engineering. It is a crucial task beyond one's skill to give an overview on all aspects of hypersonic flows in general.

Fortunately, problem areas and problems associated with hypersonic flows had been considerably clarified through the researches peaked out about the late 1960's. One may attribute this period to the first generation of the hypersonics if one may assign the zeroth generation to the era of theoretical arguments on the hypersonics traceable back to the Newton's law of fluid resistance. In the first generation, several possibilities of hypersonic flight had been seeked, regardless of the realization. It was a matter of regret that CFD could not take part in the researches in the first generation due to poor computers and unskilled algorithms available at that time. Discussions on these areas and problems can be found elsewhere as still readable survey or review articles among the stored volumes of appropriate technical journals, proceedings of conferences concerned with the hypersonics, written books with emphasis on the hypersonics, and so on. With a citation of the article by Kumar [1] as is giving an entry for current interest, repetition of the abovementioned discussions may be avoided here for a simplicity of the present argument.

From the aforementioned fact and a consideration that is required for avoiding missing of the important reseach works, the present paper will concentrate its discussion to a very restricted subject to illustrate some technical aspects of hypersonic flow computation.

It will be a technological milestone in the second hypersonic generation on which now we are going on to develope a maneuverble trans-atmospheric vehicle with propulsion engine system, i.e. a spaceplane. It is a consensus that CFD will play at this time an important role both in research and development. Particular efforts in the development of such a vehicle have focused a great deal of attention on the design of Thermal Protection System (TPS). The TPS design depends on a precise prediction capability of aerothermal loads and is among the many critical items that will determine the succes of a spaceplane. A difficult problem to evaluate accurate aerothermal loads is encountered even under low altitude flight or continuum flow regime, if a significant portion of trans-atomospheric flight profile falls into continuum flow regime. In continuum flow regime, one encounters a problem resulting severe local heat loads when disturbed flows caused by shock interference can attatch to the body surface under consideration. These disturbed flows may be shear layers, supersonic jets, expansion waves, or shock waves and the overall flow can be dominated by viscous/inviscid effects depending on the interacting geometry of primal shock waves. The interacting geometry is 3D in general and results necessarily in a complex 3D flow, though whose representative flow characteristics may be interpreted qualitatively with the aid of the results obtained by Edney [2]. However, in order to first gain more insight into the physics of the disturbed flows by shock wave interference, a simplified modeling of the flow without loss of essence would be suitable for furthering our understanding on the physics of the flows. Flows disturbed by an oblique shock impinging on a bow shock ahead of a cylinder have been analyzed with the renewed interest both in experiment and theory. Flows aroud an integrated rectangular engine inlet cowl lip of the NASP can be considered as a representative of application of such a shock interfering flow analysis.

Although such an interfering flow problem is a very localized one, it includes an enough amount of problems characterisic to the hypersonic flow. Several numerical studies have been tried by using the modern CFD technology. This paper first addresses to discussions on some computational aspects about this type of shock intereference and its flow analysis and also problems needed to be studied further to improve the solution, i.e. to obtain the more accurate prediction of heat loads. The discussions will derive an inevitability of a high speed computation in the TPS design of an engine inlet cowl in practice.

Continued efforts for obtaining more powerful computers and skilled algorithms, both of them can be considered to be a necessary requirement constituting of a tightly coupled pair to advance CFD technology, have been devoted to enhance the status of CFD with worldwide spread. As is well known, it is one of the fruits of these efforts that the recent progress in the application of CFD techniques has opened the way for a dramatic new manner in the design of an aircraft and the analysis of its performance. Aircraft designers are recognizing now that CFD has got the status of an indispensable partner to a hardware simulation or an experiment. This fact will be indispensably applicable to the design and analysis of spaceplanes of current concern.

Efforts having been undertaken in Japan will be mentioned with a particular emphasis on supercomputer development. Some disccussions on the existing Japan made supercomputers will be provided from a technological point of view on a computer. And moreover utilization of supercomputers in Japan will be briefed. An outline of forthcoming machines in Japan, maybe available in 1990, will be included.

2. Shock interference computation

Up to date, some countries are in definition study of their own hypersonic vehicle(s) according to respective circumstances. Vehicle with an airbreathing engine system would be a foremost one among them. A blended body configuration is being candidated as a baseline for such a vehicle. The blended body configuration enables a design of an airframe in which propulsion engine system may be integrated. The airframe surfaces may be shaped to act as components of the propulsion system and the great weight saving would be realized if the surfaces are properly shaped as an engine inlet as well as an exhaust nozzle of the propulsion system. The shaping is related to an engine performance to be specified. One of the problems at the inlet portion that needs to study precisely is an interaction of a shock wave emanating from the forebody compression portion with the bow shock standing ahead of the engine cowl lip. To obtain a definite inlet efficiency one has to keep inlet air capture and compression ratios within respectively prescribed levels. The requirement from the engine inlet leads to an inevitable design scheme that is bridging a shock wave between the forebody and the cowl lip portions. If the inlet design mismatches the shock bridging portion both the desired air capture and compression will not be attained. However, the matching of the shock bridging will elevate heating rate on the cowl surface to an intorerable level for the surface material to be used and brings a severe requirement into the TPS design about the cowl surface, particularly in that portion around the leading edge of the cowl surface where a space allowed for the TPS is very limited. Effects caused by the shock interference may be understood fundamentally by analyzing the following simplified but very useful model flow.

However, since this subject is among ongoing studies the most is still inconclusive situation. Note that the present argument does not make a conclusive remark. The aim of the present article therefore will be restricted to give an overview on computational aspects of the subject in general terms.

2.1 Representative model flow and its computation

To simulate the physical situation at hand, consider an interaction of a planar oblique shock with a bow shock supported by a cylinder immersed in a hypersonic stream as shown schematically in Fig. 1. That the interaction causes basically six different types of the interacting patterns according to the intersecting location of both waves, the freestream condition, and the obliquely incident shock angle was investigated systematically by Edney [2] in 1968. The interacting patterns are successively clasified as type I through type VI. Edney gave an excellent description on each type of interacting pattern. Pioneering computation for the type III interference had been conducted by Tannehill et al without [3,4] and with [5] a turbulence model, respectively.

In view of the aerothermal loads on the leading edge resulting from the shock interference, all the six types is not of equal importance.

A schematic of the type III interference is shown in Fig. 2. This type of interference occurs when the impinging shock intersected the leading edge bow shock within the subsonic region without the impingement (inside the upper sonic line). Two shear layers and one transmitted shock appear usually. One of the shear layers and the trnsmitted shock may hit confluently the leading edge, as is shown in the figure. The type III is characterized by this shear layer. A close inspection shows the shear layer attaching to the leading edge is subsonic on one side and supersonic on the other side. Note that a supersonic jet emanating from the secondary triple point may merge with the wall boundary layer, and complicates the type III flow pattern. The type IV shown in Fig. 3 occurs under the similar situation with the type III, but the shock intersection in this type was taken at more near normal portion of the bow shock, and characterized by a supersonic jet hitting the leading edge instead of a shear layer grazing the leading edge as in the type III. The supersonic jet is sorrounded with the subsonic region on both sides. When the supersonic jet impinges perpendicular to the leading edge a local stagnation results with a high pressure and an extremly high heat-flux at the point of impingement. Experimental data indicate that the maximum heat-flux can be attained a tremedously high peak greater than the maximum heat-flux without the interference. In this respect the type III interference is not severe as the type IV. The type V shown in Fig. 4 occurs when the impinging shock intersected the bow shock just below sonic line. In the type V interference, we observe a supersonic jet from the first triple point and a set of shear layers and a transmitted shock. Of which only the transmitted shock can hit the leading edge. Therefore the type V can be characterized by the shock impingement. Neither the shear layer and the supersonic jet hit the leading edge. This type of interference may be considered to be lesser importance than the type III and IV in terms of the aerothermal loads.

The remaining types (I,II, and VI, and characterized by a shock or expansion wave impingement) are of more little importance in terms of the aerothermal loads on the leading edge. We will proceed exclusively with the type III - V interferences. Computational studies on these three types have been conducted using the modern CFD technology [6-9,11-15]. The cited references in this paper are missing something of analytical, empirical, duplicated, experimental studies, by the personal judgement of the present author, and also ones unavailable to the present author. However, the missing may be restored easily from references among the quoted papers.

The general behavior on this type of shock interference computation reveals that the bow shock without the inpingement moves closer to the body on one side and away on the other side, as illustrated in Fig. 5. Therefore, computation must be carried out on such a grid that an excess amount grid point is unnecessary ahead of the retroceded bow shock while an enough amount of grid point must be provided ahead of the advanced bow shock. More precisely, grid to be used must have such a distribution that grid points are
(1) fine near the body surface to resolve the boundary layer,
(2) fine in the shock interfered region to resolve its flow structure,
(3) fine ahead of the bow shock without impingement to capture the advanced shock and resolve shock layer behind the advanced shock, and

(4) coarse ahead of the retroceded shock.

These result finally a grid with a highly skewed outer boundary is a preferable one to compute such interfering flows. Uniformly distributed grid in the circumferential direction, easy to generate, must be dense enough to satisfy the abovementioned requirement, except use is made on preliminary computation to asses a flow structure. Relying the shock capturing methods, a uniformly distributed grid is not preferable because a large number of grid point is wasted in the shock retrocedent region while it becomes insufficient to capture the interfered bow shock and to resolve the shock layer behind it in the shock advanced region. The most stringent situation is encountered when one has to analyze the type III to V flows. Shown in Fig. 6 is an illustration of such a skewed grid on which the type III to V flow analyses were conducted by using a FDM. Grids used among the typical computations are summarized in Table 1. Solution adaptive techniques are especially preferable to compute the type IV flow and to resolve structure of the jet characteristic to this type. The type IV flow exemplyfies the greatest pressures and heat-transfer rates among the interferance patterns and is of the greatest concern to TPS designer. The importance of grid will be stressed again with respect to solution convergence and more precise heat-transfer rates prediction.

The system of equations to be solved is composing of the 2D Reynolds averaged Navier Stokes equations and appropriate conservation equations according to the fluid model used, such as turbulence or chemical species transports. An equation of the gas state with functionals depending on the fluid model used completes the system. These equations are well documented, therefore may be omitted here. To solve the system, one has to discretize the given equations. Fortunately one has shared several methods those are composed of formulism by the characteristic oriented procedure. Almost of all methods were devised originally for solving the transonic or low supersonic inviscid flows, and then extended for solving inviscid dominant flows (i.e., high Reynolds number flows). These manner are extendable to the hypersonic flow computation as well without essential modification. Of cause, for example, reorganization of the so called the aerodynamics matrices is needed if one uses a method including such matrices, to enhance stability or to include gas imperfection effects, such as chemical reactions, into the characteristic variables.

The desirable solution behavior that the characteritic oriented methods have demonstrated in transonic flow regime retains still unchanged in hypersonic flow regime. That is to give us a robust algorithms and to yield sharp and smooth shock transitions. This is especially desirable to compute a flow with chemical reaction since physically unwanted oscillatory reaction due to temperature oscillation can be suppressed. These modernized methods have been discretized in a variety of forms, i.e. the FDM, the FVM, and the FEM (somewhat restrictive than the others). These forms can be applicable to the interfered flow computation with success and failure indistinguishable each other. Currently used forms are summarized in Table 1.

It is not overemphsized that the pressure causes almost of all troubles on implementation of hypersonic computation. Considering a normal shock wave in a perfect gas for a simplicity, one can understood readily from the shock relations that the pressure undergoes an unlimitted jump across the shock, as well the temperature, whereas the density reaches a plateau as the Mach number increases. A particular caution must be paid to overcome this large jump of the pressure. The pressure based sensing of shock waves or rescaling of the flow variables are typical recipes for success in hypersonic computation for preventing nonphysical solutions and also for obtaining stable solutions [6,8,15].

Initial conditions strongly affect the overall convergence rate of the solution. One has two possible strategies for obtaining an interfering flow solution. The first of which is to obtain the interfering solution via the steady state solution without the impingement. The second one is to obtain directly an impinging solution without an intermediate steady state solution. The first one is appropriate to a series of parametric study, e.g. a variation of shock intersection location under a fixed freestream Mach number, whereas the second one to a fixed set of the conditions. Here, we assume the first strategy.

The solution procedure for obtaining a steady state solution without the impingement and accerating its convergence are well known. Assumption of the Newtonian pressures along the body surface and an appropriate guess of bow shock location as well as its shape and flow variables both in the shock and wall boundary layers are being used scheme independently to enhance the convergence. These suggest that the nearer the initial setup, conventionally phrased as the initial conditions, approximates the true physical situation, the faster the convergence will be realized. In this respect, a solution obtained at nearby Mach number provides a very useful database. Among the other techniques to accerate the convergence, the local time stepping and a sequential control of the time step increment are also applicable to obtain a steady state solution without the impingement rapidly and stably. Moreover, wellknown solution accerating techniques, such as an enthalpy damping, a mesh refinement, a multi-grid, and etc. are can be applicable. In any way, the hypersonic steady state blunt body flow is not a difficult problem to obtain the solution, regardless of the explicit and implicit relaxation methods.

Once a steady state solution without the impingement can be obtained, we restart the shock impingement computation by rewriting the inflow conditions. Flow variables within the grid points along the inflow boundary and belonging to behind the given oblique shock are rewritten with those obtained by solving the oblique shock relations at the given incident angle under the prescribed Mach number or by solving a hypersonic flow past a wedge generating the incident oblique shock with Navier Stokes equations that simulates a wind tunnel experiment. Flow variables among the remaining grid points are kept unchanged.

The boundary conditions to be applied in the shock impingement computation are those on the inflow and outflow boundaries and the cylinder surface. We assume the interacted shock does not cross the inflow boundary for a easiness of treatment. If this is the case, the inflow boundary is treated as the aforementioned manner. Since the outflow is supersonic for all types of the present interference, the flow variables at the outflow boundaries can be extrapolated from the interior grid points except those within the boundary layer where a subsonic state prevails. One can specify a boundary condition over these grid points. We assume a constancy of the pressure across the boundary layer at this station. The others are extrapolated from the interior grid points. The isothermal body surface is a plausible assumption for the viscous flow since an active cooling will be instrumented into the actual engine cowl. The surface pressures may be obtained simply by solving the normal momentum equation along the surface although the conservation law may not be held exact. The velocity assumes to satisfy no slip condition for the viscous flow. The surface density is evaluated from the isothermal wall condition. The surface heat-flux may be evaluated from the Fourier law and the Prandtl number definition.

If one seeks the solution of a prescribed flow pattern, the starting of the solution with a semi-adaptive grid as shown in Fig. 6 is desirable, particularly seeking for the solution of the type IV. The solution adaptive grid method is recomendable for analyzing the type III through V flows since more complicated patterns than the other remaining types among the six possible types can appear. The weighting factor to be used for a solution adaptive method may be in general any one among the computed variables, but the pressure is one of the candidates because of its sensibility to the shock. Since for each type of the interfered flow a dominant change arises in the normal direction to the wall rather than the circumferential direction, adaption with the pressure in the normal direction and no adaption or adaption with flow variale less sensitive to a shock or a shear layer, such as the density, may be allowed in the circumferetial direction. Note that the density is a readily available variable without complex decoding in the conservation law form.

It is notable that the descrimination of a type can not often be obtained distinctly. Intermediate or merged type between two adjacent types, or early or late of a type can be observed and one encounters a difficulty to identify a type from the solution. We can observe a merged interference pattern as shown in Fig. 7 (III + IV). Distinctly justifiable patterns can be seen in [6-15]. Figure 7 shows the Mach number contours and obtained in a series of preliminary Euler solutions. It is very likely that every transition between the adjacent types is not discontinuous. A variety of graphic outputs other than the Mach contours would assist one's judgement.

One can encounter a peculiar phenomenon while solving the shock interference problems under the steady flow approach. That is an oscillatory behavior exhibited by the solution. As the interaction progressed, the solution displayed a perfectly periodic phenomenon. One obtains the periodic nature in the residual, in which the bow shock ahead of the cylinder oscillated while the point of maximum pressure, accordingly maximum heat transfer, on the cylinder surface oscillated circumferentially. The type IV interference solution shows the most pronounced behavior. The residual was dropped and in a fixed level at the 2 or 3 orders below from the initial level in average and oscillates around the average. This had been noted first by Klopfer and Yee [7]. Not a few authors since then have reported the same behavior regardless of difference of their numerical methods, i.e. irrespective to the FDM, the FVM, and the FEM. This phenomenon could be observed in the ideal as well as the equilibrium air solutions, and in viscous as well as inviscid flows, and moreover in 2D as well as 3D computations [14]. The steady state solution without the impingement obtained by using the same algorithm with the impingement has not shown such a phenomenon. Therfore, this phenomenon seems to be peculiar only to (the type IV) shock interference problem.

A plausible cause of this oscillatory behavior of the solution seems to be due to coarseness of grid used within the boundary layer. The oscillatory behavior of the type IV flow can be observed in the measurement, but the frequency of the oscillation in computation does not compare readily with the measurement until the numerical method used in the computation becomes clear, because the computation is using occasionally a steady flow approach in which time inaccurate relaxation with the large Courant number as well as other convergence accerating means may be oftenly adopted. In this case, the time in the computation will be lost its physical reality. By using finer grid and time accurate relaxation, though hefty time consuming, the oscillatory behavior can asymptote the measurement.

As is well known, grid density near the wall must be high enough, particularly for obtaining a good prediction capability of heat-transfer. The cell Reynolds number, based on the minimum grid spacing next to the wall, gives an appropriate measure expressing an adequancy of a grid for the purpose. The cell Reynolds number of order unity is preferable for heat-transfer prediction. Effect of the cell Reynolds number is investigated precisely in [7]. Table 1 includes the cell Reynolds numbers used in the existing impingement computations.

All of the existing computational results shows that they predict the pressures favorably well with the existing wind tunnel data but fails to predict the heat transfer rates, the several order of magnitude lower than the measurement. Aside the problems including problems in measurement, such as problems (e.g., turbulence or contamination levels in freestream, and etc.) inherent to facility and problems (e.g., methods of experimental setup and data reduction) dependent to operational scheme of a facility, for the moment, the problems in computation to be reexamined are several factors contributing to the mentioned discrepancy. These are

(1) laminar flow assumption, mostly used among the existing computations,

(2) grid coarsness,

(3) time accuracy of relaxation method used,

(4) the uncertainty in the experimental data used for comparison, and

(5) others.

Some of the problems stated may be in relationship of interdependence. The most difficult problem is a turbulent flow modeling (turbulence modeling) adequate to the shock interference since measurements show a localized turbulent flow prevails at the impingement region. The type IV flow exemplifies a measurable turbulence level in the impingement region, although a correlation between heat-flux versus pressure implies a validity of laminar flow assumption. Much should be said about turbulence modeling, but it will need further space and the present article may limit further discussion on the turbulence modeling.

Among the others, the most stringent problem is a question on whether a numerical scheme used is truly multi-dimensional one. Generally grid lines are hard to align with the direction of disturbance propagation and then each grid point does not contain all information about the disturbance. Conventional numerical schemes are mostly a formal extention of 1D expression and consider only a variation along a grid line. Thus the numerical scheme used does not follow the true physical feature, and would cause the peculiar solution behavior.

2.2 Extention : Illustrations

The TPS design will demand a tight interdisciplinary coupling more than the present time, e.g., in between and among the aerodynamic, structural, and heat-transfer analyses. Up to date, these analyses have been conducted with relatively independent manner. An aerodynamic analysis predicts the surface pressures and heating rates under an appropriate assumption. The aerodynamic heating rate data are used for the heat-transfer analysis to predict the structural temperature distribution. The aerodynamic pressure and structural temperature data are used for the structural analysis to predict the response, i.e. deformation and stresses of the TPS materials to be used. This analyzed result is reported to the aerodynamic analysis and thus a circulation of analysis is closed. Such a traditional approach becomes inefficient in the coming TPS design due to mainly a lack of speediness between or among mutual interactions. An approaching method to couple fluid dynamic and thermal-structural analyses has been demonstrated [10]. Such interdisciplinary couplings may arise in between and among a variety of disciplines related the TPS design. These couplings will invoke the increased computer memory as well as CPU time.

Most of current computation relyes on an assumption that a steady state holds for respective types of the interference. If this is justified, then the conventional steady flow relaxation methods are also justified. In this case, the numerical method used may be that of a lower order in time-accuracy. However, it is likely that the type IV interference is unsteady in the jet impinging region. As stated previously, computational method including a local unsteadiness such as the type IV may invoke some disccusions. Apart from this local flow unsteadiness, one shall be encountered an essentially nonlocal unsteadiness in an actual flight. In an actual flight, even under steady cruse operation, the vehicle will oscillate around some mean attitude due to vibrations from a variety of sources. In a maneuvering operation, enviroment surrounding the engine inlet should be altered unsteadyly and influences both of the shocks under the interaction, also possibly associating with the vibrations. Thus the interference will be of unsteady. That the unsteady interference are resulted in patterns not necessarily the same with those of the counterparts in the steady state was shown by Kloper and Yee [7] under a limited codition, like that of maneuvering. Since trans-atomospheric vehicles have not been built, data needed by the vibrational analysis, e.g. frequencies, particularly that of the forebody, are yet unknown. But an analysis in cluding these oscillation and vibration will need truly time-accurate computation and therefore will appear as an increased CPU time consumer.

Effects resulting from the Mach number, Reynolds number, and gas imperfection are being limited in issue or nothing and left as future studies. All requires an intense computaion.

The above arguments are only illustrative to increase the prediction capability of aerothermal loads or the interdisciplinary coupling and indicate a latent use of computer, especially supercomputer, in the TPS design.

The aforementioned arguments were described in general terms, therefore hold scheme independetly. However, the present author hopes the reader should consult the cited references for more details and prior to the actual computation.

3. Supercomputers in Japan : manufacturing and utilization

One may make no doubt that CFD must be play a key role in the upcoming space vehicle designs that demand indispensably supercomputers. Although CFD itself must not be completly supported by computer hardware only, that the developments of the computer hardware prompted the developments of CFD can not be laid outside of our attention. Recent developments of Japan made supercomputers are well worth mentioning. For a while, let us turn our eyes to Japan made supercomputers and some related problems.

In retrospect, until the end of 1981, it seemed to be very likely that no one has been being paid any attention to Japan made computers. It can be understood from the fact that performance data about the Japan made computers had been never appeared in the literatures written by persons outside of Japan. One of the reasons was perhaps because of the worldwide known language barrier, which causes a lack of the information exchange and seems to be continuing still, however the most might be nothing more or less than the poor performances of the Japan made computers. The situation that little is known about the Japan made computers was drastically altered when some information about a Japan made supercomputer was transmitted to the USA. Since 1982 the performance data of the Japan made supercomputers have been occasionally quoted in the literatures by persons ouside of Japan.

Among the factors contributing to this rise of the Japan made supercomputers, one can single an episode that none is more significant than the national projects that have been organized by MITI, who controls the international trade of Japan and the Japanese industries and one of the ministries among the Japanese government. The initial object of this project was the design and development of a high performance VLSI computer system, and completed with success. The world fastest scalar machine coded as M-series, the top of them had a comparable speed with Cray X-MP/22, was an outgrowth of this project.

The utilization of VLSI allows expanded memory handling capability and an increased number of vector processing pipelines with a significant reduction in the number of compuputer hardware components. The forenamed project was expanded into a supercomputer project to design and develop a supercomputer in which the scalar processor developed within the initial project was integrated with a vector processor. This scheme influenced the current basic design concept common to the Japan made supercomputers that differentiate basically architectures from the USA made counterparts. The design of the Japan made supercomputers have drawn performance from their in-house device technology, while the USA made supercomputers have balanced the device technology disadvantage by pushing the state of the art in parllel architecture. This basic architectural difference between the Japan and USA made supercomputers can be seen partly from the simple diagram as shown in Fig. 8. In Fig. 8, architecture is characterized in terms of the number of concurrent processor, while the device technology is described representatively in terms of the number of vector pipeline sets. Thus, Japan made supercomputer is designed as to obtain the fastest processing speed on a single job. On the other hand, the USA made supercomputers aim the system throughput.

In any way, both the Japan and USA made supercomputers have realized the limitation of the state of the art technology regardless of difference in architectrure between them, although the architectural dimension was limited.

The machine installed in 1976 to NAL was the first Japan made supercomputer, though the device technology used was not that of the M-series. It was a highly pipelined array processor operatable under a host scalar machine (I/O) and had kept a speed of 22 MFLOPS, which was the highest among the Japan made computers available at that time. Though it was inferior to Cray 1, in the sense of both the hardware and software systems, it had been in operation at NAL by 1982. To extract its performance at maximum, a Fortran callable instruction system like an assembler was devised. But, this instruction system accerated an isolation of the computer further. As a result, the machine followed a fate similar to the ILLIAC IV due to its particular language system used.

Early in 1987, NAL joined the vector processor community by installing two supercomputers, both of them are belonging to a common ancestry and operatable under almost the same operating system with the old one which was quite similar to IBM's MVS. Therefore, a troubleproof replacement of the computing system was realized without any trial operation while absorbing softwares accumulated during 1982 to 1987. The exising system at NAL is outlined in Fig. 9. The system is named locally as NS (Numerical Simulator) and Fig. 9 shows only its kernel. It is operated as a loosely coupled multi-processor system composing of two supercomputers and one large scale scalar computer (shown as FEP in the figure), one of the supercomputer is FACOM VP400 (shown as SHP1 in the figure) and the other is VP200 (SHP2). The SHP1 system keeps 1.14 GFLOPS processing speed while the SHP2 keeps 0.57 GFLOPS processing speed (then the total processing speed of the NS amounts about 1.7 GFLOPS). However, it will move to another system within the foreseeable future, maybe in 1991, since a forerun official work necessary to introduce a new computer system at NAL has been completed. This indicates that NAL is headding always a new computer system and, on the other hand, the life of a computer system is quite short. The latter situation is not unusual in Japan because of the wellestablished rental or lease services of computers. The servicing traders take over a large amount of investment associating with new computer system installations from vender. Enduser may pay for the new computer system in monthly installments. The enduser is only obliged at least one year use of the new system (changeable depending on agreement between the enduser and vendor).

Unauthorized statistics shows that Japan has 140 sets of the supercomputer at the end of 1989, most of them were installed within industrial cities distributed over the longest but narrow main island (Honshu, one of Japan's main islands) and consisted of almost the products supplied by the domestic main frame manufactures. The figure 140 indicates that about 1/3 of the supercomputers installed in the world is in Japan as well as a supercomputer is installed about every mile apart on the average (a simple arithmetic). Roughly speaking, 2/3 of the supercomputers installed in Japan are used within the industrial sector, the remainder within universities and national research laboratories, and none within the other sectors, non-scientific as well as engineering sectors, such as commercial and financial sectors, at the present. However, latent users from these last sectors will increase exhaustively within the foreseeable future considering the existing enormous amount of energy growing among their markets.

One can find three major supercomputer manufactures in Japan. They are manufacturing their own scalar computers parallel with supercomputers. Their products are expanding over main-frame- to micro- computers as well as general electronic components including computer components. In addition to these major manufactures, a number of manufactures with a variety of investments and productive schemes, among them one can find those from the American origin, are competing the mini- (scalar, super, and parallel) and micro- computers' market share. The competition between and among these manufactures is quite intense. Thus, the availability of computers in Japan is very excellent.

The supercomputers, conventional large scale scalar computers, and mini (scalar, super, and parallel) computers are consisting of Japan's computing power. However, this high density distribution seems to be an obstacle, on the other hand, to explore an efficient computer network technology.

In the mid 1989 and coincidentally, two of the major manufactures have announced that they will be able to deliver a new series of the respectively manufacturing supercomputers to the domestic market within 1990. Their new series are conceptually the same in the sense that these series are of powered up version against the respectively existing machines although seeming architectural difference exist. The some advanced technology has been introduced to realize the new series. The announced performance of the respective machines, mostly interesting also to all of the users, can be considered in basic components as that of very competitive one. Therfore, the present paper explains only the characteristic features about one of these new series.

A high-speed and density logic LSI with a 80 ps/gate propagation delay and about 15000 gates/chip used in the CPU, and a SRAM with a 35 ns access time and 1M bits /chip used in the main storage unit are the representative advanced semiconductor technologies introduced. Packaging is also advanced by developing a multi-layer glass-ceramic board. The advanced packaging technology used enables not only reducing the number of boards to be used but also attaining a faster transmission speed. The arithmetic and load/store pipelines in the vector unit were modified or enhanced against the existing machines, which aimed an enhanced parallelism of the arithmetic and logical (or mask) operations. To enhance scalar processing ability, an augmentable scalar processor which can be field-upgradable is considered. All the advanced technologies and enhancements introduced enables to realize a 5 GFLOPS speed and 2 GB maximumly loadable main memory system. An example of block diagramatic system configuration is shown in Fig. 10. System storage unit (8 GB) shown in Fig. 10 is a multi purpose swapping area for job stream control.

In the software dimension, up to now, the operating system of the Japan made supercomputers are batch-oriented. Of cause, TSS or like environments were served but were far more batch-oriented as compared with UNIX, now becoming standard in scientific computation in the current use. Both manufactures have announced support of UNIX ports for their new series which should help to make their computers more attractive to users, for both of the existence and comers. The UNIX environment should also facilitate the porting of upcoming application software to Japan made supercomputers. In fact, the scarcity of application software was a major weakness that Japan made supercomuters have been handicapped in the worldwide supercomputer race. For a typical example, the application software, such as a crush simulation, is available only for Cray systems and Japanese automobile manufactures are using Cray systems exclusively for that simulation. This weakness in the existing software dimension will be restored by the support of the UNIX.

Another weakness of the Japan made supercomputers seems to be a lack of the connectivity all kind. In fact there is no support for the high-bandwidth network at the present time, as one being comparable to the NSFnet. None of Japan made supercomputers support a high speed I/O channel like Cray HSX and caused the inability to do real time visualizarion on the superomputers. Therefore most supercomputers support local access only. As stated previously, this lack of the connectivity would cause the high density distribution of the supercomputers in Japan. However, new machines announced will be closing the gap in these aspects within the couple of years.

4. Conclusions

Some computational aspects on shock-shock interference around a cylindrical leading edge at hypersonic speed were described in general terms and with emphasis on numerical scheme, grid, and peculiar solution behavior on the type IV interference. And plausible factors contributing to the low prediction capability of computation at the present time were considered as problems suggesting to further investigation. In addition, the several demands necesitated in the upcoming application works on the TPS design were suggested. These indicated the indispensability of a supercomputer. Then a brief description on Japan made supercomputers and utilization of them in Japan were given to clarify the present Japan's status as a member of the supercomputer community, which includes a personal view of the present author.

References ;

[1] Kumar,A. Lecture Notes in Physics, Vol.323, Springer,1989.
[2] Edney,B. FFA REp.115, A.R.I.Sweden,1968.
[3] Tannehill,J.C. and Holst,T.L., AIAA J. Vol.14, 1976.
[4] Tannehill,J.C. et al, AIAA J. Vol.14, 1976.
[5] Tannehill,J.C. et al, AIAA J. Vol.17,1979.
[6] White,J.A. and Rhie,C.M., AIAA paper-87-1895, 1987.
[7] Klopfer,G.H. and Yee,H.C. AIAA paper-88-0233, 1988.
[8] Perry,K.M. and Imlay,A.T. AIAA paper-88-2904, 1988.
[9] Moon,Y.J. and Holt,M. Lecture Notes in Phy. Vol.323, 1989.
[10] Dechaumphi,P. et al, AIAA J.Spacecraft, Vol.26, 1989.
[11] Thareja,R.D. et al, Intern.J. for Numer.Methods in Fluid, Vol.9, 1989.
[12] Singh,D.J. et al, **AIAA paper-89-2184-CP**, 1989.
[13] Prabhu,R.K. et al, **AIAA paper-90--6-6**,1990.
[14] Singh,D.J. et al, **AIAA paper-90-0529**, 1990.
[15] Montagne,J.-L. et al, NASA TM100074, 1988.

R	T	G	S	M	Re or Msp	A	Mo
[6]	IV	109*110	U,V	5.94	$3.8*10^{**}-6$(m) /10.0254(m)	N	L
[7]	I~V	291*141	U,D	8.03~15	310~1.6	Y	L
[8]	IV	151*91	U,V	8.0	$1.1*10-6$(m) /10.0762(m)	N	T
[9]	III,IV	151*81 121*81	U,V				L,T
[11]	III,IV	11466$	U,E	8.03	$1.0*10^{**}-5$(in) /3(in)	Y	L
[14]	IV		U,V	5.94 8.03		Y	L
[12]	III,IV	10673$	U,E	16		Y	L

Legend :
R=Reference, T=Types, G=Grid(typical): $=total elements, S=Scheme: U=Upwind scheme; D=finite Difference; V=finite Volume; E=finite Element, M=Mach number, Re=cell Reynolds number, Msp=minimum grid spacing/diameter: meter (m); inch(in), A=Solution adaptive grid; Y=Yes, N=No, Mo=flow Model; L=Laminar, T=Turbulent.

Table 1. Summary of the modern computations.

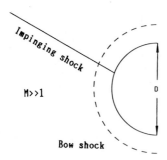

Figure 1. Simplified model of shock impingement.

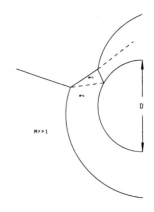

Figure 2. The type III interference.

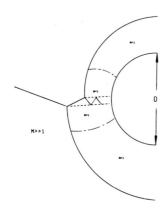

Figure 3. The type IV interference.

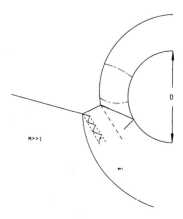

Figure 4. The type V interference.

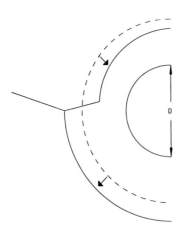

Figure 5. The general feature of shock interference.

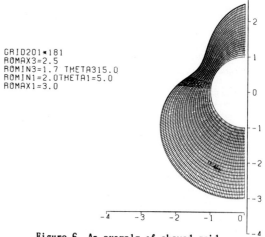

Figure 6. An example of skewed grid.

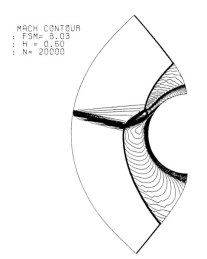

Figure 7, Merged interference.

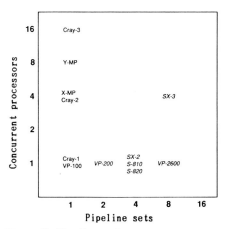

Figure 8, Pipeline sets versus processors.

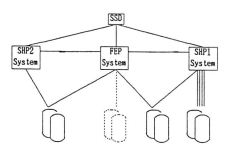

Figure 9, NAL's NS kernel.

Figure 10, Block diagram of typical 1990 machine.

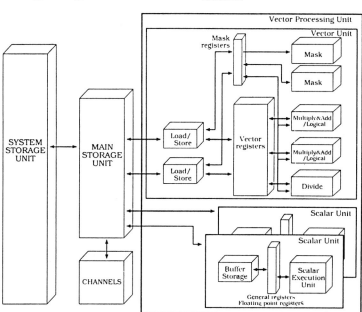

CHAOTIC FLOW OVER AN AIRFOIL

Thomas H. Pulliam, NASA Ames Research Center
and
John A. Vastano, Stanford University / NASA Ames Research Center

In recent years the topic of "chaos" (or more properly "nonlinear dynamics") has captured the imagination of scientists, engineers, and just about everyone who has seen the beautiful, intricate structures emblematic of the field. In computational fluid dynamics (CFD) we have always been faced with nonlinear systems. It is natural, therefore, to ask what the field of nonlinear dynamics can contribute to CFD, what physical insights can be gained with these new tools for analyzing complex behavior. This paper will review our attempt to apply the ideas of nonlinear dynamics to the results from a particular model problem, the two-dimensional (2D) flow past an airfoil at a high angle of attack. The three-fold purpose of these numerical experiments is to (1) determine the bifurcation sequence leading from simple periodic flow to complex aperiodic flow, (2) identify and quantify the chaos present in the aperiodic flow, and (3) verify that the observed bifurcation scenario is not an artifact of the numerical method.

The model problem is the unsteady flow over a NACA 0012 airfoil at a high angle of attack and low Reynolds numbers. The implicit centered finite difference code ARC2D [1] is used to solve the full two-dimensional Navier-Stokes equations. Details of the implementation of the code have been provided elsewhere [2,3]. The results of mesh refinement and algorithm parameter studies will be discussed below. The physical parameters of the flow were set at $M_\infty = 0.2$ and angle of attack $\alpha = 20°$. The Reynolds number was then varied in the range $800 \leq Re \leq 3000$. At each Reynolds number studied, the unsteady flow was evolved until an asymptotic regime was achieved. Extremely long time transient behavior was observed, especially for Reynolds numbers near bifurcations. Flowfield visualizations show the same basic scenario over the entire Reynolds number range: a shear layer off the leading edge, massive separation over the upper surface, and trailing edge vortices. As the Reynolds number is increased from 800, a series of bifurcations lead to chaotic behavior. The chaos in this flow arises as the unsteady vortex shedding, periodic for low Reynolds numbers, becomes irregular, primarily in the intensity of the shed vortices. The observed route to chaos is a period-doubling cascade, which occurs in the range $800 \leq Re \leq 1600$. Past $Re = 1600$, windows of periodic behavior appear in the chaos, with the last observed periodic window beginning around $Re = 1865$. Above this point, no periodic solutions were found, and the chaos began to become higher dimensional. The technique of time delay reconstruction is used in Figure 1 to visualize the asymptotic behavior of the flow. The time series are of the vertical component of velocity at a single spatial site just past the trailing edge of the airfoil.

Figure 2 summarizes the results and shows the overall picture of a period-doubling cascade into chaos. To further characterize the chaotic states, partial Lyapunov exponent spectra were computed for the attractors at $Re = 1600$, 2000, and 3000. The Lyapunov exponent spectrum of an attractor is perhaps the most complete characterization possible of the geometric properties of an attractor and the dynamical properties

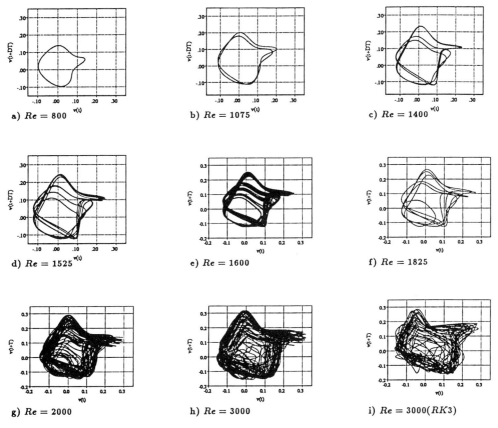

FIGURE 1. Time delay reconstructed attractors at various Reynolds numbers. The attractors are: (a) period 1, (b) period 2, (c) period 4, (d) period 8, (e) chaotic, (f) period 6, and (g-i) chaotic.

of a system as it evolves on that attractor. The Lyapunov exponents $\{\lambda_i\}$ measure the long-time average exponential growth or decay of infinitesimal perturbations to a trajectory. The number of Lyapunov exponents thus equals the number of independent phase space directions. The Lyapunov exponents are ordered so that $\lambda_1 \geq \lambda_2 \geq \ldots$, i. e. from largest to smallest. Negative Lyapunov exponents correspond to the decay of perturbations towards the attractor. If λ_1 is positive, then the attractor is defined to be chaotic. A positive Lyapunov exponent indicates that there is a direction *on the attractor* such that perturbations in that direction will grow exponentially. The Kaplan-Yorke [4] conjecture relates the Lyapunov exponent spectrum to the dimension of the attractor:

$$D_\lambda = m + \left(\frac{\sum_{i=1}^{m} \lambda_i}{|\lambda_{m+1}|}\right),$$

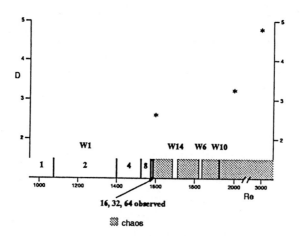

FIGURE 2. The observed bifurcation sequence. Dimension estimates for the chaotic attractors were obtained from Lyapunov exponent calculations.

where m is the largest number for which $\sum_{i=1}^{m} \lambda_i$ is positive. Estimating the Lyapunov exponents of an attractor involves following the evolution of perturbations to some fiducial phase space trajectory on the attractor. If the equations of motion for the system are known and solvable, either analytically or numerically, then as many Lyapunov exponents as desired may be estimated by solving the full set of equations for the fiducial trajectory and simultaneously evolving a set of N perturbations to estimate N Lyapunov exponents.

The estimated Lyapunov exponents for the three cases studied are presented in Table 1, along with the dimension estimate obtained using the Kaplan-Yorke conjecture. In all three cases there is one positive Lyapunov exponent, one zero exponent (to acceptable accuracy) and finally negative values. The magnitude of the positive exponent increases with Re, i. e., the rate of information loss in the system increases. The attractor dimension also increases with Re, but note that this is primarily due to a decrease in the magnitudes of exponents, but is instead caused by a decrease in the magnitudes of the first few negative exponents. This points out that a very chaotic system (high rate of loss of information) need not be high-dimensional.

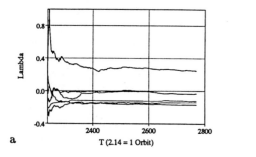
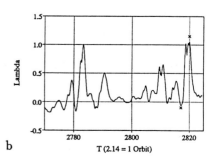

FIGURE 3. a) Convergence of λ_i for $Re = 3000$, b) $\lambda_1(t)$ for $Re = 3000$.

Lyapunov Exponents and Fractal Dimension		
$Re = 1600$	$Re = 2000$	$Re = 3000$
+0.10067	+0.19348	+0.52650
+0.00678	-0.00321	-0.00073
-0.22545	-0.16266	-0.06946
-0.27604	-0.27182	-0.26022
-0.30910	-0.29130	-0.30011
-0.34065	-0.33226	-0.36016
D_λ	D_λ	D_λ
2.48	3.10	4.65

Table 1. Estimated partial Lyapunov exponent spectra. The dimension estimates were obtained by applying the Kaplan-Yorke conjecture to the computed spectra.

While the long-time average Lyapunov exponent λ_1 is positive for a chaotic attractor, the short-time contribution to that average at time t, $\lambda_1(t)$ may be either positive or negative, and in fact the range of contributions is very large. Large positive or negative maxima in $\lambda_1(t)$ mark important moments in the evolution of the flow. Furthermore, it can be shown (see Vastano and Moser [5] for details) that the form of the perturbation corresponding to λ_1 at any given instant of time t is unique and depends only on the phase space location of the fiducial trajectory at that time. By using $\lambda_1(t)$ to locate the important times, and then studying the form of the unique perturbation corresponding to λ_1 at those times, it may be possible to discover the physical instabilities driving the flow. Figure 3a shows the convergence of the λ_i to their asymptotic values for the $Re = 3000$ case, and Figure 3b shows a part of the short-time contribution $\lambda_1(t)$. Figure 4 shows contours of the convective derivative of the perturbation field intensity (dotted curves) and the base vorticity field (solid curves) at the maximum of $\lambda_1(t)$ marked in Figure 3b. The convective derivative field is an order of magnitude stronger than the corresponding field at the minimum (also marked in Figure 3b). The perturbation field intensity is largest between the two counter rotating vortices being shed by the body. These vortices are at their largest size at this time and at other times they are not interacting as strongly. This indicates that the chaos in the system is associated with vortex interactions in the near wake of the airfoil. Work is currently in progress to use this perturbation field technique to further analyze this set of flow results.

The final question to be addressed is the effect of the numerics on the observed flows and bifurcation structure. Pulliam [2] and Pulliam and Vastano [3] show that the bifurcation sequence is sensitive to the numerics, e. g., mesh refinement, time step size, grid distribution, and other numerical parameters. The bifurcation sequence shifts in parameter space and thus the flow state observed at a fixed value of Re may depend strongly on the algorithm used. Therefore, one is cautioned not to make too much of the drastic changes in the nature of the attactors with algorithm variation. However, the same bifurcation sequence, i.e., a period-doubling cascade to chaos, exists across a wide variety of algorithm variations, mesh refinements, and other numerical variations. It has been shown that chaos can be generated numerically in discrete approximations of nonlinear equations. In particular, Stuart [6] has studied systems of nonlinear PDE's,

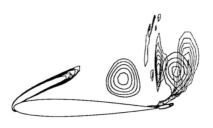

FIGURE 4. Contours of Vorticity Solid and Convective Derivative of Perturbation Metric Dashed For $Re = 3000$

and has shown that time steps larger than the linear stability limit lead to spurious periodic solutions (and hence in some cases to chaos). The time step studies performed for the airfoil cases imply that numerically generated chaos is not the source of the period-doubling cascade to chaos. In particular, a numerical experiment was performed using third order Runge-Kutta (RK3) time integration. For this method the linear stability bound corresponds to the bifurcation boundary. Computations were performed using a time step a factor of two below this limit for the $Re = 3000$ case. The observed reconstructed attractor, seen in Figure 1i, is very similar to that seen with the previous numerical method (see Figure 1h). In addition, the Lyapunov exponent spectrum is essentially unchanged. These numerical tests offer strong evidence that both the observed bifurcation to chaos and the chaotic states themselves are not simply due to numerics but are real solutions of the 2D Navier-Stokes equations.

REFERENCES

[1] Pulliam, T. H., " Efficient Solution Methods for The Navier-Stokes Equations", 1985, Lecture Notes for the von Kármán Institute For Fluid Dynamics Lecture Series : Numerical Techniques for Viscous Flow Computation In Turbomachinery Bladings, von Kármán Institute, Rhode-St-Genese, Belgium.

[2] Pulliam, T. H., 1989, " Low Reynolds Number Numerical Solutions of Chaotic Flow", *AIAA-89-0123*, AIAA 27th Aerospace Sciences Meeting, Reno, NV.

[3] Pulliam, T. H. and Vastano, J. A., "Numerical Simulation Of Chaotic Flows: Measures of Chaos", 3rd Joint ASCE/ASME Mechanics Conference Forum on Chaotic Flows, La Jolla, CA, 1989, ASME Publication FED-Vol. 77

[4] Frederickson, P., Kaplan, J. L., and Yorke, J. A., "The Lyapunov Dimension of Strange Attractors", *J. Diff. Equ.*, Vol. 49, 1983, p 185.

[5] Vastano, J. A. and Moser, R. D., "Lyapunov Exponent Analysis and The Transition To Chaos In Taylor-Couette Flow", *CTR Manuscript 108*, Stanford U., Stanford, Ca. 1990, Submitted Journal Fluid Mechanics.

[6] Stuart, A., "Linear Instability Implies Spurious Periodic Solution", *IMA J. of Num. Anal.*, Vol. 9, 1989, pp 465-486.

Modelling Shock/Turbulent-Boundary-Layer Interaction with a Cell-Vertex Scheme and Second-Moment Closure

K P Dimitriadis and M A Leschziner*
University of Manchester Institute of Science and Technology (UMIST)
Manchester M60 1QD, UK

1 Introduction

The line of separation from a curved surface − here provoked by a shock − and the shape of the ensuing recirculation zone depend sensitively on the turbulence structure in the boundary layer approaching separation and the shear layer bordering the recirculation bubble. In both, turbulence is strongly anisotropic due to the highly unequal levels of generation rates of different Reynolds stresses, combined with wall-proximity and flow-curvature effects. The interactions governing anisotropy and hence the level of the influential shear stresses are complex and cannot be described in any fundamentally sound manner by turbulence models based on the *eddy-viscosity* concept. Such models, whether based on a purely local, algebraic prescription or involving transport equations for scalar turbulence parameters, are in effect 'tuned' to yield the required level of the shear stress − the only stress of importance − in two-dimensional thin shear flow.
An arguably superior, more general modelling framework is offered by *second-moment closure* in which balance equations are solved for each and every Reynolds stress with all rates of stress production resolved without approximation. While this type of closure is difficult to implement in a numerically stable manner, because of its highly non-linear and coupled nature, it has been found to yield marked improvements in predictive accuracy in extensively separated and swirling *incompressible* flow, once stability problems have been overcome by means of special algorithmic measures. It is this experience which has motivated the present extension of the methodology to shock-induced separation.
The numerical vehicle for the present study combines a conservative cell-vertex finite-volume approach with a Lax-Wendroff time-marching scheme, originally formulated by Hall [1] for solving the Euler set. This has been extended by the present authors to apply to the Reynolds-averaged Navier Stokes equations. Previous publications have

---

*The starting point of this work has been a single-grid, Euler-equation solver developed by Dr. M Hall of RAE, Farnborough. Dr. Hall's input is gratefully acknowledged, as is that of Dr. B R Williams, also of RAE Farnborough. The work has been supported by the DTI and MoD through grant 2044/157/XR/AERO.

focused on numerical issues associated with the above extension [2], on the performance of high- and low-Reynolds-number $k-\varepsilon$ eddy-viscosity models for attached and separated transonic bump flows [3], and on multigrid convergence acceleration within the above modelling context [4]. The present paper aims to highlight numerical issues pertinent to the stable implementation of a Reynolds-stress model, and illustrate the performance of the resulting algorithm for shock-induced separation over a nominally two-dimensional bump in a channel.

2 Turbulence model and its implementation

The modelling strategy adopted herein combines the high-Re-number Reynolds-stress model of Gibson & Launder [5] in the fully-turbulent outer region with the low-Re-number turbulence-energy/length-scale $(k - \ell)$ eddy-viscosity model of Wolfshtein [6] covering the semi-viscous region very close to the wall. Attention is focused below on the Reynolds-stress model only.

The transport equation for $\rho \overline{u_i u_j}$ can be written, symbolically, as:

$$C_{ij} - D_{ij} = P_{ij} + \Phi_{ij} - \rho\varepsilon_{ij} \tag{1}$$

The processes whose balance is expressed by (1) are modelled as follows:

- <u>Transport</u> by convection and diffusion is approximated in terms of the transport of turbulence energy, as proposed by Rodi [7], for computational simplicity and in recognition of the fact that stress transport is rarely influential:

$$C_{ij} - D_{ij} \leftarrow \frac{\rho \overline{u_i u_j}}{\rho k}(C_k - D_k) = \frac{\rho \overline{u_i u_j}}{\rho k}(P_k - \rho\varepsilon)$$

- <u>Production</u> is exact and given by $\quad P_{ij} = -\left(\rho \overline{u_j u_k}\frac{\partial U_i}{\partial x_k} + \rho \overline{u_i u_k}\frac{\partial U_j}{\partial x_k}\right)$

- <u>Pressure-strain correlation</u> is modelled according to Gibson & Launder as $\Phi_{ij} = \Phi_{ij,1} + \Phi_{ij,2} + \Phi^w_{ij,1} + \Phi^w_{ij,2}$ where

$$\Phi_{ij,1} = -C_1 \rho\varepsilon\left(\frac{\overline{u_i u_j}}{k} - \frac{2}{3}\delta_{ij}\right) \quad \Phi_{ij,2} = -C_2\left(P_{ij} - \frac{2}{3}P_k\delta_{ij}\right)$$

$$\Phi^w_{ij,1} = C^w_1 \frac{\rho\varepsilon}{k}\left(\overline{u_k u_m}n_k n_m \delta_{ij} - \frac{3}{2}\overline{u_k u_i}n_k n_j - \frac{3}{2}\overline{u_k u_j}n_k n_i\right)f$$

$$\Phi^w_{ij,2} = C^w_2\left(\Phi_{km,2}n_k n_m \delta_{ij} - \frac{3}{2}\Phi_{ik,2}n_k n_j - \frac{3}{2}\Phi_{kj,2}n_k n_i\right)f$$

in which y_n is the distance from the wall; $\vec{n} \equiv (n_1, n_2, n_3)$ is the unit vector normal to the wall; the function f accounts for wall proximity and is given by $f = \frac{0.4k^{1.5}}{\varepsilon y_n}$

- <u>Dissipation</u> is assumed isotropic $\varepsilon_{ij} = \frac{2}{3}\varepsilon\delta_{ij}$

The above model is introduced into a cell-vertex scheme in which the inertial processes are approximated by integration over the primary cell at whose vertices all variables are stored. Fluxes and stresses are evaluated at cell centroids, however, and are integrated over secondary cells formed by primary-cell centroids. For further details, see [2]. The algebraic stress equations are formed in terms of normalized stresses:

$$\tau_{xx} \equiv \overline{\rho u^2}/\rho k \quad \tau_{yy} \equiv \overline{\rho v^2}/\rho k \quad \tau_{xy} \equiv \overline{\rho uv}/\rho k$$

In inviscid regions, where $k \to 0$, the normal stresses $\tau_{xx}, \tau_{yy} \to 2/3$. This enhances the stability of the solution procedure. The equation for the stress τ_{xx} may be written as

$$\begin{aligned} & \tau_{xx} \quad [P_k + \rho\varepsilon(C_1 - 1) + Q\rho k U_x] \\ + & \tau_{yy} \quad [R\rho k V_y - C_1^w f\rho\varepsilon] \\ + & \tau_{xy} \quad [Q\rho k U_y + R\rho k V_x] = \frac{2}{3}\rho\varepsilon(C_1 - 1) \end{aligned} \quad (2)$$

where $U_x = \partial U/\partial x$ e.t.c.; $Q = 2 - \frac{2}{3}C_2(2 - C_2^w f)$; $R = \frac{2}{3}C_2(1 - 2C_2^w f)$.

The system of three algebraic stress equations is solved iteratively, with one iteration performed per time step of the main marching sequence. An inversion of the algebraic system can readily be performed but results in rapid divergence of the solution sequence. For the iteration sequence to converge, care must be taken to enhance diagonal dominance of the system and to ensure positivity of the normal stresses at all times. This is achieved by appropriate rearrangement of the terms according to their sign. Terms whose sign is not known a priori are split into two parts, e.g.:

$$P_k = 0.5(|P_k| + P_k) - 0.5(|P_k| - P_k) \equiv P_k^+ - P_k^-$$

where $P_k^+, P_k^- \geq 0$. Equation (1) is finally written in the form

$$\tau_{xx}^{(m+1)} = \frac{C\tau_{xx}^{(m)} + D}{A + B/\tau_{xx}^{(m)}} \quad (3)$$

where m, $m+1$ are iteration levels.

To ensure convergence of (3) to a positive value, we demand:

$$A, B, C, D \geq 0 \quad (4)$$

$$A > C, \quad D > B \quad (5)$$

In the case of equation (1), the coefficients are:

$$\begin{aligned} A &= P_k^+ + C_1\rho\varepsilon + \rho k(QU_x)^+ \\ B &= \tau_{yy}\rho k(RV_y)^+ + \rho k\left[(\tau_{xy}QU_y)^+ + (\tau_{xy}RV_x)^+\right] + \frac{2}{3}\rho\varepsilon \\ C &= P_k^- + \rho\varepsilon + \rho k(QU_x)^- \\ D &= \tau_{yy}\left[\rho k(RV_y)^- + C_1^w\rho\varepsilon f\right] + \rho k\left[(\tau_{xy}QU_y)^- + (\tau_{xy}RV_x)^-\right] + \frac{2}{3}C_1\rho\varepsilon \end{aligned}$$

The above expressions guarantee satisfaction of (4) but not of (5). In practice, however, (5) is satisfied due to the large value of C_1 (= 1.8) and because P_k is usually positive.

It is observed from the above expressions that care has been taken to make A,B,C,D as large as possible. This is a form of underrelaxation delaying the evolution of the stress field and keeping it in pace with that of the turbulence energy and dissipation fields which are subjected to time-step restrictions.

Finally the constraints

$$\tau_{xx} + \tau_{yy} \leq 2$$

$$|\tau_{xy}| \leq q\sqrt{\tau_{xx}\tau_{yy}} \qquad (q \cong 5)$$

are always imposed to avoid explosive numerical divergence.

3 Application

Comparisons are presented here with the LDA data of Delery [8] for the strong-interaction case 'C'. This flow may reasonably be treated as 2-D, in contrast to other separated cases. The shock is provoked by a plane bump along one channel wall. The Mach number upstream of the shock is 1.4. Two sets of computations have been performed over a 193x97 mesh, one with the low-Re $k - \varepsilon$ model of Launder & Sharma [9] and the other with the stress model stated in Section 2 combined with the low-Re $k - \ell$ model of [6] in the range $y^+ < 50$. Space contraints only permit a narrow selection of mean-flow results to be included, although comparisons for shear and normal stresses have been performed.

Fig. 1 gives an overall view of the computed flow field in terms of Mach contours. Clearly recognisable is the characteristic Lambda structure of the shock as it interacts with the boundary layer. The somewhat rugged appearance of the contours downstream of the shock reflect the low level of artificial diffusion introduced by the pressure-gradient adaptive smoothing methodology used. This ensures that smoothing is only active in the shock region outside the shear layer.

Figs. 2 and 3 compare, respectively, variations of bump-wall pressure and displacement thickness, while Fig. 4 gives velocity profiles in the boundary layer. All comparisons demonstrate that the Reynolds-stress model predicts a more extensive recirculation zone, a result arguably attributable to curvature-induced attenuation of turbulence in the separated shear layer. The pressure plateau in Fig. 2 is well captured, and so is the maximum displacement thickness. There is, however, the suggestion that the stress model over-estimates displacement in the recovering wake region, although the experimental conditions in this region are affected by a pressure gradient arising from a downstream contraction which has not been accounted for in the calculation.

References

[1] Hall M G,(1986), in *Numerical Methods for Fluid Dynamics II, Clarendon Press.*

[2] Dimitriadis K P and Leschziner M A, (1989), *6th Int. Conf. on Numerical Methods in Laminar and Turbulent Flows, Swansea, pp. 861-881.*

[3] Dimitriadis K P and Leschziner M A, (1989), *Proc. Royal Aeronautical Society Conference on The Prediction and Exploitation of Separated Flow, London*, pp. 10.1-10.15.

[4] Dimitriadis K P and Leschziner M A, (1989), *Proc. 4th Copper Mountain Conference on Multigrid Methods, Colorado, SIAM* pp. 130-148.

[5] Gibson M M and Launder, B E (1978), *JFM*, 86, pp. 491-511.

[6] Wolfshtein, M W (1969), *Int. J Heat Transfer*, 12, p. 301.

[7] Rodi, W (1976), *ZAMM*, 56, p. 219.

[8] Delery, J, *ONERA Rapport Technique No 42/7078 AY 014*, Decembre 1980/Juin 1981.

[9] Launder, B E and Sharma, B I (1974), *Letters in Heat and Mass Transfer*, 1, pp. 131-138.

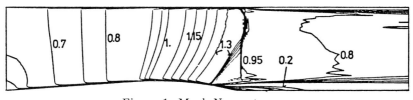

Figure 1: Mach No contours

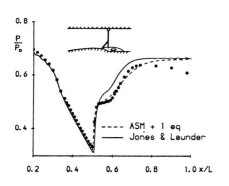

Figure 2: Wall pressure

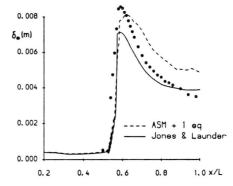

Figure 3: Displacement thickness

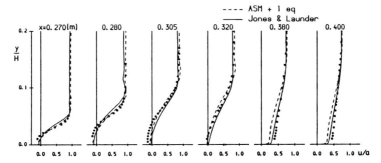

Figure 4: Axial velocity profiles

THE NONLINEAR GALERKIN METHOD FOR THE TWO AND THREE DIMENSIONAL NAVIER-STOKES EQUATIONS

Thierry Dubois, François Jauberteau, Roger Temam
Laboratoire d'Analyse Numérique, Université Paris-Sud
Bâtiment 425, 91405 Orsay, France

1. Introduction

Modern large scale computing allows the utilization of a very large number of variables/modes for spatial discretization of the partial differential equations arising in Computational Science and Applications. Because of that, the computers tend to be saturated by computations on small wavelengths that carry a small percentage of the total energy and that affect very mildly the final results.

The nonlinear Galerkin method is a new integration method that is well adapted to the solution of dissipative evolution equations on large intervals of time. An essential aspect of turbulent flow is the relation/interaction between small and large eddies. All the frequencies of the spectrum, up to the Kolmogorov dissipation frequency k_d interact ; large eddy break into small eddies and those, in turn, feed the large eddies : small eddies are negligible at each given time but their cumulative effects are not negligible on large intervals of time. Thus a proper and economical treatment of small eddies is necessary for large time computation. The novelty in this algorithm is that it combines the time and space discretizations ; it stems from recent developments in dynamical systems theory.

2. Modelling the interaction of small and large eddies

We consider, in space dimension $n = 2$ or 3, the incompressible Navier-Stokes equations :

$$\frac{\partial \vec{u}}{\partial t} - \nu \Delta \vec{u} + (\vec{u} \cdot \nabla)\vec{u} + \vec{\nabla} p = \vec{f} \tag{1}$$

$$\text{div } \vec{u} = 0. \tag{2}$$

Here $\vec{u} = \vec{u}(x,t)$ is the velocity vector, $p = p(x,t)$ the pressure and $\vec{f}$ represents volume forces applied to the fluid. We assume that $\vec{u}$, p and $\vec{f}$ are space periodic with period 2π in each direction x_1, x_2 if $n = 2$, or x_1, x_2, x_3 if $n = 3$. Hence we can expand $\vec{u}$ in Fourier series :

$$\vec{u}(x,t) = \sum_{k \in \mathbb{Z}^n} \hat{u}_k(t) e^{ik \cdot x} \tag{3}$$

where $n = 2$ (or 3), $k = (k_1, k_2)$ (or $k = (k_1, k_2, k_3)$). The nonlinear Galerkin method is based on a decomposition of u into small and large eddies :

$$\vec{u} = \vec{y} + \vec{z}. \tag{4}$$

We choose a cut-off value N and call large waves the wave vectors corresponding to $k = (k_1, k_2)$ (or $k = (k_1, k_2, k_3)$) such that

$$|k| = \left(\sum_{i=1}^{n} k_i^2\right)^{1/2} \leq N, \quad n = 2 \text{ or } 3.$$

Hence the small waves correspond to k's such that : $|k| > N$. Of course, the value N is here arbitrarily choosen, we will comment on this point hereafter. Having choosen N, we write :

$$\vec{y}(x,t) = \sum_{k \in \mathbb{Z}^n, |k| \leq N} \hat{u}_k(t) e^{ik \cdot x}$$

and

$$\vec{z}(x,t) = \sum_{k \in \mathbb{Z}^n, |k| > N} \hat{u}_k(t) e^{ik \cdot x}$$

At this point, it is convenient to remove the pressure p of equation (1), as usual by taking the divergence of (1). We obtain

$$\frac{\partial \vec{u}}{\partial t} - \nu \Delta \vec{u} + B(\vec{u}, \vec{u}) = \vec{f}. \tag{5}$$

where $B(\vec{u}, \vec{u})$ is now the sum of the inertial term and the pressure gradient expressed in terms of $\vec{u}$. Using the preceding decomposition (4) of $\vec{u}$, we project the equation (5) onto the first modes $|k| \leq N$ and onto the other modes. So we obtain a coupled system of equations for $\vec{y}$ and $\vec{z}$:

$$\frac{d\vec{y}}{dt} - \nu \Delta \vec{y} + P_N B(\vec{u}, \vec{u}) = P_N \vec{f} \tag{6}$$

$$\frac{d\vec{z}}{dt} - \nu \Delta \vec{z} + Q_N B(\vec{u}, \vec{u}) = Q_N \vec{f} \tag{7}$$

Here P_N is the projector on the modes k such that $|k| \leq N$ and Q_N is the projector on $|k| > N$. It was shown in [FMT] that $\vec{z}$ (resp. $\frac{d\vec{z}}{dt}$) is small compared to $\vec{y}$ (resp. $\frac{\partial \vec{y}}{\partial t}$). So we can approximate (7) by

$$-\nu \Delta \vec{z} + Q_N B(\vec{y}, \vec{y}) = Q_N \vec{f}. \tag{8}$$

Equation (8) provides an interaction law between small and large eddies, i.e. the small eddies $\vec{z}$ are expressed in terms of the large ones by

$$\vec{z} = \Phi(\vec{y}) = (\nu \Delta)^{-1} \left[Q_N B(\vec{y}, \vec{y}) - Q \vec{f} \right]; \tag{9}$$

with (9) we obtain the equation of a manifold that we call $\mathcal{M}_1$.

It was shown in [F.M.T.] that the attractor is closer to $\mathcal{M}_1$ than it is to the space $\vec{z} = 0$, i.e. the space spanned by the functions $e^{ik \cdot x}$ with $|k| \leq N$, which is used in the usual Galerkin method. Hence, it seems that $\mathcal{M}_1$ provides a better approximation of the attractor. In the next section we describe the approximation scheme based on (9).

3. Implementation of the nonlinear Galerkin method :

For fixed N, we are looking for an approximation u_N of u, based on the modes k such that $|k| \leq N$:

$$\vec{u}_N(x,t) = \sum_{k \in \mathbb{Z}^n, |k| \leq N} \hat{u}_k(t) e^{ik \cdot x} \tag{10}$$

Taking advantage of the fact that $\vec{z}$ is small and looking for an approximation of the attractor, that describes dynamic of the flow, we decompose u_N like u in (4) :

$$\vec{u}_N = \vec{y}_{N_1} + \vec{z}_{N_1} \tag{11}$$

where $N_1 < N$ and

$$\vec{y}_{N_1}(x,t) = \sum_{k \in \mathbb{Z}^n, |k| \leq N_1} \hat{u}_k(t) e^{ik \cdot x}$$

$$\vec{z}_{N_1}(x,t) = \sum_{k \in \mathbb{Z}^n, N_1 < |k| \leq N} \hat{u}_k(t) e^{ik \cdot x}, \; n = 2 \text{ or } 3 \tag{12}$$

The nonlinear Galerkin method is based on a different treatment of $\vec{y}_{N_1}$ and $\vec{z}_{N_1}$. The corresponding scheme is written :

$$\frac{\partial \vec{y}_{N_1}}{\partial t} + P_{N_1}\left[B(\vec{y}_{N_1}, \vec{y}_{N_1}) + B(\vec{y}_{N_1}, \vec{z}_{N_1}) + B(\vec{z}_{N_1}, \vec{y}_{N_1})\right]$$
$$- \nu \Delta \vec{y}_{N_1} = P_{N_1} \vec{f} \tag{13}$$

$$-\nu \Delta \vec{z}_{N_1} + Q_{N_1} B(\vec{y}_{N_1}, \vec{y}_{N_1}) = Q_{N_1} \vec{f}. \tag{14}$$

P_{N_1} is defined like P_N in (6) and Q_{N_1} is the projector onto the set of modes k such that $N_1 < |k| \leq N$. In the system (13)-(14), $\vec{y}_{N_1}$ corresponds to the large waves and so contains most of the energy. From a numerical viewpoint, $\vec{y}_{N_1}$ is the large term ; hence, (13) is integrated with a classical scheme i.e. the time discretization is implicit for the linear terms and explicit with a Runge-Kutta scheme of order 3 for the nonlinear terms. By contrast, $\vec{z}_{N_1}$ are small terms and appear as a perturbation (or correction) in the equation of the large waves (13). So, we impose $\vec{z}_{N_1}$ to verify equation (14) from time to time along the temporal integration.

The evaluation of the level of discretization N_1 is based on a cut-off strategy described in [DJT]. From time to time the size of the $\vec{y}_{N_1}$ grid is adjusted ; its depends on the evolution of $\vec{y}_{N_1}$ and $\vec{z}_{N_1}$.

In the next section we describe some computational tests performed with this new numerical method.

4. Numerical results :

We describe here the results of the computational tests performed with the nonlinear Galerkin method. Comparisons are also made with the usual Galerkin method. $3D$ as well as $2D$ numerical tests are reported.

First figures 1 and 2 show results on computational tests performed with the nonlinear Galerkin method in space dimension 3 ; the time step is $\Delta t = 10^{-2}$ and the viscosity is $\nu = 10^{-2}$. The solution is a priori choosen, hence the exact solution is known and it is easy to safely test the accuracy. The results show that the errors for the two methods are similar, with a gain in CPU computing time of 50%. More complicated tests are performed and will be reported later.

On another $2D$ example, we have compared the stability of the usual and nonlinear Galerkin methods. The total number of modes is $N = 128$ in each direction, the viscosity is $\nu = 10^{-3}$ and the time step $\Delta t = 7,5.10^{-2}$ is at the limit of stability for both schemes (i.e. the schemes are unstable for a larger Δt). We can note on figure 3 that the solution of the usual Galerkin method blows up much sooner than the solution of the nonlinear Galerkin method : in this sense, we can say that the usual method is less stable than the nonlinear Galerkin method and consumes more CPU time (figure 4).

To conclude we present the entrance, in the neighborhood of the manifold $\mathcal{M}_1$ of a solution computed with the $2D$ code (see figure 6). In this example, $\vec{f}$ is constant in time and is acting only on some small modes. We retain for the computation $N = 300$ modes in each direction, the time step is $\Delta t = 10^{-3}$ and the viscosity is $\nu = 7,5.10^{-3}$. Our aim is the numerical simulation of turbulence. These numerical results confirm the theoretical expectations (see [FMT] and [MT]). More results about such examples will be reported elsewhere.

5. Conclusion

In this article we have described numerical tests for a new discretization algorithm called the nonlinear Galerkin method. The results show that the algorithm is more stable and less time consuming than the usual (pseudo-spectral) Galerkin method. The algorithm is robust and well suited for large time integration because it takes into account the interaction between the small and large eddies, by neglecting the very small quantities. Also this algorithm can be adapted to other form of discretizations (finite elements, finite differences, wavelets...) ; this will be reported elsewhere.

Aknowledgement. The computations presented in this article were done on the CRAY 2 of the C.C.V.R.

References.
[DJT] T. Dubois, F. Jauberteau and R. Temam, to appear.
[FMT] C. Foias, O. Manley and R. Temam, Math. Model. Numer. Anal. (M2AN), 22 (1988) 93-114.
[JRT1] F. Jauberteau, C. Rosier and R. Temam, Appl. Numer. Math., 6 (1990).
[JRT2] F. Jauberteau, C. Rosier and R. Temam, Computer Methods in Applied Mechanics and Engineering, 78 (1990).
[MT] M. Marion and R. Temam, SIAM J. Numer. Anal., 26 (1989) 1139-1157.
[RT] R. Temam, Proc. 11th International Conf. on Numerical Methods in Fluid Dynamics, D.L. Dwoyer, M.Y. Hussaini and R. Voigt Eds., Lecture Notes in Physics 323, Springer Verlag, 1989.

Figure 1 : $G(t) = \frac{\|\vec{u} - \vec{u}_{ex}\|_{L_\infty}}{\|\vec{u}_{ex}\|_{L_\infty}}$

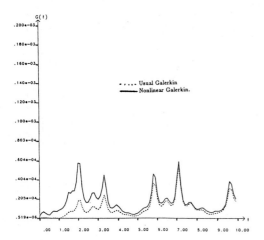

Figure 2 : $G(t)$ = gain of computing time between the usual and nonlinear Galerkin method

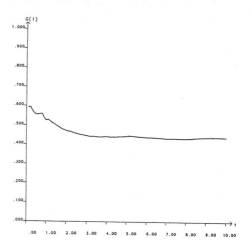

Figure 3 : $G(t) = \frac{\|\vec{u} - \vec{u}_{ex}\|_{L_\infty}}{\|\vec{u}_{ex}\|_{L_\infty}}$ (Blow-up of the solutions)

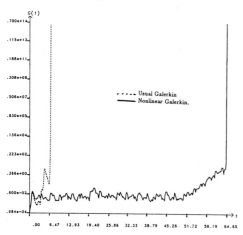

Figure 4 : $G(t)$ = gain of computing time between the usual and nonlinear Galerkin method

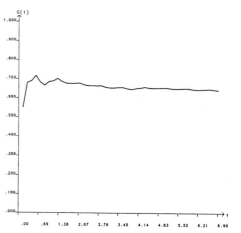

Figure 5 : $G(t)$ = evolution of the lower envelope of the levels of discretisation

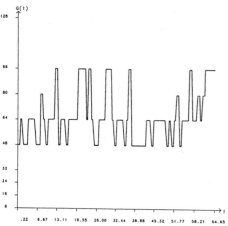

Figure 6 : $G(t) = \|\nu A\vec{z} + QB(\vec{y},\vec{y}) - Q\vec{f}\|_{L_2}$ (entrance of the orbit in the manifold $\mathcal{M}_1$)

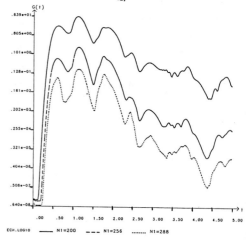

ON THE LARGE-EDDY SIMULATION OF COMPRESSIBLE ISOTROPIC TURBULENCE

G. Erlebacher, M.Y. Hussaini, and C.G. Speziale
ICASE, NASA Langley Research Center
Hampton, VA, 23665

and

T.A. Zang
NASA Langley Research Center
Hampton, VA 23665

INTRODUCTION

Direct numerical simulations of the complex turbulent flows of technological interest — at high Reynolds numbers with all scales resolved — will not be possible in the foreseeable future. The calculation of turbulent flows of practical importance will be done primarily by Reynolds stress models or by large-eddy simulations (LES). Among these two alternative approaches, large-eddy simulations appear to have the best predictive capabilities. The reason for this is clear: in LES, the large, energy containing eddies — which are strongly dependent on the geometry of the flow — are computed directly; only the small scales (which are more universal in character) are modeled. Large-eddy simulations have been applied to a variety of incompressible turbulent flows of basic engineering and geophysical interest with a considerable amount of success (cf. Moin and Kim [1], Bardina, Ferziger and Reynolds [2], Piomelli, et al. [3], Clark and Farley [4] and Smolarkiewicz and Clark [5]). However, there have been virtually no applications of LES to compressible turbulent flows. The purpose of this paper is to establish the feasibility of compressible LES — particularly for supersonic turbulent flows. The development of a general compressible LES capability could have a wealth of important aerodynamic applications involving high speed flows.

Large-eddy simulations of compressible isotropic turbulence are conducted based on the Favre-filtered equations of motion for an ideal gas. The calculations are based on compressible subgrid scale models that were recently derived by the authors [6]. The computed results from the coarse grid LES are validated by comparisons with the results of a fine grid direct numerical simulation (DNS) of isotropic turbulence for the same initial conditions. An analysis of the ability of the large-eddy simulation to capture the essential physics of compressible turbulence is provided along with a brief discussion of prospective future research.

FORMULATION OF THE PROBLEM

The large-eddy simulation of compressible isotropic turbulence is considered. Here, the Favre-filtered continuity, momentum, and energy equations are solved in the following form [6]:

$$\frac{\partial \overline{\rho}}{\partial t} + \frac{\partial}{\partial x_i}(\overline{\rho}\tilde{v}_i) = 0 \tag{1}$$

$$\frac{\partial}{\partial t}(\overline{\rho}\tilde{v}_i) + \frac{\partial}{\partial x_j}(\overline{\rho}\tilde{v}_i\tilde{v}_j) = -\frac{\partial \overline{p}}{\partial x_i} + \frac{\partial \overline{\sigma}_{ij}}{\partial x_j} + \frac{\partial \tau_{ij}}{\partial x_j}$$

$$\frac{\partial}{\partial t}(C_v\overline{\rho}\tilde{T}) + \frac{\partial}{\partial x_i}(C_v\overline{\rho}\tilde{v}_i\tilde{T}) = -\overline{p\frac{\partial v_i}{\partial x_i}} + \overline{\sigma_{ij}\frac{\partial v_i}{\partial x_j}} + \frac{\partial}{\partial x_i}\left(\overline{\kappa\frac{\partial T}{\partial x_i}}\right) - \frac{\partial Q_i}{\partial x_i} \tag{2}$$

where ρ is the density, v_i is the velocity vector, T is the absolute temperature, $p = \rho RT$ is the thermodynamic pressure, R is the ideal gas constant, C_v is the specific heat at constant volume, κ is the thermal conductivity,

$$\sigma_{ij} = -\frac{2}{3}\mu \nabla \cdot \mathbf{v}\, \delta_{ij} + \mu\left(\frac{\partial v_i}{\partial x_j} + \frac{\partial v_j}{\partial x_i}\right) \tag{3}$$

is the viscous stress tensor, τ_{ij} is the subgrid scale stress tensor, and Q_i is the subgrid scale heat flux. In (1)-(2), an overbar represents a spatial filter whereas a tilde represents a mass weighted or Favre filter, i.e.

$$\tilde{\mathcal{F}} = \frac{\overline{\rho \mathcal{F}}}{\overline{\rho}} \tag{4}$$

where $\mathcal{F}$ is any flow variable. A Gaussian filter is used for this study. The subgrid scale stress tensor and subgrid scale heat flux are modeled as follows (see Speziale et al. [6]):

$$\tau_{ij} = -\overline{\rho}(\widetilde{\tilde{v}_i\tilde{v}_j} - \tilde{\tilde{v}}_i\tilde{\tilde{v}}_j) + 2C_R\overline{\rho}\Delta_f^2 II_{\tilde{S}}^{1/2}(\tilde{S}_{ij} - \frac{1}{3}\tilde{S}_{mm}\delta_{ij}) - \frac{2}{3}C_I\overline{\rho}\Delta_f^2 II_{\tilde{S}}\delta_{ij} \tag{5}$$

$$Q_i = C_v\overline{\rho}(\widetilde{\tilde{v}_i\tilde{T}} - \tilde{\tilde{v}}_i\tilde{\tilde{T}} - \frac{C_R}{Pr_T}\Delta_f^2 II_{\tilde{S}}^{1/2}\frac{\partial \tilde{T}}{\partial x_i}) \tag{6}$$

where $II_{\tilde{S}} \equiv \tilde{S}_{ij}\tilde{S}_{ij}$, Δ_f is the filter width, and C_R and C_I are constants which assume the values of 0.012 and 0.0066, respectively. The turbulent Prandtl number Pr_T is taken to be 0.7. In the incompressible limit, the subgrid scale stress model (5) reduces to the linear combination model of Bardina, et al. [2]. Of course, in the limit as the filter width $\Delta_f \to 0$: $\tilde{\mathcal{F}}, \overline{\mathcal{F}} \to \mathcal{F}$ (where $\mathcal{F}$ is any flow variable) and hence the unfiltered continuity, momentum and energy equations are recovered.

DISCUSSION OF NUMERICAL RESULTS

Both direct and large-eddy simulations of compressible isotropic turbulence are conducted using a Fourier collocation method. The time advancement was done utilizing a third-order Runge-Kutta method and a time splitting scheme to alleviate stability problems associated with acoustic waves at low turbulence Mach numbers (see Erlebacher et al [7] for more details). This code has been tested extensively for two-dimensional and three-dimensional isotropic turbulence for a variety of Mach numbers [7].

The results of the simulations to be shown correspond to the initial conditions: $Re_\lambda = 26$, $< M_t^2 >^{1/2} \equiv < q^2/\gamma RT >^{1/2} = 0.1$, $\rho_{rms}=0$, $T_{rms} = 0.0626$ and $Pr = 0.7$ (where $\gamma \equiv C_p/C_v = 1.4$, $q^2 = v_i v_i$, and $< \cdot >$ denotes a spatial average). A direct numerical simulation was performed on a 96^3 mesh for these initial conditions. The results were then filtered and injected onto a coarse 32^3 mesh for comparison with the LES. Large-eddy simulations were performed on a 32^3 mesh with filter widths $\Delta_f = 0$ and $\Delta_f = 2$ (here the former value of $\Delta_f = 0$ corresponds to a coarse-grid direct simulation). In figures 1 (a)-(d), the integrated average rms divergence of velocity and vorticity, the average turbulence Mach number, the compressible and incompressible average turbulent kinetic energy ($E = < q^2 >$), and the partition function $\chi = E^c/E$ obtained from the LES and DNS are compared. It is clear that these results — which correspond to a filter width $\Delta_f = 2$ and are given for a few eddy turnover times — are in excellent agreement. We now demonstrate that the excellent results obtained from the LES are due to the adequate performance of the subgrid-scale models. In figures 2(a)-(d), the same turbulence statistics are shown for a filter width $\Delta_f = 0$ — namely, the case where the LES is actually a 32^3 coarse grid DNS. It is clear from figure 2 that the quantitative accuracy of the results degrades considerably. Hence, it follows that the success of the LES shown in figure 1 is largely due to the good performance of the subgrid-scale models in draining the proper amount of energy from the large scales to account for the presence of unresolved scales.

CONCLUDING REMARKS

The results obtained in this study strongly support the feasibility of LES for compressible turbulence. In fact, the results obtained are all the more impressive since — for the initial conditions considered — 20% of the turbulent kinetic energy was compressible. Future research will be directed toward extending these results to anisotropic and inhomogeneous compressible turbulent flows.

Additional refinements in the subgrid scale models are being explored in order to enhance the predictions for the dilatational turbulence statistics. With these further refinements — and with additional advancements in computer capacity — compressible large-eddy simulations could become an important tool for the analysis of high-speed compressible flows.

ACKNOWLEDGEMENTS

The first three authors (GE, MYH and CGS) acknowledge the support of the National Aeronautics and Space Administration under NASA Contract No. NAS1-18605 while in residence at the Institute for Computer Applications in Science and Engineering (ICASE), NASA Langley Research Center, Hampton, VA 23665. The authors would also like to thank Drs. R.B. Dahlburg and J.P. Dahlburg for sharing their revision to our DNS code and their LES results prior to publication.

References

[1] P. Moin and J. Kim. Numerical investigation of turbulent channel flow. *J. Fluid Mech.* **118**, 341-377, 1983.

[2] J. Bardina, J.H. Ferziger, and W.C. Reynolds. Improved turbulence models based on large-eddy simulation of homogeneous, incompressible turbulent flows. *Stanford University Technical Report, TF-19*, 1983.

[3] U. Piomelli, J.H. Ferziger, and P. Moin. New approximate boundary conditions for large-eddy simulations of wall-bounded flows. *Phys. Fluids A* **1**, 1061-1068, 1989.

[4] T.L. Clark, and R.D. Farley. Severe downslope windstorm calculations in two and three spatial dimensions using anelastic interactive grid nesting: a possible mechanism for gustiness. *J. Atmos. Sci.* **4**, 329-350, 1984.

[5] P.K. Smolarkiewicz and T.L. Clark. Numerical Simulation of a three-dimensional field of cumulus clouds. Part I. Model description, comparison with observations and sensitivity studies. *J. Atmos. Sci.* **42**, 502-522, 1985.

[6] C.G. Speziale, G. Erlebacher, T.A. Zang and M.Y. Hussaini. The subgrid scale modeling of compressible turbulence. *Phys. Fluids* **31**, 940-942, 1988.

[7] G. Erlebacher, M.Y. Hussaini, H.O. Kreiss, and S. Sarkar. The analysis and simulation of compressible turbulence. *ICASE Report No. 90-15*, NASA Langley Research Center, 1990.

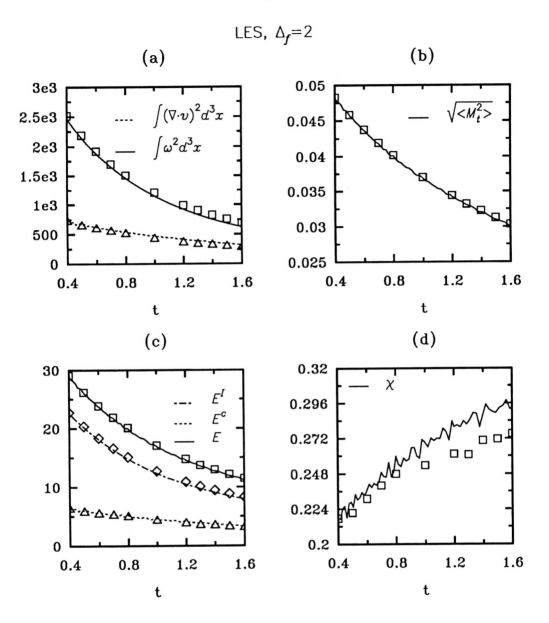

Figure 1. Comparison of the results of DNS ($\Box$, $\Diamond$, $\triangle$) and LES (———, ---, —·—) of compressible isotropic turbulence for $\Delta_f = 2$: (a) integrated average rms divergence of velocity and vorticity, (b) average turbulence Mach number, (c) incompressible, compressible and total turbulent kinetic energy, and (d) partition function χ.

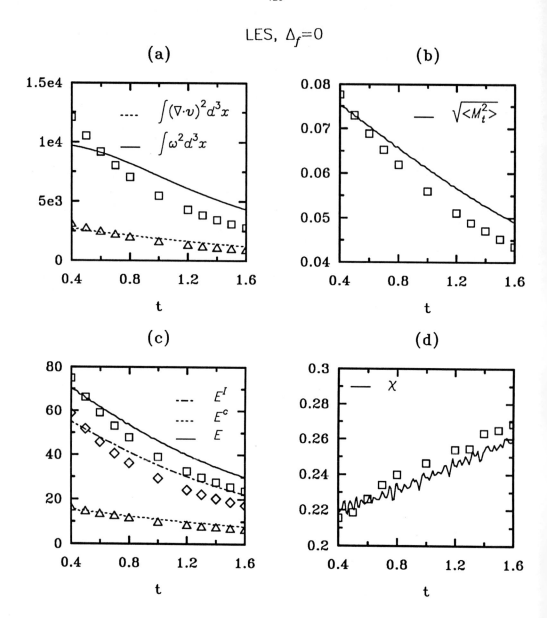

Figure 2. Comparison of the results of DNS ($\square$, $\diamond$, $\triangle$) and LES (——, - - - -, - · -) of compressible isotropic turbulence for $\Delta_f = 0$: (a) integrated average rms divergence of velocity and vorticity, (b) average turbulence Mach number, (c) incompressible, compressible and total turbulent kinetic energy, and (d) partition function χ.

A Calculation of the Compressible Navier-Stokes Equations in Generalized Coordinates by the Spectral Method

Yoshiaki Nakamura, Jian Ping Wang, and Michiru Yasuhara

Department of Aeronautical Engineering, Nagoya University,
Furocho Chikusa-ku, Nagoya 464-01, Japan

1. Introduction

The treatment of complex geometries has been a key issue for spectral methods[1]. Two approaches can be considered to overcome this difficulty: splitting the whole computing region into a number of elements[2], and using generalized curvilinear coordinates. The former makes it possible to properly adjust the construction of elements, and save CPU time by reducing the terms of Fourier transformation. In contract, the latter has the advantage of generating a smooth grid to get a uniformly distributed accuracy.

Generalized coordinates have been introduced by us into the 3D Euler equations solver by the spectral method to fit the forward part of a shuttle-like body[3],[4]. Viscous supersonic flows around a sphere were also calculated by using the axisymmetric N.S. equations[5],[6]. A common point to these problems is that the governing equations are transformed from the spherical coordinates which are very suitable to blunt bodies. To deal with arbitrary geometries, it is necessary to introduce generalized coordinates based on the Cartesian coordinates. Furthermore, in order to calculate the complex flow such as hypersonic reacting flow, a gas mixture model must be incorparated into the code. These are the objectives of the present paper.

2. Governing equations

2D compressible N.S. equations can be written in the Cartesian coordinates as follows:

$$Q_t + E_x + F_y = 0, \tag{1}$$

where

$$Q = [\rho, \quad \rho u, \quad \rho v, \quad e]^t,$$

$$E = \begin{bmatrix} \rho u \\ \rho u^2 + p - \tau^{xx}/Re \\ \rho v u - \tau^{xy}/Re \\ u(e+p) - (u\tau^{xx} + v\tau^{xy} - q^x)/Re \end{bmatrix},$$

$$F = \begin{bmatrix} \rho v \\ \rho u v - \tau^{yx}/Re \\ \rho v^2 + p - \tau^{yy}/Re \\ v(e+p) - (u\tau^{yx} + v\tau^{yy} - q^y)/Re \end{bmatrix},$$

with

$$\tau^{xx} = (2/3)\mu(2u_x - v_y), \quad \tau^{yy} = (2/3)\mu(2v_y - u_x), \quad \tau^{xy} = \tau^{yx} = \mu(u_y + v_x), \tag{2}$$

$$q^x = [(\gamma - 1)M_\infty^2 Pr]^{-1} \kappa T_x, \quad q^y = [(\gamma - 1)M_\infty^2 Pr]^{-1} \kappa T_y. \tag{3}$$

Pressure and temperature are calculated as

$$p = (\gamma - 1)[e - 0.5\rho(u^2 + v^2)], \tag{4}$$

$$T = p/\gamma M_\infty^2 \rho. \tag{5}$$

The generalized coordinates form of eq. (1) is

$$\tilde{Q}_t + \tilde{E}_\xi + \tilde{F}_\eta = 0, \tag{6}$$

$$\tilde{Q} = J^{-1}Q, \quad \tilde{E} = J^{-1}(\xi_t Q + \xi_x E + \xi_y F), \quad \tilde{F} = J^{-1}(\eta_t Q + \eta_x E + \eta_y F).$$

Thus all derivatives involved in the Cartesian coordinates are transformed to those in the computational coordinates.

3. Numerical method

3.1. Time integration The Euler explicit scheme of the first order accuracy is employed for the time integration. This scheme is sufficient to the steady problem considered in the present study.

3.2. Spatial derivatives There exist three groups of spatial derivatives in eq. (6): $(u_\xi, u_\eta, v_\xi, v_\eta, T_\xi, T_\eta)$, $(\tilde{E}_\xi, \tilde{F}_\eta)$, and $(x_\xi, x_\eta, y_\xi, y_\eta)$. A Chebyshev collocation method was utilized for calculating these values. The discrete Chebyshev polynomial expansion for a variable Q is expressed as

$$Q(\xi_i, \eta_j, t^k) = \sum_{n=0}^{N} \sum_{m=0}^{M} \hat{Q}_{mn}(t^k) \cos\frac{\pi i m}{M} \cos\frac{\pi j n}{N}, \tag{7}$$

in which the coefficients $\hat{Q}_{mn}$ can be obtained by using the FFT from the known values on the left hand side. The spatial derivatives can also be expressed in the same fashion as eq.(7):

$$Q_\xi(\xi_i, \eta_j, t^k) = \sum_{n=0}^{N} \sum_{m=0}^{M} \hat{Q}_{mn}^\xi(t^k) \cos\frac{\pi i m}{M} \cos\frac{\pi j n}{N},$$
$$Q_\eta(\xi_i, \eta_j, t^k) = \sum_{n=0}^{N} \sum_{m=0}^{M} \hat{Q}_{mn}^\eta(t^k) \cos\frac{\pi i m}{M} \cos\frac{\pi j n}{N}. \tag{8}$$

3.3. Filtering There are three reasons for using filters to remove high frequency modes; (1) the high frequency modes are emphasized in deriving the coefficients of derivatives, (2) high frequency oscillations are also generated by boundary conditions, and (3) the aliasing error mainly comes from high frequency modes.

An efficient time-varied filter which we call the Residual-Dependent Filter[7] was developed. In order to avoid sudden growth of pressure instability (eq.(4)), the four solution variables: $\rho, \rho u, \rho v, e$ were alternately filtered.

4. Gas mixture problem

Around hypersonic vehicle, chemical reactions may occur, where the air must be considered as a mixture of several species. The characteristics of the gas model of the mixture are; (1) the continuity equations for each species, (2) the thermal gas model such as the frozen specific heat, enthalpy, viscosity and thermal conductivity of each species and the mixture, (3) the influence of binary diffusion, which appears in the heat flux terms of the energy equation, and the diffusion terms of continuity equation for each species, (4) the relation between enthalpy, pressure, internal and kinetic energy which is used to calculate temperature by the Newton iteration method instead of eq.(5), and (5) the shock-fitting process tuned so as to fit the gas mixture model rather than simple Ranking-Hugoniot relations for a calorically perfect gas.

5. Results

A two-dimensional cylinder was chosen as a test problem. For the N.S. equations written in generalized coordinates, comparisons between the spectral collocation method (SCM) and the finite difference method (FDM) were carried out under the following conditions: Mach number $M_\infty = 4$, Reynolds number $Re = 1000$, Prandtl number $Pr = 0.72$, grid number 17×17, and a time increment $\Delta t = 0.009$.

Figure 1 illustrates the Mach number contours which show generally good agreement except for slight differences in the region outside the boundary layer. In Fig.2, the shock locations are exactly the same for two numerical results, and located slightly more upstream than experiment[7]. The pressure distributions along the body surface are shown in Fig.3, where the SCM and FDM results show excellent agreement, and are close to the experimental data. The time histories of the average residuals of the SCM and FDM codes are illustrated in Figs. 4a and 4b. The superiority of the former over the latter is obvious.

For the N.S. equations including a gas mixture model, we set the following free stream conditions: velocity $V_\infty = 2550 m/s$, temperature $T_\infty = 252.6 K$, pressure $p_\infty = 20.35 N/m^2$, and mass fractions $C_{O_2} = 0.2629$, and $C_{N_2} = 0.7371$.

Figure 5 shows the time history of the shock stand-off distance. Figures 6 and 7 illustrate the contours of enthalpy per unit volume and a diffusion coefficient which is difined as $D = \kappa Le/\rho C_{pf}$, where Le is the Lewis number and C_{pf} the specific heat at constant pressure for the frozen flow.

6. Conclusion

A spectral collocation method was developed in the present paper for the compressible N.S. equations written in the generalized coordinates. Comparisons between the SMC and FDM show the superiority of the former in accuracy, though it takes several times as large CPU time as the latter. Furthermore, the gas mixture model was successfully incorparated into our program code, which can become a fundation of dealing with hypersonic chemically reacting flows around blunt bodies by the spectral method.

References

1) Orszag, S. A.: Spectral Methods for Problems in Complex Geometries, *J. Comp. Phy.*, *37(1980)*, pp. *70-92*.

2) Patera, A. T.: A Spectral Element Method for Fluid Dynamics: Laminar Flow in a Channel Expansion, *J. Comp. Phy.*, *54(1984)*, pp. *468-488*.

3) Yasuhara, M., Nakamura, Y. and Wang, J. P.: Computations of the Hypersonic Flow by the Spectral Method, *Journal of the Japan Society for Aeronautical and Space Sciences*, *36(1988)*, pp. *542-549 (in Japanese)*.

4) Yasuhara, M., Nakamura, Y. and Wang, J. P.: Numerical Calculation of Hypersonic Flow by the Spectral Method, *Proc. 11th Int. Conf. on Numerical Methods in Fluid Dynamics, Lecture Notes in Physics 323, Springer-Verlag(1989)*, pp. *607-611*.

5) Wang, J. P., Nakamura, Y. and Yasuhara, M.: Several Improvements of Spectral Method in Compressible Flow Calculation, *Proc. Int. Sympo. Comp. Fluid Dynamics, Nagoya(1989), pp. 1210-1215*.

6) Wang, J. P., Nakamura, Y. and Yasuhara, M.: Accurate Computations of Compressible Navier-Stokes Equations by the Spectral Collocation Method, *Journal of the Japan Society for Aeronautical and Space Sciences*, **38***(1990), pp. 232-240 (in Japanese)*.

7) Kim, C. S.: Experimental Studies of Supersonic Flow past a Circular Cylindar, *J. Phy. Soc. Japan* **11***(1956), pp. 439-445 (in Japanese)*.

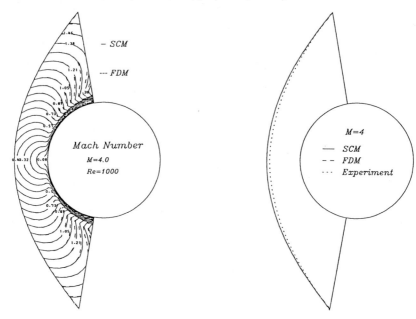

Fig. 1 Mach number contours.   Fig. 2 Schok wave locations.

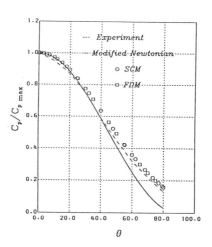

Fig. 3 Surface pressure distributions.

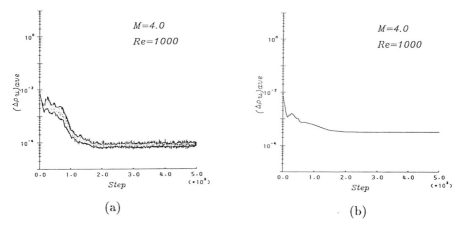

Fig. 4 Time histories of residual for ρu by SCM (a) and FDM (b).

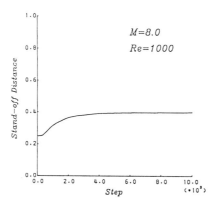

Fig. 5 Time history of shock stand-off distance.

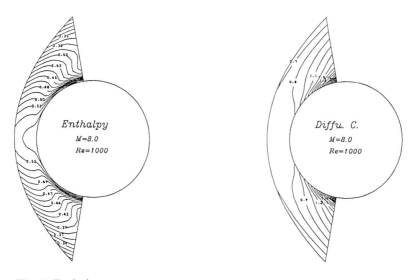

Fig. 6 Enthalpy contours.　　　　Fig. 7 Diffusion coefficient contours.

Numerical Methods for Viscous Incompressible Flows
D. Kwak, S.E. Rogers, C. Kiris and S. Yoon
NASA Ames Research Center, Moffett Field, CA, U.S.A.

I. Introduction

Unlike the compressible flow formulation, the incompressible Navier-Stokes formulation does not yield the pressure field explicitly from the equation of state or through the continuity equation. Depending on how the elliptic nature of the pressure field is computed, various methods have been developed in the past. The most frequently used primitive variable methods are based on a pressure iteration which usually results in partially implicit numerical algorithms. The pressure-based methods are easy to use. However, they are time-consuming, especially, in three-dimensional simulations. Therefore, it is particularly desirable to develop computationally efficient methods by implementing fast algorithms and by utilizing computer capabilities such as vectorization and parallel processing.

Recent advances in the state of the art in computational fluid dynamics (CFD) have been made in conjunction with compressible flow computations. Therefore, it is of significant interest to be able to use some of these compressible flow algorithms. To apply these algorithms, the method proposed by Chorin [1] can be used. In this formulation, the continuity equation is modified by adding a time derivative of a pressure term. This method, known as the pseudocompressibility method, has been extensively utilized in developing several incompressible Navier-Stokes codes at NASA Ames Research Center [2-4]. In the present paper, the pseudocompressibility formulation will be discussed followed by presentation of computed results.

II. Numerical Methods

After introducing the pseudocompressibility relation, the governing equations for steady-state solutions can be written in curvilinear coordinates, (ξ, η, ζ), as

$$\frac{\partial}{\partial \tau}\hat{D} = -\frac{\partial}{\partial \xi_i}(\hat{E}_i - \hat{E}_{vi}) = -\hat{R} \qquad (1)$$

where $\hat{R}$ is the right-hand-side of the momentum equation and can be defined as the residual for steady-state computations, and where

$\xi_i = \xi, \eta$ or ζ for $i = 1, 2,$ or 3

$$\hat{D} = \frac{D}{J} = \frac{1}{J}\begin{bmatrix} p \\ u \\ v \\ w \end{bmatrix}, \quad \hat{E}_i = \frac{1}{J}\begin{bmatrix} \beta(U_i - (\xi_i)_t) \\ (\xi_i)_x p + uU_i \\ (\xi_i)_y p + vU_i \\ (\xi_i)_z p + wU_i \end{bmatrix}, \quad \hat{E}_{vi} = \frac{\nu}{J}\nabla\xi_i \cdot (\nabla\xi_l\frac{\partial}{\partial\xi_l})\begin{bmatrix} 0 \\ u \\ v \\ w \end{bmatrix} \qquad (2)$$

$U_i = (\xi_i)_t + (\xi_i)_x u + (\xi_i)_y v + (\xi_i)_z w, \quad J = $ Jacobian of the transformation

Here, t is the time, x_i the Cartesian coordinates, u_i the corresponding velocity components, p the pressure, ν the total kinematic viscosity, and β is a pseudocompressibility parameter. All variables are nondimensionalized by a reference velocity and/or length scale.

In the steady-state formulation the equations are to be marched in a time-like fashion until the divergence of velocity vanishes. The time variable for this process no longer represents physical time. Therefore, τ can be thought of as a pseudotime or iteration parameter. Since this forms a hyperbolic-parabolic system of time-dependent equations, fast implicit schemes developed for compressible flows, such as the approximate-factorization scheme [5] and the implicit lower-upper symmetric-Gauss-Seidel (LU-SGS) scheme [4], can be implemented.

To obtain time-dependent solutions using this method, an iterative procedure can be applied in each physical time step such that the continuity equation is satisfied [3]. In this formulation the time derivatives in the momentum equations are differenced using physical time. To advance the governing equations in time for a divergence free velocity at the next time level, the following pseudocompressibility relation is introduced:

$$\frac{p^{n+1,m+1} - p^{n+1,m}}{\Delta \tau} = -\beta \nabla \cdot \hat{u}^{n+1,m+1} \tag{3}$$

where the superscript $n+1$ denotes the quantities at time $t = (n+1)\Delta t$ and a pseudotime level is denoted by a superscript m. The equations are iteratively solved such that $\hat{u}^{n+1,m+1}$ approaches the new velocity $\hat{u}^{n+1}$ as the divergence of $\hat{u}^{n+1,m+1}$ approaches zero. Combining equation (3) with the momentum equations gives the governing equations for the time-accurate formulation. Further information on the numerical algorithm can be found in [3].

III. Computed Results

Several flow solvers have been developed using the methods described above. The one based on the time accurate formulation, named the INS3D-UP code [3], uses an implicit line relaxation scheme. The other one based on the steady-state formulation, named the INS3D-LU code [4], uses an LU-SGS scheme. The results obtained using these codes are compared with experimental data and computed results obtained using a fractional-step based solver, INS3D-FS [6].

Curved Duct of Square Cross-Section

The flow through a square duct with a 90° bend offers a good test case for a full Navier-Stokes solver. This flow is rich in secondary flow phenomena both in the corner regions and through the curvature in the streamwise direction. This particular geometry was used as a steady-state test case for the INS3D family of codes. The geometry is shown in figure 1 and the Reynolds number of the flow is 790. The problem was non-dimensionalized using the side of the square cross-section. The inflow velocity was specified to be that of a fully developed, laminar, straight square duct. The velocity is normalized by the average inflow velocity. The computed results are compared to the experimental results of Humphrey et al. [7] as shown in figure 2. Overall, the comparison is quite satisfactory.

Artificial Heart Flow Simulation

Blood flow through mechanical hearts is very complicated in many respects. The fluid may exhibit significant non-Newtonian characteristics locally, and the geometry is usually very complicated. This problem is interdisciplinary, so a complete simulation would be a formidable task. However, an analysis based on a simplified model may provide much needed physical insight into the blood flow through these devices. Therefore, the primary purpose of the current application is to demonstrate that CFD technology described above can be extended to biofluid analysis. The present demonstration calculation is being performed on the Penn State artificial heart (Tarbell et al., [8]) using the time-accurate INS3D-UP code.

The computational model was made of two parts, namely, the main chamber and the valve region. The surface grid and a computer model of the main chamber are shown in figures 3a and 3b, respectively. The heart is composed of a cylindrical chamber with two openings on the side, out of which cylindrical tubes extend for valves. These tubes contain tilting flat disks which act as the valves. The pumping action is provided by a piston surface which moves up and down inside the chamber. For the present work, the main chamber and the valve region are computed separately.

The main chamber calculations were carried out using a single grid which is allowed to move to confirm to the motion of the piston. Figure 4a shows computed particle traces as the piston nears its bottom position, while, in figure 4b, an experimental photograph (J.M. Tarbell, Pennsylvania State University, private communication, 1988) shows a two-vortex system similar to the computed particle traces.

The geometry used for the valve region calculations is similar to that used in the experimental studies of Yoganathan et al. [9] (steady-state) and of Figliola and Mueller [10] (unsteady). The exact shape of the sinus region of the aorta is not known. Therefore, a slight difference in geometry may exist. The Reynolds numbers are based on channel entrance diameter and the mean velocity at the entrance of the channel. The physiological range of Reynolds numbers for these problems is in a regime where turbulence is important. For the present computations, an algebraic mixing length model was used.

Two overlapping grids were used for the present computations as shown in figure 5 where the disk is set at 30° from the centerline. Corresponding particle traces for steady-state solutions at Re= 2,390 is shown in figure 6. In figure 7, the axial velocity profile downstream of the disk in the horizontal plane through the center of the channel is plotted. The agreement with the experimental data [9] is fairly good. In figure 8, three snap shots taken from unsteady calculations during the valve opening are shown. The Reynolds number is 6,400. The numerical results generally show the same characteristics observed by experiments [10].

The present solution shows the capability of the computational procedure to simulate complicated internal flows with moving boundaries. This work represents the first step toward developing a CFD tool for this type of flow.

IV. Concluding Remarks

In the present paper, numerical solutions of the incompressible Navier-Stokes equations are discussed. Our main interest has been in three-dimensional real-world applications. Therefore, the main emphasis has been placed on a primitive variable formulation. There are several areas to be considered which will help make the current methods as efficient as possible. Multigrid acceleration is one possibility which is physically consistent with the incompressible formulation. Overall, the solution procedure should be optimized to best utilize computer characteristics such as vectorization, parallel processing, and access to memory.

REFERENCES

1. Chorin, A. J., "A Numerical Method for Solving Incompressible Viscous Flow Problems," J. Comp. Phys., vol. 2, pp.12-26, 1967.
2. Kwak, D., Chang, J. L. C., Shanks, S. P., and Chakravarthy, S., "A Three-Dimensional Incompressible Navier-Stokes Flow Solver Using Primitive Variables," AIAA J, vol. 24, No. 3, pp. 390-396, 1986.
3. Rogers, S. E. and Kwak, D., "An Upwind Differencing Scheme For the Time-Accurate Incompressible Navier-Stokes Equations," AIAA J., vol. 28, No. 2, pp. 253-262, Feb. 1990.
4. Yoon, S. and Kwak, D., "Three-Dimensional Incompressible Navier-Stokes Solver Using the Lower-Upper Symmetric-Gauss-Seidel Algorithm," AIAA J. (to appear in 1990).
5. Beam, R. M., and Warming, R. F., "An Implicit Factored Scheme for the Compressible Navier-Stokes Equations," AIAA J., vol. 16, pp. 393-402, 1978.
6. Rosenfeld, M., Kwak, D. and Vinokur, M., "A Solution Method for the Unsteady and Incompressible Navier-Stokes Equations in generalized Coordinate Systems," AIAA Paper 88-0718, 1988.
7. Humphrey, J. A. C., Taylor, A. M. K., and Whitelaw, J. H., "Laminar Flow in a Square Duct of Strong Curvature," J. Fluid Mech., vol. 83, part 3, pp. 509–527, 1977.
8. Tarbell, J. M., Gunshinan, J. P., Geselowitz, D. B., Rosenburg, G., Shung, K. K., and Pierce, W. S., "Pulse Ultrasonic Doppler Velocity Measurements Inside a Left Ventricular Assist Device," J. Biomech. Engr., Trans. ASME, vol. 108, pp. 232–238, 1986.
9. Yoganathan, A.P., Concoran, W.H., and Harrison, E.E., "In Vitro Velocity Measurements in the Vicinity of Aortic Prostheses," J. Biomechanics, vol. 12, pp. 135-152, 1979.
10. Figliola, R.S., and Mueller, T.J., "On the Hemolytic and Thrombogenic Potential Occluder Prosthetic Heart Valves from In-Vitro Measurements," J. Biomech. Engr., vol.103, pp. 83–90, 1981.

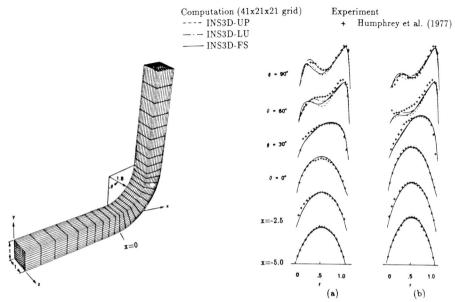

Figure 1.- Geometry and surface grid for flow through a 90° bend at Re=790.

Figure 2.- Streamwise velocity for flow through a 90° bend at Re=790 : (a) xy-plane at z=0.25, (b) xy-plane at z=0.5

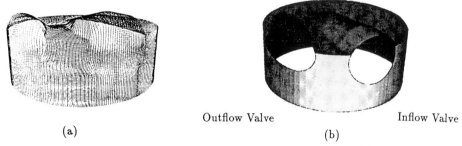

Figure 3.- Geometry of the Penn State artificial heart: (a) surface grid, (b) main chamber showing valve openings.

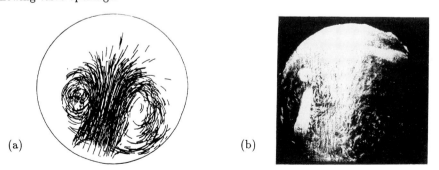

Figure 4.- Unsteady flow through an artificial heart: (a) incoming particle traces as the piston nears the bottom position (computed using INS3D-UP), (b) experiment (J.M. Tarbell, Pennsylvania State University, 1988).

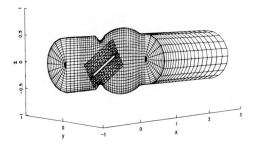

Figure 5.- Overlaid grids for prosthetic tilting-disk valve.

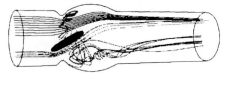

Figure 6.- Particle traces for the steady-state solution at Re=2,390 and at 30° disk angle.

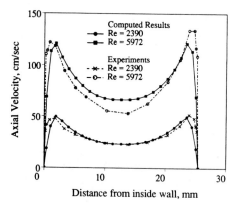

Figure 7.- Axial velocity profile downstream of the disk in the horizontal plane through the center of the channel compared with the experimental data of Yoganathan et al. [9].

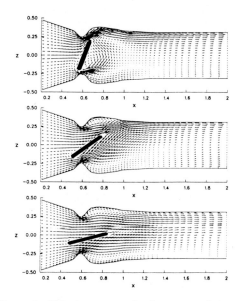

Figure 8.- Velocity vectors in the center plane showing valve opening at disk angle of 75° (top), 45° (middle), and 20° (bottom).

THREE-DIMENSIONAL SIMULATION OF VORTEX BREAKDOWN

G. Kuruvila
Vigyan, Inc.
Hampton, Virginia, USA

M. D. Salas
NASA Langley Research Center
Hampton, Virginia, USA

Introduction

The phenomenon of vortex breakdown (an abrupt change in the structure of the core of a swirling flow), first reported by Peckham and Atkinson[1], has been under investigation for over 30 years. The change in the structure of the vortex core is characterized by either a spiral deformation of the vortex axis or the formation of a stagnation point along the vortex axis followed by a bubble of recirculating flow. An example of this phenomenon is the breakdown of the vortex shed from a highly swept leading edge of a delta wing. The severe adverse effect of vortex breakdown on the performance of a wing has stimulated many experimental and theoretical studies. Comprehensive reviews of the progress made in understanding and predicting the occurrence of vortex breakdown have been given by Hall[2] and Leibovich[3].

In the past, several investigators (including the present authors) have simulated the bubble-type breakdown using various explicit numerical methods. All these investigations simplify the problem by assuming that the flow is axisymmetric, incompressible and laminar. In all these cases, bubble-type breakdown solutions were obtained for Reynolds numbers (based on vortex core radius) up to 200. However the performance of these algorithms was found to deteriorate rapidly at higher Reynolds numbers. A few investigators, including the present authors, were able to overcome this Reynolds number barrier using direct matrix inversion techniques[4]. With this method, solutions were obtained for Reynolds numbers as high as 1500. The present authors also studied the effect of three-dimensional perturbations by Fourier decomposition in the azimuthal direction and solved for the first non-axisymmetric Fourier mode[5]. Although the first Fourier component indicated a relatively small effect at low Reynolds numbers, the effect was significant at higher Reynolds numbers. Inclusion of higher modes leads to a formidably large and expensive problem. Hence the present effort was directed towards solving the full three-dimensional problem using an iterative technique.

Governing Equations

In the present study, the complete unsteady, compressible, three-dimensional Navier-Stokes equations in the primitive variable form are solved in a domain as shown in Fig.1. An isolated vortex is introduced at the $(0, y, z)$ plane by specifying the profiles of the flow variables. The three reference parameters governing the flow are the Reynolds number, Re, the swirl velocity parameter, S, and the Mach number, M_∞. The vortex-core

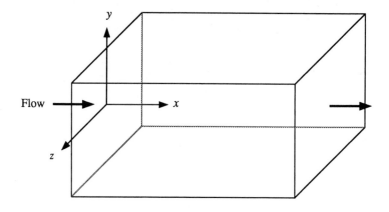

Figure 1. Computational Domain

radius along with reference values of density, speed of sound and viscosity are used to nondimensionalize the governing equations. Hence Re is based on the vortex-core radius and a reference velocity of 1. S is the swirl velocity at the edge of the vortex-core. The integral form of the nondimesionalized, full, conservative Navier-Stokes equations, cast in generalized curvilinear coordinate system, can be written as,

$$\frac{1}{J}\frac{\partial Q}{\partial t} + \frac{\partial (E - \mathcal{E})}{\partial \xi} + \frac{\partial (F - \mathcal{F})}{\partial \eta} + \frac{\partial (G - \mathcal{G})}{\partial \zeta} = 0 \qquad (1)$$

where Q is the vector of conserved variables, E, F and G are the Euler fluxes, $\mathcal{E}$, $\mathcal{F}$ and $\mathcal{G}$ are the viscous fluxes and J is the Jacobian of the transformation.

Discretization and Solution Procedure

The governing equations are integrated in time using a backward Euler AF scheme of Beam and Warming[6].

The Euler fluxes are discretized using Roe's upwind-biased flux-difference splitting[7]. First the primitive variables P are interpolated to both sides of the cell interfaces using the 2nd-order accurate fully upwind formula given by the following Eqns. 2[8].

$$\begin{aligned}(P_L)_{i+\frac{1}{2}} &= P_i + \frac{1}{4}[(1+\kappa)(P_{i+1} - P_i) + (1-\kappa)(P_i - P_{i-1})] \\ (P_R)_{i-\frac{1}{2}} &= P_{i+1} + \frac{1}{4}[(1+\kappa)(P_{i+1} - P_i) + (1-\kappa)(P_{i+2} - P_{i+1})]\end{aligned} \qquad (2)$$

where $\kappa = -1$ and P_i, P_{i+1} etc. are the cell average values. The fluxes at the cell interfaces are evaluated using the Roe's flux formula which, for example at a $i+\frac{1}{2}$ interface can be written as

$$E_{i+\frac{1}{2}}(Q_L, Q_R) = \frac{1}{2}\Big[E_L + E_R - \mathbf{T}|\mathbf{\Lambda}|\mathbf{T}^{-1}(Q_R - Q_L)\Big] \qquad (3)$$

where $\mathbf{T}$ and $\mathbf{T}^{-1}$ are the right and left eigenvector matrices respectively and $\mathbf{\Lambda}$ is a diagonal matrix containing the eigenvalues of the flux Jacobian $\partial E/\partial Q$. These matrices are evaluated using the Roe average quantities of the flow variables. The viscous fluxes are evaluated by 2nd-order central differencing. Cross-derivatives are evaluated using symmetric molecules. The resulting set of algebraic equations are linearized and they form a 5×5 block tridiagonal system of equations. These equations are solved using the Thomas algorithm. A full approximation multigrid algorithm was used to accelerate the convergence to steady state. The residuals were restricted to the coarse meshes by a volume weighted average while the corrections were prolongated to the fine meshes using tri-linear interpolation. Both the residuals and the corrections were smoothed, using a Laplacian like operator, before transferring.

Boundary Conditions

A major hurdle in simulating vortex breakdown, especially the spiral-type breakdown, is the numerical modelling of boundary conditions consistent with experiments. In the past we were able to simulate bubble-type breakdown by specifying that the flow is essentially inviscid near the outer edge of the computational domain[5]. Other investigators have simulated similar bubble-type breakdown by specifying pressure distribution along the outer edge[9]. In the present study, we assume that the outer edges of the computational domain are reasonably far away from the breakdown region. Hence the flow in this region is considered to be inviscid, steady and reasonably benign. The boundary conditions that are used are based on the characteristic variables[10].

The $(0, y, z)$ plane is treated as an inflow boundary while the $(10, y, z)$ plane is treated as an outflow boundary. The $(x, 4, z)$, $(x, -4, z)$, $(x, y, 4)$ and $(x, y, -4)$ planes are treated as mixed inflow and outflow boundaries. The appropriate boundary condition is applied after determining the direction of the velocity vector at these boundaries.

Initial Conditions

The initial velocity profiles are the same one that has been used by several previous investigators including the present authors[5] in their axisymmetric studies. They can be written in the Cartesian coordinate system as follows

$$\begin{aligned} u &= M_\infty \\ v &= -M_\infty \mathcal{V} z/r \\ w &= M_\infty \mathcal{V} y/r \end{aligned} \quad (4)$$

where u, v and w are the velocity components in the x, y and z direction and $\mathcal{V}$ is the velocity component in the yz-plane, which is given as

$$\begin{aligned} \mathcal{V} &= \mathcal{S} r \left(2 - r^2\right), & 0 \leq r \leq 1 \\ \mathcal{V} &= \mathcal{S}/r, & r \geq 1 \end{aligned} \quad (5)$$

where

$$r^2 = y^2 + z^2 \quad (6)$$

is the radial distance from the center of the vortex and $\mathcal{S}$, the swirl parameter, is the circumferential velocity at the edge of the vortex-core. Assuming a total enthalpy, based on the maximum total velocity and a temperature = 1, and a constant entropy, the profiles of ρ the density and e the total energy per unit mass were evaluated. Fig. 2 shows the vortex profile described by Eqn. 5 for $\mathcal{S} = 1$.

Results

Presented here is the solution for the case where $Re = 100$, $S = 1$ and $M_\infty = 0.1$. The computational domain, based on the vortex–core radius, of $10 \times 8 \times 8$ in the x, y and z directions respectively, had $64 \times 32 \times 32$ cells with grid clustering around the vortex-axis. Fig. 3 shows the residual history for this case with and without multigrid. The use of a three level V-cycle multigrid resulted in a machine-zero solution in about 6 hours of CPU time on Cray-2. In the same amount of time the single grid residual was over six orders of magnitude higher. About 2400 multigrid cycles were required to drive the steady state residual to machine zero. About 9 megawords of core memory was used for this case. Fig. 4 shows the bubble-type breakdown of the vortex. A tube of particles seeded at $(0, y, z)$ bulges around the breakdown region. A "tape" of particles seeded above the axis can be seen to follow the swirling velocity field, stretching and rolling-up as it travels downstream. Similar steady state bubble-type solutions were obtained for other values of Re and S.

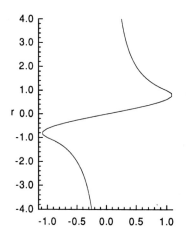

Figure 2. Vortex Profile

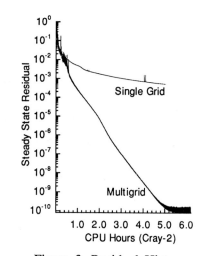

Figure 3. Residual History

Figure 4. Streamsurfaces

Concluding Remarks

A three-dimensional finite volume full Navier-Stokes code has been developed and used to study the vortex breakdown. The code is robust and its performance, both in terms of computer resources requirement and rate of convergence, is good. Steady bubble-type breakdown has been simulated for several sets of Re and S. The results agree, qualitatively, with the ones obtained in the past using an axisymmetric incompressible formulation. Attempts were made to obtain spiral-type breakdown by perturbing the flowfield or by increasing the back pressure. In the case when the flowfield was perturbed the flowfield decayed back to the unperturbed field. When a slightly higher back pressure was applied to a un-burst vortex, a bubble-type flowfield resulted. However the solution diverged when large values of back pressure were applied. These results are only preliminary and further extensive numerical experiments are required before this phenomenon can be fully understood.

References

[1] D. H. Peckham and S. A. Atkinson, "Preliminary Results of Low Speed Wind Tunnel Tests on a Gothic Wing of Aspect Ratio 1.0", Technical Report CP No. 508, Aeronautical Research Council, 1957.

[2] M. G. Hall, "Vortex Breakdown", Annual Review of Fluid Mechanics, vol. 4, pp. 195-218, 1972.

[3] S. Leibovich, "Vortex Stability and Breakdown", AIAA Journal, vol. 22, pp. 1192-1206, 1984.

[4] M. D. Salas and G. Kuruvila, "Vortex Breakdown Simulation: A Circumspect Study of the Steady Laminar Axisymmetric Model", Computers & Fluids, vol. 17, pp. 247-262, 1989.

[5] M. D. Salas and G. Kuruvila, "Study of Three-dimensional Effects on Vortex Breakdown", 16th Congress of the International Council of the Aeronautical Sciences, Jerusalem, Israel, August/September 1988.

[6] R. M. Beam and R. F. Warming, "An Implicit Factored Scheme for the Compressible Navier-Stokes Equations", AIAA Journal, vol. 16, pp. 393-402, April 1978.

[7] P. L. Roe, "Characteristic-Based Schemes For the Euler Equations", Annual Review of Fluid Mechanics, vol. 18, pp. 337-365, 1986.

[8] B. van Leer, "Upwind-difference Methods for Aerodynamics Problems Governed by the Euler Equations", 15th AMS/SIAM Seminar on Large Scale Computations in Fluid Dynamics, La Jolla, California, June/July 1983.

[9] S. Menne, "Vortex Breakdown in an Axisymmetric Flow", AIAA 26th Aerospace Sciences Meeting, Reno, Nevada, January 1988. Paper No. 88-0506.

[10] J. L. Thomas and M. D. Salas "Far-Field Boundary Conditions for Transonic Lifting Solutions to the Euler Equations", AIAA Journal, vol. 24, pp. 1074-1080.

ACCURATE FINITE-DIFFERENCE METHODS FOR SOLVING NAVIER-STOKES
PROBLEMS USING GREEN'S IDENTITIES

- by -

S C R Dennis

Department of Applied Mathematics, University of Western Ontario, London, Canada

and

J D Hudson

Department of Applied & Computational Mathematics, University of Sheffield,
United Kingdom

Abstract

An investigation is presented of a method of numerical approximation to the steady-state Navier-Stokes equations in two space dimensions using the finite-difference methods of h^4 accuracy recently published by Dennis and Hudson (1989) in conjunction with the global, or integral, methods based on Green's identities given by Dennis and Quartapelle (1989). It is shown that uniformly h^4-accurate results can be obtained using this combination of methods without the necessity of making use of local approximations to the boundary vorticity. Some detailed results of computations are given for flow past a circular cylinder in the Reynolds number range 10-100 based on the diameter of the cylinder. The problem of flow in a square cavity in which one side is moved parallel to itself with constant velocity is also considered.

1. Introduction and basic method

In obtaining numerical solutions of the Navier-Stokes equations using finite-difference methods, a problem associated with all h^4-accurate methods is to maintain equivalent accuracy at the boundaries when boundary values must be calculated there, e.g., for the boundary vorticity in the vorticity-stream function formulation. Local methods of approximation are normally used for calculations at boundaries but Dennis and Quartapelle (1989) have reviewed various applications of Green's identities which are used to transform local boundary conditions to global conditions, termed integral conditions, over the solution domain. Since integrals are easily approximated using h^4-accurate methods, it is relatively easy to complete the h^4-accurate analogue of the equations without using local conditions.

We consider only the formulation

$$R^{-1}\nabla^2\zeta = (\partial\psi/\partial y)(\partial\zeta/\partial x) - (\partial\psi/\partial x)(\partial\zeta/\partial y) \qquad (1)$$

$$\nabla^2\psi + \zeta = 0 \qquad (2)$$

in terms of the scalar vorticity ζ and the stream function ψ, although the corresponding equations for unsteady flow and also the use of other dependent variables could be considered. The variables ψ and ζ are dimensionless and R is the Reynolds number based on some representative length and velocity.

It is supposed that the solution is required inside a region bounded by a closed curve C with given conditions

$$\psi = f(x,y), \quad \partial\psi/\partial n = g(x,y) \text{ on } C, \qquad (3a,b)$$

where $\underset{\sim}{n}$ is the outward normal to C. Equations (1) and (2) are solved using the Dennis and Hudson (1989) h^4-accurate method, assuming Dirichlet boundary conditions. The condition needed for (2) is given by (3a), while that needed for (1) on C is found from one of Green's identities. If ϕ is any harmonic function and ψ satisfies (2), we easily find that

$$\iint_\sigma \phi\zeta d\sigma = \oint_C (\psi\partial\phi/\partial n - \phi\partial\psi/\partial n)dS, \qquad (4)$$

where σ denotes the plane region bounded by C and S is distance measured along C in the counter-clockwise sense.

For a given function ϕ, the right-hand side of (4) is known by (3) and thus (4) gives a condition on ζ of integral type. There is one such condition for each ϕ we choose and we may determine ζ_c by choosing $\phi = \phi_m$ (m = 1,2,3, ...) as a complete set of functions for the region interior to C. If ζ is expressed as a series of these functions, it is easy to construct a method of utilizing the corresponding set of conditions of type (4) to determine ζ_c in terms of the distribution of vorticity inside C (cf Dennis and Quartapelle, 1989).

Quadrature formulae of h^4 accuracy are used to approximate the integral on the left-hand side of (4) and thus an over-all h^4-accurate scheme is preserved which avoids using local approximations at the boundary. This is the basic principle of the method. Of course if C is a general curved boundary there may be practical problems to overcome in maintaining the h^4 accuracy but, nevertheless, we can find numerous practical examples in which the method is relatively straightforward. Two illustrative example are now given.

2. Flow past a circular cylinder

The first example is symmetrical flow past a circular cylinder treated in the usual modified polar coordinate system (ξ,θ), where

$$x = \exp(\xi)\cos\theta, \quad y = \exp(\xi)\sin\theta. \qquad (5)$$

Here the domain of the solution is $\xi \geq 0$, $0 \leq \theta \leq \pi$ and the contour C is in fact the corresponding curve bounded by $\theta = 0$, $\theta = \pi$, $\xi = 0$, $\xi = \infty$, although in practice the infinite boundary is approximated by a finite, large enough, value of ξ. The harmonic functions in (4) are $\phi_m = \exp(-m\xi) \sin m\theta$ and the integral conditions equivalent to (4) can be reduced to (cf Dennis and Quartapelle, 1989)

$$\int_0^\infty \exp\{(2-m)\xi\} g_m(\xi)d\xi = -2\delta_{m,1} \qquad (6)$$

where

$$\zeta = \sum_{m=1}^\infty g_m(\xi)\sin m\theta \qquad (7)$$

and $\delta_{m,1}$ is the Kronecker delta. The coefficients $g_m(\xi)$ in (7) are evaluated from a numerical solution obtained using two-dimensional h^4-accurate finite-differences based on the scheme of Dennis and Hudson (1989).

From this numerical solution $\zeta(\xi,\theta)$ of the equation

$$\partial^2\zeta/\partial\xi^2 + \partial^2\zeta/\partial\theta^2 = (R/2)[(\partial\psi/\partial\theta)(\partial\zeta/\partial\xi) - (\partial\psi/\partial\xi)(\partial\zeta/\partial\theta)]$$

obtained using (5) in (1), where $R = 2aU/\nu$, with a the cylinder radius, U the stream velocity and ν the coefficient of kinematic viscosity, we obtain $g_m(\xi)$ from the Fourier integral

$$g_m(\xi) = \frac{2}{\pi}\int_0^\pi \zeta(\xi,\theta) \sin m\theta \, d\theta \qquad (8)$$

This evaluation is made for all $\xi \neq 0$ and then satisfaction of (6) for each m using an appropriate quadrature formula determines $g_m(0)$ and hence $\zeta(0,\theta)$ is found from (7). The number of terms M required to approximate the infinite sum (7) is a parameter of the numerical procedure. Moreover, since M may be reasonably large, specialized formulae of Filon's type (Abranowitz and Stegan, 1968) are necessary to evaluate (8) and formulae of a similar type, modified suitably, are employed to evaluate the integrals in (6).

The only function of the expansion (7) is to determine the boundary vorticity $\zeta(0,\theta)$. The stream function is obtained by applying the same h^4-accurate method to the equation obtained by transforming (2) into the (ξ,θ) coordinate system, subject to suitable Dirichlet boundary conditions. The infinite domain of computation is limited in the ξ direction by a finite boundary $\xi=\xi_\infty$ on which suitable asymptotic expressions for ψ and ζ were assumed. The position of ξ_∞ was varied to ensure that it was sufficiently large. Calculations were carried out for R = 10, 40 and 100. In all cases the number of terms required in (7) was M = 20 and there was no significant difference when M was increased to 30.

As an illustration of some results obtained in the case of flow past a circular cylinder we show in Table 1 a comparison between present calculations and those of Fornberg (1980) for the drag coefficient C_D and front and rear stagnation point pressure coefficients $P(\pi)$ and $P(0)$ at Reynolds numbers, based on the diameter, R = 40 and 100. Three grid sizes h = $\pi/20$, $\pi/40$, $\pi/60$ were used in the present calculations and the results displayed correspond to h = $\pi/60$.

R	C_D Present	C_D Fornberg	$P(\pi)$ Present	$P(\pi)$ Fornberg	$P(0)$ Present	$P(0)$ Fornberg
40	1.508	1.498	0.571	0.570	-0.43	-0.46
100	1.053	1.058	0.530	0.532	-0.19	-0.17

TABLE 1

What is quite interesting about the present calculations, however, is the lack of grid dependence of many of the results, provided the Reynolds number is not too high. For example, at R = 10 the values of C_D obtained using the three grid sizes h = $\pi/20$, $\pi/40$, $\pi/60$ are, respectively, C_D = 2.812, 2.778, 2.763 and the corresponding values of the maximum absolute surface vorticity are 2.75, 2.71,

2.71. The corresponding values of $P(\pi)$ at $R = 100$ are 0.531, 0.530, 0.530 but there is a much greater variation of $P(0)$. In this latter case the corresponding values are -0.058, -0.140, -0.194. The maximum absolute surface vorticity values are 8.64, 9.16 and 9.20.

3. Flow in a square cavity

The second example considered is that of two-dimensional flow in a square cavity of side d when one side is moved parallel to itself with constant velocity U. We can take the governing equations in the forms (1) and (2), where (x,y) are dimensionless, C is the unit square $0 \leq x, y \leq 1$ and $R = Ud/\nu$. It is assumed that the side $y = 1$ is moved in the negative x direction and that the velocity components are scaled with respect to U. Then the boundary conditions are

$$\psi = \partial\psi/\partial x = 0 \text{ when } x = 0,1; \quad \psi = \partial\psi/\partial y = 0 \text{ when } y = 0;$$
$$\psi = 0, \partial\psi/\partial y = -1 \text{ when } y = 1. \tag{9}$$

Exactly the same h^4-accurate methods of Dennis and Hudson (1989) have been used to solve this problem, but two different methods of calculating the boundary vorticity have been used.

In the first we use a local method for calculating ζ_C, namely the h^4-accurate formula derived by Dennis and Hudson (1989) for the case of heat transfer in a square cavity but suitably adapted to the present problem. This is, at any point of C,

$$\zeta_C = 90 w_C/(23h) - 15(8\psi_1 - \psi_2)/(23h^2) - (16\zeta_1 - 11\zeta_2 + 2\zeta_3)/23, \tag{10}$$

where w_C is the velocity of the boundary moved parallel to itself in an anticlockwise sense and the subscripts 1, 2, 3 denote the first three internal grid points along the inward normal from the point considered. Calculations were carried out for $R = 10$ and 100 using grid sizes $h = 1/20, 1/40$. In Table 2, the values obtained for ζ at four grid points when $R = 100$ are compared with accurate values obtained by Dennis and Hudson (1980). The entry E indicates that the values were obtained by extrapolation from four h^4-accurate solutions with $h = 1/20, 1/30, 1/40, 1/50$; the present calculations therefore seem quite good.

Source of result	h	$\zeta(0.7,0.7)$	$\zeta(0.7,0.3)$	$\zeta(0.3,0.3)$	$\zeta(0.3,0.7)$
Dennis & Hudson (1980)	E	0.7861	-0.0914	-0.3392	3.535
Present calculations	1/20	0.7952	-0.0913	-0.3383	3.527
	1/40	0.7885	-0.0911	-0.3390	3.536

TABLE 2

The second approach uses the integral conditions based on the Green identity (4). We can only report briefly here the scope of this work and full details, including results and comparisons, will be published later. Two applications of

(4) are used, one to calculate ζ on $x = 0,1$ and the other to calculate ζ on $y = 0,1$. Thus, in the latter case, we substitute

$$\psi(x,y) = \sum_{n=1}^{\infty} f_n(y) \sin n\pi x \qquad (11)$$

and it follows from (9) that

$$f_n(0) = f_n(1) = 0; \quad f_n'(0) = 0, \quad f_n'(1) = a_n \qquad (12)$$

where $a_n = 2\{(-1)^n - 1\}/(n\pi)$. The harmonic functions $\phi(x,y)$ in (4) are taken as the terms in the series (11) for $n = 1, 2, \ldots$ and we can then deduce the conditions

$$\int_0^1 r_n(y) \exp(\pm n\pi y) dy = a_n \exp(\pm n\pi), \qquad (13)$$

where

$$r_n(y) = -2 \int_0^1 \zeta(x,y) \sin n\pi x \, dx. \qquad (14)$$

The equations (13) give two sets of conditions, obtained by taking the positive and negative signs respectively, which serve, by means of numerical quadrature, to determine $r_n(0)$ and $r_n(1)$ once $r_n(y)$ has been determined for $y \neq 0,1$ from (14). This is carried out also by numerical quadrature from the numerical solution for $\zeta(x,y)$ obtained by h^4-accurate methods. From $r_n(0)$ and $r_n(1)$ we obtain $\zeta(x,0)$, $\zeta(x,1)$ by inversion of (14).

In a similar way we determine $\zeta(0,y)$, $\zeta(1,y)$ by writing

$$\psi(x,y) = \sum_{n=1}^{\infty} F_n(x) \sin n\pi y \qquad (15)$$

and satisfying the conditions

$$\int_0^1 s_n(x) \exp(\pm n\pi x) dx = 0 \qquad (16)$$

where

$$s_n(x) = -2 \int_0^1 \zeta(x,y) \sin n\pi y \, dy. \qquad (17)$$

This is the basic method. However, there are many details requiring discussion, including the effect of the singularities in ζ; detailed results will be published in due course.

We can conclude, however, from the results obtained so far the present methods of using Green's identities give effective methods of obtaining, in many cases, accurate numerical solutions of the Navier-Stokes equations and that the h^4-accurate methods of Dennis and Hudson (1989) are capable of giving highly accurate results.

REFERENCES

M Ambranowitz and I A Stegun, Handbook of Mathematical Functions, National Bureau of Standards, Washington DC, (7th printing), p 890, 1968.
S C R Dennis and J D Hudson, J.Comput.Phys. 85, 390 (1989)
S C R Dennis and J D Hudson, J.Inst.Math.Applics, 26, 369 (1980)
S C R Dennis and L Quartapelle, Int.J.Num.Methods Fluids 9, 871 (1989).
B Fornberg, J.Fluid.Mech. 98, 819 (1980).

INFLUENCE OF LONGITUDINAL VORTICES ON THE STRUCTURE OF THE THREE DIMENSIONAL WAKE OF A PARTIALLY ENCLOSED BODY

M. Sanchez, N.K. Mitra, M. Fiebig
Institut für Thermo- und Fluiddynamik
Ruhr-Universität Bochum
Postfach 102148, 4630 Bochum, FRG

Introduction

The structure of the 3D-flow around a circular cylinder built in a rectangular channel is highly complex. Horse shoe vortices appear on the top and bottom walls in the stagnation zones. Depending on the Reynolds numbers and geometry, a Karman vortex street may appear on the midplane. Between the midplane and the bottom or the top wall, a helical vortex tube appears in the wake /1/.
The geometrical configuration of a channel with a tube models a cross flow fin-tube heat exchanger where the channel walls represent plate fins. The three dimensional structure of the wake determines the heat transfer between the fluid and the fin in the wake. The structure of the wake can be controlled and the separation on the cylinder surface can possibly be avoided by introducing longitudinal vortices in the flow. The longitudinal vortices can be generated by deforming part of the channel wall in form of a wing or winglet. In previous works, the structure of the laminar flows in a rectangular channel with a built-in circular cylinder and in a channel with built-in vortex generators were investigated /1, 2/. The purpose of the present work is a numerical investigation of the interaction of the longitudinal vortices with the wake of a cylinder in a channel.

Basic equations and method of solution

Figure 1 shows the computational domain consisting of a rectangular channel of height H, width B = 10H, with a built-in cylinder of diameter D and a pair of delta winglets of height of H and span of 0.5D in the wake of the cylinder. The flow field in this domain is calculated by solving complete unsteady Navier Stokes equations for incompressible fluid with constant viscosity.

Uniform parallel flow is used as initial condition at the channel inlet. No-slip condition is used at the top and bottom walls of the channel and the symmetry condition is used at the side boundaries. At the channel exit, the second derivatives of velocity components are taken equal to zero. Previous computations for flows in a channel with a cylinder with H/D = 1 and the Reynolds number based on H, Re_H = 1100 showed unsteady wake /1/. With H/D = 0.4 and Re_H = 1000 we obtained steady symmetric wake. In the present work with H/D = 0.2 and Re_H = 600 (typical of a fin-tube heat exchanger) the wake will be steady. Hence the flow need to be calculated only in the half width of the channel (between the side wall and the dotted line) with symmetry condition on the middle.

The basic equations are solved by a modified version of the marker and cell (MAC) technique /3/. The computation is carried out in two steps. In the first step cartesian grids are used to discretize the channel and to simulate the circular cylinder. Once a steady or a periodic solution is obtained in cartesian grids, a second step of solution, in which the flowfield in the neighborhood of the cylinder is again computed in a polar grid, is performed. Details of the computational scheme can be found in refs. 2, 3.

Results and Discussion

Computations have been performed in the channel with 56*42*12 grids for Re_H = 600 and 1000. Figure 2 shows velocity vectors at the channel cross section at a distance D downstream of the cylinder (shown in fig. 1). The horse shoe vortex which forms in front of the cylinder bends around the cylinder and forms two longitudinal vortices on two sides of the symmetry plane.

The cross section of one of these longitudinal vortex is seen on the left side of fig. 3. The next vortex is the cross section of the longitudinal vortex generated by the winglet. At the extreme right induced secondary vortices near the top and bottom walls can be seen. Figure 3 compares surface stream lines on the cylinder for the cases of (a) channel without winglet and (b) channel with winglet. The angular position 0° corresponds to the forward stagnation point. For flow without winglet, the two stagnation points (forward and rearward) are located in the middle of the channel. Between them at about θ = 109°, a saddle point appears. The streamlines coming out of the two stagnation points converge on the separation line and form two separation points at heights of 0.37H and 0.63H.

For the flow with winglet (fig. 3b) the rear stagnation point moves down to the bottom plate. The saddle point disappears and there is now only one separation point which appears on the bottom plate. The separation line meets the bottom plate at 155° from the stagnation point. The right half of fig. 3b is strongly asymmetric and has revolved in the same direction as the longitudinal vortex from the winglet rotates.

Figure 4 shows the velocity vectors on cross sectional planes in the wake at a distance of H/24 from the cylinder center. With winglet (fig. 4a), there is no recirculation inmediately behind the cylinder, but a recirculation zone appears further downstream. Without winglet (fig 4b), the flow separates from the cylinder at about 100° from the stagnation.

Conclusion

With delta winglets positioned on the channel wall in the wake of a cylinder, the separation on the cylinder can be avoided. A recirculation zone may be appear in the wake. It will be much smaller than recirculation zone without winglet.

References

/1/ Mitra, N.K., Kiehm, P., Fiebig, M., in Proc. 10th Int. Conf. on Numerical Methods in Fluid Dynamics, ed. F.G. Zhang and Y.L. Zhu, Beijing, 1986, Springer, pp. 481-487

/2/ Fiebig, M., Brockmeier, U., Mitra, N.K., Güntermann, T., in Numerical Heat Transfer, A, 15, 1989, pp. 281-302

/3/ Hirt, C.W., Nichols, B.D., Romero, N.C., Report LA - 5652, 1975, Los Alamos Scientific Lab.

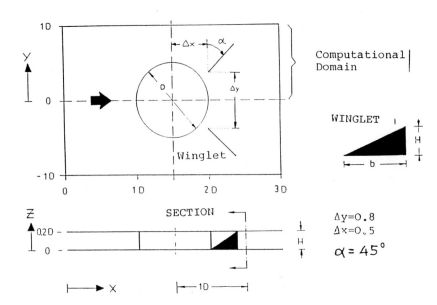

Fig.1: Geometry and computational domain

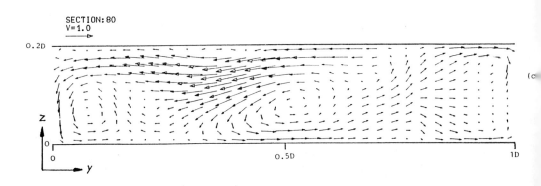

Fig. 2: Cross-stream velocity vectors on the cross section shown in fig. 1, $Re_H = 600$

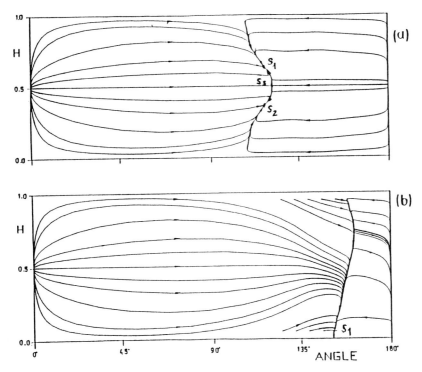

Fig. 3: Surface streamlines on the cylinder for the case channel without winglet (a) and with winglet (b). Re = 600. S_1, S_2: separation points; S_3: saddle point

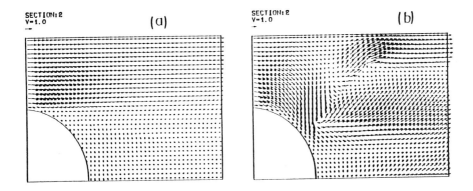

Fig. 4: Velocity vectors in the wake on lateral cross section at a distance of H/12 from the bottom plate. Re_H = 600. (a) without winglet, (b) with winglet

The ISNaS Compressible Navier-Stokes Solver; First Results for Single Airfoils[1]

F.J. Brandsma [1], J.G.M. Kuerten [2]

[1] National Aerospace Laboratory, Aerodynamics Division, P.O. Box 90502, 1006 BM Amsterdam, The Netherlands
[2] Department of Applied Mathematics, University of Twente, P.O. Box 217, 7500 AE Enschede, The Netherlands

1 Introduction

In the ISNaS project, an information system for flow simulation based on the Navier-Stokes equations is being developed. As part of this project the National Aerospace Laboratory and the University of Twente cooperate on the development of a flow solver for three-dimensional, compressible flow based on the Navier-Stokes equations. The first application will be multi-element airfoils with extended flaps and slats for take-off and landing conditions, taking into account infinite-sheared wing effects. Because of the complex geometrical shape of this application, it has been decided to develop a multi-block solver. The computational domain is divided into blocks, where in each block a boundary-conforming, structured grid is used. This choice facilitates the easy use of the Euler equations in the outer regions of the flow, where viscous effects are negligible, whereas the more expensive evaluation of the Navier-Stokes equations can be restricted to the viscous, inner regions of the flow (boundary layers and wakes). The flow solver will be based on the Reynolds-averaged Navier-Stokes equations, where initially the Baldwin-Lomax turbulence model will be used [1].

A stepwise development of the flow solver has been planned, starting from inviscid flow, via viscous, laminar flow to viscous, turbulent flow, each around a single airfoil. Subsequently, the integration scheme will be implemented in a multi-block solver. After each development step the flow solver will be validated in view of requirements with respect to numerics, performance and flow physics. In this paper results of the monoblock solver for viscous, laminar flow over a flat plate, and viscous, turbulent flow around a single airfoil will be presented. Implementation of the multi-block solver is in progress.

---

[1] The ISNaS project is partly subsidized by the Dutch Ministries of Education and Sciences, and of Transport and Public Works. Part of the necessary supercomputer time is subsidized by 'Stichting SURF' from the 'Nationaal Fonds gebruik Supercomputers (NFS)'

2 Numerical Method

The unsteady, three-dimensional, Reynolds-averaged Navier-Stokes equations for compressible flow are integrated in time towards a steady state solution. The equations are written in integral form, where density, total energy density and the three Cartesian components of the momentum density are used as dependent variables.

Since the ISNaS project aims at applications on the industrial level, the numerical method must be robust and has therefore been selected from the group of well-proven integration schemes. This has led to a time-explicit scheme, using a finite-volume discretization in physical space. In order to cope with the highly non-uniform grids which are typical for the applications of interest, the present solver employs a cell-vertex technique instead of the commonly used cell-centred approach. For each vertex the control volume is the union of the eight grid cells meeting at that vertex.

For the spatial discretization central differences are used combined with second- and fourth order artificial dissipation. The fourth order dissipation is necessary to prevent odd-even decoupling, whereas the second order dissipation makes it possible to capture shocks. This formulation is based on the work of Jameson [2] for simulation of inviscid flow. For viscous flow several modifications are necessary in order to minimize the numerical dissipation in boundary layers and wakes [3]. As boundary conditions at the solid surface the velocity components are set equal to zero, while the density and energy density at the surface are found by solving the continuity and energy equations, assuming an adiabatic wall. The treatment of the far-field boundaries is based on approximate Riemann invariants, where in case of two-dimensional applications a circulation correction (bound vortex formulation) is added to the free stream quantities [4].

Integration in time is performed using a multistage Runge-Kutta scheme. Since the objective is the calculation of the steady state solution (no time accuracy is pursued), the coefficients in the Runge-Kutta scheme are chosen in such a way that the stability region is as large as possible. For the same reason local timestepping is applied to accelerate the convergence. For the time being no other convergence acceleration techniques, like residual averaging and multigrid, are considered, in order to avoid difficulties at block interfaces. At present, both a three-stage and a five-stage scheme are used. In the three-stage scheme the dissipative terms are evaluated only at the first stage, while in the five-stage scheme they are recalculated at the third and at the fifth stage.

3 Results

3.1 Laminar Flow over a Flat Plate

The flow solver has been validated for laminar flow by comparing the results for flow over a flat plate with analytical solutions. To this end the two-dimensional subsonic laminar non-lifting flow over a finite flat plate at free-stream Mach number $M_\infty = 0.3$, and Reynolds number $Re_\infty = 10^4$ (based on the platelength), is considered. A rectangular grid with 312 cells in streamwise direction (256 on the plate) and 128 in normal direction is used. The grid points in streamwise direction are clustered around the leading and trailing edge with the smallest distance 5×10^{-4}, in order to resolve the local small scale phenomena. The grid spacing in normal direction is stretched, with the first grid line at a distance of 1.8×10^{-4} platelengths from the wall. The computational domain extends from 0.5 platelengths upstream till 25 platelengths downstream of the plate, and 1 platelength above the plate. For the low Mach number used here, the numerical solution should compare well with the analytical solutions for incompressible flow. Above the plate, away from the leading and trailing edge, the analytical solution is determined by the classical Blasius theory. Figure 1 shows that the agreement between the calculated velocity profiles and the Blasius solution is good.

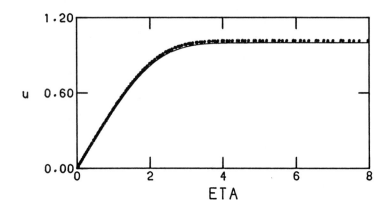

Fig. 1. Comparison of the calculated velocity profiles (nondimensional velocity u against scaled normal distance $\text{ETA} = z\sqrt{Re_\infty/2x}$) between 0.4 and 0.6 platelengths with the Blasius solution (solid line)

In figure 2 the calculated velocity at the wake centre-line is compared with the analytical solution, deduced from asymptotic theory in several flow regimes. Theoretically three regions can be distinguished here: immediately behind the trailing edge there is a small region of $\mathcal{O}(Re_\infty^{-3/4})$ platelengths where the full Navier-Stokes equations should be considered [5]; at larger distances from the trailing edge the flow is well described by the triple-deck approximation [6]; this triple-deck merges with the classical Goldstein solution, which is valid up to downstream infinity. As can be seen, in all three regimes the comparison between calculation and theory is good. This is clearly demonstrated by the asymptotic behaviour for vanishing distance x from the trailing edge. Theoretically, for $Re_\infty = 10^4$ the wake centre-line velocity behaves as $1.39\sqrt{x}$ near the trailing edge, whereas the calculation yields $u \sim 1.37\sqrt{x}$. The calculated drag coefficient ($C_D = 0.02822$) compares excellently with the analytical triple-deck value (0.02823) [6].

3.2 Turbulent Flow around an Airfoil

A good testcase for turbulent flow over a lifting airfoil is RAE2822 case 6: $M_\infty = 0.725$, $Re_\infty = 6.5 \times 10^6$, $\alpha = 2.92°(uncorr.)$. For this testcase well-documented experimental results are available [7], and this case was also included in the 'Holst workshop' [8]. The calculations have been performed on a C-type grid with 352 cells in streamwise direction (256 cells around the airfoil) and 64 cells in normal direction. The first grid line

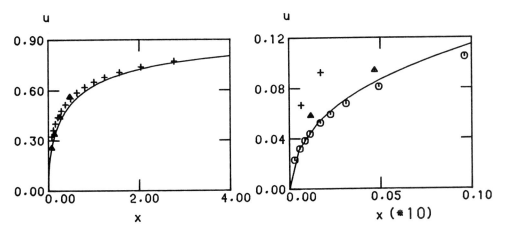

Fig. 2. Comparison of the velocity at the wake centre line (nondimensional velocity u against nondimensional distance from the trailing edge x) between calculations with the ISNaS solver (solid line) and asymptotic theory (o: Navier-Stokes; $\triangle$: triple-deck; +: Goldstein)

above the airfoil is chosen in such a way that the law-of-the-wall coordinate y^+ is close to 1 along this line. The far-field boundary is located at a distance of approximately 10 chordlenghts away from the airfoil. For the calculations the angle of incidence was chosen to be $\alpha = 2.44°$ according to the windtunnel corrections indicated in [7]. A total number of 12,000 iterations has been performed, resulting in a reduction of more than four orders of magnitude of the density residuals.

In figure 3 the results of the ISNaS flow solver for this case are compared with experimental data. For the pressure distribution the overall agreement is good. The small discrepancies on the lower surface, and on the upper surface just behind the suction peak and near the shock, are comparable with most of the other numerical results for this case presented in the Holst workshop [8]. The calculated lift coefficient ($C_L = 0.772$) is larger than the experimental value (0.743). The boundary layer on the upper surface is well resolved, as may be concluded from the agreement between the experimental and calculated displacement thickness. The calculated drag coefficient ($C_D = 0.0125$) is close to the experimental value (0.0127).

4 Conclusions

This paper describes the cell-vertex scheme for the compressible Navier-Stokes equations, to be part of the ISNaS flow simulation system. A comparison between numerical and analytical results for viscous, laminar flow shows that boundary layers and wakes are correctly resolved. Also for transonic turbulent flow accurate results have been obtained, as can be seen from a comparison with experimental results and other numerical solutions for the RAE2822 testcase.

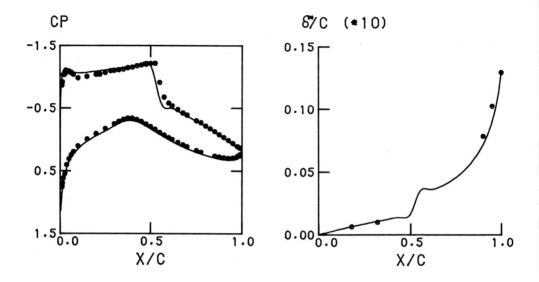

Fig. 3. RAE2822 case 6 ($M_\infty = 0.725, Re_\infty = 6.5 \times 10^6, \alpha = 2.92°$): comparison of the ISNaS results with experimental data. left: pressure distribution; right: displacement thickness on the upper surface; o: experiment; solid line: calculation

References

1. F.J. Brandsma: Mathematical Physics Aspects of Simulations based on the Navier-Stokes Equations, ISNaS 88.12.024 / NLR TP 89069 L (1989)
2. A. Jameson, W. Schmidt, E. Turkel: Numerical Solutions of the Euler Equations by Finite Volume Methods Using Runge-Kutta Time-Stepping-Schemes, AIAA Paper 81-1259 (1981)
3. J.G.M. Kuerten: Numerical Definition Document for the Time-explicit ISNaS Solver, ISNaS 88.10.031 (1988)
4. J.L. Thomas, M.D. Salas: Far-Field Boundary Conditions for Transonic Lifting Solutions to the Euler Equations, AIAA Paper 85-0020 (1985)
5. D. Dijkstra: The Solution of the Navier-Stokes Equations near the Trailing Edge of a Flat Plate, Ph. D. Thesis, University of Groningen (1974)
6. A.E.P. Veldman: Boundary Layer Flow past a Finite Flat Plate, Ph. D. Thesis, University of Groningen (1976)
7. P.H. Cook, M.A. McDonald, M.C.P. Firmin: Aerofoil RAE 2822 - Pressure Distributions, and Boundary Layer and Wake Measurements, Agard-AR-138 (1979)
8. T.L. Holst: Viscous Transonic Airfoil Workshop Compendium of Results, AIAA Paper 87-1460 (1987)

This paper was processed by the authors using the TEX macro package from Springer-Verlag.

LEADING EDGE EFFECTS ON BOUNDARY LAYER RECEPTIVITY

Thomas B. Gatski
NASA Langley Research Center
Hampton, Virginia 23665-5225 USA

Edward J. Kerschen
University of Arizona
Tucson, AZ 85721 USA

ABSTRACT

Numerical calculations are presented for the incompressible flow over a parabolic cylinder. The computational domain extends from a region upstream of the body downstream to the region where the Blasius boundary layer solution holds. A steady mean flow solution is computed and the results for the scaled surface vorticity, surface pressure and displacement thickness are compared to previous studies. The unsteady problem is then formulated as a perturbation solution starting with and evolving from the mean flow. The response to irrotational time harmonic pulsation of the free stream is examined. Results for the initial development of the velocity profile and displacement thickness are presented. These calculations will be extended to later times to investigate the initiation of instability waves within the boundary layer.

1. INTRODUCTION

The transition to turbulence in boundary layer flows is the result of the linear and nonlinear development of initially small amplitude Tollmien-Schlichting (T-S) waves into finite-amplitude disturbances which break down into random chaotic motion. The location of the transition point and the initiation of the T-S waves is closely related to the free stream disturbances; however, the mechanisms responsible for the coupling of the free stream disturbances and the spatially evolving T-S waves, that is the receptivity problem, is still a matter of some debate (see the recent review by Kerschen [5] for further details).

Numerical simulations of boundary layer receptivity have been somewhat limited. Murdock [6] calculated the flow over a parabolic cylinder in the incompressible limit (infinite sound speed and wavelength). He showed that, near the leading edge, sound waves feed energy into the T-S waves. The process was found to be sensitive to the parabola nose radius; the T-S wave amplitude for the case of a nose radius $r_n = O(\lambda_{TS})$, (λ_{TS} is the T-S wavelength), was an order of magnitude smaller than the T-S wave amplitude for the flat plate case ($r_n \ll \lambda_{TS}$).

In the initial phase of this study reported here, the steady mean flow solution for three parabolic cylinders with different nose radii is examined. We consider nose radii smaller than those studied by Murdock [6], since these body thicknesses have been shown to have boundary layers which are most sensitive to the free stream sound wave forcing. Unlike the previous numerical studies, but consistent with the theoretical analyses [4,5], the present formulation utilizes the length scale associated with the free stream forcing frequency and velocity rather than the usual viscous length scale. Results are presented for the initial development of the unsteady motion induced by the start-up of the time harmonic oscillation of the free stream velocity. A forcing amplitude of 0.1% is used to insure the linear behavior of the perturbation quantities.

2. FORMULATION

The present geometry suggests that the problem be formulated in terms of parabolic coordinates. The parabolic coordinates (ξ, η') are defined in the usual way by

$$z = x + iy = \frac{U_\infty}{2\omega^*}(\xi + i\eta')^2, \tag{1}$$

or

$$x = \frac{U_\infty}{2\omega^*}(\xi^2 - \eta'^2) \tag{2a}$$

$$y = \frac{U_\infty}{\omega^*}\xi\eta', \tag{2b}$$

where U_∞ and ω^* are the characteristic velocity and frequency (time) scales of the flow. The corresponding nondimensional divergence-curl and vorticity transport equations, in parabolic coordinates, can be written as

$$\frac{\partial u_1}{\partial \xi} + \epsilon^{-3}\frac{\partial u_2}{\partial \eta} = 0 \tag{3a}$$

$$\frac{\partial u_2}{\partial \xi} - \epsilon^{-3}\frac{\partial u_1}{\partial \eta} = K \tag{3b}$$

$$\frac{\partial K}{\partial t} + u_1\frac{\partial \Gamma}{\partial \xi} + u_2\epsilon^{-3}\frac{\partial \Gamma}{\partial \eta} = \epsilon^6\frac{\partial^2 \Gamma}{\partial \xi^2} + \frac{\partial^2 \Gamma}{\partial \eta^2}, \tag{3c}$$

where

$$\eta = \epsilon^{-3}\eta' \tag{4a}$$

$$K = (\xi^2 + \eta'^2)\Gamma \tag{4b}$$

$$\epsilon^6 = \frac{\nu\omega^*}{U_\infty^2} \equiv Re^{-1}, \tag{4c}$$

Γ is the vorticity and (u_1, u_2) are the nondimensional covariant velocities in the (ξ, η') coordinate directions, respectively. Note that the transverse coordinate η' has been stretched to augment the resolution of the solution normal to the wall. The ϵ^6 inverse Reynolds number scaling is chosen for consistency with the theoretical framework of Goldstein [4]. These equations are solved using a second-order spatially and temporally accurate vorticity-velocity numerical algorithm developed by Gatski, Grosch and Rose [3].

The focus of this study is high Reynolds number flows ($\epsilon \ll 1$) so that the results can be compared with the asymptotic analyses of boundary layer receptivity [4,5]. Of course, comparisons of the mean flow with previous steady flow calculations are also required. Fortunately, a parameter range can be found which satifies the constraints imposed by both the steady and unsteady problems.

The Reynolds number defined in Eq.(4c) is inversely proportional to the dimensionless frequency parameter associated with the stability of the Blasius boundary layer (e.g. Drazin and Reid, [2]). The asymptotic theories are valid at high Reynolds numbers, suggesting that low frequencies are desirable. Unfortunately, high Reynolds numbers place restrictions on spatial resolution for both the steady and unsteady problems. A frequency parameter of $O(10^{-4})$, which corresponds to a Reynolds number of $10^4 (=\epsilon^{-6})$, is a reasonable compromise among the varying constraints imposed on the numerical solution. The theoretical analysis of Goldstein [4] then provides the guidelines for the spatial resolution and range of the (ξ, η) coordinate directions.

In the streamwise direction, the theory gives the neutral stability point at $\xi = O(\epsilon^{-1})$, which in the present study corresponds to $\xi \approx 6$. In addition, the unsteady boundary layer region spans a range in ξ of $O(1)$ and the full Navier-Stokes equations are required in a small region near the leading edge with $\xi = O(\epsilon^{1/2})$ or $\xi \approx 0.5$. This dictates the use of a stretched grid in the ξ

direction ranging from the leading edge ($\xi = 0$) to a point downstream of the neutral stability point. The downstream limit on ξ was chosen as $\xi = \xi_B = 10$ in order to be sufficiently far downstream for the mean Blasius solution to hold and to also include a region of instability wave growth. Numerical tests indicated that a stretched streamwise computational grid of 500 cells was sufficient to obtain the desired resolution in the various regions of the flow.

In the direction normal to the boundary, Goldstein's [4] theory gives a triple-deck structure to the boundary layer region, with the main inviscid region being of $O(1)$, the outer inviscid region being of $O(\epsilon^{-1})$ and the viscous wall layer being of $O(\epsilon)$. This multi-deck structure dictates the use of a stretched grid in the normal direction. Since $\epsilon \approx 0.2$ in the present study, the viscous features of both the mean boundary layer and the unsteady motion are adequately resolved by using a stretched grid of 100 cells in the η direction and extending out to $\eta = \eta_T = 15$. The boundary condition at η_T is obtained from the potential flow solution.

In the present scaling, Murdock's [6] nose radii, R ($\equiv r_n \omega^*/U_\infty$), ranged from approximately 10^{-1} to approximately 10. We consider nose radii which range from the flat plate case, $R = 0$, to 10^{-3}. These radii are, of course, smaller than the T-S wavelength of the unsteady flow. At a Reynolds number of 10^4, and with the present scaling, the corresponding T-S frequency is 8.20×10^{-5}. Thus, the flows calculated here are well suited for the numerical forcing experiments.

3. RESULTS

In order to verify the accuracy of our calculations, the mean flow results for three different nose radii are compared to the results of Davis. The scaled surface vorticity for nose radii of $R = 0, R = 10^{-4}$ and $R = 10^{-3}$ is shown in Fig. 1. These results are in excellent agreement with those of Davis [1]. Even though the numerical procedure allows for both positive and negative values of ξ (above and below the parabolic cylinder), only the flow in the upper half of the cylinder ($\xi \geq 0$) was computed.

Another measure of the effect of cylinder radius on the boundary layer flow is the displacement thickness. In the thin boundary layer flow under study here, the displacement thickness is given by

$$\eta^* = 1.217 - \xi^{-1} \int_{\eta_0}^{\eta_T} u_1' \, d\eta, \tag{5}$$

where u_1' is the velocity perturbation to the Blasius solution in the steady problem or the velocity perturbation to the total mean flow solution in the unsteady problem and η_0 is the location of the body surface. The displacement thicknesses for the three different cylinder radii are shown in Fig. 2. As can be seen, there is a consistent thinning of the boundary layer with an increase of cylinder radius. This is to be expected since the flow in the vicinity of the leading edge is locally an accelerating stagnation point flow. Van De Vooren and Dijkstra [7] also calculated the displacement thickness for the case of an infinitely-thin flat plate; however, they indicate that the accuracy of their displacement thickness calculation was in question near the leading edge. Figure 2 shows the large deviation from the present results for values of $\xi^2 < 0.1$; however, the consistency of the present results with varying nose radii suggests that these results more correctly reflect the proper trends.

A final comparison to the steady results can be made with the surface pressure distribution. Figure 3 shows excellent agreement between the present calculations and the results of Davis [1]. Note that there are no regions of adverse pressure gradient along the parabolic cylinder surface. This is a desirable and necessary flow feature if one is to isolate the leading edge receptivity mechanisms from other receptivity mechanisms such as that due to abrupt changes in surface geometry.

The free stream forcing is a simple time harmonic oscillation starting at $t = 0$,

$$u'_1 = \epsilon_s \xi \sin t, \tag{6}$$

where $\epsilon_s (= 10^{-3})$ is the amplitude of the perturbation. This distribution induces a time harmonic pressure force on the boundary layer which is readily seen by examining the displacement thickness at different times within a forcing period. Figure 4 shows this variation, for the parabolic cylinder of nose radii $R = 10^{-3}$, at both $\pi/2$ and $3\pi/2$. The positive value for u'_1 at $t = \pi/2$ causes a thinning of the boundary layer, whereas, at $t = 3\pi/2$ the velocity u'_1 is negative and the boundary layer thickens.

It is also possible to compare these early time results with the Stokes wave solutions. Figure 5 shows this comparison at five different stations along the parabola. The early times at which these comparisons are made preclude any quantitative assessments; however, the flow does display the same qualitative features in the η direction as the exact Stokes wave solution and also shows a decrease in the thickness of the Stokes layer with downstream distance.

4. CONCLUDING REMARKS

This paper has considered the flow over parabolic cylinders with different nose radii. Both the steady and unsteady flow cases have been considered. To validate the accuracy of the calculations, the results for the mean scaled surface vorticity and surface pressure distributions have been favorably compared to previous results. It was shown that in the immediate vicinity of the leading edge, the local streamwise acceleration thins the boundary layer relative to its flat plate Blasius value. This region of favorable pressure gradient may have an important influence on the initiation of the boundary layer disturbances. In the unsteady case, the boundary layer thickness is correlated with the transient behavior of the free stream velocity. The results have shown, that even in the early stages of forcing, the flow development is qualitatively similar to the Stokes wave solutions.

References

[1] Davis, R. T., 1972, *Numerical Solution of the Navier-Stokes Equations For Symmetric Laminar Incompressible Flow Past a Parabola*, J. Fluid Mech., 51, pp. 417-433.

[2] Drazin, P. G. and Reid, W. H., 1981, *Hydrodynamic Stability*, Cambridge University Press, Cambridge.

[3] Gatski, T. B.; Grosch, C. E. and Rose, M. E., 1989, *The Numerical Solution of the Navier-Stokes Equations for 3-Dimensional, Unsteady, Incompressible Flows by Compact Schemes*, J. Comp. Phys., 82, No. 2, pp. 298-329.

[4] Goldstein, M. E., 1983, *The Evolution of Tollmien-Schlichting Waves Near a Leading Edge*, J. Fluid Mech., 127, pp. 1-10.

[5] Kerschen, E., 1989, *Boundary Layer Receptivity*, AIAA 12th Aeroacoustics Conference, April 10-12, San Antonio, Texas, Paper No. 89-1109.

[6] Murdock, J. W., 1981, *Tollmien-Schlichting Waves Generated by Unsteady Flow over Parabolic Cylinders*, AIAA 19th Aerospace Sciences Meeting, January 12-15, St. Louis, Missouri, Paper No. 81-0199.

[7] Van de Vooren, A. S. and Dijkstra, D., 1970, *The Navier-Stokes Solution For Laminar Flow Past a Semi-Infinite Flat Plate*, J. Engr. Maths., 4, pp. 9-27.

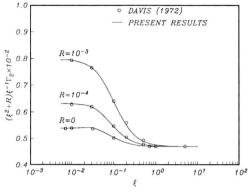

Figure 1: Comparison of steady-state scaled surface vorticity for nose radii of $R = 0$, $R = 10^{-4}$ and $R = 10^{-3}$.

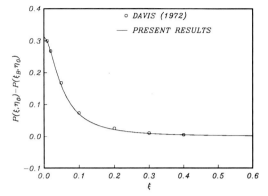

Figure 3: Surface pressure distribution for cylinder with nose radii of $R = 10^{-3}$.

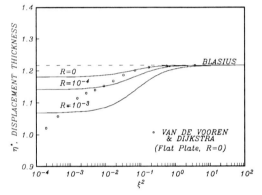

Figure 2: Variation of displacement thickness along cylinder for nose radii of $R = 0$, $R = 10^{-4}$ and $R = 10^{-3}$.

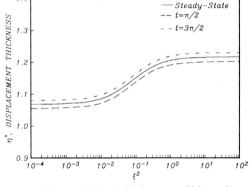

Figure 4: Time variation of displacement thickness for cylinder with nose radii of $R = 10^{-3}$.

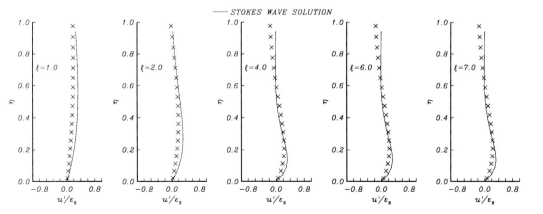

Figure 5: Comparison of unsteady streamwise velocity at $t = 2\pi$ with Stokes wave solution at various streamwise stations.

The Viscous-Inviscid Coupling via a Self-Adaptive Domain Decomposition Technique

Renzo Arina[1], Claudio Canuto[2]

[1] Dipartimento di Ingegneria Aeronautica e Spaziale,
Politecnico di Torino, C.so Duca degli Abruzzi, 24, I-10129 Torino ITALY
[2] Dipartimento di Matematica, Politecnico di Torino
and Istituto di Analisi Numerica del C.N.R., Pavia ITALY

Abstract: A new approach for the viscous/inviscid coupling is discussed. The viscous part of the Navier-Stokes equations is replaced by a monotonic function of the viscous terms themselves. The function is zero when the viscous terms are less than a prescribed value. This formulation leads to a natural viscous/inviscid splitting of the domain.

1. Introduction

The prediction of viscous compressible flows remains a major challenge of fluid mechanics. The reliable calculation of viscous flows by the full Navier-Stokes equations may not be an economic approach when the Reynolds number is high, being the viscous effects confined in narrow regions. In these cases a zonal method coupling the local solutions of the Euler equations and the Navier-Stokes equations (or boundary-layer like equations) can be very effective [1]. Among the advantages of the zonal approach is the possibility of using grids of different coarseness in the Euler and Navier-Stokes regions, or methods of different order in the two regions, or even coupling different techniques (such as finite volumes and spectral methods) in order to enhance the overall accuracy of the solution. These advantages are particularly pronounced in 3D problems and complex geometries. The potential of the zonal approach is fully exploited when the matching of the models is appropriate, and the interface is located in an optimal, or nearly optimal, way. For weak viscous/inviscid interactions, both the search of the position of the interface and the coupling conditions can be obtained by asymptotic matching concepts. However for strong interactions, which are very common in the hypersonic range, other approaches must be devised.

A new formulation of the viscous/inviscid coupling has been recently introduced in [2] under the name of χ-*formulation*, with the aim of providing an automatic detection of the viscous/inviscid interface and a smooth transition between the Euler and Navier-Stokes models. In this approach a monotone function of the viscous terms (built-in in the mathematical model) is used both as a zonal indicator and as a matching mechanism between the zones. The efficient implementation of the χ-formulation is discussed in [3] for the Burgers equation, with focus on the self-adaptive capability of the method in detecting the optimal position of the interface.

In the present paper, we will apply the χ-formulation in establishing a viscous/inviscid coupling for the conical Navier-Stokes equations. The numerical results enlight the distinctive features of our approach in providing an accurate and optimally displaced zonal decomposition.

2. Definition and basic properties

Let us describe the χ-formulation on the following model problem of incompletely parabolic type:

(2.1)
$$\begin{cases} u_t + Bu_x + Cu = Au_{xx} \ , \ x \in (0, L) \, , \, t > 0 \, , \\ + \text{ initial and boundary conditions } , \end{cases}$$

where $u = \begin{pmatrix} u_1 \\ u_2 \end{pmatrix}$, $A = \begin{pmatrix} 0 & 0 \\ 0 & \nu \end{pmatrix}$ with $\nu > 0$, and B is a symmetric matrix. Let us choose a cut-off parameter $\delta > 0$, and a further parameter $\sigma > 0$ such that $\sigma << \delta$, and let us introduce the monotone function $\chi = \mathbb{R} \to \mathbb{R}$ defined as:

$$(2.2) \quad \begin{cases} \chi(s) = \begin{cases} s & if\ s > \delta\ , \\ (s - \delta + \sigma)(\delta/\sigma) & if\ \delta - \sigma \leq s \leq \delta\ , \\ 0 & if\ 0 \leq s < \delta - \sigma\ , \end{cases} \\ \chi(s) = -\chi(-s)\ , \quad s < 0\ . \end{cases}$$

The idea of the χ-formulation consists of replacing the viscous term $Au_{xx} = (0, \nu u_{2,xx})^T$ by the modified term $(0, \nu \chi(u_{2,xx}))^T$, which for brevity will be denoted by $A\chi(u_{xx})$. Thus, we consider the χ-problem

$$(2.3) \quad \begin{cases} u_t + Bu_x + Cu = A\chi(u_{xx})\ ,\quad x \in (0, L)\ ,\ t > 0, \\ + \text{ initial and boundary conditions } . \end{cases}$$

It is shown in [4] (see also [2]) that, under suitable mathematical assumptions on the matrices B and C and the boundary conditions, the χ-problem admits a unique solution. Moreover, the maximal deviation of u from the solution of (2.1) is proportional to $\delta\nu$. This estimate yields a guideline for the practical choice of δ as a function of ν. The domain $(0, L)$ is *automatically* decomposed into three regions: the viscous region where $|\chi(u_{2,xx})| \geq \delta$ and the original system is solved, the inviscid region where $\chi(u_{2,xx}) \equiv 0$ and the reduced system without diffusion is solved, and finally a transition region where $0 < |\chi(u_{2,xx})| < \delta$. The solution u is proven to be continuously differentiable all over the domain ([4]). This remarkable feature of our approach is a consequence of the fact that the viscous/inviscid domain decomposition is not fixed a-priori, but it is free to adjust itself to give the smoothest solution of (2.3).

It is clear that solving the more expensive χ-problem throughout the domain $(0, L)$, instead of the original problem (2.1), would be a nonsense. The χ-formulation becomes attractive if the region where $|\chi(u_{2,xx})| \neq 0$ is small compared to the whole domain, and the reduced pure hyperbolic problem is solved for outside this region. Precisely, suppose that there exists a point b, $0 < b << L$, such that $|\chi(u_{2,xx})| \equiv 0$ in $(b - \epsilon, L)$ for some $\epsilon > 0$. Then, setting $w = u$ in $(0, b)$ and $v = u$ in (b, L), problem (2.3) is equivalent to the splitted problem

$(2.4 - a) \qquad w_t + Bw_x + Cw = A\chi(w_{xx})\ ,\quad in\ (0, b)\ ,\ t > 0\ ,$

$(2.4 - b) \qquad v_t + Bv_x + Cv = 0 \qquad\qquad ,\ in\ (b, L)\ ,\ t > 0\ ;$

supplemented by initial and boundary conditions, and by an *interface condition* which couples w and v at $x = b$. Since by assumption $\chi(u_{2,xx}) \equiv 0$ in a neighborhood of b, the proper interface condition is of hyperbolic type, namely, each subdomain takes from the adjacent subdomain the information running on the ingoing characteristics.

The coupled problems (2.4-a/b) are solved by an iterative procedure which can be included into the time-advancing scheme. Since the location of the interface is itself unknown, the actual position of b can be adjusted in time as discussed in [3], until a steady state is reached.

3. The χ-formulation for the conical Navier-Stokes equations

The conical Navier-Stokes equations are a simplified set of equations, appropriate for studying supersonic viscous flows that have truly conical inviscid flow counterpart, such as supersonic flows past pointed cones. A conical flow has the property that all flow quantities are invariant on rays passing through the apex of the cone. Consequently, all derivatives in the conical direction may be neglected, reducing the number of spatial dimensions by one. Moreover, in the case of axisymmetric flows, such as a flow past a circular cone with zero angle of incidence, the governing equations reduce to the one-dimensional system

$$(3.1) \qquad \frac{\partial U}{\partial t} + \frac{1}{z}(\frac{\partial F}{\partial \sigma} + B) = \frac{1}{Re}\frac{1}{z^2}(\frac{\partial S}{\partial \sigma} + G) ,$$

where $\sigma = tg^{-1}\theta$ is the angular coordinate measuring the distance from the cone surface, $U = (\rho, \rho v_r, \rho v_z, \rho e)^T$, F and B the inviscid flux and inviscid source terms, and S and G the viscous flux and viscous source terms. Further details are given in [5]. The conical assumption is exact for inviscid steady flow, whereas the axis z still remains as a length-scale factor in the steady viscous problem. System (3.1) is a simplified set of equations that greatly reduces the complexity of numerical computations, yet retaining the important features involved in the solution of viscous flows. In this way, it is possible to perform an exhaustive and representative study of the χ-formulation for the Navier-Stokes equations, at a reasonable computational effort.

In order to introduce the χ-formulation for system (3.1), and more generally for the complete Navier-Stokes equations, it is necessary to define the χ-function of the viscous terms. Being $\underline{\tau}$ the viscous stress tensor, the viscous terms of the momentum equations can be written in the form $div\underline{\tau}$. The mathematical results presented in Section 2 are valid if the function $\chi(s)$ is defined as a monotonic function of the viscous terms. Then the appropriate definition in this case is

$$(3.2) \qquad \chi_M(div\underline{\tau}) = \alpha_M(\|div\underline{\tau}\|) \cdot div\underline{\tau} ,$$

where, for a given $\delta > 0$ and $\sigma > 0$,

$$(3.3) \qquad \alpha_M(\|\mathbf{s}\|) = \begin{cases} 1 & if \ \|\mathbf{s}\| \geq \delta , \\ f(\|\mathbf{s}\|) & if \ \delta - \sigma < \|\mathbf{s}\| < \delta , \\ 0 & if \ \|\mathbf{s}\| \leq \delta - \sigma , \end{cases}$$

with $f(\|\mathbf{s}\|)$ any smooth monotonic function, with values between 0 and 1.

In the total energy equation, the scalar diffusive term can be written in the form

$$(3.4) \qquad -div\mathbf{q} + div(\underline{\tau} \cdot \mathbf{V}) = -div\mathbf{q} + grad\mathbf{V} \cdot \underline{\tau} + \mathbf{V} \cdot div\underline{\tau} ,$$

with $\mathbf{q}$ the heat conduction vector and $\mathbf{V}$ the velocity vector. Two distinct dissipative mechanisms are present: the heat diffusion $div\mathbf{q}$, and the part due to the viscous stress tensor $\underline{\tau}$. The former part can be taken into account by an appropriate χ function defined similarly to χ_M, as $\chi_Q(div\mathbf{q}) = \alpha_Q(|\ div\mathbf{q}\ |) \cdot div\mathbf{q}$. As far as the latter part is concerned, the quantity $\mathbf{V} \cdot div\underline{\tau}$ is negligible if the viscous terms of the momentum equation $div\underline{\tau}$ are negligible. Consequently, the same function χ_M given in (3.2) can be applied to these terms, when they appear in the total energy equation. On the other hand, the dissipation function $grad\mathbf{V} \cdot \underline{\tau}$ does not contain second-order derivatives and in particular circumstances it can be important even if all the second-order terms are negligible (e.g., in a Couette flow). Therefore, a third χ-function, say χ_D, has to be applied to the dissipation function. However, in the present application, we found that such a term is essentially comparable to the term $div\underline{\tau}$; hence, it is possible to control both terms by the same cut-off function α_M. In conclusion, an appropriate χ-formulation of the diffusive terms (3.4) has the form

$$(3.5) \qquad -\alpha_Q(|\ div\mathbf{q}\ |) \cdot div\mathbf{q} + \alpha_M(\|div\underline{\tau}\|) \cdot (grad\mathbf{V} \cdot \underline{\tau} + \mathbf{V} \cdot div\underline{\tau}) .$$

Equations (3.1) and their χ-formulation have been solved numerically by finite-differences. The convective fluxes are split into upwind components as proposed by Van Leer [6], and discretized by second-order backward and forward differences on a uniform grid, with the MUSCL approach [6]. The viscous terms are discretized by second-order centred differences. The time integration is obtained with a one-dimensional form of the Beam and Warming algorithm in delta form [7]. First-order upwind differencing is used on the implicit part giving rise to a block-tridiagonal matrix structure.

The test problem chosen is the supersonic flow over a circular cone (10 degree semi-apex angle, $M_\infty = 4$, Re= $.1\,10^5$, $T_w/T_\infty = 3.2$, Pr= 0.72). The radial and axial velocity components and the temperature plots are shown in Fig. 1-a/b for both the Navier-Stokes equations (solid lines), and the corresponding χ-formulation obtained with a cut-off value $\delta = 100$ (circles). Both solutions have been obtained with 251 grid points, and can be considered as grid independent. In the same Figures, shown are also the cut-off functions α_M (Fig. 1-a) and α_Q (Fig. 1-b), which indicate where the viscous terms are retained ($\alpha = 1$) or neglected ($\alpha = 0$). It can be observed that the χ-solution essentially coincides with the solution of the original problem (3.1). We found that setting the cut-off value δ to 1, the viscous terms are turned off only in the free-stream region; on the contrary, with $\delta = 10^4$, the viscous terms are turned off even in the outer part of the boundary layer, leading to an excessive deviation of the corresponding χ-solution from the exact one. This is in accordance with the theoretical error estimate in Sect. 2.

It is interesting to note the capability of the functions α_M and α_Q to detect the viscous and thermal boundary-layer regions and the shock region. Also note that the transition between the states $\alpha = 0$ and $\alpha = 1$ invariably occurs across very few grid points. Discarding the shock region, which is well captured by the Euler equations, the viscous region is formed by the points where at least one α is not zero. By slightly overestimating the viscous region, and placing the interface in a point where $\alpha_M = \alpha_Q = 0$, we obtain an optimal viscous/inviscid domain decomposition. In the inviscid region the Euler equations are solved, while the χ-Navier-Stokes equations are solved in the viscous region. At the interface, the viscous terms of the χ-Navier-Stokes equations are zero, then the matching condition is of the type Euler-Euler, which can be easily obtained by imposing the appropriate characteristic boundary conditions. Figure 2 shows the results obtained by such a coupling strategy. The number of grid points in the viscous region is fixed to $N = 62$, while the computational effort is progressively reduced in the inviscid region: there we used a second-order scheme with $M = 188$ grid points (plot (a)), or with $M = 100$ (b), and a first-order scheme with $M = 75$ (c) or with $M = 50$ (d). The χ-solution maintains its overall accuracy, as well as the good matching at the interface. This demonstrates the interest of locating the computational effort in the boundary-layer region.

References

1 R.C. Lock and B.R. Williams, *Viscous-inviscid interactions in external aerodynamics*, Prog. Aerospace Sci., vol.24, 1987, 51-171.

2 F. Brezzi, C. Canuto and A. Russo, *A self-adaptive formulation for the Euler/Navier-Stokes coupling*, Computer Meth. Appl. Mech. and Eng., vol.73, 1989, 317-330.

3 R. Arina and C. Canuto, *A Self-Adaptive Domain Decomposition for the Viscous-Inviscid Coupling. - I. Burgers Equation*, IAN-CNR Report 719, Pavia 1989.

4 C.Canuto, *A Mathematical Model for Coupling Differential Systems of Different Type*, in preparation.

5 R. Arina and C. Canuto, *A Self-Adaptive Domain Decomposition for the Viscous-Inviscid Coupling. - II. Conical Navier-Stokes Equations*, in preparation.

6 B. Van Leer et al., *A Comparison of Numerical Flux Formulas For the Euler and Navier-Stokes Equations*, AIAA Paper 87-1104, 1987.

7 R.F. Warming and R.M. Beam, *On the Construction and Application of Implicit Factored Schemes for Conservation Laws*, SIAM-AMS Proceedings, Comput. Fluid Dynamics, vol.11, 1978.

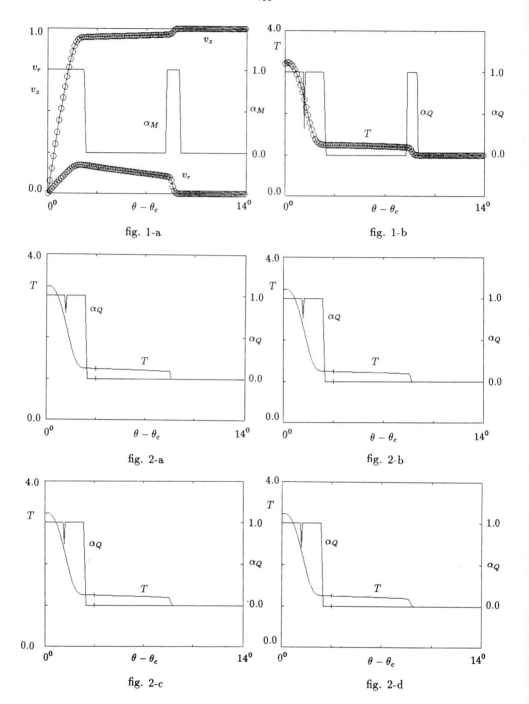

fig. 1-a

fig. 1-b

fig. 2-a

fig. 2-b

fig. 2-c

fig. 2-d

EFFICIENT COMPUTATION OF WING BODY FLOWS

by

N.–H. Cho, C.A.J. Fletcher and K. Srinivas
University of Sydney, NSW 2006, Australia

1. INTRODUCTION

The flow around a body mounted on a wall or a flat plate generates a very complex flow interaction that can be described superficially as generating a horse–shoe vortex system. The actual behaviour is greatly complicated if the flow is also turbulent and compressible. This flow configuration is both a severe test for CFD codes and of direct practical importance.

Here we describe a control–volume formulation in body–fitted coordinates with generalised four–point discretisation [1] for first derivatives and the use of limiters to control local dissipation to provide an accurate representation of severe gradients, e.g. shock waves. Turbulence is modelled by combining an algebraic Reynolds stress model (ASM) with a two–layer wall function [2,3]. This turbulence model provides the spatial distribution of all the Reynolds stresses and turbulent heat fluxes throughout the computational domain. A transport equation for a velocity potential correction is used to satisfy the mass conservation equation. The discretised equations are marched repeatedly in the main flow direction with a strongly implicit procedure [1] updating the solution in each transverse plane.

2. GOVERNING EQUATIONS

The governing equations in Reynolds–averaged form (retaining only dominant terms) can be written as

$$\frac{\partial}{\partial x_i}(\rho u_i) = 0 \qquad (1)$$

$$\frac{\partial}{\partial x_i}(\rho u_i u_j + \delta_{ij} p_j - \tau_{ij} + \rho \overline{u'_i u'_j}) = 0 \qquad (2)$$

$$\frac{\partial}{\partial x_i}(\rho c_p u_i T - k_{eff}\frac{\partial T}{\partial x_i}) = u_i \frac{\partial p}{\partial x_i}, \qquad (3)$$

where k_{eff} includes the eddy diffusivity. The Reynolds stresses are obtained via an algebraic Reynolds stress (ASM) or a k–ϵ turbulence model. At present turbulent heat fluxes are obtained via a turbulent Prandtl number/eddy viscosity construction. A two–layer wall function [2] is used to provide the solution and boundary conditions close to solid surfaces. The above ASM/wall function turbulence representation has been used previously to investigate internal swirling flows [2,3].

3. COMPUTATIONAL ALGORITHM

The governing equations are written symbolically as

$$\partial F_i^\ell / \partial x_i = S^\ell \tag{4}$$

Equation (4) is discretised using a control–volume formulation in generalised–coordinate space with the metric information expressed in terms of area vectors [4],

$$\sum_{n=1}^{6} F_n^\ell \cdot \underline{A}_n = V_c < S^\ell > \tag{5}$$

where the control volume is a unit cube in computational space. $\underline{A}_n$ is the area vector associated with the corresponding n^{th} face in physical space and V_c is the physical control volume.

In order to retain higher accuracy on relatively coarse grids first derivatives in the governing equations are evaluated with the equivalent of four–point asymmetric discretisation [1]. If the grid were uniform a typical control surface value of F would be

$$F_{i-\frac{1}{2}} = F_{i-1} + \psi_{i-\frac{1}{2}} [0.5 \, \Delta F_i - (q/3)(\Delta F_i - \Delta F_{i-1})] \tag{6}$$

For $q = 0.5$ and $\psi_{i-\frac{1}{2}} = 1.0$ the resulting scheme is third–order accurate and has excellent dispersion–suppression properties [5]. For $q = 1.5$ a three–point upwind scheme is obtained which is appropriate in regions of supersonic flow. Here we use a variable q suitable for mixed subsonic/supersonic flows,

$$q_{i-\frac{1}{2}} = 0.5 + \min[1.0, \, \gamma \, M_n^2 / \{1 + (\gamma-1) M_n^2\}]_{i-\frac{1}{2}} \tag{7}$$

where M_n is the normal Mach number. The form of (7) is obtained from symbolic analysis [6] of the governing equations. Equations (6) and (7) are effective unless very strong gradients occur, e.g. shocks. In this case the Roe–type [7] limiter, ψ, is activated.

Each governing equation in (5) is sequentially relaxed to modify a particular dependent variable. A velocity potential correction, ϕ, is introduced via a modified SIMPLEC [8] algorithm to satisfy the continuity equation. It is assumed that ϕ generates corrections to the velocity, pressure and density solutions as follows,

$$u'_i = \alpha \partial \phi / \partial x_i, \quad p' = -\beta \phi, \quad \rho' = \overset{*}{\delta} p'/(a*)^2 = -\delta \phi/(a*)^2 \tag{8}$$

Substitution into the continuity equation generates the following transport equation for ϕ,

$$\delta \frac{\partial}{\partial x_i}\left[\frac{u_i^* \phi}{(a^*)^2}\right] - \alpha \frac{\partial}{\partial x_i}\left[\rho^* \frac{\partial \phi}{\partial x_i}\right] = R_c^*, \tag{9}$$

where R_c^* is the continuity equation residual based on the current solution (*). Equation (9) is solved with $\phi = 0$ on all boundaries and in discrete form can be written as a component of eq. (5). The general transmission of information is in the main flow direction with the exception of eq. (9) which has a strong elliptic character if the flow is incompressible ($a^* \to \infty$). Reflecting this a strongly implicit procedure [1] (SIP) is solved in transverse planes of successive downstream stations to upgrade the individual dependent variable solutions.

4. RESULTS AND DISCUSSION

The present formulation has been used to compute the flow about wing body junctions (Fig. 1). Experimental data is available for a semi-infinite elliptic-nosed wing at zero incidence [9] and for a NACA-0012 airfoil at $\alpha = 10^0$ [10]. A typical convergence history on a 46×26×26 grid for the Shabaka case [9] is shown in Figure 2. The algorithm requires about 1.7×10^{-3} secs per iteration per grid point on a SUN Sparc Station 1 (1.4 Mflops).

For a downstream station the transverse velocity field is indicated in Fig. 3. The core location of the primary vortex agrees well with experimental data [9]. In addition the present computation indicates the presence of a secondary vortex at about $y^+ = 50$ which could not be resolved experimentally. The transverse distribution of the normal Reynolds stress, $\overline{\rho u'u'}$, is compared with the measurements of Shabaka in Fig. 4. The general level and fall off in moving from the wing and body surfaces to the freestream are well predicted. Transverse velocity components for the Wood and Westphal [10] test case (Fig. 1), obtained on a 101×31×21 grid, are shown in Fig. 5. Due to the 10^0 airfoil inclination the wake is characterised by a single trailing vortex whose strength and position is predicted accurately. Figure 6 shows the corresponding distribution of k (obtained from a k-ϵ turbulence model). The overall levels and the local peak aligned with the axis of the trailing vortex correspond closely to the experimental data.

In the region remote from the body the flow field is very similar to that for a two-dimensional inclined airfoil. Consequently the present algorithm (with the k-ϵ turbulence model) has been used to compare solutions with the two-dimensional experimental results of Harris [11]. The surface pressure distribution about a NACA-0012 airfoil (Fig. 7) was obtained on a 135×35 C-mesh and demonstrates the presence of an internal shock adjacent to the upper surface which is captured accurately by the present method. The overall pressure distribution is seen to be in good agreement with the experimental data [11]. The corresponding Mach number contours are given in Fig. 8.

In conclusion the present efficient sequentially–marched control volume formulation is able to provide mean flow and turbulent Reynolds stress and heat flux solutions of the steady compressible Reynolds–averaged Navier Stokes equations. Good agreement is obtained with the wing body and isolated airfoil experimental data.

References
1) C.A.J. Fletcher, *Computational Techniques for Fluid Dynamics*, Springer, Heidelberg, 1988, Vol. 1, p. 296.
2) S.W. Armfield, N.–H. Cho, and C.A.J. Fletcher, *A.I.A.A. J.*, 28, (1990) 453–460.
3) N.–H. Cho and C.A.J. Fletcher, "Computation of turbulent conical diffuser flows using a non–orthogonal grid system, *Computers and Fluids* (1990, to appear).
4) A.D. Burns and N.S. Wilkes, AERE Report R 12342, Harwell, U.K., July 1987.
5) C.A.J. Fletcher, *Comp. Maths. Applic.*, 16 (1988), 31–39.
6) C.A.J. Fletcher, *Advances in Fluid Dynamics* (eds. W.F. Ballhaus and M.Y. Hussaini), Springer, Heidelberg, 1989, 57–68.
7) P.L. Roe, *Ann. Rev. Fluid Mechanics*, 18 (1986), 337–365.
8) J.R. van Doormal and G.D. Raithby, *Num. Heat Transfer*, 7 (1984), 147–163.
9) I.M.M.A. Shabaka, Turbulent Flow in an idealised Wing–Body Junction, Ph.D. Thesis, Imperial College, U.K., 1979.
10) D.H. Wood and R.V. Westphal, "Measurements of the Flow around Lifting–Wing/Body Junction", *A.I.A.A. J.* (1990, to appear).
11) C. Harris, NASA TM 81927, April 1981.

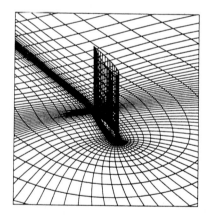

Fig. 1 Grid for inclined airfoil on plate

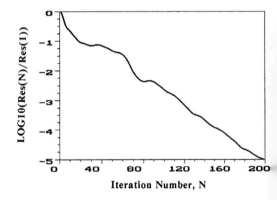

Fig. 2 Axial momentum convergence history on a 46×26×26 grid [9]

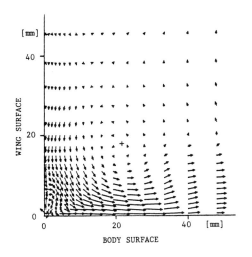

Fig. 3 Transverse velocity components, Station 5 [9]

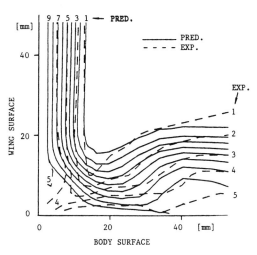

Fig. 4 Transverse distribution of $\overline{\rho u' u'}$, Station 5 [9]

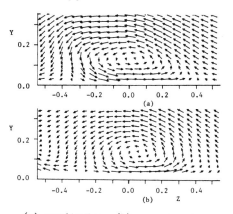

(a) prediction (b) experiment
Fig. 5 Transverse velocity components in wake ($\Delta x_{te} = 2.5$ C), [10]

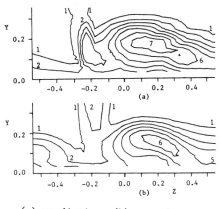

(a) prediction (b) experiment
Fig. 6 Transverse distribution of k ($\Delta x_{te} = 2.5$ C), [10]

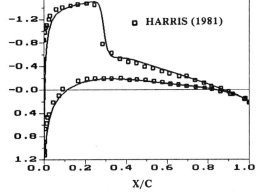

Fig. 7 Pressure coefficient distribution, $\alpha = 3.86$, $M_\infty = 0.70$ [11]

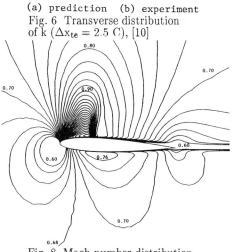

Fig. 8 Mach number distribution, $\alpha = 3.86$, $M_\infty = 0.70$ [11]

STAGGERED AND NON-STAGGERED FINITE-VOLUME METHODS FOR NONSTEADY VISCOUS FLOWS: A COMPARATIVE STUDY

H.I. ANDERSSON
Division of Applied Mechanics, The Norwegian Institute of Technology,
N-7034 Trondheim, Norway

J.T. BILLDAL
Division of Fluid Dynamics, SINTEF, N-7034 Trondheim, Norway

P. ELIASSON and A. RIZZI
FFA, the Aeronautical Research Institute of Sweden,
P.O. Box 11021, S-161 11 Bromma, Sweden

Background
A number of different methods exist today for the numerical solution of the incompressible and steady Navier-Stokes equations. One of the most successful numerical methods in a generalized coordinate system with the primitive variable formulation has been based on Chorin's artificial compressibility technique [1], in which an extra time-dependent pressure term is added to the continuity equation. The discretisation in space has often been done with a finite volume scheme [2]. Since the finite-volume method can be applied to complex three-dimensional geometries, a non-staggered grid with the dependent variables defined in the center of the cells has usually been used.

There are only a limited number of time-dependent incompressible Navier-Stokes solvers today for arbitrary geometries in two and three dimensions. It would be nice to retain the advantages with the finite volume methods also for unsteady flows, but the artificial compressibility technique can not be used however. One might of course consider a subiteration technique, where several intermediate iterations are taken for the pressure at each time step [3], but it seems unnecessary expensive. Instead, we have considered an explicit fractional step approach with finite-volume discretisation, which decouples the solution of the time-dependent momentum equations from the solution of the mass continuity equation. This leads to a Poisson equation for the pressure to be solved at each time step.

In this paper we consider two different finite-volume methods for the fractional step approach. The two methods use staggered and non-staggered grids which is the major difference between the methods. Other important differences are the time discretisation and the solution method of the Poisson equation. The objective is to show the differences between the staggered and non-staggered methods and to compare the accuracy as well as efficiency for the two methods.

Outline of the finite-volume methods
The explicit fractional-step approach using the integral formulation of the governing equations for a control volume V with surface S, is written in semi-discretised form as:

$$\int_V (\vec{u}^* - \vec{u}^n)dV = \Delta t \oint_S ((1+\alpha)\mathbf{T}^n - \alpha \mathbf{T}^{n-1} - \beta p^n \mathbf{I}) \cdot \vec{dS} \qquad (1)$$

$$\oint_S \nabla P^{n+1} \cdot \vec{dS} = \frac{1}{\Delta t} \oint_S \vec{u}^* \cdot \vec{dS} \qquad (2)$$

$$\int_V \vec{u}^{n+1} dV = \int_V \vec{u}^* dV - \Delta t \oint_S P^{n+1} \vec{dS} \qquad (3)$$

where the tensor **T** and the potential function P are defined as

$$\mathbf{T} = -\vec{u}\vec{u} + \nu \nabla \vec{u} \quad , \quad P^{n+1} = p^{n+1} - \beta p^n \qquad (4)$$

respectively, **I** is the identity tensor, and α and β are dimensionless control parameters.

A tentative, generally not divergence free, velocity field is calculated explicitly from the equation (1) using velocities and pressure at time levels n and $n-1$. The solution of the integral form of the Poisson equation (2) for the potential function P at the new time level $n+1$ is then obtained by an efficient Poisson solver, and the new velocity field at time level $n+1$ is finally evaluated from (3).

Two different finite-volume methods are used to produce the results reported in the paper, NSE (Non-Staggered Euler) and SAB (Staggered Adams-Bashforth). NSE uses a non-staggered grid-system and a first-order accurate Euler scheme for the time discretisation (i.e. $\alpha = 0$). The potential function P in (4) is taken as the pressure change from time level n to $n+1$ (i.e. $\beta = 1$). SAB, on the other hand, uses a staggered grid-system for arbitrary geometries with the cell-face fluxes and mid-point pressure as dependent variables. The integrator is a second order Adams-Bashforth scheme (i.e. $\alpha = 1/2$), and the potential function P equals the pressure p (i.e. $\beta = 0$). Detailed descriptions of the methods can be found in refs. [4], [5], respectively.

It is of crucial importance to use an efficient Poisson solver for the Poisson equation (2) to reduce the computer time. The NSE-method [4] uses Gauss-Seidel iterations by line, and the resulting tridiagonal matrix is solved by the Thomas algorithm. The SAB algorithm [5] uses a multigrid algorithm to solve the Poisson equation effectively. Successive-line-over-relaxation (SLOR) is used as a smoother, the interpolation between the mesh levels is linear. The use of a staggered grid for the solution of the Poisson equation with multi-grid is important. In order to obtain a discretely divergence-free velocity field at each time-step, the use of a non-staggered grid leads to a coarser Laplace-stencil. This is the basic difference between the two methods. A discrete Fourier analysis has been applied to a one-dimensional test problem to show that the coarse Laplace stencil can not be solved effectively with multi-grid, it is also shown that multi-grid might even diverge for this stencil [5].

The same type of boundary conditions have been used for the two methods. On boundaries where the velocity is given, a homogeneous Neumann condition for the pressure is used. On an outflow boundary, on which the velocity should be extrapolated from the interior, a constant normal pressure gradient is used. This must be done to conserve mass and to ensure the existence of a solution to the Poisson equation with Neumann boundary conditions [5].

Validation

In order to validate the different methods, we have considered two different two-dimensional flow cases. The first case is a test case where the analytical solution is known, the second case is flow past a circular cylinder.

The following two-dimensional flow is a solution to the Navier-Stokes equations:

$$u(x,y,t) = -cos(x)sin(y)e^{-2\nu t}$$
$$v(x,y,t) = sin(x)cos(y)e^{-2\nu t} \qquad (5)$$
$$p(x,y,t) = -0.25(cos(2x) + cos(2y))e^{-4\nu t}$$

Computations have been carried out on an equidistant, Cartesian mesh $[0 \leq x, y \leq \pi]$. The time-dependent velocity was prescribed on the boundaries according to (5), homogeneous Neumann boundary conditions for the pressure were used. In fig. 1 the difference between the numerical and the analytical velocity is plotted versus grid size after 100 time steps with $\nu = 1$, $t = 5 \cdot 10^{-3}$ and $\Delta t = 5 \cdot 10^{-5}$. Exactly second order accuracy is obtained for the SAB algorithm which could be expected. The accuracy is even higher for the (NSE) algorithm even though second order accuracy is expected.

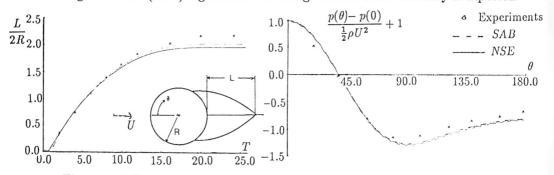

Figure 1. Difference between numerical and analytical velocity versus cell length.

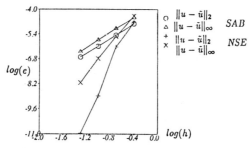

Figure 2. a) Time development of the wake length behind a cylinder. b) Pressure distribution along the cylinder. $Re = 40$.

The flow past a circular cylinder has been computed for two different Reynolds numbers, $Re = 40$ and $Re = 100$. The cylinder, with radius R, is impulsively set in motion with constant speed U at the dimensionless time $T \equiv tU/R = 0$. The resulting time-dependent flow pattern strongly depends on the Reynolds number $Re = 2RU/\nu$. For $Re = 40$, the flow is transient and a closed attached wake which grows with time develops behind the cylinder. The predicted time evolution of the wake length L and the resulting static pressure distribution along the cylinder are shown in fig. 2 for a 129×65 O-mesh with the outer boundary 10 cylinder diameters from the cylinder. The numerical results of the two different codes are compared with experimental results [6][7], and it is seen that the numerical predictions compare favourably with the experimental findings.

The slightly underpredicted wake length L for $T > 10$ is most likely an effect of the numerical boundary conditions on the outer boundary of the computational domain.

Some other characteristic data of the ultimate steady-state solution are given in table 1 together with corresponding numerical data of two other papers [8][9]. Here, p_∞ denotes the constant pressure of the incoming stream, ρ the constant density, $C_{D,f}$ and $C_{D,p}$ the friction drag and pressure drag, respectively.

Method	$180 - \theta$	L/R	$p(0) - p_\infty / .5\rho U^2$	$p_\infty - p(180)/ .5\rho U^2$	$C_{D,f}$	$C_{D,p}$
SAB	52.9	4.05	1.14	0.66	.28	.55
NSE	53.4	3.96	1.15	0.63	.28	.55
Dennis & Chang	53.8	4.69	1.14	0.51	.26	.50
Collins & Dennis	53.6	4.30	1.16	0.53	.27	.51

Table 1. Characteristics of the steady solution for $Re = 40$.

The efficiency of the two codes was compared for this flow case. The computational data can be seen in table 2. The two codes have used a CRAY-XMP and the computational data are shown at the dimensionless time $T = 20$ for different tolerances of the divergence of the velocity $\|\nabla \cdot \vec{u}\|_2$. It can be noted that the SAB algorithm can take a larger time-step due to a larger area of stability in the complex plane which results in shorter CPU-time. It is also interesting to see that even though the SAB algorithm uses multi-grid for the solution of the pressure equation, SAB spends more time solving the pressure equation than the NSE algorithm compared to the total CPU-time. One explanation is that it is not possible to vectorise the multi-grid cycle.

$\|\nabla \cdot \vec{u}\|_2$	Time step	No. of time steps	No. of points	Tot. CPU time (s)	CPU Press./ CPU Tot.	Method
10^{-3}	$3.612\ 10^{-3}$	5536	8385	1162.5	81 %	SAB
10^{-4}	$3.612\ 10^{-3}$	5536	8385	1328.6	83 %	SAB
10^{-5}	$3.612\ 10^{-3}$	5536	8385	1599.9	86 %	SAB
10^{-3}	$2.441\ 10^{-3}$	8192	8385	1532.8	64 %	NSE
10^{-4}	$2.441\ 10^{-3}$	8192	8385	1648.4	66 %	NSE
10^{-5}	$2.441\ 10^{-3}$	8192	8385	1792.3	68 %	NSE

Table 2. Computational data for the two codes, flow over a cylinder, $Re = 40$, $T = 20$.

As the Reynolds number becomes greater than 40, some of the perturbations that are always present can not be damped out leading to the well known Kármán vortex streets. To simulate such a flow, a perturbation is introduced to the flow field by rotating the cylinder just after the start up of the cylinder.

Only the SAB code was used to simulate the flow past an impulsively started cylinder at $Re = 100$. The NSE algorithm turned out to have numerical instabilities on the outer boundary, which was not discovered for the transient case at $Re = 40$. This outflow boundary condition is under investigation but unfortunately no numerical results are included herein for the NSE algorithm at $Re = 100$.

The periodic properties of the flow can clearly be seen in fig. 3 where the velocity vectors are shown for a specific time. Note the alternating vortices from the lower and upper side of the cylinder. The Strouhal number compares well with other numerical data and experiments [5].

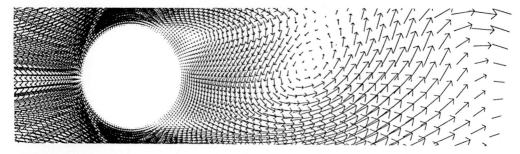

Figure 3. a) Instantaneous velocity vectors behind a cylinder, $Re = 100$.

Conclusions

Two different finite-volume methods for the solution of the unsteady, incompressible Navier-Stokes equations have been compared. The two methods are spatially discretised using staggered (SAB) and non-staggered (NSE) grids and they use different time integration. The basic differences between the methods are the solution methods of the Poisson equation for the pressure and its discretisation. SAB solves the Poisson equation with multi-grid, NSE uses Gauss-Seidel iteration.

Both methods predict the transient flow over a cylinder very well. They are both second order accurate and their efficiency is approximately the same as far as CPU-time is concerned. The investigation of the boundary conditions for the NSE algorithm needs to be carried out further to handle more complicated flow cases.

References

[1] CHORIN, A.J.: "A numerical method for solving incompressible viscous flow problems", J. Comp. Phys. 2 (1967), pp 12-26.

[2] ELIASSON, P., RIZZI, A. and ANDERSSON H.: "Time-marching method to solve steady incompressible Navier-Stokes equations for laminar and turbulent flow", Notes on Numerical Fluid Dynamics 24 (1989), pp 105-112.

[3] PEYRET, R. and TAYLOR, T.: "Computational methods for fluid flow", Springer (1983).

[4] ANDERSSON, H.I. and BILLDAL, J.T.: "A finite-volume method for unsteady viscous flows in complex geometries", in preparation.

[5] ELIASSON, P.: "A solution method for the time-dependent Navier-Stokes equations for laminar incompressible flow", FFA Report 146 (1990).

[6] COUTANCEAU, M. and BOUARD, R.: "Experimental Determination of the Main Features of the Viscous Flow in the Wake of a Circular Cylinder", J. Fluid Mech. 79 (1977), pp. 231-257.

[7] GROVE, A.S., SHAIR, F.H., PETERSEN, E.E. and ACRIVOS, A.:"An Investigation of the Steady Separated Flow past a Circular Cylinder", J. Fluid Mech. 19 (1964), pp. 60-80.

[8] DENNIS, S.C.R. and CHANG, G.Z.:"Numerical Solutions for Steady Flow past a Circular Cylinder at Reynolds Number up to 100", J. Fluid Mech. 42 (1970), pp. 471-489.

[9] COLLINS, W.M. and DENNIS, S.C.R.:"Flow past an Impulsively Started Circular Cylinder", J. Fluid Mech. 60 (1973), pp. 105-127.

DIRECT NUMERICAL MODELLING OF TURBULENCE: COHERENT STRUCTURES AND TRANSITION TO CHAOS

BELOTSERKOVSKII O.M.

INSTITUTE OF COMPUTER AIDED DESIGN, ACADEMY OF SCIENCES, MOSCOW, U.S.S.R.

The direct numerical modelling (without using the semiempirical theories) of the complete unsteady dynamical equations is realized by means of the method of splitting over physical processes, within the frame of a computational experiment. The problem under investigation is that of the evolution of coherent structures for the limiting regime (Reynolds number $Re \to \infty$) of the free shear turbulence in the near wake behind the body moving in the fluid, as well as the process of transition of a laminar flow into irregular state ("chaos") by the study of oceanic vortical structures, emerging under the influence of the wind-induced stresses.

A discrete dissipative model, obtained from the non-stationary Euler equations by averaging over an elementary cell and a time interval, is proposed for the modelling of ordered motions and large scale vortices on the basis of ideas concerning structural turbulence [1]. By means of appropriate representations the dissipative mechanism provides a stability of solutions "as a whole" and reflects the contribution of small scale ("subgrid") pulsations for different scales of resolution. The stochastic component of turbulence is modelled at the kinetic level by a statistical Monte-Carlo method; such an approach sharply reduces the level of demand on computer resources. The efficiency of the algorithm proposed is demonstrated by solution of the problems of the sparated flows about bodies and of the evolution of a turbulent wake for two- and threedimensional flows in supercritical and transonic regimes of the motion [2,3].

The investigation of the vortices production in oceanic flow is conducted on the basis of nonlinear equations taking into account a turbulent dissipation and an Earth's rotation (β-effect, connected with a latitudinal alteration of Coriolis parameter), by the presence of wind-induced stresses. Here at the highest derivatives appear two small parameters, which characterize the effects of Earth's rotation ($\varepsilon = \sqrt{(\tau/\beta)}/L$) and of viscosity ($\gamma = \varepsilon/Re^{1/3}$); this leads to the formation at the western boundary of the perturbed area of an inertial-viscous boundary layer.

By the growth of Reynolds number, at the small values of $\varepsilon \sim 0.02$, one obtains from the calculations the following succession of the flow regimes: (1) stationary (Re ~ 0.25), (2) periodical (Re $\sim 1 \div 2$), (3) stable irregular ($\varepsilon \sim 0.0126$, Re ~ 1), and (4) unstable irregular (Re ~ 3). The last of these regimes is characterized by a sharp growth of the kinetic energy of the whole system [4] (it is worth to note that by a numerical study of "Kolmogoroff flow" [5] one failed to obtain the transition to chaos).

When using the medium-power computers, the models listed above permit to obtain the results, well agreeing with experimental data.

References

1. Cantwell B.J. Organized Motion in Turbulent Flows (Russian translation), Mir, Moscow, 1984.
2. Belotserkovskii O.M. Direct Numerical Modelling of Free Induced Turbulence. Zh. Vychisl. Mat. & Mat. Fiz., 25, 12, 1856-1882, 1985
3. Babakov A.V. Numerical Simulation of Unsteady Vortex Structures in Near Wake. Zh. Vychisl. Mat. & Mat. Fiz., 28, 2 267-272, 1988.
4. Belotserkovskii S.O., Pastushkov A.R. Numerical Study of the Oceanic Vortical Structures Originated by the Wind-Induced Stresses (in publication).
5. Belotserkovskii S.O., Gushchin V.A. New Numerical Models in Continuum Mechanics. Uspekhi Mekhaniki, 8, 1, 97-150, 1985.

CONFORMING CHEBYSHEV SPECTRAL COLLOCATION METHODS FOR THE SOLUTION OF THE INCOMPRESSIBLE NAVIER-STOKES EQUATIONS IN COMPLEX GEOMETRIES

Andreas Karageorghis, Mathematics Department,
Southern Methodist University, Dallas, Texas 75275-0156, USA

Timothy N Phillips, Department of Mathematics,
University College of Wales, Aberystwyth SY23 3BX, United Kingdom

ABSTRACT

This paper examines the numerical simulation of steady planar two-dimensional, laminar flow of an incompressible fluid through an abruptly contracting channel using spectral domain decomposition methods. The flow domain is divided into a number of conforming rectangular subregions. Within each of the subregions the solution is approximated by a truncated expansion of Chebyshev polynomials. With a judicious choice of collocation strategy we show that the resulting approximations are pointwise C^0 and C^1 continuous across the interfaces [1].

The governing equations are the incompressible Navier-Stokes equations, which in the stream function formulation reduces to:

$$\nabla^4 \psi - Re \left[\frac{\partial \psi}{\partial y} \frac{\partial}{\partial x} (\nabla^2 \psi) - \frac{\partial \psi}{\partial x} \frac{\partial}{\partial y} (\nabla^2 \psi) \right] = 0 \qquad \text{(A)}$$

where Re is the Reynolds number. The introduction of the stream function ensures that mass is conserved identically. Equation (A) is nonlinear in the stream function and is linearized by means of a Newton-type technique [2,3].

At each Newton step the collocation of the linearized equation in conjunction with the interface patching conditions and the imposition of the boundary conditions (in a collocation sense) produces a system of linear equations for the expansion coefficients. The linear systems resulting from spectral discretizations suffer from not being sparse thus requiring large amounts of storage and being costly to invert. To some extent this can be overcome if the block structure of the spectral domain decomposition matrix is exploited. We explore two techniques: a capacitance matrix technique [4,5] and an algorithm for almost block diagonal systems [6,7].

Numerical results are presented demonstrating the convergence of the numerical solution for different numbers of degrees of freedom and for different values of the Reynolds number. For values of $Re \leq 100$ the Newton process converges from a zero solution after six steps. For higher values of Re continuation in Re in increments of 50 is used. The use of conforming subregions enables solutions to be obtained for much larger values of Re than is possible for nonconforming subregions [2]. Quantitative agreement is reached with previous work [8] on the salient corner vortex. Further, our spectral discretizations are able to resolve the flow sufficiently enough to detect a vortex downstream of the contraction which appears at a value of Re around 200 and then continues to grow as the Reynolds number is increased further. A full discussion of these results is given. Contours of the stream function are presented in Figs 1 and 2 for $Re = 100$ and 500, respectively.

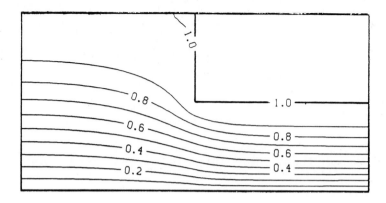

Fig 1. Contours of the stream function for $Re = 100$.

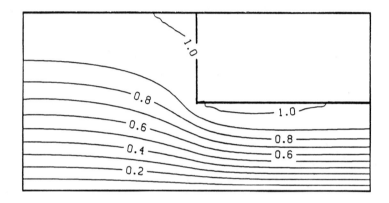

Fig 2. Contours of the stream function for $Re = 500$.

REFERENCES

[1]   T N Phillips and A Karageorghis, ICASE Report 89-60, 1989.
[2]   A Karageorghis and T N Phillips, J Comput Phys 84, 114-133 (1989).
[3]   T N Phillips, J Comput Phys 54, 365-381 (1984).
[4]   B L Buzbee, F W Dorr, J A George and G H Golub, SIAM J Numer Anal 8, 722-736.
[5]   T N Phillips and A Karageorghis, SIAM J Sci Statist Comput 10, 89-103 (1989).
[6]   J C Diaz, G Fairweather and P Keast, ACM Trans Math Software 9, 358-375 (1983).
[7]   R W Brankin and I Gladwell, Comp Math Applics, to appear.
[8]   S C R Dennis and F T Smith, Proc Roy Soc Lond A 372, 393-414 (1980).

DIRECT SIMULATION OF TRANSITION AND TURBULENT BOUNDARY LAYER

I.N.Simakin, S.E.Grubin
Central Aero-Hydrodynamic Institute
TsAGI, Zhukovsky-3, 140160, USSR

The incompressible viscous fluid flows in the boundary layer on a flat plate with a zero external pressure gradient are investigated. The evolution of a finite amplitude disturbances is simulated by numerical integration of the three-dimensional time-dependent Navier-Stokes equations. The laminar flow is assumed to be local-parallel and the disturbances to be periodic in streamwise (x) and spanwise (y) directions. A spectral method with Fourier expansions is used for approximating the Navier-Stokes equations in x,y-directions, and a collocation method with Laguerre polynomials expansions in z-direction. The time integration is perfomed using the Crank-Nicolson type scheme of the second order of accuracy. The "neutrality" property of nonlinear terms is valid for the problem in question, i.e. the time-dependent equation of the total energy of disturbances does not contain nonlinear terms of the Navier-Stokes equations. A discrete analog of this property in the method constructed is valid.

Fig.1 gives plots showing time variations of local friction coefficient c'_f, normalized to c'_{fo} for laminar flow, and fluctuation energy E' calculated at a Reynolds number of $R = (R_x)^{1/2} = 580$, periodicity intervals of $X = 2\pi/\alpha_0$ ($\alpha_0 = 0.179$), $Y = 2X$. Initial energy $E(t=0) = E_0 = 0.7 \times 10^{-3}$ (case 1) and $E_0 = 1.5 \times 10^{-2}$ (case 2). A comparatively small number of basic functions was used in the representation of the approximate solution, i.e. 21x21 Fourier-modes in x,y-directions, and 33 polynomials in z-direction, the total number of grid points being $S \simeq 3.4 \times 10^4$. At the initial stage there is a weak nonlinear interaction, as well as a slow growth of the friction coefficient and fluctuation energy. This regime is absent in case 2. When the disturbance energy attains its critical value of $E \sim 10^{-2}$, there occurs a transition to a developed turbulent flow accompanied by a drastic increase in friction and fluctuation energy, particularly in small-scale disturbances. The boundary layer thickness rises approximately threefold. It was found that mean characteristics of turbulent flow(velocity profile, Reynolds stress etc.) are not depend on the initial amplitude of disturbances and are in satisfactory agreement with experiments. The result was established proves to some degree the possibility of the boundary layer transition simulation in the framework of local-parallel model. It should be noted that if a moderate number of basic functions is used in the solution representation, it is possible to obtain the transition from the laminar flow to the fully developed turbulent one using only the algorithm for which the analog of the "neutrality" property of nonlinear terms is valid. Fig.2 shows the velocity profile and the Reynolds stress, averaged in time and variables x,y, for a turbulent flow, as well as the calculated results with a very great number of basic functions $\sim 10^6$ [1] and the experimental data [2,3]. The results show satisfactory agreement. A similar agreement takes place also for the rms pulsation velocity.

Thus, it is shown that within the framework of the model considered for the Navier-Stokes equations it is possible to simulate a transition from a laminar to a developed turbulent flow in the boundary layer at comparatively small number of basic functions in the solution representation.

References
1. Spalart P.R. J.Fluid Mech.,1988,v.187,p.61-98.
2. Erm L.P.,Smits A.J.,Joubert P.N. Proc. 5th Symp.Turbul.Shear Flows. 1985, Ithaca, N.Y.
3. Klebanoff P.S. NACA TN-3178, 1954.

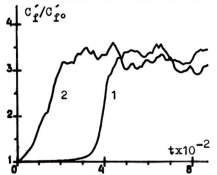
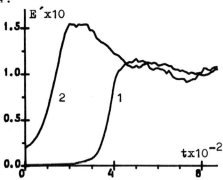

Fig.1.

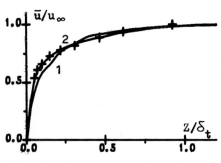
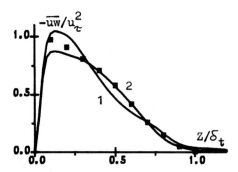

Fig.2. 1 - our results for R_θ = 690, 2 - Spalart[1] R_θ = 670, + - Erm et al.[2] R_θ = 617, ■ - Klebanoff[3] $R_\theta \approx$ 7500. δ_t - turbulent boundary layer thickness.

NUMERICAL SIMULATION OF TRANSITION TO TURBULENCE USING HIGHER ORDER METHOD OF LINES

Hiroshi Tokunaga, Kencihi Ichinose and Nobuyuki Satofuka

Department of Mechanical and System Engineering
Kyoto Institute of Technology, Matsugasaki, Sakyo-ku, Kyoto 606, Japan

1 Introduction

The pseudo-spectral method is accurate and efficient so that it is widely used. The transition to the turbulence and large eddy simualtions of turbulent flows are treated by the spectral method [1-2]. In claculating turbulent flows with finite difference method, however, there exist a number of difficulties although these methods are widely used in the computational fluid dynamics. The high accuracy and high efficiency have to be assured. The formualtion is also important in calculating three-dimensional incompressible flows. As is well known, it is difficult to satisfy the solenoidal condition for the velocity field in the primitive variable formulation. From these points of views we propose a new computational method, for calculating turbulent shear flows, which is composed of the higher order method of lines combined with the vorticity-vector potential formulation. Although the solenoidal condition is satisfied automataically, the Poisson equations have to be solved for three components of the vector potential. The multi-grid mehtod is therefore used in order to accelerate the convergence of the Poisson eqiuations. The computations are carried out with the super computer Fujitsu FACOM VP-400E at the Data Processing Center of Kyoto University.

2 Governing Equations and Computational Methods

In two-dimensional flows the vorticity and the stream function formulation is often used. The vector potential ψ, is however, introduced on three-dimensional flows from $\mathbf{u} = rot\psi$ where $\mathbf{u}$ denotes the velocity field. The Navier-Stokes equations are then transformed into the vorticity transport equations for the vorticity $\zeta = rot\mathbf{u}$ and the Poissson equations are defined for the vector potential ψ. In the present formulation the equation of continuity is automatically satisfied and the present method is expected to be an efficient computational method for solving incompressible flows.

$$\frac{\partial \zeta_x}{\partial t} = -\nabla \bullet (\mathbf{u}\zeta_x) + (\zeta \bullet \nabla)u + \frac{1}{Re}\nabla^2 \zeta_x, \tag{1}$$

$$\frac{\partial \zeta_y}{\partial t} = -\nabla \bullet (\mathbf{u}\zeta_y) + (\zeta \bullet \nabla)v + \frac{1}{Re}\nabla^2 \zeta_y, \tag{2}$$

$$\frac{\partial \zeta_z}{\partial t} = -\nabla \bullet (\mathbf{u}\zeta_z) + (\zeta \bullet \nabla)w + \frac{1}{Re}\nabla^2 \zeta_z, \tag{3}$$

$$\nabla^2 \psi = -\zeta. \tag{4}$$

The method of lines is adopted in the present investiagtion. In this method spatial discretizations and the time integration are treated separately. For spatial discretizations the eighth order modified differential quadrature (MDQ) method is used. Partial derivatives of ζ for the x-direction, for instance, are approximated in terms of vorticity values $\zeta_{ijk} = \zeta(x_i, y_j, z_k)$ at the grid point (x_i, y_j, z_k), according to the following expressions:

$$\frac{\partial \zeta}{\partial x}\bigg|_{ijk} = \sum_{l=-4}^{4} a_{il}\zeta_{i+l,j,k}, \tag{5}$$

$$\frac{\partial^2 \zeta}{\partial x^2}\bigg|_{ijk} = \sum_{l=-4}^{4} b_{il}\zeta_{i+l,j,k}, \quad b_{il} = \sum_{l=-4}^{4} a_{im}a_{ml}, \tag{6}$$

where a_{il} are numerical coefficients specified adequately [3]. The same discretizations are yielded for derivatives in the $y-$ and $z-$direction. After spatial discretizations, the partial differential equations (PDEs) are reduced to a set of ordinary differentail equations (ODEs) for the vorticity values at all inner grid points.

$$\frac{d\vec{\zeta}}{dt} = \vec{F}(\vec{\zeta}), \tag{7}$$

$$\vec{\zeta} = (\zeta_{x,2,2,2}, \zeta_{x,3,2,2}, ..., \zeta_{z,imax-1,jmax-1,kmax-1})^T. \tag{8}$$

In expressions (8) $imax, jmax$ and $kmax$ represent the number of grid points in the $x-, y-$ and $z-$direction, respectively. The fourth step Runge-Kutta-Gill Method is used for the time integration of ODEs (7) with (8). The Poisson equations are discretized by the MDQ method as (4) and the higher order multi-grid method is used for the relaxation.

3 Computational Results

The present computational method has been already validated for the steady and unsteady flow [4]. In the present paper, therefore, the numerical simulation of the subcritical transition to the turbulence in a plane channel is treated with the eighth accuracy. The computational domain is shown in Fig. 1. The initial condition is composed of the Benney-Lin type flow [3]. The computaion is carried out at $Re = 1500$ and in the number of grid points $65 \times 65 \times 129$. The Figure 2 shows the spanwise vorticity contour on the plane $z = 0$ at $t = 21$. The horseshoe vortices are elongated, bent extremely and touch another newly generated vortices. This is the perspective view of the transient turbulent shear flows. In Fig. 3 we show constant enstropy surfaces with four levels. The structure of the horseshoe vortex is clearly visible. The three-dimensional vortices are extremely bent in both the spanwise direction and the normal direction to the wall. In order to inspect feasibility of the direct numerical simulation (DNS) of the turbulent shear flow, the energy spectrum on the $x - z$ plane are shown in Fig. 4. At time $t = 18.75$, the energy spectrum does not attain the maximum wave number region. At $t = 21$, howvere, the energy spectrum covers all wave number regions, so that DNS is not feasible after this time.

4 Conclusions

The higher order method of lines approach proposed here has several important advantages to calculate shear flow turbulence directly.
(1) The present method adopts the vector potential method and satisfies the divergent-free condition for the velocity field automatically so that the spurious error is excluded.
(2) The eighth order accyrate method is used, and therefore numerical dissipations is eliminated completely.
(3) The transition process of the shear flow turbulence is captured till the time 21 in the $65 \times 65 \times 129$ grid.
(4) The computer graphics displays show that the transient turbulent shear flow is composed of extremely elongated and rolled up horseshoe vorticies.

References

1. T.A.Zang, S.A.Krist, G.Erlebacher and M.Y.Hussaini, AIAA Paper **87-1204**(1987).
2. P.Moin and J.Kim, J. Fluid Mech. **118**, 341(1982).
3. H.Tokunaga, N.Satofuka and H.Miyagawa, Lecture Notes in Physics **264**, 617(1986).
4. H.Tokunaga, K.Ichinose and N.Satofuka, Presented to *Int. Symp. Comp. Fluid Dynam.-Nagoya*, (1989).

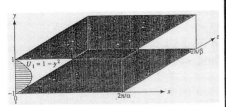

Fig.1 Computational domain.

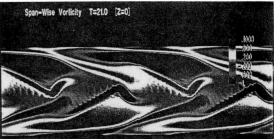

Fig.2 Spanwise vorticity contour at $t=21$.

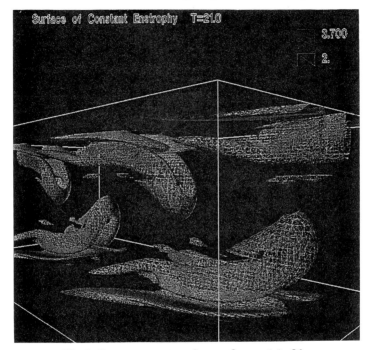

Fig.3 Constant enstrophy surfaces at $t=21$.

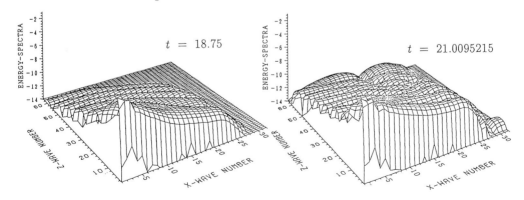

Fig.4 Energy spectra on the $x-z$ plane at $t=18.75$ and 21.

A New Fully-coupled Method for the Solution of Navier-Stokes Equations

M. Ferry & J. Piquet
CFD Group ; ENSM, URA 1217 CNRS

1. Depending on the way the solenoidality constraint is satisfied, two approaches can be used for the solution of the steady incompressible Navier-Stokes equations : (i) the segregated approach in which the continuity and the momentum equations are solved sequentially ; this approach leads the most often to a so-called pressure correction method. (ii) the fully-coupled approach where the pressure-velocity coupling is retained implicitly, then the need of a pressure correction equation is removed : velocity and pressure are written in a block form and are simultaneously updated in a linear sense, iterations being performed to solve for the non linearity. In contrast to [1][2] where the continuity is retained in its primitive form, the present work rather follows [3] in the use of an approximate pressure equation but it departs from [3] in its two most significant aspects.

2. First, following e.g.[4][5][6], colocated grids are used where all (node-centered) variables share the same grid location. Another significant aspect of the present work lies in the solution retained for the so-called closure problem (fig.1a). In order to integrate the continuity and momentum equations over *the control volume* around C, the fluxes through the surfaces like Ae and An have to be expressed with respect to the values of the unknowns at points C, E, N, NE. Such points are the vertices of the so-called *closure volume* V_A. Each closure volume, like V_A, is built with four triangles T_S, T_N, T_E, T_W (fig.1b) whose associated quantities may be computed with four parallel processors.

3. The difference between the technique used here and other works is best explained on the following onedimensional example (fig.2). Given the unknowns at points W, C, E, the closure problem consists in expressing fluxes like u_e under the form : $u_e = f(U_C, U_E) + d_e (P_E - P_C)$. On staggered grids, the closure problem is reduced to a substitution, into the continuity equation, of u_e resulting from the discrete form of the momentum equation between C and E. On colocated grids, the most common practice follows [4] where d_e and $f(U_W, U_C, U_E, U_{EE})$ result from a weighted mean of the discrete form of the momentum equation written around C and E. In [5][6], either shape functions satisfying locally the momentum equations are used [5], or the convective form of the momentum equations is discretized at the face center [6]. Both approaches end up with a result equivalent to a substitution of the momentum equations into the continuity equation. The present work departs from [3][4][5][6] in that the introduced shape function satisfies the continuity equation *as well as* the momentum equation. On each triangle of V_A, each velocity component is defined by a shape function : $\Phi = \lambda_1 + \lambda_2 \exp(\text{Re } V \, l) + \lambda_3 \, n - S_\Phi \, l / V$. where l and n are the axis defined along and normally to the local advection velocity V, while $\lambda_1, \lambda_2, \lambda_3$ involve the values of Φ at points A, C, E. The pressure is assumed to behave linearly over each triangle. Integrating the continuity and the momentum equations over the closure volume allows one to eliminate the unknowns at point A. Consequently, the fluxes, like u_e, can be analytically found along each face Ae and An inside V_A, and expressed only with respect to the vertice unknowns of V_A (fig.1b). This computation is performed for every closure volume and one is left with a nine point stencil, each node of which involves the three simultaneous U, V, P. The corresponding (block-3x3) nanodiagonal system is solved with a linear multigrid method using a block-3x3 line Jacobi method as the (highly vectorizable) relaxation solver. Given the fine grid operator L_h, simple injection R and bilinear prolongation P are used to define the coarse grid operator $L_H = R L_h P$. The boundary condition for the pressure results from the continuity equation written on the half control volume (fig.1c).

4. Fig.3 demonstrates the efficiency of the multigrid method when applied to the square driven cavity problem (Re=100). The non linearity is updated on the finest grid any four cycles, the efficiency of the smoother is improved (especially for high Re) by underrelaxation (.9). The case Re = 10000 is presented in fig.5. The solution on a 257^2 grid is obtained without any particular difficulty, although the residuals cannot be driven to machine zero. The effect of the grid regularity is studied, at Re = 1000, on a strongly randomly disturbed grid (fig.6) and no accuracy is apparently lost, because of the quality of the closure. The present method appears also to have a good behavior when computing instationnary flow over a cicular cylinder Re=40 as shown in fig.4.

5. References.
[1] Vanka, S.P. ; J. Comp. Phys. 65, 138-158, 1986.
[2] Bruneau,C.H., Jouron,C. & Zhang,L.B.; ICNMFD 11, Lect. Note in Phys. 323, 172-177, Springer Verlag, 1988.
[3] Schneider, G.E. & Zedan, M. ; AIAA Paper 84-1743 (19th. Thermophysics Conf.)
[4] Rhie, C.M. & Chow, W.L. ; AIAA Journ. 21, 1525-1532, 1983.
[5] Prakash, C. ; Num. Heat Transfer 11, 401-406, 1987.
[6] Schneider, G.E. & Raw, M.J. ; Num. Heat Transfer 11, 363-400, 1987.
[7] Ghia, U. & Ghia, K.N. & Shin, C.T. ; J. Comp. Phys. 48, 387-411, 1982.

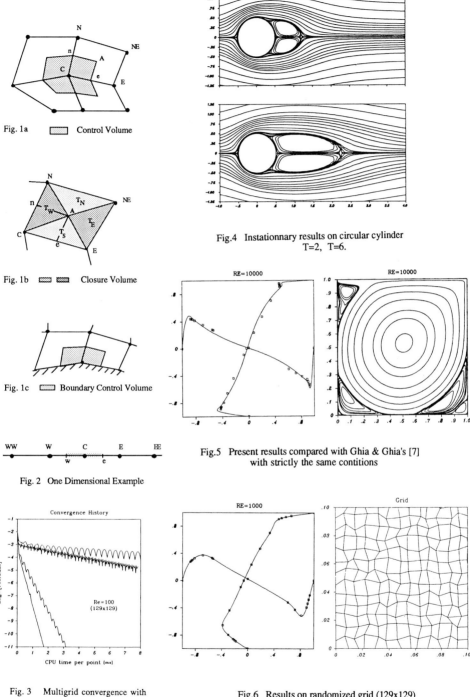

Fig. 1a Control Volume

Fig. 1b Closure Volume

Fig. 1c Boundary Control Volume

Fig. 2 One Dimensional Example

Fig. 3 Multigrid convergence with 1, 2, 3 and 4 grids.

Fig. 4 Instationnary results on circular cylinder T=2, T=6.

Fig. 5 Present results compared with Ghia & Ghia's [7] with strictly the same contitions

Fig. 6 Results on randomized grid (129x129) Compared with Ghia & Ghia's [7]

A MULTIGRID SCHEME WITH SEMICOARSENING FOR ACCURATE COMPUTATIONS OF VISCOUS FLOWS

R. Radespiel and N. Kroll
DLR, Institute for Design Aerodynamics
Am Flughafen, D-3300 Braunschweig

Multigrid schemes based on explicit Runge-Kutta time stepping and central differencing have been shown to yield good convergence rates for both inviscid and viscous flows. The principal reason for this is that the coefficients of the Runge-Kutta schemes can be chosen in a way that a good high-frequency damping behaviour is obtained, which is necessary to drive the multigrid process. For computations of viscous flows, however, one has to use meshes with high-aspect-ratio cells. There, the explicit time step is limited by the length of the short cell face. Thus, the explicit time step is much smaller than the characteristic time scale in the direction of the long side of the cell, which has an adverse effect on the damping characteristics of the scheme. Numerical experiments with a conventional multigrid scheme [1] have shown that the damping behaviour of the explicit time stepping scheme has to be improved by increasing the scaling factor of the fourth-difference dissipation terms. For the scaling in the i-coordinate direction

$$\bar{\lambda}_i = \lambda_i \left[1 + \left(\frac{\lambda_j}{\lambda_i}\right)^\omega\right]$$

is used where λ_i and λ_j are the spectral radii of the flux Jacobian in the coordinate directions i and j. Obviously, accuracy improves with decreasing ω whereas convergence of the conventional multigrid scheme slows down. To overcome this problem and to allow for the implementation of upwind schemes the conventional multigrid scheme is replaced by a multigrid scheme with semicoarsening, which is similar to Mulder's scheme [2]. As shown in figure 1 the finest mesh is successively coarsened in both coordinate directions. Details of the restriction and prolongation process are given in [3]. By using the semicoarsening strategy the aspect ratio of the cells is varied on the coarse meshes. Hence, for a fine mesh with high-ascpect-ratio cells, there are always coarse meshes with smaller aspect ratios and better damping characteristics. Numerical computations of 2-D high-Reynolds-numbers flows using multigrid with semicoarsening fulfill our expectations. The convergence rates given in figure 2 are almost insensitive to the exponent ω. For the computation of flows with strong shocks we have implemented the upwind TVD scheme [4] into the code [3]. Results for a hypersonic laminar high-Reynolds-number flow with strong viscous-inviscid interaction are shown in figures 3-4. Convergence is considerably accelerated using the multigrid scheme and computing times are reduced by factors around 7.

[1] Radespiel, R.; Rossow, C.-C.; Swanson, R.C.: An Efficient Cell-Vertex Multigrid Scheme for the Three-Dimensional Navier-Stokes Equations. AIAA Paper 89-1953.
[2] Mulder, W.A.: A New Multigrid Approach to Convection Problems. Journ. Comp. Phys. 83, 1989, pp. 303-323.
[3] Radespiel, R.; Kroll, N.: A Multigrid Scheme with Semicoarsening for Accurate Computations of Viscous Flows. DLR-IB 129-90/19, 1990.
[4] Yee, H.C.; Harten, A.: Implicit TVD Schemes for Hyperbolic Conservation Laws in Curvilinear Coordinates. AIAA Paper 85-1513.

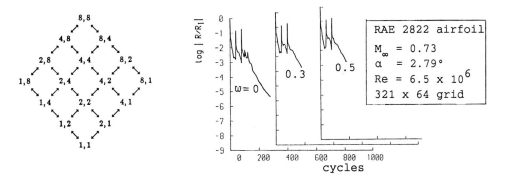

Fig. 1: Semicoarsening scheme for 8 x 8 grid

Fig. 2: Convergence history for central-difference scheme

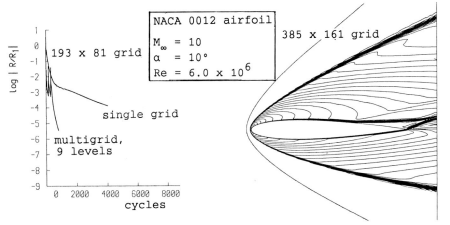

Fig. 3: Convergence history for upwind TVD scheme

Fig. 4: Mach contours

NUMERICAL SIMULATION OF SPATIALLY-EVOLVING INSTABILITY IN THREE-DIMENSIONAL PLANE CHANNEL FLOW

G. Danabasoglu[a], S. Biringen[b]
Department of Aerospace Engineering Sciences,
University of Colorado, Boulder, Colorado 80309.
and
C. L. Streett[c]
Theoretical Aerodynamics Branch,
NASA/Langley Research Center, Hampton, Virginia 23665

In this work, a three dimensional computational study concerning the spatial stability and transition in plane Poiseuille flow is presented. For this purpose, a Navier-Stokes solver which accurately resolves inflow/outflow boundary conditions in wall-bounded flows was developed. Experimentally, instabilities in wall-bounded flows are generated by a localized periodic disturbance that evolves in space (e. g. generated by a vibrating ribbon). For such flows, the computationally efficient "temporal" approach which follows the time evolution of a single wavelength of the disturbance becomes at best a good approximation to the laboratory flow. The difference between temporal and spatial evolution may become more prominent during the later stages of the transition process.

The accuracy of a Navier-Stokes solver for tracking the spatial evolution of a disturbance field depends on several factors. Among these are (a) phase-accurate discretization of the convective terms, (b) viscous resolution close to solid boundaries and (c) outflow boundary conditions that will remain non-reflective even in the presence of nonlinear wave interactions. The first requirement can be met by the use of spectral collocation methods or high (fourth) order finite differences, whereas the implementation of the Chebyshev spectral technique should suffice to meet the second condition. Finally, the outflow boundary conditions can be imposed by the buffer domain technique[1]. The attractive features of these boundary conditions are the absence of any restrictions on the wave number content of the outgoing disturbances and the elimination of any explicit definition for a convection velocity.

The present numerical scheme employs a time-splitting method[2] to integrate the full Navier-Stokes equations using spectral collocation/finite-difference discretization on a non-staggered mesh. The eigenvalue decomposition procedure is applied for the solution of the Poisson equations after the transformation of the homogeneous Neumann boundary conditions to nonhomogeneous Dirichlet conditions using the capacitance matrix technique. The buffer domain method is incorporated for the outflow boundary conditions. The input perturbation velocities are obtained by solving the Orr-Sommerfeld equation for the non-linear (spatial) eigenvalue problem employing the companion matrix method[3]. The above numerical scheme requires about 5.3 s. per time step on the CRAY 2 at NASA/Langley Research Center on a $41 \times 561 \times 17$ mesh.

We first present a test case obtained by the application of the above numerical scheme to the simulation of plane Poiseuille flow stability using only two-dimensional disturbances. The appended buffer domain was of equal length and had the same number of grid points as the physical domain. Computations were performed on a $41 \times 129 \times 17$ mesh in the physical domain for an aspect ratio of 8, $Re = 5000$ and $\omega_R = 0.33017$. Here, Re and ω_R are the Reynolds number and the real frequency, respectively. For these parameters, the Orr-Sommerfeld solution gives a streamwise wavenumber of $\alpha_{2d} = 1.1557 + i0.0106$. Furthermore, the perturbation amplitude was set to $A_{2d} = 0.0012$. In Fig. 1, the streamwise distribution of the u perturbation

[a] Graduate Student, [b] Associate Professor, [c] Senior Research Scientist.

velocity is compared with the linear solution at a vertical position close to the upper wall ($y = 0.891$). In this figure, no amplitude and phase errors (including at the inflow/outflow boundaries) are observed and the numerical solution is indistinguishable from the linear (exact) solution.

Next, we consider three-dimensional inflow disturbances with $Re = 5000$ and $\omega_R = 0.33698$ using the conditions of the experiments of Nishioka et al.[4] The spanwise wavenumbers were chosen as $\beta = \pm 1$. For these parameters, the Orr-Sommerfeld solution gives $\alpha_{2d} = 1.17249 + 0.0128073i$ and $\alpha_{3d} = 1.02570 + 0.0684559i$. For the first case, we set the amplitudes of the two- and three-dimensional disturbances to $A_{2d} = 0.02$ and $A_{3d} = 0.001$, respectively, and we increased the amplitudes to $A_{2d} = 0.03$ and $A_{3d} = 0.002$ for the second case. In these simulations, the length of the buffer domain was 40% of the physical domain and a $41 \times 401 \times 17$ mesh in the physical domain for an aspect ratio of 27.5 was used. In Fig. 2, the spanwise component of vorticity is plotted at a vertical location ($y = 0.809$) close to the critical layer for both cases. Figure 2a reveals the formation of lambda vorticies with a progressive increase in perturbation amplitudes along the streamwise direction. The sudden formation of a lambda vortex is observed in the high amplitude case (Fig. 2b).

References:
1. Streett, C. L. and Macaraeg, M. G., "Spectral Multi-Domain for Large-Scale Fluid Dynamics Simulations," *Appl. Numerical Math.* (in press).
2. Biringen, S. and Danabasoglu, G., "Oscillatory Flow with Heat Transfer in a Square Cavity," *Phys. Fluids A*, 1, p. 1796, 1989.
3. Danabasoglu, G. and Biringen, S., "A Chebyshev Matrix Method for Spatial Modes of the Orr-Sommerfeld Equation," *Int. J. Num. Meth. Fluids* (in press).
4. Nishioka, M., Asai, M. and Iida, S., "Wall Phenomena in the Final Stage of Transition to Turbulence," in *Transition and Turbulence* (ed. R. E. Meyer), p. 113, Academic Press (1981).

Acknowledgement: This work was in part supported by NASA/Langley Research Center, under grant NAG-1-798.

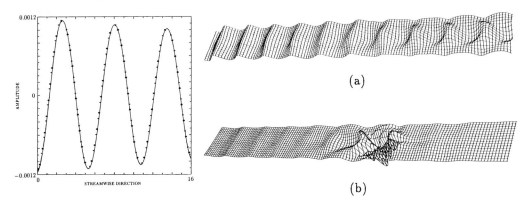

Figure 1. The streamwise perturbation velocity after 7 Tollmien-Schlichting (TS) periods ($y = 0.891$).

Figure 2. The spanwise perturbation vorticity surface plots on an xz plane a) low amplitude case after 11 TS periods and b) high amplitude case after 6.3 TS periods ($y = 0.809$).

SECONDARY INSTABILITY AND TRANSITION IN COMPRESSIBLE BOUNDARY LAYERS

D. Pruett*, G. Erlebacher**, and L. Ng***
NASA Langley Research Center
Hampton, Virginia, USA 23665

INTRODUCTION

It is well known that transition to turbulence in incompressible boundary layers often occurs via primary and secondary instability mechanisms. Little is known however, either experimentally or theoretically, regarding the paths to transition in compressible boundary layers. In this paper we combine direct numerical simulation and secondary instability theory to investigate and compare transition mechanisms in compressible flat plate and axisymmetric boundary layers, focusing on the influences of transverse curvature on transition along cylindrical bodies.

We find linear theory and direct simulation to complement one another in several ways. In the absence of experimental data, direct numerical simulation provides a means to validate theory and to establish the limits of that validity. Within its range of validity, secondary instability theory is preferable to computationally expensive numerical simulation for exploration of the parameter space. When nonlinear stages of transition beyond the reach of linear theory are simulated numerically, initial conditions derived from secondary instability theory provide a "jump start" thereby reducing significantly the integration time necessary to attain nonlinear states.

DIRECT NUMERICAL SIMULATION (DNS)

The 3D, compressible Navier-Stokes equations are solved by a highly accurate spectral collocation method developed by Erlebacher and Hussaini [1]. The algorithm is adapted for either flat plate or cylindrical geometry, depending on the value of a switch parameter. A temporal model of stability is assumed in which the mean flow is regarded as locally parallel and subject to temporally-evolving, spatially-periodic perturbations. The spectral collocation method employs Fourier basis functions in the periodic streamwise (x) and spanwise (y) (azimuthal-θ) directions and Chebyshev polynomial basis functions in the aperiodic wall-normal (z) (radial-r) direction. The governing equations and appropriate boundary conditions define an initial value problem which is integrated in time explicitly by means of a 3rd-order low-storage Runge-Kutta scheme. Initial conditions are defined by superposition of a parallel mean flow, derived from a spectrally accurate boundary layer code, and eigenfunctions of primary and secondary disturbances, generated from a temporal linear stability code, also of spectral accuracy.

SECONDARY INSTABILITY THEORY (SIT)

A linear theory is developed governing secondary instabilities of both fundamental (K) and subharmonic (H) type in compressible flat plate and axisymmetric boundary layers. Following a Floquet analysis similar to that of Herbert [2], the mean flow is assumed to be locally parallel and to be perturbed by a "saturated" primary instability wave characterized by wave numbers (α_1, β_1) and finite amplitude ε_1. The perturbed mean flow is then subjected to a pair (H) or triad (K) of secondary instability waves characterized by wave numbers (α_2, β_2) and amplitude ε_2. The mathematical formulation leads to an eigenvalue problem whose complex eigenvalue $\omega = \omega_r + i \omega_i$ defines the

---
*National Research Council Associate, Theoretical Flow Physics Branch
**Senior Scentist, ICASE
***Research Scientist, Analytical Services and Materials

growth rate ω_i and the frequency ω_r of the secondary instability wave ensemble. The theory is further refined by including additional wave components to the secondary disturbance. For example, a generalized form of a secondary disturbance of subharmonic type is the following:

$$e^{-i\omega t} e^{i\beta R\theta} \sum_{j=-n (j \neq 0)}^{n} \phi_j(r) e^{ij\frac{\alpha}{2}x} + \text{complex conjugate}$$

RESULTS

Whereas the boundary layer of a flat plate is self-similar (in the absence of streamwise pressure gradient), that of the cylinder is not, as shown in Fig. 1. Thus, transverse curvature affects the stability of flows along axisymmetric bodies both directly, through the "curvature terms" in the governing equations, and indirectly, through modification of the underlying mean flow velocity and temperature profiles. Figure 2, obtained through temporal linear theory, presents the direct, indirect, and combined influences of increasing transverse curvature $C = \delta^*/R$ on primary (a) and secondary (b) instability waves in the boundary layer of a cylinder, where δ^* and R are the displacement thickness and radius respectively. The parameters of the flow are $M_\infty = 4.5$, $Re = 10000$, and $T_\infty = 110R$, and the reference length is δ^*. The primary instability wave is of "second mode" type with $(\alpha_1, \beta_1) = (2.52, 0.0)$ and $\varepsilon_1 = 0.085$. (Eigenfunctions are normalized so that the maximum amplitude in the temperature fluctuation is unity relative to freestream temperature T_∞.) The secondary instability is of H-type with $(\alpha_2, \beta_2) = (1.26, 2.09)$. The growth rate of the primary is strongly influenced by both the direct and indirect effects of increasing transverse curvature. However, for this particular parameter set, these effects are in opposite directions and the combined influence is small. Our experience with other parameter sets is that the direct and indirect influences can be in the same direction, with substantial combined effect on the growth rate of the primary. In contrast, the growth rate of the secondary is only weakly influenced by increasing curvature, for fixed ε_1. It is however strongly affected by the amplitude of the primary; in the absence of the primary, the secondary is in fact damped. The shape of the secondary eigenfunction (peaked well away from the wall), its relative insensitivity to changes in the mean flow, its large (convective) growth rate, and the persistence of instability as $Re \to \infty$, identify the secondary instability mechanism as predominantly inviscid.

Figure 3 compares the results of SIT ($n=2$) and DNS for flow along a cylinder of curvature $C \cong 0.1$. The primary (1,0) wave grows slowly at the rate predicted by linear theory for more than 30 periods of oscillation. From an initial amplitude $\varepsilon_2 = 0.0085$, the secondary (1/2,1) mode also grows initially at a rate in agreement with SIT, and deviates slightly toward higher growth rates once its energy is roughly equal that of the primary. Although strictly valid only for neutrally stable primary waves, the theory appears to be quantitatively valid to relatively large amplitudes whenever the growth rate of the primary is incorporated, as denoted by squares in Fig. 3. Growth rates do not begin to depart severely from their theoretically predicted values until the amplitude of the (1,2) mode, formed by the self-interaction of the (1/2,1) component of the secondary, is large, at which time (period 38), the secondary amplitude is 12% relative to T_∞.

Acknowledgement The authors are especially grateful for the contributions of Dr. Tom Zang and Dr. Michele Macaraeg to this work.

REFERENCES

1) Erlebacher, G. and M. Y. Hussaini (1987), "Stability and Transition in Supersonic Boundary Layers", AIAA-87-1416.

2) Herbert, Th. (1988), "Secondary Instability of Boundary Layers", *Ann. Rev. Fluid Mech.* **20**, 487-526.

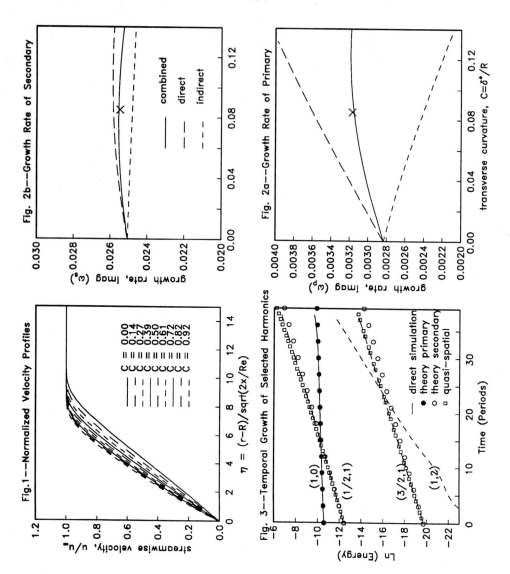

Fig. 1—Normalized Velocity Profiles

Fig. 2a—Growth Rate of Primary

Fig. 2b—Growth Rate of Secondary

Fig. 3—Temporal Growth of Selected Harmonics

THREE-DIMENSIONAL INTERNAL WAVE BREAKING AT A CRITICAL LEVEL

Kraig B. Winters, Eric A. D'Asaro and James J. Riley
University of Washington, Seattle WA USA

We present a numerical study of an important wave/mean flow interaction that occurs in stably stratified fluids; the breaking of internal gravity waves at critical levels. A critical level is a location in the flow field where the horizontal advection velocity of the large scale ambient flow is equal to the horizontal phase speed of a propagating internal wave. In the linear limit of this interaction, the wave becomes trapped at a critical level as the vertical wave scale shrinks to zero. The linear limit is a non-physical limit as the shrinking wave scale implies an infinite wave energy density. In geophysical flows such as in the atmosphere and the oceans, the appropriate limit is highly nonlinear and can result in turbulent wave breaking. The long term objective of this research is to understand, quantify, and parameterize the enhanced mixing associated with small scale wave breaking events in the stratified ocean.

We investigate the nonlinear interaction by formulating an initial value problem in which a wave packet propagates downward into an accelerating shear flow towards a critical level. We wish to resolve the transition dynamics of the breaking waves in the nearly inviscid limit. The high degree of scale separation inherent to the problem makes this difficult. We introduce a sub-grid scale model which maintains numerical stability during the cascade of energy to small scales while confining the dissipative/diffusive effects to only the smallest scales of motion in the numerical model. This model is simple in form, we merely replace the operator $Re^{-1}\nabla^2$ with $Re_*^{-1}\nabla^6$. A similar replacement is used for the diffusive term. This form of the operator allows the coefficient to be set to a much smaller value, with the result that a much greater proportion of wavenumber space is treated essentially inviscidly. We solve the equations using a pseudo-spectral method with second order explicit time stepping on a 32 x 32 x 200 mesh in the x, y (transverse) and z (vertical) directions.

In this paper we present results concerning two aspects of wave breaking. We first investigate the importance of three-dimensionality to the dynamics of wave instability. To do this, we prescribe initial conditions which are two-dimensional (x,z) except for a broad spectrum of small amplitude three-dimensional noise. The propagating wave packet is then free to evolve in R^3. As the waves propagate, they maintain a two-dimensional character until shortly before they break. Near the critical level, the waves exhibit a great deal of regular transverse structure. The wave breaking can be described as a rapidly growing instability driven by both shear and buoyancy. Visualization of the flow field shows the formation of counter-rotating convective cells oriented in the transverse direction, tilted in the (x,z) plane.

Our second aim is to identify means of diagnosing the mixing associated with isolated turbulent patches. Our first approach is to calculate an energy balance in an open volume, fixed in space, containing the wave breaking region. Crucial to this balance is the decomposition of potential energy into background and available potential energy, where the background potential energy is defined as the minimum potential energy state achievable through adiabatic resorting. Changes in the background state are then due only to diffusive mixing. The analysis is complicated in an open volume by the lack of known boundary conditions and the need to correct for changes in potential energy due to mass flux across the boundaries. A second approach is based on the idea that the volume of water between any two isodensity surfaces can change only as a result of mixing. By computing the volumes for a given discretization of the density range, the fine grain details of the small scales are meaningfully averaged and an estimate of net density flux obtained. This flux, plotted as a function of time and density, is shown in Figure 1. A third approach is to calculate the production of mean square potential vorticity, i.e. vorticity in the direction of the local density gradient, between isodensity surfaces. Potential vorticity variance can only be produced through diffusion and dissipation and hence can be used as a tracer for mixed fluid. This diagnostic is particularly well suited to this initial value problem because neither the waves nor the ambient shear have potential vorticity associated with them. It is generated by the wave breaking which excites small scale three-dimensional motions.

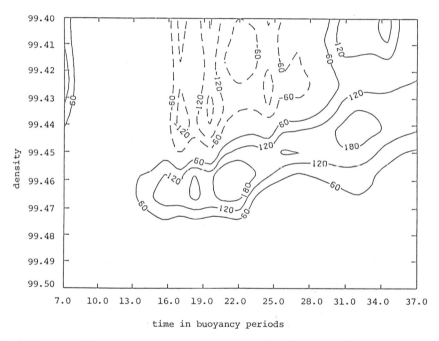

Figure 1: Dimensionless net density flux. The initial critical level location corresponds to a density value of 99.46 Lighter water is located above and heavier water below. Flow visualization and energy diagnostics showed wave breaking to occur near $t = 20$ just above the critical level. The flux estimates reveal regions of both positive (solid contours) and negative (dashed) fluxes. The strongest positive fluxes occur in the breaking region, as expected, but there was also significant positive flux associated with the nonlinear waves prior to breaking. A region of negative flux was formed above the critical level, possibly the result of reflecting waves.

NUMERICAL SIMULATION OF TRANSITIONAL BLUFF-BODY NEAR-WAKE SHEAR FLOWS

F.F. Grinstein and J.P. Boris
Laboratory for Computational Physics & Fluid Dynamics
US Naval Research Laboratory
Washington, D.C. 20375, USA

We present results of time-dependent numerical simulations of subsonic, spatially evolving, near-wake planar shear flows behind a rectangular cylinder, in the limit of high Reynolds numbers. The dynamics and topology of large-scale coherent structures (CS) in the transitional plane wake are investigated by examining its response to transverse spanwise excitation, and studying the flow evolution from the region of the initial thin shear layer to the region of the large-scale instability and vortex formation. From the practical point of view, we are interested in the effects of the near wake on parameters such as drag, lift, and base pressure coefficients.

The streams on each side of the bluff-body are regarded as different species, and passive scalars associated with the species are convected separately to evaluate entrainment and mixing. The unsteady conservation equations are solved numerically using the Flux-Corrected Transport (FCT) algorithm [1], direction- and timestep-splitting techniques, and appropriate inflow and outflow boundary conditions [2]. FCT is a non-linear, high order, explicit, finite-difference algorithm. A fourth-order phase-accurate FCT algorithm is used in the present simulations for Reynolds numbers (based on the bluff-body thickness and free-stream velocity), $Re \gtrsim 14,000$. An LES approach is used to simulate the large scale features of the flow dynamics. The FCT high-frequency filtering combined with the conservative, causal and monotone properties of the algorithm, effectively provide a minimal subgrid model, maintaining the large scale structures without aliasing while numerically smoothing the scales with wavelengths smaller than a few computational cells. The codes are fully vectorized and optimized for processing on the CRAY-2 and CRAY-YMP computers. The approach has been shown to be adequate for simulating the high Reynolds number vorticity dynamics in transitional free shear flows [2-4].

The systems modeled involve laminar streams of air with free-stream Mach numbers in the range 0.3–0.6. The flows are initially constrained to be strictly two-dimensional and allowed to develop in this way. The three-dimensional simulations are initialized with the two-dimensional Karman vortex street. The 3D instabilities are triggered through a spatial (spanwise) sinusoidal perturbation of the inflow velocity field. Such spanwise excitation introduces streamwise vorticity in the flow and simulates an excitation which can be introduced in laboratory experiments by putting a spanwise indentation in the edge of the thick splitter plate. The perturbations are sinusoidal, with a level typically of the order of a few percent of the initial mean streamwise velocity.

A typical instantaneous 3D flow visualization is shown in the figure below, where the flow direction is from the lower right to the upper left. The figure shows surfaces of constant vorticity magnitude equal to 15% of the peak value. High strain rates in the saddle regions between spanwise rollers are responsible for stretching and concentrating the transverse and streamwise vorticity introduced through the spanwise excitation. The resulting streamwise structures appear in counterrotating pairs distinctly superimposed on the spanwise rollers, and can initially be interpreted as legs of horseshoe vortices. The observed CS dynamics is closely related to the topology of the high-strain regions of the flow. As rollers of opposite sign are alternatively shed, the high-strain region initially lying in I, between rollers of the same sign, shifts to II, between rollers of opposite sign. As this topological change occurs, the legs of a horseshoe vortex with head wrapping roller '3', reconnect with each other on the newly shed roller '2' of opposite sign, thus forming a closed vortex loop.

The present simulations give direct evidence of CS interactions leading to the formation and development of unconnected closed vortex loops in the spanwise-excited plane-wake; the vortex loops are distributed on each side of the wake following a staggered pattern. The loops can partially overlap in the fully developed vortex street, as the region where a vortex loop wraps a roller becomes very close to that where the next loop downstream wraps a roller of opposite sign. Further details are discussed, including the effects of initial conditions.

This work is sponsored by ONR, NRL, and by DARPA. The calculations were performed at the computational facilities of the NAS at NASA Ames Research Center.

[1] Boris, J.P. & Book, D.L., in Methods in Computational Physics, Vol. 16, pp. 85-129 (1976), Academic Press, New York.
[2] Grinstein, F.F., Oran, E.S. & Boris, J.P., *J. Fluid Mech.* **165**, 201 (1986); *AIAA J.* **25**, 92 (1987); Grinstein, F.F., Hussain, A.F, & Oran, E.S., to appear in Europ. J. Mech. B / Fluids.
[3] Grinstein, F.F., Hussain, A.F, & Oran, E.S., Proc. 6th Symp. Turb. Shear Flows, Toulouse (1987); AIAA Paper 88-0042, Reno (1988); AIAA Paper 89-0977, Tempe.
[4] Grinstein, F.F., Guirguis, R.S., Dahlburg, J.P. & Oran, E.S., Proc. 11th Int. Conf. Num. Meth. Fluid Mech., ed. by D.L. Dwoyer and M.Y. Hussaini, Springer Verlag, pp. 283–287 (1989).

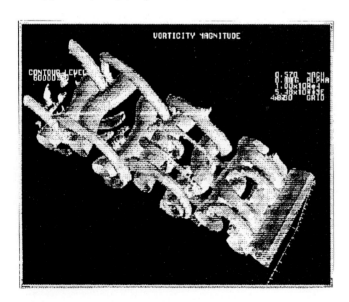
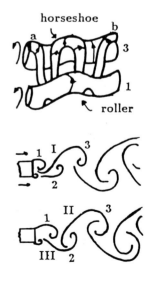

On the stability of the infinite swept attachment line boundary layer

V. Theofilis*,   P. W. Duck [†] &   D. I. A. Poll [‡]

Abstract

A number of numerical schemes were employed in order to gain insight in the stability problem of the infinite swept attachment line boundary layer. The basic flow was taken to be the classical Hiemenz flow. A number of assumptions for the perturbation flow quantities were considered. In all cases a pseudo- spectral approach was used; the chordwise and spanwise directions were treated spectrally, while an implicit Crank-Nicolson scheme was used temporally. Extensive use of the FFT algorithm has been made.

The significance of the attachment line boundary layer which is formed on the windward surface of an infinite swept cylinder, with respect to transition to turbulence has been recognized and studied analytically, numerically and experimentally by a number of investigators. Attention has been generally restricted, however, to specific assumed forms of the time-dependence and chordwise variation of the solution, in particular to those appropriate to instability waves which occur in boundary layers.

In this work we study the flow using a spectral method and an initial-value-problem approach to tackle the problem numerically, thus departing from the specified spatial and time dependence of a number of previous works.

The Navier-Stokes equations have been written using the velocity-vorticity formulation and the basic flow is taken to be of the classical Hiemenz class. The perturbed flow has been considered linearly dependent on the chordwise component x, compatible with stagnation point/Hiemenz flows. A spectral method has been used in the spanwise direction z and the resulting equations

---
*Dept. of Aeronautical Engineering, Univ. of Manchester, U.K.
[†]Dept. of Mathematics, Univ. of Manchester, U.K.
[‡]Dept. of Aeronautical Engineering, Univ. of Manchester, U.K.

are approximated using a standard second-order finite-difference scheme in the transverse direction, while a Crank-Nicolson scheme is used in time marching. A forcing is applied to the transverse velocity component on the wall which is akin to a jet discharge out of the wall (and as such has the potential for comparison with experiment). This forcing is taken to be a short-lived prescribed function of z and time which may trigger the onset of instability. Overall the scheme may be described as a pseudo-spectral approach.

The numerical results for the linearized case of the perturbed flow equations (i.e. for small amplitudes of the forcing) in spectral space have been compared with previous works (*ICASE Rep.No.84-64/NASA Contr.Rep.172504* and *NASA Contr.Rep.172300*) and the agreement is excellent. This justifies the assumptions of previous investigators on the time-dependence of the instability waves. Results for the velocity and the vorticity components have also been obtained in real space.

In the non-linear case of the governing equations we make the same assumptions as in the linear case, except that the smallness condition is relaxed. The Fast Fourier Transform (FFT) method has been used to back-transform the numerical spectral space results at any time-step into real space where the non-linear terms are calculated. These are then transformed into spectral space again using the FFT and the iterative process continues until convergence is achieved. The solution then proceeds to the next time step. The advantage of using the FFT method is that it requires only $O(N log_2 N)$ operations for the transformation of a set of data defined on N points, instead of $O(N^2)$ which would be required for the evaluation of the convolutions resulting from the non-linear terms in the equations in spectral space.

A time-periodic dependence of the flow quantities has also been implemented and the resulting equations have been solved subject to the same linear dependence of the chordwise velocity component but (of course) without the need of the time-marching procedure. Results obtained using the above scheme were compared with the time-marching results and the agreement at times sufficiently large, when the transients in the time-marching scheme have decayed, was excellent. A number of linear results have been presented at the *4th IUTAM Symposium on Transition and Turbulence* in Toulouse, France and are due to be published in the Conference Proceedings.

Three-dimensional perturbations have also been introduced in the basic flow (which is again taken to be the Hiemenz flow) and their time evolution is currently being investigated using a linear time-marching scheme. The (strictly non-rational) approximations on the form of the perturbation quatities which we have made are analogous to the parallel-flow approximations , namely that there exist two length-scales in the chordwise direction x, a slow one indicating the slow boundary layer acceleration, along which the basic flow varies and a fast one, upon which perturbation quantities are dependent

The previously studied two-dimensional cases are incorporated in our three-dimensional scheme, which enables us to study the stability of the flow off the attachment line. Results at a number of x−stations for different streamwise wavenumbers are currently being obtained.

References

[1] Burggraf, O. R. & Duck, P. W. in *"Numerical and Physical Aspects of Aerodynamical Flows"*, (1981) Springer

[2] Cooley, J. W. & Tuckey, J. W. *Math. Comp.* **19** (1965), p.297

[3] Crank, J. & Nicolson, P. *Proc. Camb. Phil. Soc.* **43** (1946), p.50

[4] Dennis, S. C. R., Ingham, D. B. & Cook, R. N. *J. Comp. Phys.* **33** (1979), p.325

[5] Duck, P. W. *J. Fluid Mech.* **160**, (1985), p.465

[6] Duck, P. W. & Burggraf, O. R. *J. Fluid Mech.* **162** (1986), p.1

[7] Hall, P., Malik, M. R. & Poll, D. I. A. *Proc. Roy. Soc.* **395**, (1984), p.229

[8] Hall, P. & Malik, M. R. *J. Fluid Mech.* **163**, (1986), p.257

[9] Hall, P. & Seddougui, S. *NASA Contr. Rep. 181653*, (1988)

[10] Hiemenz, K. *Dingl. Polytechn. J.* **326**, (1911), p.321

[11] Poll, D. I. A. *The Aeronautical Quart.* **30**, (1979), p.607

[12] Poll, D. I. A. *J. Fluid Mech.* **150**, (1985), p.329

[13] Rosenhead, L. *"Laminar Boundary Layers"*, O.U.P., Oxford, 1963

[14] Spalart, P. R. *AGARD CP-438*, (1988), p.5-1

[15] Schlichting, H. *"Boundary Layer Theory"*, McGraw-Hill, N.Y., 1979

LARGE EDDY SIMULATION OF TURBULENT FLOW IN A DUCT OF SQUARE CROSS SECTION

Takeo Kajishima, Yutaka Miyake and Toshiyuki Nishimoto
Department of Mechanical Engineering, Osaka University
2-1 Yamadaoka, Suita, Osaka, 565 Japan

Introduction The turbulence-driven secondary flow is observed in the fully developed turbulent flow in a straight non-circular duct. In order to obtain the near wall behavior of the Reynolds stress components which cause the secondary flow, we carried out the large eddy simulation(LES) for fully developed turbulent flows in a duct of square cross section using 160,000 (50^2 in the cross section and 64 in the streamwise direction) grid points. Reynolds numbers based on the duct width H and the bulk velocity u_m are 6,200(Case 1) and 67,400(Case 2). In this report, a high speed scheme of LES for the duct flow, an explanation to the origin of the secondary flow and the discussion on the turbulence modeling are outlined.

Numerical Scheme The coordinate system is as follows: x_1, the mean flow direction; x_2 and x_3, the cross stream directions. In the cross section, the grid spacings are $0.0040H \leq h_{2(,3)} \leq 0.038H$ for Case 1 and $0.0016H \leq h_{2(,3)} \leq 0.049H$ for Case 2, respectively. The no-slip condition is applied at the wall. In the direction x_1, in which the periodic boundary condition is applied, the computational domain is $H_1 = 6.4H$ with uniform grid spacing of $h_1 = 0.1H$. The effect of the subgrid scale(SGS) turbulence is approximated by the Smagorinsky model.

For the finite difference calculation, the staggered grid is adopted. The basic equations are expressed by the conservative form and spatial derivatives are approximated by the 2nd order central difference. Time marching scheme is SMAC type: (1) Prediction by the 2nd order Adams-Bashforth time marching; (2) Solving the Poisson equation for the scaler potential; (3) Correction to satisfy the continuity equation at the new time step. In the 2nd step, the FFT is applied for the reduction of CPU time. The SOR procedure to solve the Poisson equation

$$\left(-k_1'^2 + \frac{\partial^2}{\partial x_2^2} + \frac{\partial^2}{\partial x_3^2}\right)\phi(k_1, x_2, x_3) = \varphi(k_1, x_2, x_3), \quad k_1'^2 = \frac{2}{h_1^2}\left(1 - \cos\frac{2\pi k_1}{N_1}\right) \tag{1}$$

parallel for each wavenumber(k_1) is suited for the vector processor. About 1.5sec CPU time to advance one time step with 120 SOR iterations is required on NEC SX-2N.

Numerical Results The mean flow field and turbulence statistics have been obtained by averaging for $20H_1/u_m$ integrating time, after the statistically steady state is reached. CPU time are 17.5hr for Case 1 and 27.5hr for Case 2. Figure 1 shows the time averaged flow field and the turbulence kinetic energy in the lower-left quarter section. Numerical values are nondimensionalized by the centerline velocity u_c. Figure 2 compares the numerical results with the experimental data collected in the literature[1],[2],etc. along the corner bisector, z_1. The mean velocity and turbulence quantities are in reasonable agreement with measurements.

The budget equation of the vorticity $\overline{\omega}_1 (= \partial \overline{u}_3/\partial x_2 - \partial \overline{u}_2/\partial x_3)$ is expressed as:

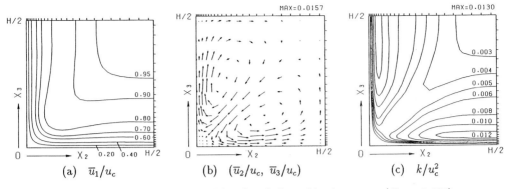

Figure 1: Time averaged flow field and turbulence kinetic energy ($R_m = 6,200$)

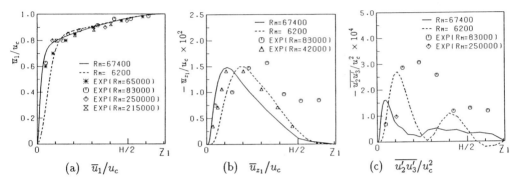

Figure 2: Comparison of LES with the measurement along the corner bisector z_1

$$\overline{u}_2 \frac{\partial \overline{\omega}_1}{\partial x_2} + \overline{u}_3 \frac{\partial \overline{\omega}_1}{\partial x_3} = \Omega_N + \Omega_S + \Omega_V,$$
$$\Omega_N = \frac{\partial^2}{\partial x_2 \partial x_3}\left(\overline{u_2'^2} - \overline{u_3'^2}\right), \quad \Omega_S = -\left(\frac{\partial^2}{\partial x_2^2} - \frac{\partial^2}{\partial x_3^2}\right)\overline{u_2'u_3'}, \quad \Omega_V = \nu\left(\frac{\partial^2 \overline{\omega}_1}{\partial x_2^2} + \frac{\partial^2 \overline{\omega}_1}{\partial x_3^2}\right). \quad (2)$$

The numerical results shown in figure 3 indicate that the normal stress term Ω_N is positive for the production of $\overline{\omega}_1$ and that the shear stress term Ω_S is negative. In this sense, the major aspect of the conclusion by Demuren and Rodi[1] is acceptable. But these effects are prominent in the near wall region in which experimental data have not been available. $\Omega_N + \Omega_S$ results in a small positive effect which is balanced almost with the destruction effect of the viscosity Ω_V

We consider now the balance of forces to get the complete description on the origin of the secondary flow. Figure 4 shows the effect of normal stress $\vec{F}_N(=\vec{F}_{RN} + \vec{F}_{PN})$, that of shear stress $\vec{F}_S(=\vec{F}_{RS} + \vec{F}_{PS})$ and the total $\vec{F}_N + \vec{F}_S$ in the momentum equation of the secondary motion. Reynolds stress gradients are represented as:

$$\vec{F}_{RN} = \left(-\frac{\partial \overline{u_2'^2}}{\partial x_2}, -\frac{\partial \overline{u_3'^2}}{\partial x_3}\right), \quad \vec{F}_{RS} = \left(-\frac{\partial \overline{u_2'u_3'}}{\partial x_3}, -\frac{\partial \overline{u_2'u_3'}}{\partial x_2}\right). \quad (3)$$

Pressures which appear in

$$\vec{F}_{PN} = \frac{1}{\rho}\left(-\frac{\partial \overline{p}_N}{\partial x_2}, -\frac{\partial \overline{p}_N}{\partial x_3}\right), \quad \vec{F}_{PS} = \frac{1}{\rho}\left(-\frac{\partial \overline{p}_S}{\partial x_2}, -\frac{\partial \overline{p}_S}{\partial x_3}\right) \quad (4)$$

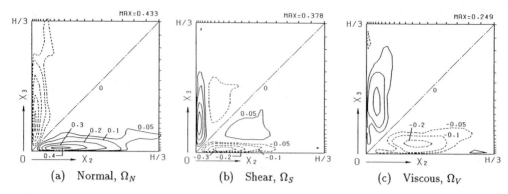

Figure 3: Contributions of Reynolds stresses and the viscosity on the $\overline{\omega}_1$ budget ($R_m = 6,200$)

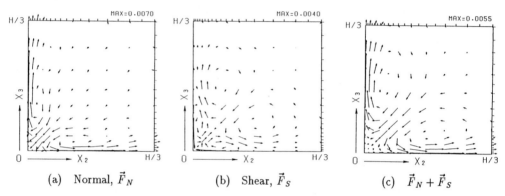

Figure 4: Effects of Reynolds stresses on the production of the secondary flow ($R_m = 6,200$)

are the solutions of the following Poisson's equations:

$$\frac{1}{\rho}\left(\frac{\partial^2 \overline{p}_N}{\partial x_2^2} + \frac{\partial^2 \overline{p}_N}{\partial x_3^2}\right) = -\frac{\partial^2 \overline{u_2'^2}}{\partial x_2^2} - \frac{\partial^2 \overline{u_3'^2}}{\partial x_3^2}, \quad \frac{1}{\rho}\left(\frac{\partial^2 \overline{p}_S}{\partial x_2^2} + \frac{\partial^2 \overline{p}_S}{\partial x_3^2}\right) = -2\frac{\partial^2 \overline{u_2'u_3'}}{\partial x_2 \partial x_3}. \quad (5)$$

The fact that each contour plot in the figures 3(a),(b) corresponds to the vortex pattern in figures 4(a),(b) indicates that the origin of the secondary flow is the imbalance between the gradient of the turbulence stress field and that of the corresponding pressure field near the wall.

Conclusion The near wall behavior of the Reynolds stress and the corresponding pressure field calculated by LES have elucidated the mechanism of the turbulence-driven secondary flow. The essential factor is the rapid attenuation of the turbulence stresses due to the wall effect. The pressure, on the other side, has the gradient along the wall. The imbalance between them causes the secondary motion. Therefore, the turbulence model should have the reasonable behavior in the near wall region. In addition, since the opposite effects of normal and shear components are almost balanced in the region very close to the corner wall, the consistency among the components of the model is important.

References (1) Demuren,A.O. and Rodi,W., J. Fluid Mech., 140 (1984), p.189.
(2) Fujita,H., et al., Trans JSME (in Japanese), 53-492B (1987), p.2370.

DIRECT NUMERICAL SIMULATION OF TURBULENT FLOW IN AN OPEN CHANNEL

Richard I. Leighton
Science Applications International Corporation
McLean, Virginia

Thomas F. Swean, Jr.
Laboratory for Computational Physics and Fluid Dynamics
Naval Research Laboratory
Washington, D.C.

The incompressible three-dimensional Navier-Stokes equations are solved for initial and boundary conditions approximating a turbulent open-channel flow of water at $Re_h = U_{fs}h/\nu = 2340$ where h is the channel depth and U_{fs} is the mean free-surface velocity or $Re_* = 134$ based on the friction velocity, $u_* = \sqrt{\frac{\tau_{wall}}{\rho}}$. The governing equations are recast into a 4th order equation for the vertical velocity and a 2nd order equation for the vertical vorticity and the continuity equation is solved explicitly in the recovery of the streamwise velocity. The numerical algorithm used follows that described in Kim, Moin and Moser (1987). This method involves the use of the homogeneous solutions of the time discretized 4th order equation to satisfy all the required boundary conditions. The equations are numerically solved after they are Fourier transformed in the streamwise (x) and spanwise (z) directions and Chebychev transformed in the vertical direction (y). The calculations are preformed on a grid of 48 × 65 × 64 nodes in x,y,z respectively for a resolution of $26.3 l_*$ in the streamwise direction and $13.2 l_*$ in the spanwise direction, where $l_* = \nu \sqrt{\frac{\rho}{\tau_{wall}}}$. The total box size is $1684 l_* \times 134 l_* \times 632 l_*$. The boundary conditions are periodic on all dependent variables in the streamwise and spanwise directions. No slip conditions are used at the channel bottom while the free surface is approximated as a rigid free slip surface with vanishing shear. The approximations for the free-surface boundary condition are

$$v = 0; \quad \omega_y = 0; \quad \text{and} \quad \frac{\partial^2 v}{\partial y^2} = 0 \quad \text{on the free surface.}$$

The assumption that the surface be rigid is equivalent to $v = 0$ on the free-surface. The remaining two can be derived from continuity and the shear free boundary conditions.

An estimate of the error produced by replacing the free-surface boundary conditions with the above condition can be obtained by considering what elevation of the free-surface would be produced by the fluctuating normal stresses. A dynamic free-surface boundary condition is produced requiring a balance of normal stress across the interface (nondimensionized by U, h and ρ):

$$(Fn^2 Re)p - \eta Re - 2Fn^2 \frac{\partial v}{\partial y} = 0 \text{ at } y = 0$$

where $Fn = U/\sqrt{gh}$ is the Froude number, and the ambient external pressure is zero. By assuming a reasonable channel depth, $h = 4$cm, similar to published physical experiments the following estimate can be made:

$$\eta_{rms} = Fn^2 \left\langle \left(\frac{2}{Re}\frac{\partial v}{\partial y}\right)^2 + p^2 - \frac{2}{Re}\frac{\partial v}{\partial y}p \right\rangle^{1/2}$$
$$= 0.4\% \text{ of } h,$$

where $\langle \ \rangle$ implies averaging occurs over planes of data parallel to the top boundary. This corresponds to a displacement of approximately $2 - 3$ gridpoints off the top boundary. Based on this estimate, we assume that the statistics generated are not adversely affected by the assumption of no free-surface displacement.

A large number of turbulence statistics are computed in the vicinity of the free surface and complete determinations of the balances of the exact Reynolds stress, turbulence kinetic energy, and isotropic dissipation rate equations are reported for the first time. The results show that while the turbulence kinetic energy is preserved in the vicinity of the free surface, the turbulence is redistributed from the vertical component into the two horizontal components. The balances of the streamwise and normal components of the turbulence kinetic energy reveal a reversal in sign of the pressure-velocity gradient correlations in this region, indicating a transfer of energy out of the normal component. Further examination reveals that the pressure due to the turbulent-turbulent interactions (related to the so-called return to isotropy term) is the most important term in mediating the transfer of kinetic energy from the normal to the planar directions. It is apparent from the kinetic energy and isotropic dissipation rate balances that there exists two separate regions near the free surface, a thin region, made apparent by large local changes in the isotropic dissipation function and a thicker zone wherein the redistribution of turbulence is more pronounced. Near surface expansions of the turbulence kinetic energy and isotropic dissipation rate are determined for use in Reynolds-averaged turbulence models.

This work was supported by the Fluid Dynamics Task Area at the Naval Research Laboratory

References:
Kim, J., Moin, P. and Moser, R.D. 1987 Turbulence Statistics in fully developed channel flow at low Reynolds nomber. J. Fluid Mech. **177**, 133-166

Direct Numerical Simulation of a Complicated Transition-Breakdown of a Circular Pipe Flow through Axisymmetric Sudden Expansions

H.L.CHEN* and K.OSHIMA**

*Department of Aeronautical Engineering, University of Tokyo
**The Institute of Space and Astronautical Science, 3-1-1 Yoshinodai, Sagamihara 229, Japan

With recent high performance computers and advanced numerical methods, it has become possible to solve the detailed dynamical behavior of the Navier-Stokes equations. This is a new and important way to study transition and other complicate physical phenomena. The purpose of this work is to examine stability problems of circular pipe flows by a numerical method. We try to utilize the maximum capacity of modern supercomputer to calculate the flow at high Reynolds number and to perform long time scale calculations of unsteady problems.

In this study, fluid flows inside a circular pipe which has an axisymmetric sudden expansion were numerically simulated for a wide range of Reynolds numbers. Axisymmetric, time-dependent, incompressible, Navier-Stokes equations were numerically solved by the finite difference method, in which second-order upwind schemes were applied to the convection terms. In order to calculate unsteady behavior at high Reynolds number, fine grid systems 31×601, 76×1501 and small time steps are used. The numerical results show that after impulsively starting from the rest condition with a plug inlet velocity profile, the flow field damps into a steady, laminar one at small Reynolds number. A laminar, recirculation region forms and the velocity distribution is of the Poiseuille type in the downstream, as shown in Fig.1(a). As the Reynolds number is increased, Taylor-Helmholtz type instability causes velocity oscillations along the shear layer. These oscillations are simple periodic wavy motions close behind the step, as seen in the phase diagrams Fig.2(a) of the momentum flux of cross-section. Development of the oscillations downstream depend strongly on the Reynolds number. At relatively small Reynolds number, they dissipate either to a steady flow field or to weak intermittent waves. As the Reynolds number is further increased, velocity oscillations develop into complicated waves and a street of vortex rings shed nonperiodically along the shear layer, as shown in Fig.1(b)-Fig.1(d). The numerical results show that the interaction of the vortex rings plays a dominant role in downstream, and the velocity fluctuations evolve into chaotic ones. A complicated bifurcation process of solution of the Navier-Stokes equations are recognized on the phase diagrams Fig.2(b) and Fig.2(c). Detailed dynamical processes of the onset of shear layer instability and transition from the steady-state solution to a chaotic one were revealed by this simulation.

Numerical simulation based on the axisymmetric Navier-Stokes equation including the swirling flow term were also carried out. If the Reynolds number and/or, the swirling velocity at the pipe inlet exceeds a certain value, stagnation point appears on the symmetric axis and bubble-like breakdown of vortex takes place. Numerical results show that the presence of vortex rings will provoke the vortex breakdown at lower Reynolds or smaller swirling velocity, because the extension of the stream tube between two vortex rings causes the pressure increase, then reverse flow appear on the axis, as shown in Fig.3. When considering the instability of axisymmetric pipe flow without the sudden expansion geometry, swirling velocity is also a crucial factor. Simulation results show that, regardless of the flow Reynolds number, the velocity distribution tends to be of the Poiseuille type without swirling velocity, that is, this type of flow is stable. When swirling velocity is present at the inlet of the pipe and the flow Reynolds number exceeds a certain value, the velocity distribution does not tend to be of the Poiseuille type. Reverse flow will appear on the symmetric axis, after impulsively starting from rest. That is, this type of flow is unstable.

REFERENCE

1. K.Oshima, H.Kanda, Y.Ishii and H.L.Chen, "Numerical Simulations of Unsteady Pipe Flows", Presented at the ISCFD-Nagoya, August 28-31, 1989.
2. H.L.Chen and K.Oshima, "Numerical Simulation of Complicate Transition-Breakdown of Vortical Flow Inside a Circular Pipe", Proceedings of the Symposium on Mechanics for Space Flight-1986, Institute of Space and Astronautical Science, Report S.P.12.

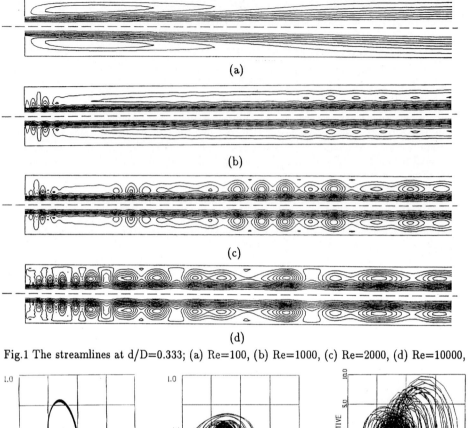

Fig.1 The streamlines at d/D=0.333; (a) Re=100, (b) Re=1000, (c) Re=2000, (d) Re=10000,

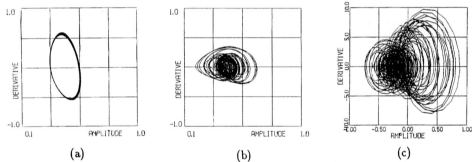

Fig.2 The phase diagrams of momentum flux of cross-section,
(a) Re=2000, at 0.05L, (b) Re=2000, at 0.667L, (c) Re=10000, at 0.2L

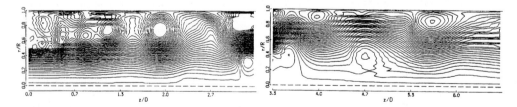

Fig.3 The streamlines with swirling inlet velocity profile at Re=10000.

Numerical Simulation of Unsteady Incompressible Viscous Flows by Improved GSMAC-FEM

Y. KATO*, T. SATO**, T. SAWADA* and T. TANAHASHI*

*Department of Mechanical Engineering, Keio University
3-14-1 Hiyoshi, Kohoku-ku, Yokohama 223, Japan
**Nagasaki Research & Development Center, Mitsubishi Heavy Ind., Ltd.
1-1 Akunoura-machi, Nagasaki 850-91, Japan

Introduction

A new FEM scheme using simultaneous relaxation of velocity and Bernoulli function is presented for unsteady incompressible viscous flow. This is an improved version of the GSMAC(Generalized Simplified Marker and Cell) FEM[1]. To verify the scheme, transient 2-D Poiseuille flow and 3-D flow in a lid-driven cubic cavity are numerically calculated.

Numerical method

The present method employs the following predictor-corrector equations.

$$\tilde{u}_i = u_i^n + \Delta t(-H_i^n + \epsilon_{ijk} u_j^n \omega_k^n + \frac{1}{Re} u_{i,jj}^n) \tag{1}$$

$$\phi_{,jj} = \tilde{u}_{j,j}, \qquad u_i^{n+1} = \tilde{u}_i - \phi_{,i} \tag{2}$$

$$H^{n+1} = H^n + \phi/\Delta t \tag{3}$$

where $\tilde{u}_i$, ω_i, H and ϕ are predictor of velocity vector, vorticity vector, Bernoulli function and velocity potential, respectively. All variables are dimensionless and Re is the Reynolds number. Finite element discretization is performed by the Galerkin method. u_i and $\tilde{u}_i$ are defined at the node point and other variables are constants in each element. To avoid iterative calculation of Poisson equation, Poisson equation of potential ϕ is rewritten by

$$\phi = \beta \tilde{u}_{j,j}, \tag{4}$$

where the parameter β depends on geometric shape of the element. Algorithm of the present method is as follows: (1) calculation of $\tilde{u}_i$, (2) simultaneous relaxation of $\tilde{u}_i$ and ϕ to satisfy the equation of continuity, (3) correction of H.

Numerical results

Velocity profiles of transient Poiseuille flow are in good agreement with exact solutions (see Figure 1). In the case of flow in a lid-driven cubic cavity at $Re = 5000$, time averaged velocity profiles on the symmetry plane are in good agreement with the experimental results by Prasad and Koseff[2] (see Figure 2). Complex flow structures are clearly simulated and non-physical pressure oscillations are not observed (see Figures 3).

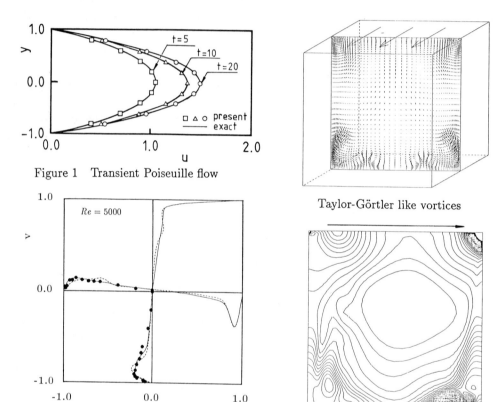

Figure 1 Transient Poiseuille flow

Taylor-Görtler like vortices

Pressure contours on the symmetry plane

Figure 3 Flow in a lid-driven cubic cavity ($Re = 5000$, $t = 50$)

---- numerical (at $t = 100$)
—— numerical (mean value)
● experiment (Prasad et al.[2])

Figure 2 Velocity profiles on the symmetry plane

References

[1] H.Kawai, Y.Kato, T.Sawada and T.Tanahashi, GSMAC-FEM for Incompressible Viscous Flow Analysis (A Modified GSMAC Method), JSME Int. J. 33-1 (1990), 17.

[2] A.K.Prasad and J.R.Koseff, Reynolds number and end-wall effects on a lid-driven cavity flow, Phys. Fluids A, 1-2 (1989), 208.

Investigation of 2-D and 3-D instability of the driven cavity flow

P. Le Quéré, A. Dulieu and Ta Phuoc Loc

LIMSI-CNRS, BP 133, 91403 Orsay Cedex, France

The square driven cavity configuration is one of the classical test-problems to test numerical algorithms developed for the integration of the Navier-Stokes equations of incompressible recirculating flows. Due to improvements of algorithms and to increasing computing resources, it is possible today to compute 2-D solutions for this problem for values of the Reynolds number, Re, based on cavity width as large as or larger than 10^4 [1,2,3,4]. There is a general agreement on the fact that, up to a critical value of the Reynolds number in the vicinity of 10^4, there exists stable steady state solutions to the 2-D equations and that, for larger values of Re, the 2-D solutions seem to undergo a transition to unsteadiness. On the other hand, reliable experimental results [5,6] have shown that the flow is already turbulent for much smaller values of Re. One can then safely argue that the first instability of this flow is certainly not two-dimensional and it has been proposed that the flow becomes first unstable to a 3-D Taylor-Görtler type instability, which has promoted the use of full 3-D algorithms [7,8].

In this work, we consider the "regularized driven cavity" problem. The unsteady governing equations are integrated with a pseudo-spectral Chebyshev spatial discretisation coupled to a second-order finite-difference time stepping scheme. We first show that the 2-D solution undergoes a transition to unsteadiness for values of Re slightly larger than 10^4, in agreement with previous results for the classical configuration (see figure 1).

We then consider the stability of the 2-D solutions with respect to 3-D perturbations. To this aim, we use a 3-D algorithm in which all the variables (3 components of velocity, pressure) are expanded according to

$$f(x,y,z,t) = \sum_{n=0}^{N} \sum_{m=0}^{M} \sum_{k=-K/2}^{K/2-1} f_{nmk}(t) T_n(2x-1) T_m(2z-1) exp(\frac{2i\pi ky}{L})$$

where L is the periodicity length in the y-direction normal to the plane (x-z) of the 2-D solution. By trial and error, we were then able to show that the 2-D solution is indeed already unstable to 3-D disturbances of infinitesimal amplitude for values of Re as low as 1.2×10^3 (see figure 2). (For the time, our best lower and upper bounds of the critical Reynolds number are 10^3 and 1.2×10^3 respectively.) Surprisingly enough, this instability is characterized by a typical wavelength which is of the same order as the size of the cavity (for Re equal to 1.4×10^3, the solution remains stable if L is chosen equal to half of the cavity-width). Furthermore the asymptotic solution is found time-periodic. These results were cross-checked with a 3-D finite-difference code in which periodicity in the y-direction was assumed.

References

[1] Ghia U., Ghia K.N. and Shin C.T., J. Comp. Phys., 1982, 48, 387-411
[2] Schreiber R. and Keller H.B., J. Comp. Phys., 1983, 49, 310-333
[3] Vanka S.P., J. Comp. Phys., 1986, 65, 138-158
[4] Bruneau C.H. and Jouron C., J. Comp. Phys, in press
[5] Koseff J.R. and Street R.L., J. Fluids Engineering, 1984, 106, 21-29
[6] Koseff J.R. and Street R.L., J. Fluids Engineering, 1984, 106, 390-398
[7] Ku H.C., Hirsh R.S. and Taylor T.D., J. Comp. Phys., 1987, 70, 439-462
[8] Perng C.Y. and Street R.L., Int. J. Num. Meth. Fluids, 1989, 9, 341-362

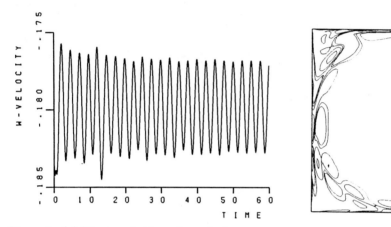

Figure 1: (a) Time evolution of w-velocity component at sampling point for Re=1.05×10^4, Integration carried out with N=M=80; time-step $\Delta t = 7.5 \times 10^{-4}$; period of asymptotic solution is approximately 2.6 (b) instantaneous plot of fluctuating vorticity; the time-periodic solution is made of 5 structures that circulate around the cavity (1 structure is made of two consecutive zones of positive and negative vorticity fluctuation).

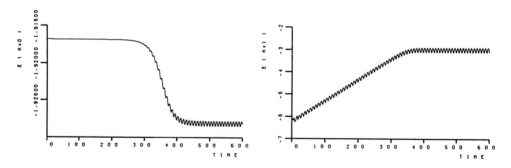

Figure 2: Time-evolution of logarithm of "energy" of u-velocity component contained within 2 first Fourier modes for Re = 1.4×10^3. Integration carried out with N=M=24, K=16; $\Delta t = 1.5 \times 10^{-2}$; Initial condition is the steady 2-D solution for Re = 1.4×10^3; Periodicity length L is equal to cavity width. Note that asymptotic solution is time-periodic with period equal to 17 approximately.

Numerical simulation of three-dimensional unsteady incompressible viscous flow for separated configurations

B. Troff (*), T.H. Lê (*), Ta Phuoc Loc (**)

* : ONERA, B.P. 72, 92322 Châtillon Cedex, France.
** : ONERA and LIMSI (CNRS), B.P. 30, 91406 Orsay Cedex, France.

The analysis of a three-dimensional separated flow around a wing is of primary importance to the evaluation of global aerodynamic performances of aircrafts.

This paper deals with the numerical simulation of three-dimensional unsteady incompressible viscous flowfields around a wing between walls.

The Navier Stokes equations in velocity-pressure formulation are used to model such flows in generalized coordinate systems. The governing equations are discretised by a second order finite difference scheme in time and space such as in the Euler approach presented in [1]. They are written as follows :

$$L(U^{n+1}-U^n) + \Delta t \, G \, Q^{n+\frac{1}{2}} = \Delta t \, [\, A(U^n, U^{n-1}) + K(U^n) \,]$$

$$D \, U^{n+1} = 0.$$

L is an implicit operator containing the boundary conditions. Its adding allows to improve the stability properties of the scheme.

$Q = \frac{1}{2} U^2 + P$ denotes the total pressure.

A , K , G and D are the matrices representing advection, diffusion, gradient and divergence, respectively.

The total pressure, computed by solving a Poisson equation, is obtained by minimising the norm of the velocity divergence. This is carried out by using the Multi-gradient algorithm developed at ONERA [2]. Grid generation is based on conformal mapping which gives a two-dimensional grid of "O" type. The three-dimensional mesh is constructed in the usual way by stacking the two-dimensional grids, generated around selected spanwise sections of the wing . No-slip boundary conditions are used on the body and on the two walls. A freestream condition is imposed at infinity. In the 2-D case, numerical results obtained with this method have been compared favorably with experiment and ψ - ω formulation [3].

Numerical results presented here in the 3-D case are related to the computation of separated flow around a wing between walls, for which there are few experimental and numerical data. This wing, rectangular planform with an aspect ratio of 3, is equipped with a NACA0012 airfoil. Results are obtained with a 15^0 angle of attack and for a Reynolds number equal to 500 based on chord length. The figure below shows the contours of the z-component of the vorticity in 3 different cross sectional planes (the first two are near the wall, the third is the midspan section) . One can see the leading edge separation vortex and the starting vortex at the trailing edge. Near the wall, these structures are dampened due to wall effect.

References

[1] Mège P., Lê T.H., Morchoisne Y. : "Numerical simulation of vortex breakdown via 3-D Euler equations", Turbulence and coherent structures, Conference proceedings, Grenoble, France, September 18-21 1989.

[2] Ryan J., Lê T.H., Morchoisne Y. : "Panel code solvers", 7th GAMM Louvain, Belgium, 1987.

[3] Daube O., Ta Phuoc Loc, Monnet P., Coutanceau M.: "Ecoulement instationnaire décollé d'un fluide incompressible autour d'un profil", Agard C.P. n° 386, 1986.

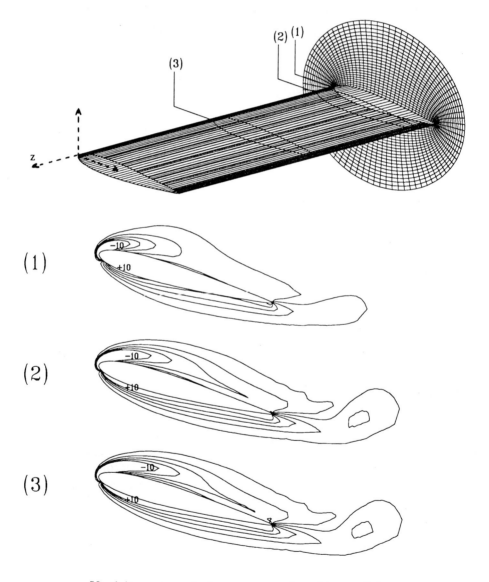

Vorticity contours in three cross sections ($\Delta\omega_z = 2.$)

FLOW OVER ARBITRARY AXISYMMETRIC MULTI-BODY CONFIGURATIONS, USING DIRECT NAVIER-STOKES SIMULATIONS

G.A. Osswald[*], K.N. Ghia[*] and U. Ghia[**]
[*]Department of Aerospace Engineering and Engineering Mechanics
[**]Department of Mechanical, Industrial and Nuclear Engineering
University of Cincinnati, Cincinnati, Ohio

An axisymmetric unsteady full Navier-Stokes direct numerical simulation analysis is extended for application to multiple body configurations. Currently, application is made to a generic submarine hull/shrouded propulsor configuration. The analysis supports arbitrarily specified hull and shroud geometries, utilizing analytically orthogonal boundary-aligned generalized coordinate grids. A generalized Schwarz-Christoffel mapping technique is employed to generate these grids. This grid-generation technique, coupled with analytical contraction mappings, permits the infinite physical domain external to this multibody configuration to be mapped to a single unit square computational domain, with the primary body being mapped to the lower edge and all secondary bodies mapped to internal slits. Zonal grid attributes are provided by clustering control capabilities, while the computational plane topology provides a single block structure very well suited to the application of the efficient direct inversion unsteady Navier-Stokes solvers previously developed by the authors; see Refs. [1-3].

The axisymmetric unsteady flow analysis employs the contravariant vorticity/stream function (ω^3, ψ) formulation of the unsteady, incompressible Navier-Stokes equations. The governing equations are discretized using only central difference operators. No artificial viscosity is introduced. The vorticity transport equation is solved using a modification of the Douglas-Gunn [4] version of the alternating-direction implicit (ADI) technique. The elliptic stream function equation is solved using a completely vectorizable efficient direct inversion technique previously developed by the authors for single body configurations but presently modified to permit multiply-connected multibody flow domains. Figure 1 provides a typical (548,101) C-grid distribution for a hull/shrouded propulsor configuration. Only the near-body grid distribution is shown. Through the use of an analytical coordinate transformation, the calculations span the entire infinite flow domain.

Figure 2 provides a typical viscous flow result for a hull/shrouded propulsor configuration. To simulate an active propulsor, the mass flow rate between the hull and shroud has been prescribed in Fig. 2 at a level greater than that which would have resulted were the area between hull and shroud simply an unobstructed duct. The instantaneous stream function, vorticity distribution, velocity field and pressure field are shown for a Reynolds number of 10,000 based on hull chord length at a characteristic time of 4 following an impulsive start from rest. This instant corresponds to a steady-state solution, as can be seen from the drag history given in Fig. 3. A small separation zone is observed to occur at the stern along the centerline of the hull.

Figure 4 presents the passive propulsor case, with the area between the hull and shroud treated as an unobstructed duct. The mass flow rate between hull and shroud is solved implicitly as a part of the direct inversion for the stream function. A small separation zone now occurs at the leading edge of the shroud.

Figure 5 presents a reverse flow solution. Massive reingestion of propulsor-generated vorticity back into the propulsor unit is now observed.

A unified numerical methodology has been developed for unsteady viscous flow simulation over multiple body configuration using efficient direct inversion techniques. This permits the attempt to achieve direct numerical simulation of highly unsteady flows over complex realistic vehicle configurations.

ACKNOWLEDGEMENTS

This research was supported, in part, by SAIC Contract No. 15-890044-65, with supercomputer resources provided by the Ohio Supercomputer Center.

REFERENCES

1. Osswald, G.A., Ghia, K.N. and Ghia, U., "Direct Simulation of Unsteady Two-Dimensional and Axisymmetric Incompressible Navier-Stokes Equations," to appear in Applied Numerical Methods, 1990

2. Osswald, G.A., Ghia, K.N. and Ghia, U., "Direct Method for Solution of Three-Dimensional Unsteady Incompressible Navier-Stokes Equations," in Lecture Notes in Physics, Editors: D.L. Dwoyer, M.Y. Hussaini and R.G. Voight, Springer-Verlag, New York, Vol. 323, 1989, pp. 454-461.

3. Osswald, G.A., and Ghia, K.N. and Ghia, U., "A Direct Algorithm for the Solution of Incompressible Three-Dimensional Unsteady Navier-Stokes Equations," AIAA CP-874, 1987, pp. 408-421.

4. Douglas, J. and Gunn, J.E., Numerishe Mathematik, Vol. 6, 1964, pp. 428-453.

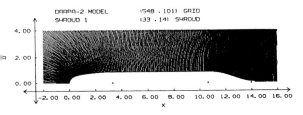

FIG. 1. NEAR BODY GRID DISTRIBUTION FOR SUBMARINE HULL SHROUDED PROPULSOR CONFIGURATION USING (548,101) POINTS: 33 POINTS ALONG SHROUD CHORD.

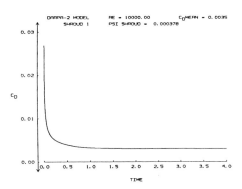

FIG. 3. DRAG HISTORY FROM IMPULSIVE START WITH ACTIVE PROPULSOR.

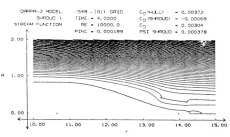

(a) STREAM FUNCTION CONTOURS

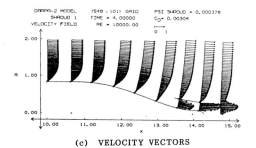

(c) VELOCITY VECTORS

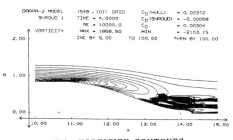

(b) VORTICITY CONTOURS

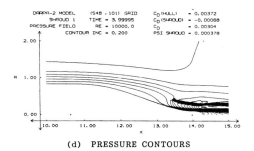

(d) PRESSURE CONTOURS

FIG. 2. STEADY STATE RESULTS FOR Re = 10,000 SHOWING NEAR SHROUD REGION: PROPULSOR ACTIVE.

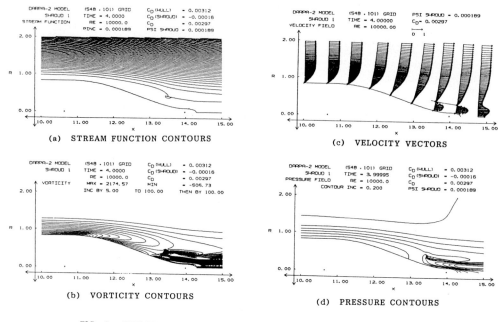

FIG. 4. STEADY STATE RESULTS FOR Re = 10,000 SHOWING NEAR SHROUD REGION: PROPULSOR PASSIVE.

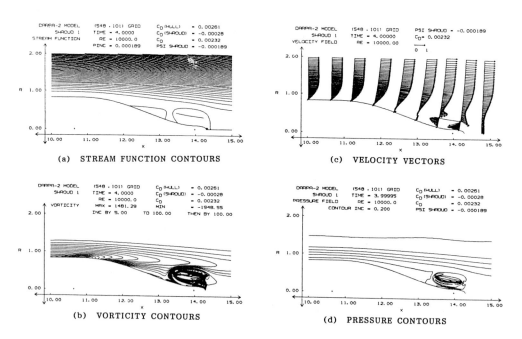

FIG. 5. STEADY STATE RESULTS FOR Re = 10,000 SHOWING NEAR SHROUD REGION. REVERSE FLOW THROUGH PROPULSOR.

A.A.Samarsky, A.P.Favorsky, V.F.Tishkin,

V.F.Vasilevsky, K.V.Vyaznikov

Keldysh Institute of Applied Mathematics,

Moscow, USSR Academy of Sciences

Construction high order monotonous difference schemes on nonregular grids.

We consider method of construction high order monotonous schemes on a nonregular grid consistent of Dirichlet cells. This is actual as soon as using nonregular grid allows easy adaptation for singularities of solution.

For computing hydrodynamic values in a cell on next time step it's nesessary to calculate numerical fluxes for each edge of given cell and then using scheme written in form

$$S_j \frac{dU_j}{dt} + \sum_{i=1}^{M} (H l)_{ij} = 0$$

where S_j - square of Dirichlet cell j, H_{ij} - numerical flux between i and j cell, and l_{ij} - length of cell's edge common for i and j cell. For choosing numerical flux H_{ij} we use such function that choosing $H_{ij} = H(U_i, U_j)$ we obtain first order monotonous scheme. We will consider following fluxes: Godunov flux [2], Lax-Friedrix flux [3], kinetic consistent flux of Chetverushkin and Elizarova [4].

To improve the resolution of this scheme we generalize the method of construction high order monotonous scheme based on the notion of TVD concept, described in papers [5] for Dirichlet cells. We briefly describe this concept.

Firstly we consider nonlinear advection equation

$$\frac{\partial u}{\partial t} + \frac{\partial f(u)}{\partial x} = 0; \quad f''(u) \geq q > 0$$

Let us write first order monotonous scheme in form

$$\frac{\partial u_i}{\partial t} + \frac{h_{i+1/2} - h_{i-1/2}}{\Delta x} = 0$$

To construct second order scheme, based on this scheme at each point of staggered grid we calculate two values $u^L_{i+1/2}$ and $u^R_{i+1/2}$ by the use of special interpolation formula

$$u^L_{i+1/2} = u_i + \frac{u_{i+1} - u_i}{\Delta x} \alpha(R^+_{i+1/2}) \frac{\Delta x}{2}; \quad R^+_{i+1/2} = \frac{u_i - u_{i-1}}{u_{i+1} - u_i}$$

$$u^R_{i+1/2} = u_{i+1} - \frac{u_{i+1} - u_i}{\Delta x} \alpha(R^-_{i+1/2})\frac{\Delta x}{2}; \quad R^-_{i+1/2} = \frac{u_{i+2} - u_{i+1}}{u_{i+1} - u_i}$$

and then use scheme

$$\frac{\partial u_i}{\partial t} + \frac{h(u^L_{i+1/2}, u^R_{i+1/2}) - h(u^L_{i-1/2}, u^R_{i-1/2})}{\Delta x} = 0$$

For linear equation this scheme coincides with scheme from [4]. For one dimensional gas dynamic equations in Euler form

$$\frac{\partial U}{\partial t} + \frac{\partial F(U)}{\partial x} = 0$$

we consider first order monotonous scheme in form

$$\frac{dU_i}{dt} + \frac{H_{i+1/2} - H_{i-1/2}}{\Delta x} = 0; \quad H_{i+1/2} = H(U_i, U_{i+1})$$

and then use the following scheme

$$\frac{dU_i}{dt} + \frac{H(U^L_{i+1/2}, U^R_{i+1/2}) - H(U^L_{i-1/2}, U^R_{i-1/2})}{\Delta x} = 0$$

To calculate values of $U^L_{i+1/2}, U^R_{i+1/2}$ we define

$$A_{i+1/2} = \frac{\partial F(U)}{\partial U}\left(\frac{U_i + U_{i+1}}{2}\right)$$

and $L_{i+1/2}$ — matrix of left eigen vectors of $A_{i+1/2}$.

Then in the grid points $k=i-1, i, i+1, i+2$ we calculate values of so called "acoustic invariants" $S^k_{i+1/2} = L_{i+1/2} U^k$ and obtain interpolation values $S^L_{i+1/2}$ and $S^R_{i+1/2}$ as described above. Finally values $U^L_{i+1/2}$ and $U^R_{i+1/2}$ we obtain using matrix $L^{-1}_{i+1/2}$.

Also when solving equations on proposed schemes it is necessary to take stock on entropy condition, written in form

$$\frac{\partial \rho S}{\partial t} + \frac{\partial \rho u S}{\partial x} \geq 0$$

where S — entropy on unity of mass.

Write approximation of entropy unequality in the form

$$\frac{\partial (\rho S)_i}{\partial t} + \frac{\Phi_{i+1/2} - \Phi_{i-1/2}}{\Delta x} \geq 0$$

where $\Phi_{i+1/2}$ — approximation of entropy flux, which must be valid for each point of grid.

Draw graphically a part of grid and using it explain the way in which we will obtain entropy unequality in numerical form

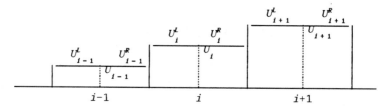

By chosing appropriate entropy flux $\Phi_{i+1/2}$ we can obtain folowing equality

$$\frac{\partial(\rho S)_i^L}{\partial t} + \frac{\Phi_{i+1/2} - \Phi_i}{0.5\Delta x} = \frac{\partial(\rho S)_{i+1}^R}{\partial t} + \frac{\Phi_{i+1} - \Phi_{i+1/2}}{0.5\Delta x}$$

If we require last expression to be positive, we come to Osher's condition

$$\left[\left(\frac{\partial \rho S}{\partial U}\right)_{i+1} - \left(\frac{\partial \rho S}{\partial U}\right)_i\right] H_{i+1/2} - F(U_{i+1})\left(\frac{\partial \rho S}{\partial U}\right)_{i+1} + F(U_i)\left(\frac{\partial \rho S}{\partial U}\right)_i +$$

$$(\rho u S)_{i+1} - (\rho u S)_i = \Delta S_{i+1/2} \geq 0$$

If this condition is not valid we can correct numerical flux $H_{i+1/2}$ by adding a diffusive term $\mu_{i+1/2}(U_{i+1} - U_i)$.

We can obtain more weak condition than this.

Calculate entropy generation for cells i and $i+1$ and claim that following conditions should be valid

$$\Delta S_{i+1/2} + \Delta S_{i-1/2} \geq 0$$

$$\Delta S_{i+3/2} + \Delta S_{i+1/2} \geq 0$$

If this conditions are satisfied we set $\mu_{i+1/2} = 0$. Else we calculate minimal $\mu_{i+1/2}$ for which this unequalities are valid.

To generalise technique of construction high order monotonous scheme for Diriclet cells, in each cell and in each neibour of given cell we need to calculate values of acoustic invariants, using matrix

$$n_x \frac{\partial F}{\partial U} + n_y \frac{\partial G}{\partial U}$$

where n_x and n_y - components of normal vector to the side between two neibour cells. Then we calculate value of $\partial S/\partial l$ for each triangle, created by the nodes of neibour cells. Further we interpolate values of invariants to the center of edge of cell, using following formula.

$$\langle\frac{\partial S}{\partial l}\rangle = \langle\frac{\partial S}{\partial l}\rangle_0 \min_k \alpha\left(\frac{\langle\frac{\partial S}{\partial l}\rangle_k}{\langle\frac{\partial S}{\partial l}\rangle_0}\right)$$

where S stands for Riemann invariants and α - antidiffusive limiter, obtained earlier, l - direction between center of cell and center of edge, $\langle\partial S/\partial l\rangle_0$ obtains from triangle, where center of edge lies. Using this interpolated values we can then obtain hydrodynamic values on edges of cell and calculate there numerical fluxes.

So, using this procedure we can obtain monotonous high order scheme, generalized for nonregular grid.

For numerical example we consider computation of the expansion of the region of high temperature gas. Computational grid was combined of Dirichlet cells of two sizes: cells in the central area of domain are two times smaller than in the other area. Cells are moved with gas with additional equidistribution of cell sizes. Presented picture shows 3-D view of energy distribution.

References

1. Samarsky A.A., Theory of difference schemes.- Moscow, USSR: Nauka, 1977. - 616 p.
2. Godunov S.K., Zabrodin A.V., Ivanov M.Ja., Kraiko A.N. et al. Numerical solution multidimensional problems of gas dinamics., Moscow, USSR, Nauka, 1976 - 200 p.
3. Osher S. and Chakravarthy S.R., High resolution schemes and the entropy condition. SIAM J. Numer. Anal., 21(1984), pp. 955-984.
4. Elizarova T.G., Chetverushkin B.N., Kinetic-consistent difference schemes for modelling of viscous heat-conductive flow., USSR, Moscow, 1988, v. 28(11), p. 695-710.
5. Vyaznikov K.V., Tishkin V.F., Favorsky A.P., Construction high order monotonous difference schemes for the systems of linear hyperbolic differential equations with constant coefficients. - USSR, Moscow, 1989, Mathematical modelling, v. 1(5), p. 95-120.

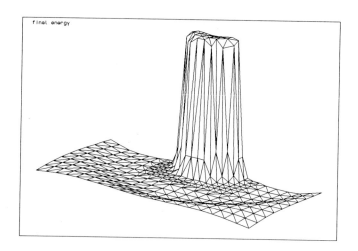

High-Order Time Integration Scheme and Preconditioned Residual Method for Solution of Incompressible Flow in Complex Geometries by the Pseudospectral Element Method

Hwar-Ching Ku, Allan P. Rosenberg and Thomas D. Taylor[*]

Johns Hopkins University Applied Physics Laboratory

Johns Hopkins Road, Laurel, MD 20707

1 Introduction

A fourth-order Runge-Kutta time integration scheme based on the Chorin's time-step splitting technique was developed for solution of incompressible flow by the isoparametric pseudospectral element method. For each intermediate step the divergence free condition on the velocity field is exactly satisfied. The preconditioned conjugate residual method is used to iteratively solve the nonseparable pressure Poisson equation which occurs in curvilinear coordinates, and the inverse of three-dimensional preconditioner can be decomposed to the simplest "algebraic" forms ($O(N^3)$) by the proposed eigenfunction expansion technique. For flow in complex geometries, implementation of the Schwarz alternating procedure (SAP), dividing the computational domain into a number of overlapping subdomains, permits an iterative procedure between subdomains. Numerical experiments are also made to examine the tested problems.

2 Primitive variable formulation

In tensor notation, the time-dependent Navier-Stokes equations in dimensionless form can be described as

$$\frac{\partial u_i}{\partial t} + u_j \frac{\partial u_i}{\partial x_j} = -\frac{\partial p}{\partial x_i} + \frac{1}{\text{Re}} \frac{\partial^2 u_i}{\partial x_j^2} \tag{1a}$$

$$\frac{\partial u_i}{\partial x_i} = 0 \tag{1b}$$

Here u_i is the velocity component and Re is the Reynolds number.

The method applied to solve the Navier-Stokes equations is a fourth-order Runge-Kutta time integration scheme based on the Chorin's [1] splitting technique. According to this scheme, the equations of motion read

$$\frac{\partial u_i}{\partial t} + \frac{\partial p}{\partial x_i} = F_i \tag{2}$$

where $F_i = -u_j \, \partial u_i/\partial x_j + 1/\text{Re} \, \partial^2 u_i/\partial x_j^2$.

At each stage, the first step is to split the velocity into a sum of predicted and corrected values. The predicted velocity is determined by time integration of the momentum equations without the

---

[*]Center for Naval Analyses

pressure term and the second step is to develop the pressure and corrected velocity fields that satisfy the continuity equation.

1st stage:
$$\bar{u}_i^1 = u_i^n + \frac{\Delta t}{2} F_i(u_i^n) \tag{3a}$$
$$u_i^1 = \bar{u}_i^1 - \frac{\Delta t}{2} \frac{\partial p}{\partial x_i} \tag{4a}$$
$$\frac{\partial u_i^1}{\partial x_i} = 0 \tag{3c}$$

2nd stage:
$$\bar{u}_i^2 = u_i^n + \frac{\Delta t}{2} F_i(u_i^1) \tag{4c}$$
$$u_i^2 = \bar{u}_i^2 - \frac{\Delta t}{2} \frac{\partial p}{\partial x_i} \tag{4d}$$
$$\frac{\partial u_i^2}{\partial x_i} = 0 \tag{4e}$$

3rd stage:
$$\bar{u}_i^3 = u_i^n + \Delta t F_i(u_i^2) \tag{5e}$$
$$u_i^3 = \bar{u}_i^3 - \Delta t \frac{\partial p}{\partial x_i} \tag{5f}$$
$$\frac{\partial u_i^3}{\partial x_i} = 0 \tag{5g}$$

4th stage:
$$\bar{u}_i^{n+1} = u_i^n + \Delta t \{ \frac{F_i(u_i^n)}{6} + \frac{F_i(u_i^1)}{3} + \frac{F_i(u_i^2)}{3} + \frac{F_i(u_i^3)}{6} \} \tag{6g}$$
$$u_i^{n+1} = \bar{u}_i^{n+1} - \Delta t \frac{\partial p}{\partial x_i} \tag{6h}$$
$$\frac{\partial u_i^{n+1}}{\partial x_i} = 0 \tag{6i}$$

The main features of this method include: (i)for a given accuracy the time step size is larger than that of the first-order scheme and (ii) the most promising time integration scheme conserves the total energy during the evolution of inviscid flow. The approach is very effective at high Reynolds numbers because the gain in time step size more than offsets the costs of four pressure solvers.

3 Preconditioned method

The pressure Poisson equation appearing in curvilinear (non-orthogonal) coordinates contains a non-separable operator for which there is no easy way to use a direct solver; this is especially more difficult in three-dimensional cases. The iterative scheme used to solve the pressure field is the preconditioned conjugate residual method [2]. A certain separable operator L_{ap} [3] is chosen and constructed from the original operator L_{sp}, which is obtained by taking the divergence operator of the equation for the corrected velocity at each stage.

The iterative procedure by the preconditioned conjugate residual method should read as follows:

Given p^0, compute $r^0 = S - L_{sp}p^0, z^0 = L_{ap}^{-1}r^0, h^0 = z^0$. Then, for $k = 0, 1, 2,...$, until $\| r^k \| < \epsilon$, do

$$p^{k+1} = p^k + \alpha^k h^k \quad (7a)$$

$$r^{k+1} = r^k - \alpha^k L_{sp} h^k \quad (7b)$$

$$z^{k+1} = L_{ap}^{-1} r^{k+1} \quad (7c)$$

$$h^{k+1} = z^{k+1} - \beta^k h^k \quad (7d)$$

where

$$\alpha^k = \frac{(r^k, L_{sp}h^k)}{(L_{sp}h^k, L_{sp}h^k)}, \quad \beta^k = \frac{(L_{sp}z^{k+1}, L_{sp}h^k)}{(L_{sp}h^k, L_{sp}h^k)} \quad (8)$$

Here (,) denotes the inner product.

With an eigenfunction expansion to the preconditioner, the solution of the three-dimensional prconditioner thus can be reduced to a simple algebraic problem (O(N^3 storage required). However, eigenvalues may not be real due to the complexity of a preconditioner. Without putting any restriction on eigenvalues, complex eigenvalues and their associated eigenvectors [4] are permissible if the pressure gradient at the imaginary part vanishes. This is true because only the pressure gradient drives flow instead of the pressure itself.

4 Domain decomposition with SAP

The solution of flow in complex geometry via the domain decomposition approach consists of first dividing the computational domain into a number of blocks (or subdomains) with inter-overlapping areas, where the grids inside the overlapping area are located at the same places. Next implement the SAP for exchanging data among different blocks, i.e., solving the problem on each block separately and then updating the boundary values on the overlapped interfaces.

The SAP iterative solution of the incompressible Navier-Stokes equations in primitive variable form for a two-dimensional flow over a square sketched in Fig. 5 is summarized by the following algorithm:

1. First assume $\mathbf{u}^{n+1}$ on $\overline{AB}$. Usually $\mathbf{u}^n$ will be a good initial guess.

2. Solve (the "O" type) domain I employing the boundary conditions derived from the divergence of velocity field on $\overline{AB}$, where the pressure solution is obtained by the preconditioned method.

3. With the solution of $\mathbf{u}^{n+1}$ on $\overline{AE}, \overline{EF}$ and $\overline{FB}$ from step (2), solve domain II∪III employing the same type boundary conditions on $\overline{AE}, \overline{EF}, \overline{FB}$ to update $\mathbf{u}^{n+1}$ on $\overline{AB}$.

4. Repeat steps (2) & (3) until the velocity $\mathbf{u}^{n+1}$ on $\overline{AB}, \overline{AE}, \overline{EF}, \overline{FB}$ does not change.

In order to guarantee that consistent values of velocity and hence pressure gradient be generated in the overlapping domains III, the divergence of the velocity field, $\nabla \cdot \mathbf{u}$, needs to be actually computed for whichever domain, I or II∪III, is counted. Since $\mathbf{u}$ on domain III is not known a priori, the divergence of velocity is only set to zero at the first SAP iteration for step (2).

5 Results and discusions

Fig. 1 plots streamlines of two-dimensional driven cavity flow calculated by the forth-order Runge-Kutta time integration scheme at Re = 3200. Six elements (6 points per element) are used in each direction and the time step size $\Delta t = 0.01$ which is five times larger than the largest stable time step size in the first-order scheme. The secondary vortex at the upper corner as well as ψ_{min} = -0.1164 agrees well with those found by the most accurate results [5].

When an inviscid density-stratified fluid in a square box with each side length 6 is disturbed by a gaussian density perturbation, $\rho' = -(\rho_0)_y(y - y_c)exp(-0.693(r/r_c)^2)$, where $r_c = 0.5, y_c = 3, (\rho_0)_y = -0.003134$, the division of the total energy into the potential, $PE = -\frac{g}{2}(\rho_0)_y \int \int_A (\rho')^2 dxdy$, and the kinetic energy, $KE = \frac{1}{2} \int \int_A (\rho_0 + \rho')(u^2 + v^2)dxdy$, will vary with time. Fig. 2 sketches the energy versus the Brunt-Väisäi (BV) period, $Nt/2\pi$ (N is the BV frequency). The total energy is conserved with 0.28% error at 8 BV periods, while for the first-order scheme, even with $\frac{1}{32}\Delta t$ of the fourth-order method, the total energy increases with BV periods.

Figs. 3 and 4 explain the three-dimensional configuration of driven cavity flow and flow direction vectors for Re = 1000 (31 * 31 *31 grids with 5 elements in each direction) at upstream plane $x = 0.3$. The preconditioned conjugate residual method is used to iteratively solve the 3-D pressure Poisson equation. The vortex phenomenon is very similar to the standard cubic driven cavity flow, a pair of strong vortices at the lower corners and weak vortices at the upper corners.

Fig. 6 describes the vortex shedding behind a square in a channel at Re = 100. The Strouhal number is found to be 0.144 in this numerical experiment.

Acknowledgement

This work was partially supported by the Office of Naval Research under Contract Number N00039 - 89 - C - 5301.

References

[1] A. J. Chorin, *Math. Comp.* **22** (1968) 745-762.

[2] Y. S. Wong, T. A. Zang and M. Y. Hussaini, *Computers Fluids* **14** (1986) 85-95.

[3] H. C. Ku and T. D. Taylor, *Advances and Applications in Computational Fluid Dynamics* (ASME, Dallas, 1990).

[4] H. C. Ku, R. S. Hirsh and T. D. Taylor, in *Proceeding of the Tenth Reservoir Simulation* (SPE, Houston, 1989) 481-492.

[5] U. Ghia, K. Ghia and C. A. Shin, *J. Comput. Phys.* **48** (1982) 387-411.

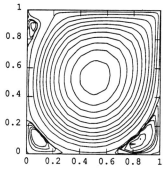

Fig. 1. Streamline pattern for Re = 3200

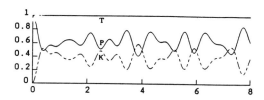

Fig. 2. Energy partition verse BV periods, $Nt/2\pi$. The solid line T indicates the normalized total energy, while the solid line P and dash line K are the potential and kinetic energy, respectively.

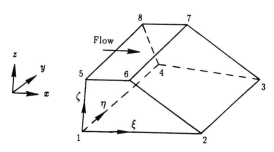

Fig. 3. Three-dimensional cavity flow configuration and coordinate system

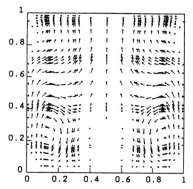

Fig. 4. Flow direction vectors at time $t = 24$ for Re = 1000 in the $x = 0.3$ plane

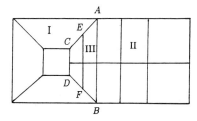

Fig. 5. Configuration of domain decomposition

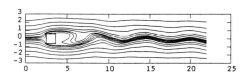

Fig. 6. Vortex shedding behind a square at Re = 100

TURBULENT FLOW CALCULATIONS USING UNSTRUCTURED AND ADAPTIVE MESHES

Dimitri J. Mavriplis
ICASE, NASA Langley Research Center
Hampton, VA, USA

INTRODUCTION

This paper describes a method for computing compressible turbulent viscous flows about arbitrary two-dimensional configurations using fully unstructured meshes and incorporating adaptive meshing techniques. The generation of meshes with highly-stretched triangular elements in the boundary-layer and wake regions is accomplished with a method based on a modified Delaunay triangulation technique [1]. The full Navier-Stokes equations are discretized and solved for on these meshes using an efficient finite-element solver which converges rapidly to steady-state using an unstructured multigrid strategy [2]. Turbulence modeling is achieved using an inexpensive algebraic model which has been devised specifically for use on unstructured and adaptive meshes [3].

MESH GENERATION

The generation of an initial unstructured mesh is accomplished in three essentially independent stages. First, a distribution of mesh points and associated stretching vectors are generated throughout the flow field. This is accomplished by generating a set of structured quadrilateral meshes about each geometry component. The resulting set of points from these multiple overlapping meshes is then employed as the basis for the triangulation procedure. A stretching vector is also constructed at each point, the magnitude of which is taken as the ratio of the local streamwise to normal spacing of the generating structured mesh, and the direction of which coincides with that of the streamwise family of structured mesh lines. These points are then joined together in a manner influenced by the local stretching values to form a set on non-overlapping triangular elements which completely fill the domain. A method for joining an arbitrary set of points together in such a manner is given by the Delaunay triangulation [1]. However, Delaunay triangulations produce the most equiangular (low aspect ratio) triangles possible, for a given set of vertices, and thus are not well suited for the generation of highly stretched meshes. Hence, a modified Delaunay triangulation is employed, whereby at each point, a locally mapped space is constructed, using the value of the local stretching vector, within which the local mesh-point distribution appears more isotropic. The Delaunay triangulation is then constructed in this mapped space, and the resulting triangulation is mapped back to physical space, thus achieving the desired stretching. The construction of a Delaunay triangulation in the locally mapped space may be performed using any standard Delaunay triangulation algorithm which exhibits local properties. Thus, in this work, Bowyer's algorithm is employed [4]. Bowyer's algorithm assumes the existence of an initial coarse triangulation, and a set of additional mesh points to be inserted. Each new mesh point is inserted into the existing triangulation, and the triangles whose circumcircles contain this new point are then tagged for restructuring. The union of these intersected triangles forms a polygonal region which contains the new point. A new triangulation is then constructed by joining the new point to all the vertices of the polygonal region. Since Bowyer's algorithm is formulated as a sequential point insertion and local restructuring process, it may easily be employed within the locally mapped space. Finally, in a post-processing stage, the mesh is smoothed out by slightly repositioning the points according to an elliptic smoothing operator discretized on the mesh.

Once the initial stretched unstructured mesh has been generated and the flow-field has been solved for on this mesh, a new adaptively refined mesh may be constructed by adding new points to the initial mesh in regions where large flow gradients or discretization errors are detected, and locally restructuring the mesh, using Bowyer's algorithm in the stretched space, thus obviating the need for global mesh regeneration. In this work, the refinement criterion is based on the undivided difference of pressure and Mach number. Pressure gradients provide a good indication of inviscid flow phenomena, such as shocks and expansions, while Mach number variations can be used to identify viscous phenomena such as boundary layers and wakes.

FLOW SOLUTION

The full Navier-Stokes equations are discretized using a Galerkin finite-element approach, where the fluxes are taken as piecewise linear functions in the construction of the convective terms, and the flow variables are taken as piecewise linear functions in the construction of the viscous terms. Additional artificial dissipation terms are required to ensure stability and to capture shocks without producing numerical oscillations. Dissipative terms are thus constructed as a blend of a Laplacian and a biharmonic operator in the conserved flow variables [2]. The spatially discretized equations are integrated in time to obtain the steady-state solution using a five-stage time-stepping scheme, where the convective terms are evaluated at each stage within a time step, and the dissipative terms (both physical and artificial) are only evaluated at the first, third, and fifth stages. This particular scheme has been designed to maintain stability in regions where the flow is dominated by viscous effects, and to rapidly dampen out high-frequency error components, which is an essential feature for a scheme intended to drive a multigrid algorithm. Convergence is accelerated by making use of local time-stepping, implicit residual averaging, and an unstructured multigrid algorithm [2], which operates on a sequence of coarse and fine meshes. In the context of unstructured meshes, a sequence of coarse and fine meshes is best constructed by generating the individual meshes independently from one another (as opposed to subdividing a coarse mesh). Thus, in general, the coarse and fine meshes of a given sequence do not have any common mesh points or nested elements. Thus, the patterns for transferring the variables, residuals, and corrections back and forth between the various meshes of the sequence must be determined in a preprocessing operation, where an efficient tree-search algorithm is employed [2]. Since no relation is assumed between the various meshes of the sequence, when adaptive meshing is employed new adaptively generated meshes may simply be added on to the top of the stack of meshes in the sequence, and thus incorporated into the multigrid cycle.

TURBULENCE MODELING

For external aerodynamic flows, algebraic turbulence models have proven to be the most reliable and inexpensive models. However, turbulence length scales must be determined by scanning profiles of flow variables at specified streamwise stations. In the context of unstructured meshes, mesh points do not naturally occur at regular streamwise locations. Thus, lines normal to the walls and viscous layers must be created, and flow variables interpolated onto these lines in order that turbulence length scales may be determined. A smooth distribution of normal mesh lines can be obtained by generating a structured hyperbolic mesh about each geometry component, based on the boundary-point distribution of the global unstructured mesh on each component [3]. These normal mesh lines are terminated if they intersect a neighboring geometry component, thus ensuring that turbulence quantities in any given region of the flow-field are only dependent on the viscous layers and walls which are directly visible from that location (c.f Figure 3). At each time-step in the flow solution phase, flow variables are interpolated onto the background turbulence mesh stations, and the Baldwin-Lomax [5] algebraic model is employed to compute eddy viscosity values along these stations. The eddy viscosities are then interpolated back onto the unstructured mesh for subsequent use in the flow solver. The patterns for interpolating back and forth between the turbulence stations and the unstructured mesh are determined in a preprocessing stage. The turbulence station points are first triangulated, and the same search routines used in the unstructured multigrid algorithm are then employed to determine the interpolation addresses and coefficients. When an adaptive meshing strategy is employed, the background turbulence meshes must be adapted in a manner analogous to the refinement of the global unstructured mesh. Hence, in regions where the unstructured mesh is refined, new points are added to the turbulence mesh stations. When new boundary points are introduced in the unstructured mesh, entire new turbulence stations must also be constructed. After each adaptation process, the transfer patterns for interpolation between the newly refined global unstructured mesh and background meshes must be recomputed. In the context of the multigrid strategy, the turbulence model is only executed on the finest grid of the sequence. The computed eddy viscosity values are then interpolated up to the coarser unstructured meshes where they are employed in the multigrid correction equations. The whole process is very efficient, and in general, the turbulence model is found to require less than 10% of the total time required within a single multigrid cycle.

RESULTS

Flow over a two-element airfoil, at a Mach number of 0.5, a chord Reynolds number of 4.5 million, and an incidence of 7.5 degrees has been computed. At these conditions, the flow becomes supercritical, and a small shock is formed on the upper surface of the slat. The adapted mesh used to compute this flow is depicted in Figure 1, and contains a total of 35,885 points. The minimum normal spacing at the wall is 0.00001 chords, and cells of aspect ratios up to 1000:1 are observed. A total of 7 meshes were employed in the multigrid sequence with the last 3 meshes generated adaptively. The computed Mach contours in the flow field are depicted in Figure 2 where a crisp resolution of the small localized shock provided by the adaptive meshing technique is observed. A good correlation between the computed and experimental surface pressure coefficients is displayed in Figure 4. The solution of this case required 15 minutes on a single processor of a CRAY-YMP, during which the fine grid residuals were reduced by 3 orders of magnitude over 200 multigrid cycles.

The next case consists of a four-element airfoil configuration involving regions with sharp corners. A multigrid sequence of 6 meshes was employed to compute the flow over this configuration with the last 2 meshes generated adaptively. The finest mesh of the sequence, which contains a total of 48,691 points, is depicted in Figure 5. The computed Mach contours for this case are depicted in Figure 6. The freestream

Mach number is 0.1995, the chord Reynolds number is 1.187 million, and the corrected incidence is 16.02 degrees. At these conditions, the flow remains entirely subcritical. Compressibility effects are nevertheless important due to the large suction peaks generated about each airfoil. The computed surface pressure coefficients are compared with experimental wind tunnel data in Figure 7, and good overall agreement, including the prediction of the height of the suction peaks is observed. The convergence of the lift and the density residuals versus the number of multigrid cycles is depicted in Figure 8. A total of 400 multigrid cycles were executed, which required 35 minutes of single processor CRAY-YMP time, and 14 Mwords of memory.

REFERENCES

1. Mavriplis, D. J., "Adaptive Mesh Generation for Viscous Flows Using Delaunay Triangulation" *ICASE Rep. 88-47, NASA CR 181699*, To appear in Journal of Comp. Physics, 1990.

2. Mavriplis, D. J., Jameson, A., and Martinelli L., "Multigrid Solution of the Navier-Stokes Equations on Triangular Meshes", *AIAA paper 89-0120*, January, 1989.

3. Mavriplis, D. J., "Algebraic Turbulence Modeling for Unstructured and Adaptive Meshes", *AIAA Paper 90-1653*, June, 1990.

4. Bowyer, A., "Computing Dirichlet Tessalations", *The Computer Journal*, Vol. 24, No. 2, 1981, pp. 162-166

5. Baldwin, B. S., Lomax, H., "Thin Layer Approximation and Algebraic Model for Separated Turbulent Flows", *AIAA paper 78-275*, 1978

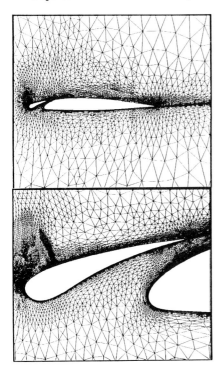

Figure 1
Adaptively Generated Unstructured Mesh about Two-Element Airfoil
Number of Nodes = 35,885

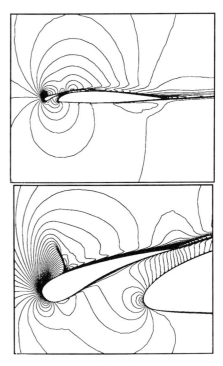

Figure 2
Computed Mach Contours for Flow over a Two-Element Airfoil Configuration
Mach Number = 0.5, Reynolds Number = 4.5 million, Incidence = 7.5 degrees

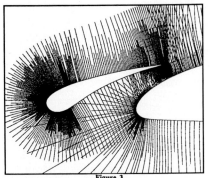

Figure 3
Illustration of Turbulence Mesh Stations Employed in Algebraic Model
Total Number of Points = 43,566

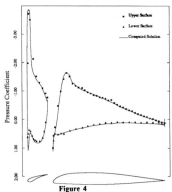

Figure 4
Comparison of Computed Surface Pressure Distribution with Experimental
Wind-Tunnel Data for Flow Over Two-Element Airfoil Configuration

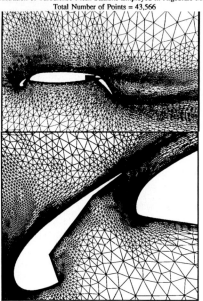

Figure 5
Adaptively Generated Unstructured Mesh about Four-Element Airfoil
Number of Nodes = 48,691

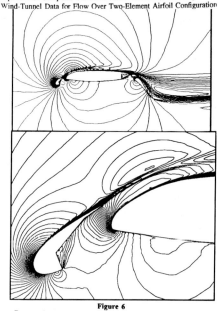

Figure 6
Computed Mach Contours for Flow over Four-Element Airfoil
Mach = 0.1995, Reynolds Number = 1.187 million, Incidence = 16.02 degrees

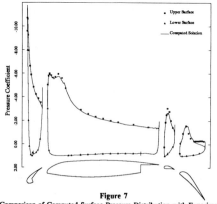

Figure 7
Comparison of Computed Surface Pressure Distribution with Experimental
Wind-Tunnel Data for Flow Over Four-Element Airfoil Configuration
Mach = 0.1995, Reynolds Number = 1.187 million, Incidence = 16.02 degrees

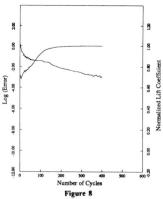

Figure 8
Convergence as Measured by the Computed Lift Coefficient and the Density
Residuals Versus the Number of Multigrid Cycles for Flow Past a Four-Element Airfoil

PARALLEL PRECONDITIONED ITERATIVE METHODS FOR THE COMPRESSIBLE NAVIER-STOKES EQUATIONS

V. Venkatakrishnan
A.S. & M., Inc.
Hampton, VA, USA

J.H. Saltz and D.J. Mavriplis
ICASE, NASA Langley Research Center
Hampton, VA, USA

ABSTRACT. The thin layer Navier-Stokes equations are solved for two-dimensional airfoil problems by preconditioned conjugate gradient-like iterative methods. These methods use applicable to structured and unstructured grids. We characterize the performance of these fully implicit schemes on vector parallel machines. We compare a domain decomposition (DD) and a parallel version of the iterative algorithm. Results are presented for a two-dimensional turbulent flow calculation on a structured mesh and an inviscid transonic case on an unstructured mesh. Performance statistics from the optimized linear algebra kernels are also presented.

INTRODUCTION. Recently, preconditioned conjugate gradient-like methods have been used with considerable success for the solution of linear systems arising from using implicit schemes in the solution of the compressible Navier-Stokes equations on structured and unstructured grids [1,2,3]. Venkatakrishnan [1] found that the Generalized Minimum Residual technique (GMRES) [4] with Incomplete LU (ILU) factorization as the preconditioner performed the best, providing an efficient implicit scheme for two-dimensional airfoil flow computations on structured meshes. Since an algebraic viewpoint is adopted, the scheme is easily applicable to unstructured meshes, thus providing a good implicit algorithm. In this paper, we show examples from structured and unstructured grids. We also analyze the parallel-vector performance of this algorithm. In particular, we analyze the performance of the powerful ILU preconditioners.

DISCRETIZATION. The equations governing fluid flow are

$$\frac{\partial W}{\partial t} + \frac{\partial f}{\partial x} + \frac{\partial g}{\partial y} = \frac{\partial R}{\partial x} + \frac{\partial S}{\partial y} \qquad (1)$$

where w, f, g, R and S are respectively, the vector unknown, convective flux vectors and viscous flux vectors in the x and y directions. The governing equations are solved in a integral form. The unknowns are stored at the centers of the cells for the structured grid code. The numerical fluxes for inviscid terms are computed using the approximate Riemann solver of Roe[5]. The construction of the numerical fluxes is done in two stages: first, a high order interpolation is used to determine the state variables on either side of a cell face and second, these are interpreted as initial data for an approximate Riemann solver to obtain the numerical fluxes. The viscous terms are discretized using central differences. The algebraic turbulence model of Baldwin and Lomax is also incorporated. In the unstructured grid code, the variables are stored at the cell vertices. A central difference approximation augmented by a blend of a Laplacian and a biharmonic dissipative terms is used [6].

IMPLICIT SCHEME. Application of Euler implicit integration and linearization in time leads to a system of linear equations

$$(\frac{I}{\delta t} + \frac{\partial R}{\partial W}) \delta W_{i,j} = -R_{i,j} \qquad (2)$$

$$\delta W_{i,j} = (W^{n+1} - W^n)_{i,j}$$

Here R represents the residual and vanishes at steady state. Eqn. (2) represents a large linear system of equations for the vector of unknowns and needs to be solved at each time step. As δt tends to infinity, the method reduces to the standard Newton's method. The term $\frac{\partial R}{\partial W}$ symbolically represents the implicit side upon linearization and involves the Jacobian matrices of the flux vectors. In the structured grid case we employ second order spatial discretizations on both sides of Eqn. (2). For the unstructured grid, however, we use the higher order discretization on the explicit side, but only a first accurate discretization on the implicit side due to storage considerations. The system of linear equations is solved by the GMRES technique developed by Saad and Schultz [4]. GMRES solves efficiently a minimization problem for the residual of the linear system over the Krylov subspace of the associated matrix. It can also be thought of as an optimal polynomial acceleration scheme. Preconditioning greatly improves the performance of GMRES. It decreases the size of the spectrum of eigenvalues so that the optimal polynomial generated by GMRES can annihilate the errors associated with each eigenvalue. A family of preconditioners arises out of incomplete LU factorization and is denoted by ILU(n), where n denotes the level of fill-in. ILU(0) (which allows no fill-in) has been found to be an effective preconditioner and has been used in the present work. The matrix produced by the ILU factorization depend on the ordering of the unknowns and affects the convergence of the method. The unknowns are ordered lexicographically in the direction normal to the airfoil. A streamwise ordering has been observed to lead to a deterioration in convergence. For the unstructured grid computations, an x-y ordering, wherein the nodes are sorted by their x and y coordinates and a reverse Cuthill-Mckee ordering have been implemented, the latter yielding the best convergence rates. A block diagonal preconditioner has also been tested, but does not work as well as the ILU preconditioner, especially for stiff problems.

Upper and lower sparse triangular systems need to be solved repeatedly when using the ILU preconditioner. By permuting the equations using wavefront methods [see 7,8] the unknowns belonging to each wavefront can be eliminated simultaneously. It is thus possible to get good vector performance out of the triangular solves and if the vectors are long enough, the work within each wavefront can be distributed across multiple processors. This ordering is derived by interpreting the triangular matrix as a directed graph and analyzing the dependencies. The off-diagonal rectangular matrix is stored in a compact form (similar to ITPACK) which enables an efficient implementation of a sparse matrix-vector multiplication. Even in the case of multiple unknowns to a grid point, the dependency graph is determined only by the grid and the stencil used. Since the operations are done on blocks (of size 4x4 in the present work) the performance of the solver is enhanced. Overall, the preconditioned GMRES algorithm performs very well in comparison with the existing schemes, while maintaining the robustness of a Newton's method [1]. The structured code runs at 100 mflops. for a 320x64 mesh and the unstructured grid code runs at 85 mflops. on the Cray-YMP for mesh with 4480 nodes. The codes run only at about 20-30 mflops. with the unvectorized triangular solves.

PARALLEL IMPLEMENTATION.

The algorithm outlined above lends itself easily to domain decomposition techniques. We adapt ideas from the work of Keyes [9]. The preconditioning phase is carried out independently for the sub-domains. In these two-dimensional airfoil problems, the sub-domains consist either of strips (C-shaped rings around the airfoil) or slabs in the streamwise direction. The incomplete LU factorization is carried out independently for each of the sub-domains, the boundary information being taken into account during the GMRES procedure. The factorization is thus carried out for each domain by zeroing out the entries which lie outside the domain, i.e. with zero Dirichlet boundary conditions. This is a good approximation since we are solving for δW and at steady state $\delta W \equiv 0$ everywhere. Still, extra operations are incurred due to the lack of coupling in the preconditioning phase. The triangular solves are also done independently for each of the domains, vector performance being achieved by using the wavefront ordering.

Two ways to achieve parallelism have been tested. In both instances the parallel solution of the linear system at each time step is the crucial issue. The rest of the code, comprising of residual calculation, forming the matrix, computing the maximum allowable time step etc. is completely parallelizable. As indicated above, one way to achieve parallelism is by partitioning the domain into sub-domains. Another way to achieve parallelism in triangular solves is to distribute the work in a wavefront across multiple processors. With domain decomposition, it is possible to achieve parallelism by concurrently performing all the work associated with the sub-domains. The performance of the triangular solves is affected by the size and shape of the sub-domains used. For the stencil used in the structured grid case, a problem solved on a mxn ($m \geq n$) domain will require n+m−1 wavefronts. For example, in the case of a 320x64 grid with 8 sub-domains, using a strip-wise decomposition leads to an average vector length of 7.83 for each sub-problem, while a slab-wise decomposition leads to a vector length of 24.85. If we partition a single triangular solve across 8 processors we obtain an average vector length of 6.68.

RESULTS In this section we present results for transonic inviscid and turbulent flows. The first flow considered is AGARD 09 test case of turbulent flow over an RAE2822 airfoil ($M_\infty = 0.73$, $\alpha = 1.25°$, $Re = 6.5 \times 10^6$) computed on a 320x64 grid. Fig. 1a shows the pressure profile and Fig. 1b, the skin friction distribution. The next case considered is the inviscid transonic flow over an NACA0012 airfoil ($M_\infty = 0.8$, $\alpha = 1.25°$). Fig. 2a and 2b. show the surface pressure profile and the convergence history for the computation done on an unstructured triangular mesh with 4480 nodes. The cpu time taken on the Cray-YMP is 85 seconds. Fig. 3 shows the convergence histories on a single CPU for AGARD 09 test case as the number of sub-domains is varied. With slab-wise DD we incur an overhead of 20% for up to 8 sub-domains while with strip-wise DD, overhead is much higher. Even when viewed in terms of iterations, slab-wise DD performs better. With the parallelized version of the GMRES-ILU algorithm, we obtain much higher overheads due to shorter vectors and synchronization. In Table 1 we show the parallel performance of the DD code on the Cray-YMP for the transonic turbulent flow case in a dedicated environment. Including the extra operations with DD, we get with 8 processors a speed-up of 4.37 (the 8 CPU run has been optimized better than the other two cases). In Table 2 we present the performance of the parallel triangular solve for the problem sizes of interest. The overheads incurred are quite high.

CONCLUSIONS A robust implicit scheme suitable for structured and unstructured meshes has been developed. The scheme has good convergence properties and has inherent parallelism, which can be exploited for implementation on parallel computers. Another means to achieve coarse-grained parallelism is via domain decomposition, which has been found to be more efficient for vector parallel machines.

REFERENCES

1. Venkatakrishnan, V., AIAA Paper 90-0586, Reno, NV, Jan. 1990.
2. V.Venkatakrishnan and Saltz, J.H., Paper presented at the 2nd SIAM Conf. on Parallel Processing, Chicago, Ill, Dec. 1989.
3. Whitaker, D.L., Slack, D. and Walters, R., AIAA Paper 90-0697, Reno, NV, Jan. 1990.
4. Saad, Y. and Schultz, M., SIAM. J. Sci. Stat. Comp., Vol. 7, No. 3, 1986.
5. Roe, P.L., Ann. Rev. Fluid Mech., 1986, 18, 337-365.
6. Mavriplis, D.J., AIAA J., Vol. 26, No. 7, July, 1988, pp. 824-831
7. Saltz, J.H., SIAM J. Sci. Stat. Comp., Jan. 1989.
8. Anderson, E. and Saad, Y., CSRD Report no. 794, Univ. of Ill., Urbana, Ill, 1988
9. Keyes, D.E., computer Physics Communications, Vol. 53, 1989.

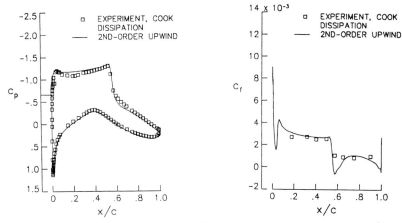

Fig. 1. Turbulent transonic flow - Case 9 (a) Surface pressure profile (b) Skin friction distribution

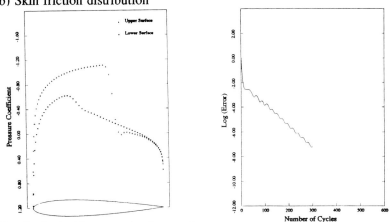

Fig. 2. Inviscid transonic flow - Unstructured grid (a) Surface pressure profile (b) Skin friction distribution

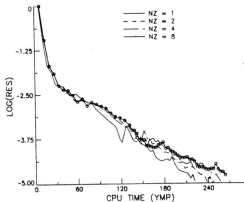

Fig. 3. Convergence histories for Case 9 with DD

NCPUS/NZ	ST/PT	η
2	1.77	1.61
4	2.94	2.60
8	5.43	4.37

Table 1. Parallel performance of the DD code; ST- Sequential time; PT - Parallel time; η = Speedup

SIZE	NCPUS	MFLOPS
96x96	4	223
192x192	4	340
96x96	1	124
192x192	1	130

Table 2. Parallel performance of triangular solves on Cray-YMP

A MULTIDIMENSIONAL CELL-CENTERED UPWIND ALGORITHM BASED ON A DIAGONALIZATION OF THE EULER EQUATIONS

Peter Van Ransbeeck, Chris Lacor and Charles Hirsch
Vrije Universiteit Brussel, Department of Fluid Mechanics
Pleinlaan 2, 1050 Brussels

Summary

A new method for solving the multidimensional Euler equations is developed. Characteristic flow directions are determined by diagonalizing the Euler system. The resulting non-linear scalar convection equations are discretized using a conservative cell-centered upwind scheme, with the numerical fluxes based on the characteristic flow properties. The scheme is combined with a MUSCL extrapolation along the characteristic propagation directions. First and second order accurate schemes are developed and tested. The first order schemes show an improved accuracy in regions of strong gradients compared to classical grid-aligned upwind schemes. Preliminary second order results are promising but do not seem to give a significant improvement over second order classical schemes.

1. Introduction

In contrast with the one-dimensional case, information in two or three dimensional Euler flows is propagated in infinitely many directions, each one corresponding to an arbitrary wave front normal. The transformation that diagonalizes the multidimensional Euler equations is derived from characteristic theory. It is shown that a complete decoupling can be obtained by an appropriate choice of two wave front propagation directions, related to the local pressure gradient and to the strain rate tensor [1]. A three-dimensional extension of this diagonalization procedure is straightforward. Further, this approach can be extended to the Navier-Stokes equations by expressing the viscous terms in the local propagation direction. Several numerical schemes based on this approach can be defined. A conservative cell-centered and cell-vertex algorithm is developed in respectively [2] and [3]. In this paper the characteristic system of non-linear scalar convection equations is discretized with a cell-centered method, using a first order upwind scheme for the convective part and a central scheme for the eventual coupling terms [2]. The conservative variables are re-introduced using the transformation between the conservative and characteristic variables. Conservation is ensured if the corresponding eigenvectors and eigenvalues of the diagonalization procedure are evaluated at the cell-faces. The resulting numerical flux is very similar to that of Roe's flux-difference splitting scheme but uses the characteristic directions instead of the mesh directions. In order to fully exploit the advantages of the present approach, the scheme is combined with a MUSCL extrapolation into the characteristic convection directions. The variations of the characteristic variables (in the numerical flux) at a cell face are evaluated using upwind and downwind extrapolation along the characteristic propagation direction associated with each characteristic component. The characteristic MUSCL extrapolation leads to a family of first order and higher order schemes.

2. Decomposition of the Euler Equations

The 2D Euler equations can be written in differential form,

$$\frac{\partial U}{\partial t} + \frac{\partial F}{\partial x} + \frac{\partial G}{\partial y} = \frac{\partial U}{\partial t} + A\frac{\partial U}{\partial x} + B\frac{\partial U}{\partial y} = 0 \qquad (1)$$

where U is the state vector of the conservative variables, F and G represent the flux vectors and A,B are the corresponding Jacobians. Any linear combination of the Jacobians $\vec{A}\cdot\vec{\kappa}$ (with $\vec{A} = (A,B)$ and $\vec{\kappa}$ a wave front normal) can be diagonalized, using a similarity transformation matrix P. Defining a set of characteristic variables

$$\partial W = P^{-1} \partial U \qquad (2)$$

the Euler equations (1) are transformed into the following form

$$\frac{\partial W}{\partial t} + \Lambda_x \frac{\partial W}{\partial x} + \Lambda_y \frac{\partial W}{\partial y} = -C_x \frac{\partial W}{\partial x} - C_y \frac{\partial W}{\partial y} \tag{3}$$

where Λ_x, Λ_y and C_x, C_y represent respectively the diagonal matrices and coupling matrices. Since the Jacobian matrices do not commute, the Euler equations are not diagonalized. Hirsch et al. [1] show that the decomposition is complete (or the right-hand-side of (3) vanishes) if two characteristic directions $\vec{\kappa}$ are chosen such that

$$\vec{\kappa}_1 \times \vec{\nabla} p = 0 \tag{4}$$

$$\vec{\kappa}_2 \cdot (\vec{\kappa}_2 \cdot \vec{\nabla}) \vec{v} - \vec{\nabla} \cdot \vec{v} = 0 \tag{5}$$

The first propagation direction satisfying eq. (4), aligned locally with the pressure gradient, can always be determined, except in uniform flow regions where the pressure gradient direction is not well defined. In this case the first characteristic direction is taken along the local velocity vector. The equation of the second characteristic direction (5) does not always have a solution. In the latter case, no complete decomposition is possible and the second characteristic direction is chosen to minimize the left-hand-side of (5) or the coupling terms of (3).

3. A Conservative Cell-Centered Algorithm

The decomposed system of 4 scalar convection equations (3), with eventual coupling terms, is discretized in a conservative way with a cell-centered method, using a first order upwind scheme for the convective part and a central scheme for the eventual coupling terms. Because the equations are non-linear, Roe averages are introduced in the transformation matrix P and the corresponding eigenvalues and eigenvectors are evaluated at the cell-face centers. The resulting numerical fluxes are based on the characteristic directions and characteristic variables, e.g. for a cartesian mesh

$$F^*_{i+1/2, j} = \frac{1}{2}(F_{i,j} + F_{i+1,j}) - \frac{1}{2} P_{i+1/2, j} |\Lambda_x|_{i+1/2, j} (W_{i+1,j} - W_{i,j}) \tag{6}$$

The resulting numerical flux (6) is similar to that of Roe's flux difference splitting

$$F^*_{i+1/2, j} = \frac{1}{2}(F_{i,j} + F_{i+1,j}) - \frac{1}{2} |A|_{i+1/2, j} (U_{i+1,j} - U_{i,j}) \tag{7}$$

The eigenvalues, eigenvectors and characteristic variables in the present scheme depend on the characteristic directions, instead of the geometry. Note also that e.g.

$$P|\Lambda_x| \Delta W \neq |A| \Delta U \tag{8}$$

since P does not diagonalize A. The conservative scheme described above is only first-order accurate. No improvement in accuracy as compared to the classical schemes is found [2]. The present scheme can easily be extended to higher-order accuracy using a classical MUSCL extrapolation in the expressions for the numerical fluxes (6). The resulting scheme has an improved resolution but the accuracy is comparable to that obtained with classical higher order methods. Since within the present approach intrinsic directions of information propagation are known, the classical MUSCL extrapolation may be replaced by a MUSCL extrapolation into the characteristic directions. The numerical flux (6) is rewritten for a general mesh as

$$(\overline{\overline{F}} \cdot \vec{n})_{i+1/2, j} = \frac{1}{2} (\overline{\overline{F}}^+_{i+1/2, j} + \overline{\overline{F}}^-_{i+1/2, j}) \cdot \vec{n}_{i+1/2, j} - \frac{1}{2} P_{i+1/2, j} |\overline{\overline{\Lambda}} \cdot \vec{n}|_{i+1/2, j} (W^-_{i+1/2, j} - W^+_{i+1/2, j}) \tag{9}$$

where the superscripts + and - indicate upwind and downwind extrapolations with respect to the direction of the normal $\vec{n}$, and $\overline{\overline{F}} = (F, G)$ and $\overline{\overline{\Lambda}} = (\Lambda_x, \Lambda_y)$. The variations of the characteristic variables in equation (9) are obtained using the characteristic MUSCL extrapolation. The extrapolation of the component w^k of W is into the corresponding characteristic propagation direction $\vec{a}_k$,

$$\begin{aligned} \vec{a}_1 &= \vec{a}_2 = \vec{v} \\ \vec{a}_3 &= \vec{v} + c \vec{\kappa}_2 \\ \vec{a}_4 &= \vec{v} - c \vec{\kappa}_2 \end{aligned} \tag{10}$$

with c the velocity of sound.

4. First order Characteristic MUSCL Schemes

If one restricts the characteristic MUSCL extrapolation to the first cell-centers nearest to the cell face, a first-order accurate scheme results. Referring to figure 1,

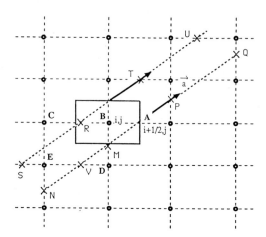

Figure 1 : MUSCL extrapolation along a characteristic propagation direction.

one obtains for the upwind and downwind values at cell face i+1/2,j

$$w^+_{i+1/2, j} = w_M$$
$$w^-_{i+1/2, j} = w_P \tag{11}$$

The values w_M and w_P are obtained by linear extrapolation between respectively nodes (i,j) ; (i,j-1) and (i+1,j) ; (i+1, j+1). This scheme is not monotone. A simplified version results if - instead of interpolating between the nodes i,j and i,j-1 - one assigns the value w_B to w_M if M lies closer to node i,j than to node i,j-1, and w_D if not. The same procedure is applied for w_P and the intersections resulting from the remaining cell sides. This clipping procedure results in a monotone scheme. Both schemes are less diffusive than the classical first-order upwind scheme and give the exact solution for a flow angle of 45°.

5. Second order Characteristic Schemes

The proper way to introduce a higher order characteristic MUSCL scheme is to involve two columns of nodes, upstream and downstream of the cell face, into the extrapolation. Referring to figure 1, the nodes M , N are used to determine the upwind value and P , Q for the downwind value at cell face i+1/2,j. One finds

$$w^+_{i+1/2,j} = w_M + \left|\frac{AM}{MN}\right| (w_M - w_N) \tag{12}$$

$$w^-_{i+1/2,j} = w_P + \left|\frac{AP}{PQ}\right| (w_P - w_Q) \tag{13}$$

where the values at M, N, P and Q are interpolated between the nearest cell-center values. This second order scheme using characteristic MUSCL extrapolation along the convection direction is expensive because of the large number of interpolations and nodes. Therefore a second order algorithm which is more compact is sought. In order to design an alternative scheme, a theoretical study of molecules for linear convection equations is conducted. A general 25 point stencil is rewritten in a Taylor expansion. Provided that this molecule is a valid discretization of the 2D linear convection equation, some consistency conditions are derived. The constraint of 2nd order accuracy limits the number of free parameters. The molecule is split in a symmetric and anti-symmetric part. The advantage of this transformation is that it becomes very easy to make some conclusions concerning conditions for monotonicity, accuracy, minimal cross diffusion, etc. The general 25 point molecule is studied by introducing several classes such as the nine-point stencils and fully-upwind schemes.

From this analysis a less expensive 2nd order accurate characteristic scheme is designed. In contrast with the characteristic MUSCL scheme, where the extrapolation depends on the convection direction $\vec{a}$ (=(a,b)), the present extrapolation coefficients are solely dependent on the geometry and not on the characteristic convection direction. This alternative scheme is based on a weighted average of 2 characteristic independent extrapolations, where the characteristic dependency is introduced through the weight coefficients. The 'flow angle' β (=b/a) is used as parameter in the weighting factors. Referring to figure 1 the upwind value at cell face i+1/2,j

$$w^+_{i+1/2,j} = \frac{1}{1+\beta} (w^+_{i+1/2,j})^{CLASS.\,MUSCL} + \frac{\beta}{1+\beta} (w^+_{i+1/2,j})^{\Delta\,EXTR.} \quad (14)$$

$$\text{with} \quad (w^+_{i+1/2,j})^{CLASS.\,MUSCL} = w_B + \frac{1}{2}(w_B - w_C) \quad (15)$$

$$(w^+_{i+1/2,j})^{\Delta\,EXTR.} = w_B + \frac{1}{2}(w_D - w_E) \quad (16)$$

is a weighted average (14) of a second order fully-upwind MUSCL extrapolation (15) along the i-gridline and an extrapolation from the three points of triangle BDE to the cell-face center A (16). Independent of the weight factors this scheme is a second order accurate fully-upwind 8-point stencil. The weights are chosen such that for zero flow angle ($\beta = 0$) the extrapolation reduces to classical MUSCL extrapolation, whereas when the flow angle goes to infinity, only the triangle extrapolation is accounted for.

6. Results

In ref. [2] first order accurate monotone and non-monotone results were given for several testcases. Fig. 2 shows a first order monotone result for a hypersonic ramp flow in comparison with first and second order classical results. The characteristic result shows a better resolution near the shock than the classical 1st order solution, and is comparable with the second order result.

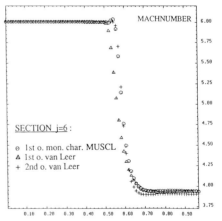

Figure 2 : First order mon. characteristic MUSCL result : Machnumber distribution for a hypersonic ramp flow (M = 6, 15 deg, mesh 81x41) with a section along nodeline j = 6.

The mesh dependency of the first order monotone characteristic MUSCL scheme has been tested for an oblique shock reflection by using three different grids, defined by a different orientation of the straight i=constant lines relative to the y-axis (0, 45 and -45 deg.). Fig. 3 shows the results. The characteristic MUSCL scheme shows less dependency than the classical scheme, though it is not completely insensitive.

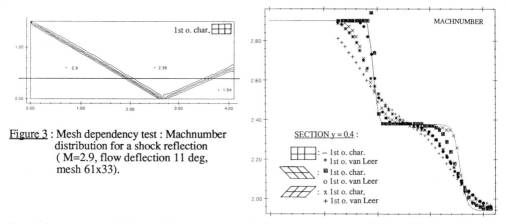

Figure 3 : Mesh dependency test : Machnumber distribution for a shock reflection (M=2.9, flow deflection 11 deg, mesh 61x33).

Fig. 4 shows some preliminary 2nd order characteristic results - using eq.'s (14) to (16) - for a channel with 5 deg. wedges. On the compression side of the shock-expansion interaction the characteristic scheme gives a crisper resolution than the 2nd order classical scheme. Overshoots are noticeable because no limiters are used. This 2nd order comparison does not seem to give an additional improvement of the same order as for the first order characteristic results.

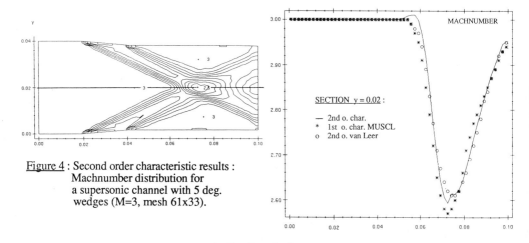

Figure 4 : Second order characteristic results : Machnumber distribution for a supersonic channel with 5 deg. wedges (M=3, mesh 61x33).

7. Conclusions

First and second order accurate cell-centered characteristic schemes, based on a diagonalization of the multidimensional Euler equations, are developed and tested. First order characteristic MUSCL results give a significantly better resolution in regions of strong gradients than the first order classical results, and are comparable to second order classical results. Preliminary second order characteristic results are promising but do not seem to give a significant improvement over second order classical results.

References

[1] Hirsch Ch., Lacor C., Deconinck H., 'Convection algorithms based on a diagonalization procedure for the multidimensional Euler equations', AIAA Paper 87-1163-CP, 1987.
[2] Hirsch Ch., Lacor C., 'Upwind algorithms based on a diagonalization of the multidimensional Euler equations', AIAA Paper 89-1958, 1989.
[3] Powell K.G., van Leer B., 'A Genuinely Multi-Dimensional Upwind Cell-Vertex Scheme for the Euler Equations', AIAA Paper 89-0095, 1989.

Cell Vertex Multigrid Methods for the Compressible Navier–Stokes Equations

P.I. Crumpton, J.A. Mackenzie, K.W. Morton, M.A. Rudgyard, G.J. Shaw

ICFD, Oxford University Computing Laboratory,
11 Keble Road, Oxford OX1 3QD, ENGLAND.

1 Introduction

In the last few years, many authors (several of whom have presented papers at this conference) have adopted a cell vertex form of finite volume method for both inviscid and viscous steady flows: but their actual schemes vary greatly. Some combine the cell vertex formulation for the inviscid fluxes with a cell centre scheme for the viscous fluxes (see Dimitriadis and Leschziner, present volume); some include their particular choice of artificial damping terms in their update procedure as an integral part of the formulation (see O'Brien and Hall, present volume); and some (see Kroll and Rossow, present volume) abandon the cell vertex ideas at such an early stage that their schemes are closer to the vertex-centred scheme of Dick [1]. This variation in approach should not be surprising because the cell vertex formulation suffers from a fundamental difficulty and many ways are sought to overcome it. The basic and distinctive difficulty is that the finite volume equations are associated with the cells while the unknowns are associated with the vertices: so there is a counting problem that cannot always be removed by a proper choice of boundary conditions.

In this paper we set up a general framework and notation that distinguishes between the cell equations that we would like to satisfy and those satisfied when the vertex values have converged. This opens up many possibilities in the choice both of cell equations and of iterative procedures. We present results in two dimensions for a particular combination which looks very promising. The cell equations use simple edge fluxes in which the inviscid and viscous terms are treated consistently; and the iteration uses a cell-based multigrid acceleration.

2 Definition of Cell Vertex Residual $\mathbf{R}_c$

Consider the system of steady conservation laws

$$\mathbf{F}_x + \mathbf{G}_y = \mathbf{0}, \qquad (x,y) \in \Omega, \tag{1}$$

where, for the Euler equations $\mathbf{F} = \mathbf{F}(\mathbf{q})$, $\mathbf{G} = \mathbf{G}(\mathbf{q})$ are vector flux functions of the state vector $\mathbf{q}$. Let Ω be divided into convex quadrilateral cells. Integration over a typical cell c of area V_c and application of Gauss' theorem gives

$$\frac{1}{V_c}\oint_{\delta c} \mathbf{F}dy - \mathbf{G}dx = \mathbf{0}. \tag{2}$$

If the approximate state vectors are stored at the vertices of the cell c, the integral along each side of the quadrilateral may then be calculated by the trapezoidal rule. This yields the residual $\mathbf{R}_c$ as an approximation to the left side of (2). $\mathbf{R_c}$ is a compact four point cell residual, which is a function of the approximate state vectors at the four vertices of c.

In the case of the Navier–Stokes equations the flux functions contain derivatives of the state vector $\mathbf{q}$. We approximate these derivatives at the vertices of c and include the relevant stress terms in the flux functions. In this way the approximation to both viscous and inviscid terms is consistently based at cells, particularly on the cell edges. There are many ways to form the approximate nodal derivatives, some of which are introduced by Mackenzie and Morton [3]. Here we describe only the simplest. Consider some differentiable function $\phi(x,y)$ and write the x–derivative as $\phi_x = \text{div}(\phi, 0)$. Application of the divergence theorem over a quadrilateral formed by the cell diagonals surrounding a vertex then yields as in (2) an approximation to ϕ_x at the vertex. Applying this to all derivatives in the viscous flux functions yields a twelve point Navier-Stokes residual $\mathbf{R}_c$.

3 Solution of the Cell Residual Equations

We would like a procedure which would set to zero all the residuals for cells that are not crossed by shocks, for two reasons. Firstly, both the truncation error arguments which motivate the choice of the cell vertex scheme [5] and the rigorous error analysis that can be carried out for simple model problems [3], [8] are based on such an equation system. Secondly, our numerical experience of the cell vertex scheme shows that the best results have always coincided with having set the l_2 norm of the residuals very close to zero. To achieve this end in the presence of severe changes in flow direction, sonic lines etc., it may be necessary to split the cells and either combine the corresponding parts of the residuals with those of neighbouring cells or to set the individual parts to zero. This is discussed in [4] but will not be considered further in this paper: instead, like most authors, we shall assume that the residuals will have to be combined in some way in order to solve for the unknowns.

Furthermore, setting the cell residuals to zero still leaves present the accuracy-destroying chequerboard mode. In [6] various modifications to the definition of the residual are discussed which in some cases obviate this problem. But at the present time there is no simple remedy of this type that we are prepared to advocate: so in this paper we consider only the simple definition described above.

To address the issue of the unknowns being associated with the vertices and residuals with the cells we distinguish between the space $\overset{\square}{\mathcal{G}} = \overset{\square}{\mathcal{G}}(\Omega)$ of cell based quantities, such as the cell residual $\mathbf{R}_c$, and the space of nodal variables $\overset{\bullet}{\mathcal{G}} = \overset{\bullet}{\mathcal{G}}(\Omega)$, such as the unknowns $\mathbf{q}$. The calculation of cell residuals defines a map between these spaces. The complete non–linear system of cell residuals and boundary conditions may be denoted

$$\mathbf{R}(\mathbf{q}) = \mathbf{f}, \quad \mathbf{R} : \overset{\bullet}{\mathcal{G}} \to \overset{\square}{\mathcal{G}}. \qquad (3)$$

If we are to solve this system in an iterative fashion, a mapping from $\overset{\square}{\mathcal{G}}$ to $\overset{\bullet}{\mathcal{G}}$ is required. This involves the combination of adjacent cell residuals to form a nodal equation. It is usually achieved via a global distribution matrix $\mathbf{D} = \mathbf{D}(\mathbf{W})$, $\mathbf{D} : \overset{\square}{\mathcal{G}} \to \overset{\bullet}{\mathcal{G}}$ which leads naturally to the relaxation method

$$\mathbf{W} := \mathbf{W} + \omega \mathbf{J}^{-1} \mathbf{D} \left[\mathbf{f} - \mathbf{R} \right], \qquad (4)$$

where ω is a relaxation parameter and $\mathbf{J} = \mathbf{J}(\mathbf{W})$, $\mathbf{J} : \overset{\bullet}{\mathcal{G}} \to \overset{\bullet}{\mathcal{G}}$ is some approximation to the Jacobian.

An example of a method which fits into this general framework is the Taylor–Galerkin pseudo-time stepping scheme. As formulated in [5], [6] it gives for node 1

$$q_1^{n+1} = q_1^n - \Delta t_1(\mathbf{D}_a\mathbf{R}_a + \mathbf{D}_b\mathbf{R}_b + \mathbf{D}_c\mathbf{R}_c + \mathbf{D}_d\mathbf{R}_d), \tag{5}$$

where the local distribution matrices at the surrounding cells a,b,c,d are given by

$$\mathbf{D}_c = \frac{1}{V_a + V_b + V_c + V_d}[V_c\mathbf{I} - \Delta t_c[(y_4 - y_2)\mathbf{A}_c - (x_4 - x_2)\mathbf{B}_c]]; \tag{6}$$

2,3,4 are the other nodes of cell c, and $\mathbf{D}_a$, $\mathbf{D}_b$ and $\mathbf{D}_d$ are similarly defined. The complete algorithm may be regarded as a simple Richardson iterative procedure applied to the modified problem

$$(\mathbf{F}_x + \mathbf{G}_y) - \frac{\Delta t_c}{2}[(\mathbf{A}(\mathbf{F}_x + \mathbf{G}_y))_x + (\mathbf{B}(\mathbf{F}_x + \mathbf{G}_y))_y] = 0 \tag{7}$$

discretised on a nine–point stencil. Whether solutions of (7) are also solutions of the original equation (1) depends on the boundary conditions: if so the distribution matrices provide a mapping from a system of cell–based equations to node–based equations for which there is a one to one correspondence between equations and unknowns.

The term Δt_1 appearing in equation (5) is effectively a relaxation parameter for the Richardson iteration, whilst Δt_c in (6) alters the ratio of the terms in the modified equation. An energy analysis for a scalar one–dimensional problem shows that the error norm is reduced by the iteration provided $\nu_1 \geq \nu_c$ and $\nu_1\nu_c < 1$ where ν_1, ν_c are CFL numbers corresponding to the two time steps. Furthermore, the damping of high frequency error modes is improved by taking larger values of ν_c. This also improves the shock–capturing ability of the scheme and no artificial viscosity is required for large enough values of ν_c. Indeed, the steady state solution for inviscid problems closely resembles that obtained by Lerat's implicit method [2].

The Richardson scheme for the modified equations may be replaced by more sophisticated iterations, e.g. the non-linear Jacobi method given by

$$\mathbf{J} = \frac{\partial}{\partial q_1}(\mathbf{D}_a\mathbf{R}_a + \mathbf{D}_b\mathbf{R}_b + \mathbf{D}_c\mathbf{R}_c + \mathbf{D}_d\mathbf{R}_d). \tag{8}$$

For scalar problems, the restrictions on ω and on Δt_1 are equivalent and hence the Richardson method is already a Jacobi scheme. The major difference is for systems of p conservation laws, for which the time step in the Jacobi method is a $p \times p$ full matrix, whilst that in the Richardson method is a constant. We may expect the Jacobi scheme to be superior to the standard method as the ratio of the speed of the fastest and slowest waves in the system increases. Indeed, we may view this 'matrix time stepping' as a means of normalising the speed of separate wave components.

Gauss-Seidel procedures are well-known to be superior to the Jacobi method, particularly if they can be contrived to march in a downstream direction. Indeed, if the mesh is stretched strongly in one direction, as will be the case for Navier-Stokes computations, it may be beneficial to consider line–relaxation schemes. An alternative is the four-colour scheme, which applies the Jacobi scheme successively on four distinct subgrids such that each update is independent of the states on its own subgrid.

A second obvious generalisation of the Richardson scheme (5) is to define the distribution matrices differently. Some recent procedures employ a multi–dimensional upwinding approach (e.g [7]), to give a shock–capturing scheme free of the arbitrary parameters associated with artificial viscosity. As yet, these have not proved to be as robust as the modified version of the Lax–Wendroff schemes which we have described.

4 Multigrid Methods

The multigrid scheme currently in use is a standard FAS scheme applied to the cell–based equations, but with a generalised iterative procedure (4) applied to node–based equations to smooth the errors. The FAS scheme therefore requires two distinct restriction operators:

$$\dot{I}_h^H: \dot{\mathcal{G}}(\dot{\Omega}_h) \to \dot{\mathcal{G}}(\dot{\Omega}_H)$$
$$\Box I_h^H: \Box \mathcal{G}(\Box \Omega_h) \to \Box \mathcal{G}(\Box \Omega_H) \tag{9}$$

where h denotes the fine grid and H the coarse grid. This leads to the coarse grid problem being defined as

$$\mathbf{R}_H(\mathbf{q}_H) = \mathbf{f}_H, \qquad \mathbf{f}_H = \Box I_h^H [\mathbf{f}_h - \mathbf{R}_h(\mathbf{q}_h)] + \mathbf{R}_H(\dot{I}_h^H \mathbf{q}_h). \tag{10}$$

If care is taken in applying boundary conditions, this procedure works well, especially when $\nu_c \gg \nu_1$ in the Taylor–Galerkin update (6),(5). Although the effect of the chequerboard mode is minimised in this way, its presence means that we cannot achieve true mesh–independent convergence rates: this will need either a better definition of the cell residual or frequency decomposition methods.

5 Results

The first test of the Navier–Stokes algorithm has been to compute the subsonic flow over a semi–infinite flat plate. The accuracy of the discretisation has been tested by comparison with the Blasius solution. Away from the leading edge the velocity profiles compare well and Fig.1 shows good agreement between the Blasius and computed skin friction coefficient along the plate.

Results have also been obtained for subsonic flow over a a blunt forebody — Fig.2 shows the computed velocity vectors for a specific test case. Mesh refinement studies have demonstrated that the steady solution is remarkably unaffected by changes in the number of grid points in the boundary layer. Indeed, successful calculations have been made on meshes more suitable for inviscid flows. The reasons for this are explored in [3], where the cell vertex scheme is applied to scalar convection–diffusion equations.

Although the present results are extremely encouraging, the l_2 norm of the cell residuals could not be driven to an arbitrarily small level, nor could that of the nodal updates. It is believed that the iteration enters a limit cycle close to the required solution and that this is a direct consequence of the existence of the chequerboard mode. Evidence of this phenomenon may also be seen in Euler calculations, where the (unconverged) results compare favourably with those obtained with the addition of artificial viscosity. Nonetheless, the multi–grid algorithm accelerated the convergence to the limit cycle by a factor of about six.

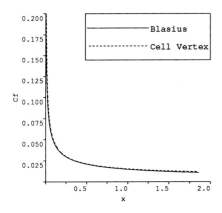 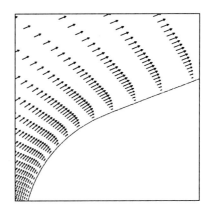

Fig.1 Comparison of Blasius and computed skin friction coefficient along a flat plate; $M_\infty = 0.5$, $Re = 2 \times 10^3$.

Fig.2 Velocity vectors for two-dimensional laminar flow over a forebody; $M_\infty = 0.5$, $Re = 1 \times 10^4$.

6 References

[1] E. Dick (1988), *A flux–difference splitting method for the steady Euler equations*, J. Comp. Phys., Vol. 76, No. 1, p. 19.

[2] A. Lerat and J. Sidès, *Efficient solution of the steady Euler equations with a centred implicit method*. Numerical Methods for Fluid Dynamics III (ed. K. W. Morton and M. J. Baines), Clarendon Press, Oxford, p. 65.

[3] J. A. Mackenzie and K. W. Morton (1990), *Finite volume solutions of convection diffusion test problems*. Oxford University Computing Laboratory, Numerical Analysis Group Report 90/7 .

[4] K. W. Morton (1990), *Finite volume methods and their analysis*. To appear in the Proceedings of the Conference on The Mathematics of Finite Elements and Applications, Brunel 1990.

[5] K. W. Morton and M. F. Paisley (1989), *A finite volume scheme with shock fitting for the steady Euler equations*. J. Comp. Phys. Vol. 80 No. 1 p. 168.

[6] K. W. Morton and M. A. Rudgyard (1989), *Finite volume methods with explicit shock representation* . To appear in the Proceedings of the Conference on Computational Aeronautical Fluid Dynamics, Antibes 1989.

[7] K. G. Powell and B. van Leer (1989), *A genuinely multi–dimensional upwind cell vertex scheme for the Euler equations*, AIAA Paper 89-0095.

[8] E. Süli (1989), *Finite volume methods on distorted meshes: stability, accuracy, adaptivity*. Oxford University Computing Laboratory, Numerical Analysis Group Report 89/6 .

A NUMERICAL STUDY OF THIRD-ORDER NONOSCILLATORY SCHEMES FOR THE EULER EQUATIONS

J. Y. Yang, Chang-An Hsu and T. H. Lee
Institute of Applied Mechanics, National Taiwan University
Taipei 10764, Taiwan.

1. INTRODUCTION

In this paper, following van Leer [1-3] and Harten, Osher, Engquist and Chakravarthy [4-7], we describe and study two third order non-oscillatory shock capturing schemes for the Euler equations of gas dynamics. These schemes are obtained by applying the characteristic flux difference splitting to an appropriate modified flux vector which may have high order accuracy and nonoscillatory property. Third order schemes are constructed using upstream interpolation and ENO interpolation. The accuracy and stability of schemes of various possible choices of stencils selected by the limiter functions in the ENO schemes are examined and compared with the Leonard's QUICKEST scheme [8] and other third-order schemes. Both explicit and implicit schemes are tested. Numerical experiments indicate that good results can be obtained for unsteady problems using explicit ENO scheme but rather slow convergence rate was experienced using implicit ENO schemes. Such slow convergence may attribute to the moving stencils employed in the ENO schemes.

2. EULER EQUATIONS AND NUMERICAL ADVECTION

Consider the one-dimensional Euler equations of inviscid gas dynamics in conservation law form

$$\partial_t Q + \partial_x F(Q) = 0 \tag{1}$$

where $Q = (\rho, \rho u, e)^T$ is the conservative state vector and $F = (\rho u, \rho u^2 + p, u(e+p))^T$ the flux vector. Here ρ is the fluid density, u, the fluid velocity, p the fluid pressure and e, the total internal energy. For a perfect gas, the pressure is related to other fluid properties by the equation of state $p = (\gamma - 1)(e - \rho u^2/2)$, where γ is the ratio of specific heats. Eq.(1) can be expressed in quasi-linear form as

$$\partial_t Q + A(Q)\partial_x Q = 0, \quad A = \partial F/\partial Q \tag{2}$$

Due to the hyperbolicity property of Eq.(1), flux Jacobian matrix A has real eigenvalues $a_1 = u$, $a_2 = u+c$, and $a_3 = u-c$. Here $c = \sqrt{\gamma p/\rho}$ is the sound speed. One can find the similarity transformation matrix T such that $A = T\Lambda T^{-1}, \Lambda = \text{diag}\{a_i\}$. We can consider the following scalar wave equation as a model equation for Eq.(2).

$$\partial_t v + a\partial_x v = 0, \tag{3}$$

Define a uniform computational mesh $\{x_j, t^n\}$, with mesh size Δx and Δt. The discrete representation of $v(x,t)$ on the mesh is v_j^n and $\lambda = \Delta x/\Delta t$ is the mesh ratio. Define the Courant number $\sigma = a\Delta t/\Delta x$ then

$$v_j^{n+1} = v_{j-\sigma}^n \tag{4}$$

where $v_{j-\sigma}^n$ is an approximation for $v((j-\sigma)\Delta x, n\Delta t)$. Since the values of $v_{j-\sigma}$ are not given, one must express it in terms of known values given at the nodal points by some interpolation.

Upstream Interpolation

Using Lagrange's interpolation formula with equal points on each side of point $j - \sigma$, we have:

$$v_{j-\sigma}^n = \sum_{m=-p+1}^{p} \prod_{\substack{i=-p+1 \\ i \neq m}}^{p} \frac{(i - \sigma + [\sigma])}{(i - m)} v_{j-[\sigma]-m}^n \tag{5}$$

Here $[\sigma]$ is the largest integer not exceeding σ. The class of schemes defined by Eq.(5) are stable for $|\sigma| \leq 1$. For $0 \leq \sigma \leq 1$, $[\sigma] = 0$, and for $-1 \leq \sigma \leq 0$, $[\sigma] = -1$.

Here we consider the scheme defined by Eq.(5) for $p = 2$ which is the "QUICKEST" method of Leonard [8]. For the purpose of present framework we consider the following form.

$$v_j^{n+1} = v_j^n - (\sigma - [\sigma])\Delta_- v_{j-[\sigma]}^n - (\sigma - [\sigma])\Delta_-\{(\frac{2 - 3(\sigma - [\sigma]) + (\sigma - [\sigma])^2}{6})\Delta_+ v_{j-[\sigma]}^n$$

$$+ (\frac{1 - (\sigma - [\sigma])^2}{6})\Delta_+ v_{j-[\sigma]-1}^n\} \quad (6)$$

Here $\Delta_\pm$ denote the forward and backward difference operators, respectively. The Taylor series expansion for equation (6) is

$$v_j^{n+1} - v_j^n = -\sigma \Delta x (\partial_x v)_j + \frac{1}{2}\sigma^2 \Delta x^2 (\partial_{xx} v)_j - \frac{1}{6}\sigma^3 \Delta x^3 (\partial_{xxx} v)_j$$

$$- \frac{1}{12}|\sigma|(1 - |\sigma|/2 - \sigma|\sigma|)\Delta x^4 (\partial_{xxxx} v)_j \quad (7)$$

The amplification factor for Eq.(7) is given by

$$g_h(\theta) = 1 - |\sigma|(1 - \cos\theta)[1 + 3|\sigma| - \sigma^2 - (1 - |\sigma|^2)\cos\theta]/3$$

$$+ i\sigma \sin\theta[(1 - \sigma^2)\cos\theta - (4 - \sigma^2)]/3 \quad (8)$$

A conservative scheme for Eq.(1) can be expressed as

$$Q_j^{n+1} = Q_j^n - \lambda[F_{j+\frac{1}{2}}^N - F_{j-\frac{1}{2}}^N] \quad (9)$$

where $F_{j+\frac{1}{2}}^N$ is the numerical flux and is given by

$$F_{j+\frac{1}{2}}^N = F_{j+1}^M - \hat{A}_{j+\frac{1}{2}}^+ \Delta_+ F_j^M = F_j^M + \hat{A}_{j+\frac{1}{2}}^- \Delta_+ F_j^M \quad (10)$$

Here F^M is the modified flux to be defined later.

<u>Leonard's Scheme Made Nonoscillatory</u>

A third order nonoscillatory scheme based on Eq.(6) can be expressed as

$$F_j^M = F_j^{TVD3} = F_j^n + D_j^n \quad (11)$$

The components of D_j are given by

$$d_j^l = (1 - S(\theta_j^l))\tilde{d}_{j+\frac{1}{2}}^l + (1 + S(\theta_j^l))\overline{d}_{j-\frac{1}{2}}^l \quad (12)$$

where $\tilde{d}_{j+\frac{1}{2}}^l$ and $\overline{d}_{j+\frac{1}{2}}^l$ are components of $\tilde{D}_{j+\frac{1}{2}}$ and $\overline{D}_{j+\frac{1}{2}}$ given respectively by

$$\tilde{D}_{j+\frac{1}{2}} = \text{sgn} A_{j+\frac{1}{2}}(\lambda^2|A_{j+\frac{1}{2}}|^2 - 3\lambda|A_{j+\frac{1}{2}}| + 2I)\Delta_+ F_j/6 \quad (13a)$$

and

$$\overline{D}_{j+\frac{1}{2}} = \text{sgn} A_{j+\frac{1}{2}}(I - \lambda^2|A_{j+\frac{1}{2}}|^2)\Delta_+ F_j/6 \quad (13b)$$

and $S(\theta_j^l)$ is the smoothness monitor given by van Leer [1,2] as

$$S(\theta_j^l) = 0, \quad \text{if} \quad |\Delta_+ q_j^l| + |\Delta_- q_j^l| = 0$$

$$= \frac{|\Delta_+ q_j^l| - |\Delta_- q_j^l|}{|\Delta_+ q_j^l| + |\Delta_- q_j^l|}, \quad \text{otherwise} \tag{14}$$

where q_j^l are components of the conservative state vector Q_j. We refer the scheme defined by Eq.(11)-(14) as TVD3 (see also [9]).

3. ESSENTIALLY NON-OSCILLATORY SCHEMES

Using reconstruction via the primitive function (RP) approach and for $N = 3$ [6], with some rearrangement, one has a numerical scheme as following:

$$v_j^{n+1} = v_j^n - \sigma \Delta_- v_j^n - \sigma \Delta_- (m[(\frac{1-\sigma}{2})\Delta_- v_j^n, (\frac{1-\sigma}{2})\Delta_+ v_j^n]$$

$$-\sigma \Delta_- \overline{m}[\Delta_- (\frac{2 - 3\sigma + \sigma^2}{6})\Delta_- v_j^n, \Delta_- (\frac{2 - 3\sigma + \sigma^2}{6})\Delta_+ v_j^n], \text{ if } |\Delta_- v_j^n| \le |\Delta_+ v_j^n|$$

$$-\sigma \Delta_- \overline{m}[\Delta_- (\frac{\sigma^2 - 1}{6})\Delta_+ v_j^n, \Delta_+ (\frac{\sigma^2 - 1}{6})\Delta_+ v_j^n], \text{ if } |\Delta_- v_j^n| > |\Delta_+ v_j^n| \tag{15}$$

In Eq.(15), m is the minmod function defined by

$$m(a,b) = s \min(|a|, |b|), \quad \text{if} \quad \text{sgn} a = \text{sgn} b = s,$$
$$= 0, \quad \text{otherwise} \tag{16}$$

and $\overline{m}$ function is defined by

$$\overline{m}(a,b) = a, \quad \text{if} \quad |a| \le |b|$$
$$= b, \quad \text{if} \quad |a| > |b| \tag{17}$$

The accuracy and Fourier stability of scheme defined by (15) can be analyzed by looking at different possible combination of the arguments in the m and $\overline{m}$ functions. We list only one of the possible schemes below, others can be similarly studied. The Taylor series expansion for one of such variations is

$$v_j^{n+1} - v_j^n = -\sigma \Delta x (\partial_x v)_j + \frac{1}{2}\sigma^2 \Delta x^2 (\partial_{xx} v)_j - \frac{1}{6}\sigma^3 \Delta x^3 (\partial_{xxx} v)_j$$

$$+ \frac{1}{24}\sigma(2 + \sigma - 2\sigma^2)\Delta x^4 (\partial_{xxxx} v)_j \tag{18}$$

The amplification factor for Eq.(18) is given by

$$g_h(\theta) = 1 - \sigma(1 - \cos\theta)[1 + 3\sigma - \sigma^2 - (1 - \sigma^2)\cos\theta]/3$$
$$-i\sigma \sin\theta(4 - \sigma^2 - (1 - \sigma^2)\cos\theta)/3 \tag{19}$$

This is identical to that given by the Rusanov-Burstein-Mirin scheme if $\omega = 2\sigma^3 + 3\sigma^2 - 2\sigma$ (JCP 5, 1970, 507 & 547). Fig. 1 shows the comparison of phase error for several schemes.

Extending the scheme Eq.(15) to system (1) as follow:

$$F_j^M = F_j^{UNO3} = F_j + E_j + D_j \tag{20}$$

In Eq.(20), the components of column vector E_j are given by

$$e_j^l = m[\tilde{e}_{j+\frac{1}{2}}^l, \tilde{e}_{j-\frac{1}{2}}^l] \tag{21}$$

where $\tilde{e}_{j+\frac{1}{2}}^l$ are given by

$$\tilde{E}_{j+\frac{1}{2}} = \text{sgn} A_{j+\frac{1}{2}}(I - \lambda|A|)\Delta_+ F_j/2 \tag{22}$$

And the components of column vector D_j are given by

$$d_j^l = m[\Delta_- \tilde{d}_{j-\frac{1}{2}}^l, \Delta_+ \tilde{d}_{j-\frac{1}{2}}^l] \quad \text{if} \quad |\Delta_{j-\frac{1}{2}} q^l| \leq |\Delta_{j+\frac{1}{2}} q^l| \qquad (23a)$$

or

$$d_j^l = m[\Delta_- \hat{d}_{j+\frac{1}{2}}^l, \Delta_+ \hat{d}_{j+\frac{1}{2}}^l] \quad \text{if} \quad |\Delta_{j-\frac{1}{2}} q^l| > |\Delta_{j+\frac{1}{2}} q^l| \qquad (23b)$$

where $\tilde{d}_{j+\frac{1}{2}}^l$ and $\hat{d}_{j+\frac{1}{2}}^l$ are components of $\tilde{D}_{j+\frac{1}{2}}$ and $\hat{D}_{j+\frac{1}{2}}$, respectively.
$\tilde{D}$ is given by Eq.(13a) and $\hat{D}$ is given by

$$\hat{D}_{j+\frac{1}{2}} = \text{sgn} A_{j+\frac{1}{2}} (\lambda^2 |A_{j+\frac{1}{2}}|^2 - I)\Delta_+ F_j / 6 \qquad (24)$$

Extension to multi-dimensional problems has been attempted using Strang-type dimensional splitting [10] for explicit schemes and Beam-Warming approximate factored ADI method [11] for implicit schemes (see [12,13]).

4. NUMERICAL RESULTS AND DISCUSSIONS

Interaction of Blast Waves

We first present numerical experiments with the above two schemes for the problem of two interacting blast waves suggested by Woodward and Colella [14]. The initial states are given by $Q(x,0) = Q_L$ for $0 \leq x < 0.1$, $Q(x,0) = U_M$ for $0.1 \leq x < 0.9$ and $Q(x,0) = U_R$ for $0.9 \leq x < 1$ and $\rho_L = \rho_M = \rho_R = 1, u_L = u_M = u_R = 0$, and $p_L = 10^3, p_M = 10^{-2}, p_R = 10^2$.

Fig. 2 shows the solutions of density obtained using TVD3 and UNO3 schemes with the modified flux approach. Using a uniform grid system of J=400 points, the CPU time required for 683 time integrations to reach time $t = 0.038$ is 99.5 seconds for the ENO2 scheme (not shown), 106.63 seconds for TVD3 scheme and 158.73 for UNO3 scheme on a Convex C-1 computer. Slight oscillation is observed for the ENO result which is allowed by the essentially non-oscillatory property.

Starting process in a nozzle

In this problem we consider a plane shock wave located initially at certain distance ahead of throat of the nozzle with shock Mach number $M_s = 20$ and propagating toward the downstream. A detailed study of this problem shall provide some information about the flow conditions for a high enthalpy shock tunnel. Here only the UNO3 results are reported. A grid system of 351x101 was used. Fig. 3 shows a sequent of density contours plots for several different times as the primary shock propagate downstream the nozzle into the test section. Very good capturing of the main flow structures such as several contact surfaces containing vortices are displayed. Mach waves from the walls show the supersonic flow established downstream of the throat.

Transonic flow over a circular arc

We consider an internal two dimensional transonic flow through a parallel channel having a 4.2% thick circular arc at the lower wall. The ratio of static downstream pressure to total pressure is 0.623512, corresponding to $M_\infty = 0.85$ in the isentropic flow. In Fig. 4, the C_p and entropy distributions at the lower surface for the ADI solution with UNO3 scheme are shown. All results were calculated with CFL=7. For the case of 85x21 grid system used, the CPU time required for the ENO2 implicit method is 5.73 seconds per step on a Convex C-1 computer and it takes 6.75 seconds for the UNO3 scheme.

ACKNOWLEDGEMENTS

This work has been supported by the National Science Council, the Republic of China under contract NSC 78-0210-D002-11.

REFERENCES

[1] B. van Leer, J. Comp. Phys., 14, (1974), pp. 361-370.
[2] B. van Leer, J. Comp. Phys., 23 (1977), pp. 263-275.
[3] B. van Leer, J. Comp. Phys., 32 (1979), pp. 101-136.
[4] A. Harten and S. Osher, SIAM J. Numer. Anal. Vol. 24, No.2 (1987) pp. 279-309.
[5] A. Harten,"Preliminary Results on the Extension of ENO Schemes to Two-Dimensional Problems," Proceedings of the International Conference on Hyperbolic Problems, Saint-Etienne, January, 1986.
[6] A. Harten, B. Engquist, S. Osher, S. R. Chakravarthy, J. Comp. Phys. 71 (1987), pp. 231-303.
[7] A. Harten, S. Osher, B. Engquist, S. R. Chakravarthy, J. Appl. Numer. Math. 2, (1986), pp. 347.
[8] B.P. Leonard, Comp. Meth. Appl. Mech. Eng. 19 (1979), pp. 59-98.
[9] J. Y. Yang, NASA TM-85959 July, 1984.
[10] G. Strang, SINUM, Vol. 5 (1968), pp. 506-517.
[11] R. F. Warming and R. M. Beam, J. Comp. Phys. Vol. 22, (1976) pp. 234-245.
[12] J. Y. Yang, Y. Liu, and H. Lomax, AIAA J. Vol. 25, No. 5, 1987, pp. 683-689.
[13] J. Y. Yang, "A Numerical Study of Uniformly Second Order ENO Schemes for the Euler Equations," AIAA Journal (to appear).
[14] P. Woodward and P. Colella, J. Comp. Phys., Vol. 54, 1984, pp. 115-173.

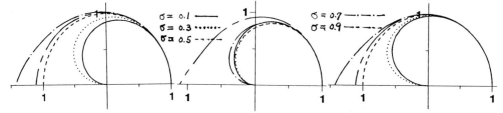

Fig. 1 Comparison of phase error for several third-order schemes.
(a) QUICKEST (b) Eq.(19) (c) RBM (minimum dispersion).

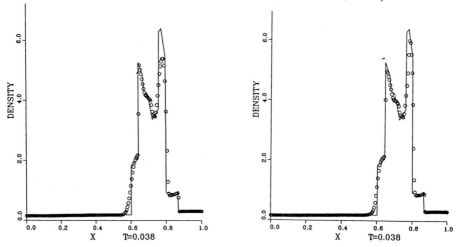

Fig. 2. Solution of one-dimensional interacting blast waves.
(a) TVD3 (b) UNO3

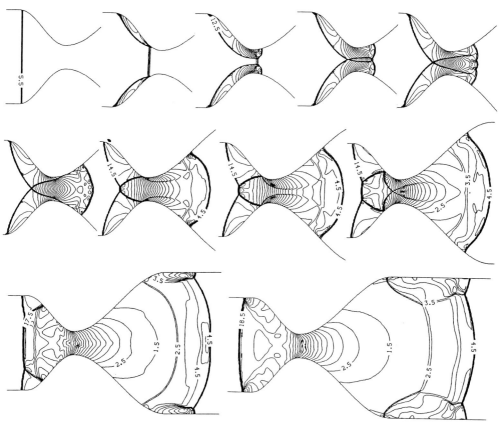

Fig. 3. Computed solution by UNO3 for nozzle starting process with $M_s = 20$. A sequence of computed density contours

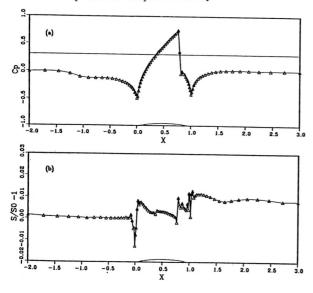

Fig. 4. Solution of transonic channel flow. UNO3 scheme.
(a) C_p distribution
(b) entropy distribution.

A HIGH-ORDER COMPACT NUMERICAL ALGORITHM FOR SUPERSONIC FLOWS

M. H. Carpenter

NASA Langley Research Center
Hampton, Virginia 23665-5225 USA

ABSTRACT

A dissipative compact two-four scheme (second-order time, fourth-order space) similar to the original MacCormack scheme has been developed, that exhibits greater accuracy than conventional fourth-order schemes. The dissipative nature of the scheme allows it to resolve weak discontinuities without artificial damping. A derivation of the scheme is presented, as well as the theoretical stability characteristics. The temporal scheme is then generalized into a steady-state formulation which achieves fourth-order spatial accuracy at steady-state. Several test problems are used to show that the scheme is more accurate than the traditional MacCormack scheme, and is nearly as efficient.

INTRODUCTION

Discontinuities in supersonic flows present a major test of a numerical algorithm's accuracy and robustness. The MacCormack scheme has over the years shown to be a simple, accurate, and reliable method for solving weakly non-linear problems in up to three spatial dimensions. The method is dissipative enough to be stable for weak non-linearities without artificial viscosity, and can achieve a steady-state without fourth-order smoothing.

The recent use of finite-difference methods to perform direct numerical simulations has renewed interest in compact numerical algorithms. Such methods have the advantage of being implementationally compact, therefore requiring a more convenient set of grid-points than non-compact schemes for a desired degree of accuracy (three instead of five for fourth-order accuracy). Thus, they are easily implemented near boundaries. In addition, their discretization error can be a factor of two to four smaller than comparable non-compact schemes [1,2,3].

Although highly accurate and efficient in smooth regions of the flow non-dissipative compact schemes have difficulties with non-linear discontinuities, and they lack the spatial coupling necessary to converge a numerical solution to steady-state. To rectify these problems, a one parameter family of compact two-four schemes has been developed and named appropriately the Dissipative Compact Parameter Schemes (DCPS). This family of schemes yields 4th order spatial accuracy in the smooth regions of the flow, and develops its own dissipation in the neighborhood of weak discontinuities. One scheme in particular differs only slightly from the basic structure of the original MacCormack formulation and will be the concentration of this work.

DEVELOPMENT OF THE COMPACT MACCORMACK SCHEME

The Compact MacCormack scheme (C-Mac) is implemented on the model equation $U_t + F_x = 0$ in the standard predictor-corrector format as:

$$U_i^* = U_i^n - \lambda \frac{\Delta^+}{(1 + \frac{\Delta^+\Delta^-}{3!})} F_i \quad U_i^{**} = U_i^* - \lambda \frac{\Delta^-}{(1 + \frac{\Delta^+\Delta^-}{3!})} F_i^* \quad U_i^{n+1} = \frac{1}{2}[U_i^n + U_i^{**}] \quad (1)$$

with $\lambda = \frac{\Delta t}{\Delta x}$, Δ^+ and Δ^- being the forward and backward difference operators, respectively, and the reciprocal operator $(1 + \frac{\Delta^+\Delta^-}{3!})$ denoting the inverse of the *Pade'* operator matrix. Thus,

$\frac{\Delta^+}{(1+\frac{\Delta^+\Delta^-}{3!})}F_i$ is an abbreviated way of writing the *Pade'* operation $\frac{1}{6}S^n_{i-1} + \frac{4}{6}S^n_i + \frac{1}{6}S^n_{i+1} = \frac{F^n_{i+1}-F^n_i}{\Delta x}$, where S_i is the numerical approximation for F_x at the grid-point i.

It can be shown that any symmetric predictor-corrector scheme implemented similar to the MacCormack scheme, will produce an amplification matrix of the form: $G = (1 - i\lambda|Im\Psi| - \frac{\lambda^2|\Psi|^2}{2})$ where Ψ is the Fourier image of either the forward or backward spatial operator. (In MacCormack-like schemes, the forward and backward spatial operators $\frac{\Delta^+}{(1+\frac{\Delta^+\Delta^-}{3!})}$ and $\frac{\Delta^-}{(1+\frac{\Delta^+\Delta^-}{3!})}$ produce Fourier images which are negative complex conjugates.) Simplifying the expression for G, with the stability condition $|G| \le 1$ produces: $\lambda \le \frac{2|Re\Psi|}{|\Psi|^2}$ where $Re\Psi$ and $Im\Psi$ are the real and imaginary parts of Ψ.

By taking the Fourier Transform of equation (1) it can be shown that the Fourier image of the spatial operator Ψ becomes:

$$\Psi = \frac{\frac{3}{2}\{(\cos\xi - 1) + i\sin\xi\}}{1 + \frac{1}{2}\cos\xi} \tag{2}$$

The stability bound for the scheme in one dimension becomes $\lambda \le \frac{1}{3}$. Standard third order extrapolation techniques can be used for the explicit boundary condition values. For the derivative specification at the boundary, a third order implicit relation can be used without losing the tridiagonal character of the scheme[3].

Extension of the C-Mac scheme to two or three spatial dimensions is accomplished by implementing a series of 1-D operators. On the model equation $U_t + F_x + G_y = 0$, the predictor stage of the C-Mac scheme becomes:

$$U^*_{i,j} = U^n_{i,j} - \lambda \frac{\Delta^+_x}{(1+\frac{\Delta^+_x\Delta^-_x}{3!})}F_{i,j} - \lambda\beta \frac{\Delta^+_y}{(1+\frac{\Delta^+_y\Delta^-_y}{3!})}G_{i,j} \tag{3}$$

As is the case with the MacCormack scheme in 2-D, the C-Mac scheme is Von Neumann conditionally unstable to certain combinations of 2-D modes. By cycling through all four of the possible permutations for the predictor-corrector sequence (FF-BB, BB-FF, FB-BF, BF-FB in two-dimensions), this instability is damped[4], similar to how it is damped in the 2-D MacCormack formulation.

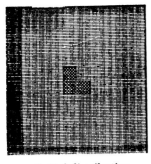

Initial distribution
Figure 1.

MacCormack
Figure 2.

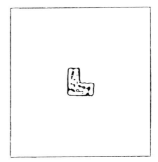

Compact MacCormack
Figure 3.

To test the stability and accuracy of the C-Mac in 2-D, a "color" problem was devised which would have the unstable modes present in the formulation. In addition, discontinuous initial data were used to test the C-Mac discontinuity tracking capabilities. The equation being solved in this study is the two-dimensional model wave equation defined by: $\frac{\partial\Omega}{\partial t} + \frac{\partial(U\Omega)}{\partial x} + \frac{\partial(V\Omega)}{\partial y} = 0$ with $\Omega(x,y,0) = \Omega o(x,y)$ and suitable boundary conditions. The velocity vector was chosen to be solid body rotation defined by $U = -\alpha y$ and $V = \alpha x$. All calculations were integrated in time until the initial distribution had rotated exactly once. The distribution was then compared with the initial distribution to assess the quality of the algorithm.

Figure (1) shows the initial distribution on the grid used in this study. The initial Ω_0 was chosen to be 1 in the interior of the "L-shaped" body, and 0 elsewhere. Different numerical algorithms were used to track the initial distribution in time. The numerical results after one complete revolution as calculated with the 1) MacCormack, and 2) C-Mac, are shown in figures (2-3). Plotted is the spatial distribution of the scalar field, with contour levels ranging from zero to one, in increments of 0.05. It is apparent that the C-Mac algorithm produces more accurate results than the MacCormack algorithm on this problem. No apparent instabilities were experienced with the C-Mac scheme (or the MacCormack) in these calculations. The unstable modes were damped by implementing the scheme in a cyclic manner, and thus did not destabilize the solution.

TWO DIMENSIONAL VISCOUS PROBLEM

A two-dimensional viscous problem for which an extremely accurate solution is known, was used as a test of the C-Mac's efficiency and to determine the formal accuracy of the method. Linear stability theory predicts that the temporally developing compressible 2-D mixing layer, is unstable to a velocity profile which is initially specified as a hyperbolic tangent axial velocity distribution. For these calculations, accurate eigenmodes for the instability were provided from a spectral linear stability code developed by Macaraeg[5]. Nondimensional growth rates and characteristic frequencies were calculated from these modes, and were used as the measure of accuracy achieved by the finite-difference algorithm.

A schematic of the mixing layer configuration is shown in figure (4). The stream-wise direction is assumed periodic at a wavelength of x = 0.6283mm (the mode which grows most rapidly, as determined from the linear analysis). The initial velocity distribution is specified as $U(x,y,0) = U_\infty \tanh(\frac{y}{\delta})$, V(x,y,0) = 0.0, Mach = 0.3, Re = 187. Source terms were added to the Navier-Stokes equations so that the momentum and energy equations would preserve free-stream.

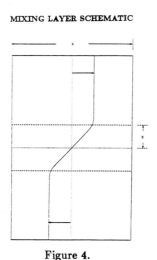

Figure 4.

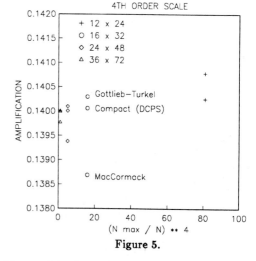

Figure 5.

The grid in the stream-wise direction is uniform, while the grid in the cross-stream direction is highly stretched. Characteristic boundary conditions were used in the transverse direction in all cases. The viscous terms for the fourth-order methods were cast in a fourth-order explicit manner. A grid convergence history of the numerical methods was used to compare the different numerical methods and to determine the formal accuracy of each.

The most unstable mode in this problem grows at an exponential growth rate with an exponent determined from linear stability theory to be 0.14000 in non-dimensional units. This rate was used as the "exact" growth rate for these conditions. A series of four grids was then defined, each having a grid density which was a constant multiple of the previous grid. Presented in figure (5) are the results obtained from: 1) MacCormack at a CFL of 1.0, 2) C-Mac at a CFL of $\frac{1}{3}$, and 3) Gottlieb-Turkel at a CFL of 0.5, all run on identical grids.

The amplification rates of all the methods are plotted against $(\frac{N_{max}}{N})^4$ to show quartic accuracy. The linear slope of the convergence towards the predetermined solution, shows the C-Mac algorithm (and the Gottlieb-Turkel) to be fourth-order spatially accurate. It is also apparent that on identical grids, the C-Mac scheme is more accurate than the Gottlieb-Turkel, and considerably more accurate than the MacCormack algorithm.

The question of efficiency is more difficult to address for the C-Mac scheme. Since the scheme requires two tri-diagonal matrix inversions at each stage of the scheme, it requires more computational effort. For these calculations, the matrix inversions required about 15 percent more computational time than comparable explicit methods, but required negligibly more memory. In addition, the maximum allowable CFL for the C-Mac is lower than either of the other algorithms. If the C-Mac scheme is consistently more accurate, then the additional cost is justified. This is certainly problem dependent, but generally has been the case.

THE STEADY-STATE FORM

For steady-state calculations, the algorithm can be reformulated in 2- or 3-D to overcome the small CFL constraint [6]. The predictor step in 2-D on the model equation $U_t + F_x + G_y = 0$, can be re-written in the form (by matrix multiplying the equation by the Pade' matrix operator: $(1 + \frac{\Delta^+ \Delta^-}{3!})$ in each of the two spatial directions)

$$(1 + \frac{\Delta_x^+ \Delta_x^-}{3!})(1 + \frac{\Delta_y^+ \Delta_y^-}{3!})\Delta U_{i,j}^* = -\lambda_x[(1 + \frac{\Delta_y^+ \Delta_y^-}{3!})\Delta_x^+ F_{i,j}] - \lambda_y[(1 + \frac{\Delta_x^+ \Delta_x^-}{3!})\Delta_y^+ G_{i,j}] \quad (4)$$

with a similar equation for the corrector step. U^{n+1} is found from $U_{i,j}^{n+1} = U_{i,j}^n + \frac{1}{2}[\Delta U_{i,j}^* + \Delta U_{i,j}^{**}]$. For steady-state calculations the tridiagonal terms on the LHS of the predictor and corrector steps can be ignored. The resulting numerical scheme is the Steady State C-Mac algorithm (SS C-Mac), which is 4th order spatially accurate at steady-state, and has a stability bound comparable with that of the standard MacCormack.

To test the efficiency and robustness of the SS C-Mac algorithm on a realistic engineering problem, a two-dimensional inviscid problem was chosen. A Mach 5, 10° compression followed by a 10° expansion provides a good test of the scheme's robustness as a shock capturing algorithm. Figure (6) shows the solution of the problem as calculated with the constant gamma SEAGULL[7] code (an accurate shock fitting code). The lower and upper wall pressures, plotted against the axial location are shown. A square pulse occurs in the lower wall pressure as the flow encounters the compression and expansion. The upper wall pressure remains constant until the shock-expansion information propagates across the domain and encounters the wall.

Figures (7-8) show a comparison between the 1) 2nd order MacCormack, and the 2) 4th order SS C-Mac algorithms. All calculations are run on identical grids. The grid distribution was 51 uniformly spaced grid points in the stream-wise direction, and 51 grid points in the cross-stream direction, with exponential compression at each wall.

In general, the global features of the flow (peak heights and shock locations) are reasonably well resolved with both the algorithms. Note that the curvature of the upper wall pressure and the location of the discontinuities (compared with the SEAGULL results) is better resolved with the SS C-Mac scheme than with the MacCormack. Also note the more monotone character of the C-Mac results for this case. The run-times for each case were approximately equal.

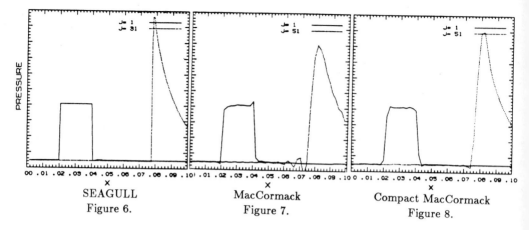

SEAGULL
Figure 6.

MacCormack
Figure 7.

Compact MacCormack
Figure 8.

CONCLUSIONS

A fourth-order compact numerical algorithm similar to the original MacCormack formulation is one member of a one parameter family of compact dissipative two-four schemes. The scheme has a theoretical stability bound of $\lambda \leq \frac{1}{3}$. It is second-order temporally and fourth-order spatially accurate for smooth flows. In addition it develops inherent numerical dissipation which stabilizes the formulation in the neighborhood of discontinuities. Extension to 2-D and 3-D is straightforward.

For steady-state calculations, the C-Mac algorithm can be reformulated into a more computationally efficient version. The resulting scheme is fourth-order spatially accurate, but only at steady-state. The SS C-Mac algorithm is as robust as the original MacCormack formulation, but provides more accurate solutions in two or more spatial dimensions. Both forms of the C-Mac algorithm have been validated against a variety of test problems to demonstrate their accuracy, efficiency and robustness.

References

[1] Orszag, S. A., Israeli, M. "Numerical Simulation of Viscous Incompressible Flows," Annual Review of Fluid Mechanics, Vol. 6, pp 281-318, 1974.

[2] Hirsh, R. S., "Higher Order Accurate Difference Solutions of Fluid Mechanics Problems by a Compact Differencing Technique," Journal of Computational Physics, Vol 19, 1 (1975), 90.

[3] Adam, Y., "Highly Accurate Compact Implicit Methods and Boundary Conditions," Journal of Computational Physics Vol 24, 10-22, 1977.

[4] MacCormack, R.W., "Numerical Solution of the Interaction of Shock Wave with a Laminar Boundary Layer," Proceeding of the Second International Conference on Numerical Methods in Fluid Dynamics, Lecture Notes in Physics, Springer-Verlag, Sept. 15-19, 1970, pp 151-163.

[5] Macaraeg, M.G., Streett, C.L., and Hussaini, M.Y., "A Spectral Collocation Solution to the Compressible Stability Eigenvalue Problem," NASA Technical Paper 2858, December 1988.

[6] Carpenter, M.H, "Three-Dimensional Computations of Cross-Flow Injection and Combustion in a Supersonic Flow," AIAA-89-1870, 20th Fluid Dynamics, Plasma Dynamics and Lasers Conference, Buffalo, NY, June 12-14, 1989.

[7] Salas, M.D., "Shock Fitting Method for Complicated Two-Dimensional Supersonic Flow," AIAA Journal, Vol 14, No. 5, May, 1976.

A STUDY OF SPURIOUS ASYMPTOTIC NUMERICAL SOLUTIONS OF NONLINEAR DIFFERENTIAL EQUATIONS BY THE NONLINEAR DYNAMICS APPROACH

H.C. Yee[1]

NASA Ames Research Center, Moffett Field, CA 94035, USA

P.K. Sweby[2]

University of Reading, Whiteknights, Reading RG6 2AX, England

and

D.F. Griffiths[3]

University of Dundee, Dundee, DD1 4HN, Scotland

ABSTRACT

The goal of this paper is to present some new results and to give insight and guidelines on the application of nonlinear dynamics theory to the better understanding of asymptotic numerical solutions and nonlinear instability in finite difference methods for nonlinear differential equations that display genuinely nonlinear behavior. Our hope is to reach researchers in the fields of computational sciences and, in particular, computational fluid dynamics (CFD). Although the study of nonlinear dynamics and chaotic dynamics for nonlinear differential equations and for nonlinear discrete maps (nonlinear difference equations) have independently flourished for the last decade, there are very few investigators addressing the issue on the connection between the nonlinear dynamical behavior of the continuous systems and the corresponding nonlinear discrete map resulting from finite-difference discretizations. This issue is especially vital for computational sciences since nonlinear differential equations in applied sciences can rarely be solved in closed form and it is often necessary to replace them by finite dimensional nonlinear discrete maps. It is important to realize that these nonlinear discrete maps can exhibit a much richer range of dynamical behavior than their continuum counterparts. Furthermore, it is important to ask what happens when linear stability in numerical integrations breaks down for problems with genuinely nonlinear behavior. Here our objective is <u>neither</u> to provide theory nor to illustrate with realistic examples the connection of the dynamical behavior of practical fluid dynamics equations with their discretized counterparts, but rather to give insight into the nonlinear features unconventional to this type of study and to concentrate on the fundamental ideas.

I. INTRODUCTION

While some applied computational fluid dynamicists are channeling their energy toward developing numerical solver computer codes, grid generation codes, three-dimensional graphical stereo displays, and stretching the limits of the faster supercomputers in the world to numerically simulate the various three-dimensional (3-D) complex aerodynamic configurations [1], there is a

---

[1] Staff Scientist, Fluid Dynamics Division.
[2] Lecturer, Department of Mathematics.
[3] Senior Lecturer, Department of Mathematical Sciences.

group of applied mathematicians, physicists, chemists, biologists, applied mechanicians, and meteorologists who are channeling their energy toward a new science called "chaotic dynamics" (or nonlinear dynamics). The science of chaotic dynamics has cut across many traditional scientific disciplines for the last decade. It offers a way of seeing order and pattern where formerly only the random, the erratic, and the unpredictable were present. It explains much of the genuinely nonlinear phenomena that were once unexplainable. See references [2-10] for an introduction to this subject.

Nonlinear Dynamics & Chaotic Dynamics: Before the birth of chaotic dynamical theory, traditional study of nonlinear dynamics belonged to the applied mechanics disciplines of mechanical engineering. Modern nonlinear dynamics (since the late seventies) includes chaotic dynamics. Strictly speaking, chaotic dynamics is a branch of nonlinear dynamics. But, for the purpose of the present discussion, the terms nonlinear dynamics and chaotic dynamics are used interchangeably. That is, nonlinear dynamics includes chaotic dynamics and vice versa.

Loosely speaking, the study of asymptotic behavior (and steady-state solutions) of nonlinear differential equations (DEs) and nonlinear discrete maps and how the asymptotes change as parameters of the system are varied is most often referred to as nonlinear dynamic analysis. Topics in this area include bifurcation theory, period doubling cascades resulting in chaos, etc. Stable chaotic solutions (chaotic attractors) may be defined loosely and simply as stable asymptotes that have infinite period and yet are still bounded and are sensitive to initial data [2-10]. It is emphasized here that unless otherwise stated, all DEs and discrete maps are nonlinear and consist of system parameters, and the terms discrete maps and difference equations are used interchangeably.

Types of Dynamical Systems: Consider an ordinary differential equation (ODE) of the form

$$\frac{du}{dt} = \alpha S(u), \tag{1.1}$$

where α is a parameter and S is a nonlinear function in u and is independent of α. An ODE of this form in which t does not appear explicitly in S is called an autonomous dynamical system. One can also consider a function S which is nonlinear in u and depends explicitly on t. ODEs of this type are called nonautonomous dynamical systems and they are more difficult to analyze; see references [5,8] for a discussion. The analysis would be more complicated if $S = S(u, t, \alpha)$ is nonlinear in both u and α. In this case, the DE is not only nonlinear in the dependent variable u (and independent variable t), but also nonlinear in the parameter space α. One can also consider systems that depend on more than one parameter and/or systems of equations of the above type.

A partial differential equation (PDE) counterpart of (1.1) might be

$$\frac{\partial u}{\partial t} + \frac{\partial f(u)}{\partial x} = \epsilon \frac{\partial^2 u}{\partial x^2} + \alpha S(u), \tag{1.2}$$

where ϵ is a parameter and the function $f(u)$ can be linear or nonlinear in u. The source term S in (1.2) can be a similar nonlinear function as the ODE (1.1) except S can depend explicitly on x as well as t and α.

Next consider nonlinear discrete maps of the forms

$$u^{n+1} = u^n + D(u^n, u^{n-1}, r), \tag{1.3}$$

and

$$u_j^{n+1} = u_j^n + G(u_j^n, u_{j\pm 1}^n, r). \tag{1.4}$$

Here r is a parameter, and D is nonlinear in u^n and u^{n-1} and linear or nonlinear in the parameter space r. The situation is similar for the function G. One can also consider discrete systems that depend on more than one parameter. A typical example is a discrete map arising from a finite-difference approximation of DEs such as (1.1) or (1.2). For the ODE, the resulting discrete maps might be nonlinear in α (eventhough the ODE is linear in α) as well as the time step Δt, depending on the ODE solvers. For the PDE, again depending on the differencing scheme, the resulting discretized counterparts can be nonlinear in α, ϵ, Δt, the grid spacing Δx and the numerical dissipation parameters even though the PDE consists of only one parameter or none. It is the introduction of new parameters due to finite discretization of genuinely nonlinear DEs that add a new dimension on the implication and interpretation of stability and convergence rate on asymptotic numerical solutions.

One can also consider discrete maps (scalar or system) of the forms

$$u^{n+1} = u^n + D(u^{n+k}, ..., u^n, ..., u^{n-l}, r_1, r_2, ..., r_m), \tag{1.5}$$

where k, l, m are positive integers and $r_1, r_2, ..., r_m$ are parameters, and

$$u_j^{n+1} = u_j^n + G(u_{j\pm 1}^{n+k}, ..., u_{j\pm 1}^n, ..., u_{j\pm 1}^{n-l}, u_j^{n+k}, ..., u_j^n, ..., u_j^{n-l}, r_1, r_2, ..., r_m). \tag{1.6}$$

Again, (1.6) can depend on more than the three indices $j, j \pm 1$. Systems (1.4) and (1.6) are sometimes referred to as partial-difference equations. Analysis of the dynamical behavior of (1.4) and (1.6) can be many orders of magnitude more difficult than that of (1.3) and (1.5). Any of the systems (1.1)-(1.6) are examples of dynamical systems. It is emphasized here that discrete maps, regardless of their origin, are dynamical systems in their own right.

II. Typical Difference in Dynamical Behavior of ODEs and Discrete Maps

The study of discrete maps is the discrete analog to the study of ODEs, as the study of recursion formulas is a discrete analog to the study of series expansions of functions. Much of the theory of ODEs can carry over to discrete maps with some slight modifications. However, there are new phenomena occurring in discrete maps which are absent in differential systems [14,15,12,13].

With repect to the topographical behavior, there are new kinds of behavior of trajectories in the neighborhood of fixed points (or equilibrium points) of discrete maps. The behavior of separatrices associated with a saddle type of fixed point for a nonlinear difference system is far more complicated than the behavior of separatrices for the corresponding differential system. See Yee [12], Hsu [13] and Hsu et al. [16,17] for details and examples.

With respect to similar equation types, the minimum number of first-order nonlinear autonomous ODEs is three for the existence of chaotic phenomena. However, a simple scalar first-order difference equation like the logistic map [18-22]

$$v^{n+1} = \mu v^n (1 - \frac{v^n}{4}), \qquad \mu \text{ a parameter}, \tag{2.1}$$

or its piecewise linear approximation [23]

$$\begin{aligned} v^{n+1} &= \mu v^n, & v^n &\leq 1 \\ &= \mu, & 1 \leq v^n &\leq 3 \\ &= \mu(4 - v^n) & 3 &\leq v^n \end{aligned} \qquad (2.2)$$

possesses very rich dynamical behavior such as period-doubling cascades resulting in chaos. Equation (2.2) has the same behavior as (2.1) except that simple closed form asymptotic solutions of all periods can be obtained. These characteristic trait differences between ODEs and discrete maps are very general. The discrete maps can arise naturally from physical and mathematical modelling as well as finite discretization of DEs. It is in this spirit that we say that discrete maps can exhibit a much richer range of dynamical behavior than DEs.

III. Background and Motivation

Spurious asymptotic numerical solutions such as chaos were observed by Ushike [24] and Brezzi et al. [25] when they used the leapfrog method of discretization for the logistic ODE

$$\frac{du}{dt} = \alpha u(1 - u). \qquad (3.1)$$

In reference [26], Schreiber and Keller discussed the existence of spurious asymptotic numerical solutions for a driven cavity problem. Some related studies are reported in [27].

Spurious solutions of Burgers' equation and channel flows have been studied and computed in [28-30]. Many other investigators in the computational sciences (e.g. [31-35]) have observed some kind of strange or chaotic behavior introduced by the numerical methods but were not able to explain precisely the source of the behavior, or most of all the implication and impact in practical applications in computational sciences.

Due to the popularity of searching for chaotic phenomena, it is very trendy to relate inaccuracy in numerical methods with the onset of chaos. It is emphasized here that inaccuracy in long time integration of discrete maps resulting from finite discretization of nonlinear DEs comes in other forms prior to the onset of chaotic phenomena. Stable and unstable spurious steady states and spurious periodic numerical solutions set in before chaotic behavior occurs. These spurious asymptotes of finite period are just as inaccurate as chaotic phenomena as far as numerical integration is concerned. In other words, the prelude to chaotic behavior is the key element that we want to stress (i.e., before the onset of chaos or a divergent solution). As can be seen from our study, the result of operating the time step beyond the linearized stability limit is not always a divergent solution in genuinely nonlinear behavior; spurious stable and unstable steady-state numerical solutions and spurious stable and unstable asymptotic numerical solutions of higher period can occur. In addition, the result of operating the time step below the linearized stability of the scheme does not always lead to the true stable steady-state solution of the DEs; spurious stable and unstable asymptotes of any period can occur, depending on the scheme and initial conditions and boundary conditions. These types of behavior are, in general, more difficult to detect than chaotic phenomena in practical computations.

Recently, it has been realized by numerical analysts that discrete maps resulting from finite discretization of ODEs and PDEs can be considered as dynamical systems. Several papers [36,37] on numerical methods as dynamical systems have appeared in recent years. These investigators studied the dynamical behavior of the different ODE solvers per se without relating their close tie with the ODEs themselves.

Why is there such a need to study the connection between the dynamical behavior of the continuum and its discretized counterparts for CFD applications? The necessity stems from the fact that current supercomputer power can perform numerical simulations on virtually any simple 3-D aerodynamic configuration and, due to the limited available experimental data, some of the applied engineers are forced to rely on the numerical simulations to help design the next generation of aircraft and spacecraft. Many of these applied scientists are still using linear analysis as their guide to study highly nonlinear equations, and often they are not aware of the limitations and pitfalls of the numerical procedures. Furthermore, most of the numerical algorithms in use operate under the accuracy and stability limit guidelines of linearized model equations. It is only appropriate to analyze nonlinear problems with the nonlinear approach; i.e., by the nonlinear dynamics approach.

IV. Connection Between the Dynamical Behavior of the Continuum and Its Discretized Counterpart

Aside from truncation error and machine round-off error, a more fundamental distinction between the continuum and its discretized counterparts is new behavior in the form of spurious stable and unstable asymptotes created by the numerical methods. This is due to the fact that nonlinear discrete maps can exhibit a much richer range of dynamical behavior than their continuum counterparts as discussed in our companion paper [38], section 2.2. Some instructive examples are given in our companion paper, section III. These new phenomena were partially explored by the University of Dundee group [39-47], Sanz Serna and Vadillo [48], Iserles [49,50,51] and Stuart [52-56]. Their primary emphasis was on phenomena beyond the linearized stability limit. The main contributions of our study are (1) the occurrence of spurious steady-state numerical solutions below the linearized stability limit of the scheme for genuinely nonlinear DEs, (2) the strong dependence of numerical solutions on the initial data, as well as other system parameters such as physical as well as numerical boundary conditions, and numerical dissipations terms, and (3) the implications for practical computations in combustion and hypersonic CFD.

The unique dynamical property of the separate dependence of solutions on initial data for the individual nonlinear DE and its discretized counterpart is especially important for employing a "time-dependent" approach to the steady state with given initial data in hypersonic CFD. In many CFD computations, the steady-state equations are PDEs of the mixed type and a time-dependent approach to the steady state can avoid the complication of dealing with elliptic-parabolic or elliptic-hyperbolic types of PDEs. However, this time-dependent approach has created a new dimension of uncertainty. This uncertainty stems from the fact that in practical computations, the initial data are not known and most often a freestream condition or an intelligent guess for the initial conditions is used. In particular the controversy of the "existence of multiple steady-state solutions" through numerical experiments will not be resolved until there is a better understanding of the separate dependence on initial data for both the PDEs and the discretized equations.

Various numerical examples are discussed in our two companion papers [38,57]. An overall summary of our findings (integrated with other relevant recent results) will be given in the next two subsections. The discussion is devoted first to steady-state solutions and asymptotes of any period, and second to transient solutions.

Steady-state Solutions and Asymptotes of Period Higher Than One: Table 4.1 shows the possible stable asymptotic solution behavior between DEs and their discretized counterparts. Some of the phenomena are supported by simple examples in references [38,57]. The main connection between the DEs and their discretized counterparts is that steady-state solutions of the continuum are usually solutions of the discretized counterparts but not the reverse. The main difference is that new phenomena are introduced by the numerical methods in the form of spurious stable and unstable asymptotic solutions of any period. In the past, the phenomena of spurious asymptotes were observed largely beyond the linearized stability of the schemes. Some numerical analysts and applied computational scientists were not alarmed and were skeptical about these phenomena since, theoretically, one is always guided by the linearized stability limit of the scheme. However, this reasoning is only valid if one is solving a scalar nonlinear ODE with known initial data using a variable step size control.

Another important factor is that associated with the same (common) steady-state solution, the basin of attraction (domain of attraction) of the continuum might be vastly different from the discretized counterparts. This is due entirely to the separate dependence on and sensitivity to initial and boundary conditions for the individual system. The situation is compounded by the existence of spurious steady states and asymptotes of period higher than one and possibly chaotic attractors. Here the basin of attraction of a dynamical system is the domain of the set of initial conditions whose solution curves (trajectories) all approach the same asymptotic state.

Intuitively, in the presence of stable spurious asymptotes, the basin of the true stable steady states (steady states of the DEs) can be separated by the basins of attraction of the stable spurious asymptotes (introduced by the numerical methods) and interwoven by unstable asymptotes, whether due to the physics (i.e., present in both the DEs and the discretized counterparts) or spurious in nature (i.e., introduced by the numerical methods).

For PDEs, there is an additional difficulty in that even with the same time discretization but different spatial discretizations or vice versa, the basins of attraction can also be extremely different. However, mapping out the basins of attraction for any nonlinear continuum dynamical system other than the very simple scalar equations relies on numerical methods. The type of nonlinear behavior and the dependence and sensitivity to initial conditions for both the PDEs and their discretized counterparts make the understanding of the true physics extremely difficult when numerical methods are the sole source. Under this situation, how can one delineate the numerical solutions that approximate the true physics from the numerical solutions that are spurious in nature? With our simple illustrations in references [38,57], we were able to demonstrate the importance of the subject and, most of all, the importance of knowing the general dynamical behavior of the schemes for genuinely nonlinear scalar DEs before applying these schemes in practical computations.

Transient or Time-Accurate Solutions: It is a common misconception that inaccuracy in long-time behavior poses no consequences on transient or time-accuate solutions. This is not the case when one is dealing with genuinely nonlinear DEs. For genuinely nonlinear problems, due to the possible existence of spurious solutions, larger numerical errors can be introduced by the numerical methods than one can expect from local linearized analysis or weakly nonlinear behavior. The situation will get more intensified if the initial data of the DE is in the basin of attraction of a chaotic transient [58-60] of the discretized counterpart. In fact, it is possible the whole solution trajectory is erroneous.

Our preliminary studies on PDEs also show that much of nonlinear dynamic phenomena have a direct relation for problems containing nonlinear source terms such as the reaction-diffusion, the reaction-convection or the reaction-convection-diffusion equations. Due to space limitations, the interested reader is encouraged to refer to our companion papers [38,57] for a more comprehensive discussion of the subject along with various examples. The intent of reference [38] is not only to present a study on the state-of-art of nonlinear dynamical behavior of ODE solvers, but more importantly also to serve as an introduction and to present new results to motivate this new yet unconventional concept to algorithm development in CFD. The following summarizes the study concluded in references [38,57].

Spurious stable as well as unstable steady-state numerical solutions, spurious stable and unstable asymptotic numerical solutions of higher period, and even stable chaotic behavior can occur when finite-difference methods are used to solve nonlinear DEs numerically. The occurrence of spurious asymptotes is independent of whether the DE possesses a unique steady state or has additional periodic solutions and/or exhibits chaotic phenomena. The form of the nonlinear DEs and the type of numerical schemes are the determining factor. In addition, the occurrence of spurious steady states is not restricted to the time steps that are beyond the linearized stability limit of the scheme. In many instances, it can occur below the linearized stability limit. Therefore, it is essential for practitioners in computational sciences to be knowledgeable about the dynamical behavior of finite-difference methods for nonlinear scalar DEs before the actual application of these methods to practical computations. It is also important to change the traditional way of thinking and practices when dealing with genuinely nonlinear problems.

In the past, spurious asymptotes were observed in numerical computations but tended to be ignored because they all were assumed to lie beyond the linearized stability limits of the time step parameter Δt. As can be seen from the study in [38,57], bifurcations to and from spurious asymptotic solutions and transitions to computational instability not only are highly scheme dependent and problem dependent, but also initial data, and physical and numerical boundary conditions dependent, and not limited to time steps that are beyond the linearized stability limit.

The symbiotic relation among all of these various factors makes this topic fascinating and yet extremely complex. The main fundamental conclusion is that, in the absence of truncation and machine round-off errors, there are qualitative features of the nonlinear DE which cannot be adequately represented by the finite-difference methods and vice versa. The major feature is that convergence in practical calculations involves fixed Δt as $n \to \infty$ rather than $\Delta t \to 0$ as $n \to \infty$. It should be emphasized that the resulting discrete maps from finite discretizations can exhibit a much richer range of dynamical behavior than their continuum counterparts. A typical feature is the existence of spurious numerical asymptotes that can interfere with stability, accuracy and basins of attraction of the true physics of the continuum.

References

[1] NASA Computational Fluid Dynamics Conference, NASA Conference Publication 10038, Vol. 1 and 2, March 7-9, 1989.

[2] R.L. Devaney, *An Introduction to Chaotic Dynamical Systems*, Addison Wesley, New York, 1987.

[3] R. Seydel, *From Equilibrium to Chaos*, Elsevier, New York, 1988.

[4] J. Guckenheimer and P. Holmes, *Nonlinear Oscillations, Dynamical Systems, and Bifurcations of Vector Fields*, Springer-Verlag, New York, 1983.

[5] J.M.T. Thompson and H.B. Stewart, *Nonlinear Dynamics and Chaos*, John Wiley, New York, 1986.

[6] C.S. Hsu, *Cell-to-Cell Mapping*, Springer-Verlag, New York, 1987.

[7] M. Kubíček and M. Marek, *Computational Methods in Bifurcation Theory and Dissipative Structures*, Springer-Verlag, New York, 1983.

[8] T.S. Parker and L.O. Chua, *Practical Numerical Algorithms for Chaotic Systems*, Springer-Verlag, New York, 1989.

[9] E. A. Jackson, *Perspectives of Nonlinear Dynamics*, Cambridge, Cambridge, 1989.

[10] E. Beltrami, *Mathematics for Dynamic Modeling*, Academic Press, Orlando, 1987.

[11] R.M. May, J. Theoret. Biol., Vol. 51, 1975, pp. 511-524.
[12] H.C. Yee, Ph.D. Dissertation, University of Calif., Berkeley, Calif., USA, 1975.
[13] C.S. Hsu, *Advances in Applied Mechanics*, Academic Press, New York, Vol. 17, 1977, pp. 245-301.
[14] A.M. Panov, Uch. Zap. Ural. Gos. Univ. vyp, Vol. 19, 1956, pp. 89-99.
[15] O. Perron, J. Reine Angew. Math. Vol. 161, 1929, pp. 41-64.
[16] C.S. Hsu, H.C. Yee and W.H. Cheng, J. Appl. Mech., Vol. 44, pp., 1977, pp. 147-153.
[17] C.S. Hsu, H.C. Yee and W.H. Cheng, J. Sound Vib., Vol. 50, 1977, pp. 95-116.
[18] R.M. May, Nature, Vol. 261, 1976, pp. 459-467.
[19] R.M. May, Science, Vol. 186, No. 15, 1974, pp. 645-647.
[20] T.Y. Li and J.A. Yorke, Am. Math. Monthly, Vol. 82, 1975, pp. 985-992.
[21] E.N. Lorenz, Tellus, Vol. 16, 1964, pp. 1-11.
[22] M.J. Feigenbaum, J. Stat. Phys., Vol. 19, 1978, pp. 25-52.
[23] C.S. Hsu and H.C. Yee, J. Appl. Mech., Vol. 44, 1975, pp. 870-876.
[24] S. Ushiki, Physica 4D, 1982, pp. 407-424.
[25] F. Brezzi, S. Ushike and H. Fujii, *Numerical Methods for Bifurcation Problems*, T. Kupper, H.D. Mittleman and H. Weber eds., Birkhauser-Verlag, Boston, 1984.
[26] R. Schreiber and H.B. Keller, J. Comput. Phys., Vol. 49, No. 1, 1983.
[27] W.J. Beyn and E.J. Doedel, SIAM J. Sci. Statist. Comput., Vol. 2, 1981, pp. 107-120.
[28] R.B. Kellogg, G.R. Shubin, and A.B. Stephens, SIAM J. Numer. Anal. Vol. 17, No. 6, 1980, pp. 733-739.
[29] A.B. Stephens and G.R. Shubin, SIAM J. Sci. Statist Comput., Vol. 2, 1981, pp. 404-415.
[30] G.R. Shubin, A.B. Stephens and H.M. Glaz, J. Comput. Phys., Vol. 39, 1981, pp. 364-374.
[31] P.G. Reinhall, T.K. Caughey and D.W. Storti, Trans. of the ASME, J. Appl. Mech., 89-APM-6, 1989.
[32] E.N. Lorenz, Physica D, Vol. 35, 1989, pp. 299-317.
[33] A.J. Lichtenberg and M.A. Lieberman, *Regular and Stochastic Motion*, Appl. Math. Sci. Bd. 38, Springer-Verlag, New York, 1983.
[34] R.H. Miller, "A Horror Story about Integration Methods," J. Comput. Phys., to appear.
[35] W.A. Mulder and B. van Leer, AIAA-83-1930, July 1983.
[36] M. Prüffer, SIAM J. Appl. Math. Vol. 45, 1985, pp. 32-69.
[37] W.-J. Beyn, SIAM J. Numer. Anal., Vol. 24, No. 5, 1987, pp. 1095-1113.
[38] H.C. Yee, P.K. Sweby and D.F. Griffiths, "Dynamical Approach Study of Spurious Steady-State Numerical Solutions for Nonlinear Differential Equations, Part I. The ODE Connection and Its Implications for Algorithm Development in Computational Fluid Dynamics," submitted to J. Comput. Phys., March 1990, also NASA TM-102820, April 1990.
[39] A.R. Mitchell and D.F. Griffiths, Report NA/88 July 1985, Department of Mathematical Sciences, University of Dundee, Scotland U.K.
[40] A.R. Mitchell and J.C. Bruch, Jr., Numerical Methods for PDEs, Vol. 1, 1985, pp. 13-23.
[41] A.R. Mitchell, P. John-Charles and B.D. Sleeman, Numerical Analysis Report 93, May 1986, Department of Mathematical Sciences, University of Dundee, Scotland.
[42] V.S. Manoranjan, A.R. Mitchell, and B.D. Sleeman, J. Comput. App. Math., Vol. 11, 1984, pp. 27-37.
[43] B.D. Sleeman, D.F. Griffiths, A.R. Mitchell and P.D. Smith, SIAM J. Sci. Stat. Comput., Vol. 9, No. 3, May 1988, pp. 543-557.

[44] D.F. Griffiths and A.R. Mitchell, Report NA/113, Jan. 1988, Dept. Math. and Compt. Science, University of Dundee, Scotland.

[45] A.R. Mitchell, G. Stein and M. Maritz, Comm. Appl. Num. Meth., Vol. 4, 1988, pp. 263-272.

[46] D.F. Griffiths and A.R. Mitchell, Inst. Math. Applics., J. Num. Analy., Vol. 8, 1988, pp. 435-454.

[47] A.R. Mitchell and S.W. Schoombie, J. Comp. Appl. Math., Vol 25, 1989, pp. 363-372.

[48] J.M. Sanz-Serna and F. Vadillo, Proceedings Dundee, 1985, G.A. Watson and D.F. Griffiths, eds., Pitman, London.

[49] A. Iserles, International Conference on Numerical Mathematics, Singapore, R.P. Agarwal, ed., Birkhauser, Basel, 1989.

[50] A. Iserles and J.M. Sanz-Serna, "Equilibria of Runge-Kutta Methods," Numerical Analysis Reports, DAMTP 1989/NA4, Univeristy of Cambridge, England, May 1989.

[51] A. Iserles, A.T. Peplow and A.M. Stuart, "A Unified Approach to Spurious Solutions Introduced by Time Discretisation, Part I: Basic Theory," DAMTP 1990/NA4, Numerical Analysis Reports, University of Cambridge, March 1990.

[52] A.M. Stuart, IMA J. Num. Anal., Vol. 9, 1989, pp. 465-486.

[53] A. Stuart, "The Global Attractor Under Discretisation," to appear, Proc. NATO Conference on Continuation & Bifurcation, 1989.

[54] A. Stuart and A. Peplow, "The Dynamics of the Theta Method," to appear, SIAM J. Sci. Stat. Comput.

[55] A. Stuart, SIAM Review, Vol. 31, No. 2, 1989, pp. 191-220.

[56] A. Stuart and M.S. Floater, "On the Computation of Blow-up," submitted to the European J. Appl. Math., May 1989.

[57] P.K. Sweby, H.C. Yee and D.F. Griffiths, "On Spurious Steady-State Solutions of Explicit Runge-Kutta Schemes," Numerical Analysis Report 3/90, U. of Reading, March 1990, also NASA TM-102819, April 1990.

[58] C. Grebogi E. Ott and J. Yorke, Science, Vol. 238, 1987, pp. 585-718.

[59] S.W. McDonald, C. Grebogi E. Ott and J. Yorke, Physica 17D, 1985, pp. 125-153.

[60] C. Grebogi E. Ott and J. Yorke, Physics 7D, 1983, pp. 181-200.

SOLUTION TYPE	ODEs or PDEs	DISCRETIZED COUNTERPARTS
# OF ASYMPTOTES OR STEADY-STATE SOLUTIONS	SINGLE	SINGLE
	SINGLE	MULTIPLE
	MULTIPLE	SAME # OF MULTIPLE
	MULTIPLE	ADDITIONAL # OF MULTIPLE
PERIODIC SOLUTIONS	NO	YES
	YES	YES (+ EXTRA)
CHAOS	NO	YES
	YES	YES (+ EXTRA)

Table 4.1 Possible stable asymptotic solution behavior for DEs and their discretized counterparts.

Simulation of unsteady Flows

M. Meinke, D. Hänel

Aerodynamisches Institut, RWTH Aachen
Wüllnerstr. zw. 5 und 7, 5100 Aachen, Germany

Introduction

Viscous flows at sufficiently high Reynolds numbers tend to unsteady motion, self induced by flow separation. An example for that is the formation of a Karman vortex street behind an obstacle in a flow. Such a flow reacts very sensitive to small disturbances and is characterized by different scale lengths in time and space. The numerical simulation of this kind of flow requires much more care, but also a larger computational effort than an equivalent stationary problem.

The aim of the present paper is to study the influence of the numerical method on the solutions for unsteady flows. The Navier-Stokes equations for compressible, laminar flows are solved in two dimensions with an explicit Runge-Kutta time stepping method. The spatial derivatives are discretized with a node-centered and a cell-vertex scheme. One central difference scheme and two upwind schemes are used with the node-centered and one central difference scheme is used with the cell-vertex formulation. The results of steady test cases and unsteady flow simulations with the different schemes are compared with each other and experimental or analytical data.

Governing Equations

The unsteady motion of a viscous, compressible fluid is described by the Navier-Stokes equations, here presented for a two-dimensional curvilinear coordinate system (ξ, η, t) and in conservative form:

$$Q_t + E_\xi + F_\eta = 1/Re \left(E_{v\xi} + F_{v\eta} \right)$$

with the conservative variables $Q = J(\rho, \rho u, \rho v, e)^T$, the Euler fluxes E, F and the viscous fluxes E_v, F_v:

$$E, F = J \begin{bmatrix} \rho W \\ \rho u W + \omega_x p \\ \rho v W + \omega_y p \\ W(e+p) \end{bmatrix} \qquad E_v, F_v = J \left[\omega_x \begin{pmatrix} 0 \\ \tau_{xx} \\ \tau_{xy} \\ e_4 \end{pmatrix} + \omega_y \begin{pmatrix} 0 \\ \tau_{xy} \\ \tau_{yy} \\ f_4 \end{pmatrix} \right] \qquad \begin{array}{l} W = U, V \\ \omega = \xi, \eta \end{array}$$

J is the Jacobian of the metric terms $x_\xi y_\eta - x_\eta y_\xi$, which can be interpreted as the cell volume. U and V are the contravariant velocities and $\tau_{xx}, \tau_{xy}, \tau_{yy}, e_4, f_4$ are the viscous terms. Laminar flow and ideal gas behaviour is assumed for all flow simulations. The viscosity is computed by a function of the temperature: $\mu = T^{\frac{3}{4}}$.

Discretization

In the node-centered scheme the Euler fluxes are approximated by two upwind and one central difference scheme. For both upwind schemes the Euler fluxes are split according to van Leer's flux-vector splitting:

$$E^\pm, F^\pm = J \begin{bmatrix} \pm \rho \hat{a}/4 (\bar{M} \pm 1)^2 \equiv P_1^\pm \\ P_1^\pm [u + \frac{p\omega_x}{\rho \hat{a}}(-\bar{M} \pm 2)] \\ P_1^\pm [v + \frac{p\omega_y}{\rho \hat{a}}(-\bar{M} \pm 2)] \\ P_1^\pm H_t \end{bmatrix} \qquad \begin{array}{l} \hat{a} = a\sqrt{\omega_x^2 + \omega_y^2} \\ \bar{M} = (u\omega_x + v\omega_y)/\hat{a} \\ \\ \omega = \xi, \eta \end{array}$$

where a is the speed of sound and H_t the total enthalpy of the flow. The split energy flux is taken from [1]. For the second upwind scheme the following modification is added from the same paper [1], in order to reduce the numerical dissipation in viscous zones. The cartesian velocities in the split fluxes are there proposed to be computed as follows:

$$u, v = f(U^\pm, V^\pm) \qquad \text{e.g. for Flux } E: \quad V^\pm = f(\tilde{Q}) \begin{cases} \tilde{Q} = Q^+ \text{ for } U^\pm \geq 0 \\ \tilde{Q} = Q^- \text{ for } U^\pm < 0 \end{cases}$$

A MUSCL type discretization is applied to the Euler fluxes. The viscous fluxes are discretized with central differences for both the upwind and the central difference schemes.

The cell-vertex scheme is used with central differences for the Euler and viscous fluxes. All spatial derivatives are discretized with second order accuracy of the spatial step.

Method of Solution

The Navier-Stokes equations are integrated in time with an explicit Runge-Kutta time stepping scheme using 5 stages. Fourth order damping terms are added in the central difference schemes to ensure stability of the algorithm. The coefficients in the Runge-Kutta stages are optimized for maximum stability of the algorithm and are 0.059, 0.14, 0.273, 1/2, 1 for the upwind scheme and 1/4, 1/6, 3/8, 1/2, 1 for the central difference scheme, respectively.

Boundary conditions

For rigid walls the no slip condition was imposed on the flow, the pressure gradient normal to the wall is assumed to be zero and the wall is assumed to be adiabatic. For the free stream boundaries the characteristic variables were extrapolated along the characteristic directions. In simulations of unsteady flows around a circular cylinder the flow field was pertubed by a short clock- and then counterclockwise rotation of the cylinder. This pertubation is not necessary to initiate the vortex shedding, but a lot of computer time can be saved with this technique.

Results

For the validation of the algorithm several steady and unsteady flow problems were computed. The steady flow around a circular cylinder at low Reynolds numbers and the flow around an impulsively started circular cylinder are presented here as typical examples.

The lengths of the closed wake behind a circular cylinder in stationary flow obtained with the different schemes are shown in Fig. 1 as a function of the Reynolds number. The results are in good agreement with experimental data of [2] and numerical results of [3]. Small deviations between the original van Leer flux-vector splitting and the other numerical results can be seen at a Reynolds number of 40. This indicates, that less accurate results are obtained with this scheme using the same grid.

The time history of the length of the closed wake is shown in Fig. 2a for an impulsively started circular cylinder at a Reynolds number of 3000 and a Mach number of 0.3. The solution done with 225x225 grid points is in good agreement with experimental data of [2] and with a numerical result of [4]. Solutions on a coarser grid (113x81 grid points) show increasing deviations for larger extension of the wake. The vorticity distribution on the surface for the same case is presented in Fig. 2b at a dimensionless time of 2.5. Here deviations become apparent between the solution with the fine grid and all other solutions with the coarser grid, especially in the secondary separation zones, where the scale lengths are small. With the same grid, however, no solution scheme shows significant better results.

The following figures refer to the unsteady flow around a circular cylinder with a Reynolds number of 3000 and a Mach number of 0.3. An impression of this flow is given in Fig. 3, where the computed streaklines of the Karman vortex street are plotted. A small part of the time history of the lift coefficient and the lines of constant pressure of the solution with the cell-vertex scheme are shown in Fig. 4a,b. The Strouhal number deduced from these results is 0.215, which is in the range of experimental data. Further computations, however, have revealed a strong influence of the numerical method and the computational grid. A number of simulations were therefore carried out with varying numerical parameters.

The results of two computations with a different magnitude of the fourth order damping terms in a node-centered central difference scheme are shown in Fig. 5a-d. Similar to [5] it is found that there is a strong influence of the damping terms on the solution. Both the Strouhal numbers and the amplitude of the lift coefficient differ substantially. The comparison of the figures with the lines of constant pressure reveal, that the small scale structures are damped out, which are, however, essential for the self induced separation process.

The comparison of the solution with the node-centered (Fig. 5a,b) and the cell-vertex scheme (Fig. 4a,b) shows qualitatively agreement for the pressure field, but the Strouhal number and the amplitude of the lift coefficient differ considerably. The same computation carried out using the van Leer flux-vector splitting, results in a much different flow structure in the wake as shown in Fig. 6a,b. The Strouhal number and the amplitude of the lift coefficient tend to values as achieved by the solution with the the node centered scheme and the larger damping terms in Fig. 5c,d.

The preceeding computations were done with a mesh, equidistant in circumferential direction. In practical computations meshes are usually stretched in all directions, to reduce the amount of grid points. Therefore comparisons were made with solutions obtained with stretched grids in circumferential direction, but with the same number of grid points. In Fig. 7a,b two time histories of the lift coefficients are presented, which were obtained with the node-centered central difference scheme. The solution done with an equidistant grid, Fig. 7a, and that with a stretched grid in circumferential direction, Fig. 7b, show a remarkable influence of the variable step sizes. Not only the Strouhal number and the amplitudes of the lift coefficients differ, but also the time averaged lift coefficient of the solution with the stretched grid is no longer zero. This unphysical behaviour may be explained by the drift of the vortices into regions of different step sizes, where the truncation error and the damping properties change. This results in a nearly unstaggered vortex street, shifted additionally above or below the axis of symmetry. The same effect in a weaker form can be observed using a scheme with a large amount of numerical dissipation as in Fig. 5d, 6b.

Conclusions

The results presented in this paper show, that simple types of unsteady flows, like accelerated body problems, where the unsteadyness is enforced or initiated by the boundary conditions are not very sensitive to small changes in the solution scheme. Unsteady flows due to self induced separation, however, react very sensitive to a change of the numerical viscosity in the solution scheme. This is due to the fact, that the numerical dissipation acts as a filter on the small scale vortices proportional to the step sizes. The filtering plays an essential role in the temporal development of the flow structure. The numerical dissipation in the solution scheme should therefore be reduced to a minimum. It was also shown that grid stretching has a strong influence on the solution and that it can lead to unphysical results for the same reasons. Concluding from that, numerical simulations of such flows require very accurate methods of solution and a grid, able to resolve the smallest significant scalings.

References

1 Schwane, R., Hänel, D.: *An implicit flux-vector splitting scheme for the computation of viscous hypersonic flow*, AIAA-paper No. 89-0274,1989

2 Bouard, R.,Coutanceau, M.: *The early stage of development of the wake behind an impulsively started cylinder for $40 < Re < 10^4$*, JFM, Vol. 101, pp.583-607, 1980

3 Dennis, S.C.R., Chang, G.: *Numerical solutions for steady flows past a circular cylinder at Reynolds numbers up to 100*, JFM, Vol. 42, pp.471-489, 1970

4 Ta Phuoc Loc, Bouard, R.: *Numerical solution of the early stage of the unsteady flow around a circular cylinder: a comparison with experimental visualization and measurements*, JFM, Vol. 160, pp.93-117, 1985

5 Dortmann, K.: *Computation of viscous unsteady Compressible flows about airfoils*, Proc. of Conf. on Low Reynolds Number Aerodynamics, Notre Dame, Indiana, 1989

Figures

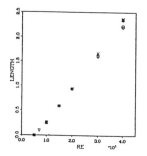

▽ Dennis, Chang, num. (1970)
◇ Coutanceau, Bouard, exp. (1980)
Navier-Stokes solution with:
O Van Leer flux-vector splitting
△ Van Leer, Hänel flux-vector splitting
+ central difference scheme, node centered
× cell-vertex scheme

Fig. 1 Steady flow around a circular cylinder, Navier-Stokes solution, Ma=0.3, 113x81 grid points

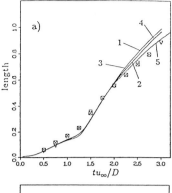

Fig. 2 Unsteady flow around an impulsively started circular cylinder, Navier-Stokes solution, Re=3000, Ma=0.3, 113x81 grid points, different solution schemes

a) time history of the length of the closed wake

b) vorticity distribution on the surface of the cylinder, $tU_\infty/D=2.5$ ($\theta=0$ for the rear stagnation point)

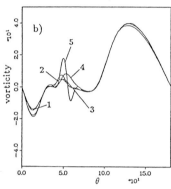

⊠ numerical result by
  Ta Phuoc Loc, Bouard, (1985)
▽ experiment by
  Coutanceau, Bouard, (1980)

1 Van Leer flux-vector splitting
2 Van Leer, Hänel flux-vector splitting
3 central difference scheme, node-centered
4 cell-vertex scheme
5 central difference scheme, node-centered (225x225 grid points)

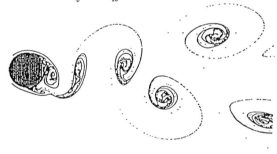

Fig. 3 unsteady flow around a circular cylinder, Re=3000, Ma=0.3, 177x113 grid points, computed streaklines of a Navier-Stokes solution

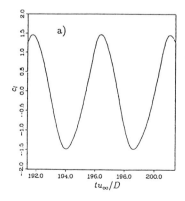

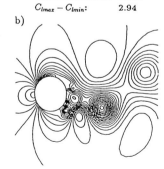

Strouhal-number: 0.215
$C_{lmax} - C_{lmin}$: 2.94

Fig. 4 legend see Fig. 3, solution with a cell-vertex scheme

a) time history of the lift coefficient

b) lines of constant pressure

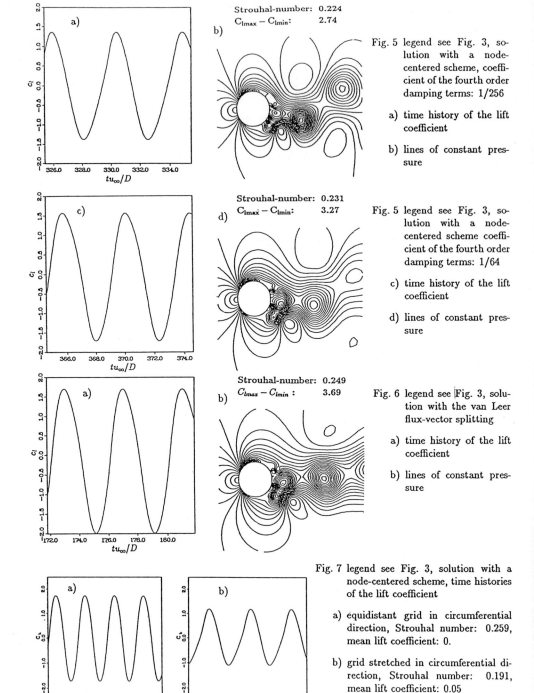

Fig. 5 legend see Fig. 3, solution with a node-centered scheme, coefficient of the fourth order damping terms: 1/256

a) time history of the lift coefficient

b) lines of constant pressure

Fig. 5 legend see Fig. 3, solution with a node-centered scheme coefficient of the fourth order damping terms: 1/64

c) time history of the lift coefficient

d) lines of constant pressure

Fig. 6 legend see Fig. 3, solution with the van Leer flux-vector splitting

a) time history of the lift coefficient

b) lines of constant pressure

Fig. 7 legend see Fig. 3, solution with a node-centered scheme, time histories of the lift coefficient

a) equidistant grid in circumferential direction, Strouhal number: 0.259, mean lift coefficient: 0.

b) grid stretched in circumferential direction, Strouhal number: 0.191, mean lift coefficient: 0.05

RECENT PROGRESS IN MULTIDIMENSIONAL
UPWINDING

P.L. Roe
Department of Aerospace Engineering
University of Michigan, Ann Arbor MI 48109, USA.

H. Deconinck, R.J. Struijs
von Karman Institute for Fluid Dynamics
B-1640 Rhode-St. Genese, Belgium.

I. INTRODUCTION

A satisfactory advection scheme is an essential ingredient of every fluid transport code. To the extent that wave propagation resembles advection, it is also the basis for numerical treatment of all hyperbolic problems. The furious rate at which new advection schemes are still published demonstrates however that no completely satisfactory scheme has yet been found, particularly for the multi-dimensional case.

The schemes reported in this paper do not fit easily into the conventional finite difference/volume/element categories. They are multi-dimensional generalisations of schemes first reported to the 1980 conference in this series [1]. They are designed for use on unstructured triangular or tetrahedral meshes, and make no assumptions relating propagation direction to mesh geometry. A rich theoretical framework is emerging, and many directions have yet to be explored. However, enough has been done to demonstrate a substantial degree of promise.

All the schemes will be presented in their two-dimensional forms; the three-dimensional extensions are straightforward.

II THE LINEAR SCALAR CASE

The governing equation is $u_t + \vec{a} \cdot \nabla u = 0$ where $\vec{a} = (a,b)$ is a constant flow vector, and the mesh is composed of arbitrary triangles, of which Fig.1 is an example. As in finite-element methods, the solution is thought of as the piecewise linear function that interpolates a given set of nodal values. With such an interpretation, the integral of u_t over each element T is

$$\varphi_T = \iint_T u_t \, dxdy = - \iint_T \vec{a} \cdot \nabla u \, dxdy$$
$$\oint_{\partial T} u\vec{a} \cdot \vec{dn} \qquad (2.1)$$

where $\vec{dn}$ is the inward normal to an element of the boundary. Since u is assumed to vary linearly along each edge, (2.1) can be expressed as

$$\varphi_T = -\sum_{1}^{3} k_i u_i \qquad (2.2)$$

where
$$k_i = \tfrac{1}{2} \vec{a} \cdot \vec{n}_i \qquad (2.3)$$

and $\vec{n}_i$ is the inward vector normal to edge E_i, whose magnitude is the length of E_i. Note that $k_1 + k_2 + k_3 = 0$, and that <u>positive</u> k: signifies flow <u>inward</u> through E_i. The quantity φ_T is called the fluctuation in triangle T.

The computational strategy will be to update each vertex value of T by some fraction of φ_T; e.g.

$$S_i u_i^{n+1} = S_i u_i^n - \alpha_i \Delta t \varphi_T + (\cdot) , \quad i = 1,2,3, \qquad (2.4)$$

where $(\cdot)$ denotes terms coming from other triangles in the set $\{T_i\}$ of triangles having V_i as a vertex. Here S_i is the area that weights u_i in the integration of u; it is one-third the area of all triangles in $\{T_i\}$. The scheme is conservative, in the sense that the integral of u is changed only by boundary terms, if, in each triangle, $\alpha_1 + \alpha_2 + \alpha_3 = 1$.

"Positive" schemes [2] are defined by every new value u_i^{n+1} being a convex combination of old values u^n. Such schemes produce no overshoots and are necessarily stable. Positivity in the one-dimensional case can be achieved by upwinding. It seems plausible to seek positivity in two dimensions by considering the weights α to depend on the flow direction and the geometry of the trinagles. Unfortunately it is easily shown [3] that no such choice of weights can lead to positivity; the weights must also depend on the data being processed. This is a fundamental difference between one and more-dimensional cases.

A simple positive scheme, S1, can be obtained as follows. In those triangles for which only one side, S_j, is an inflow side, set $\alpha_j = 1$ with the remaining weights zero. This step by itself corresponds to updating u_j using the characteristic through V_j. In some ways this would be an optimal method, but it is not conservative, because fluctuations on those triangles with two inflow sides are being ignored. For those triangles: if, say k_i and k_j are positive, then a positive scheme is

$$S_i u_i^{n+1} = S_i u_i^n - \Delta t k_i (u_i - u_k) + (\cdot) \qquad (2.5a)$$

$$S_j u_j^{n+1} = S_j u_j^n - \Delta t k_j (u_j - u_k) + (\cdot) \qquad (2.5b)$$

where V_k is the third vertex of this triangle, and the changes made sum to $\Delta t \varphi_T$ because $\sum k = 0$. The scheme is positive under the Courant restriction.

$$\Delta t \leq \min\left(\frac{S_i}{k_i}, \frac{S_j}{k_j}\right)$$

Results from this scheme are shown in Fig.2b, compared with those from a standard finite-volume upwind scheme using the same mesh (Fig.2a). The test is the transport of a square pulse in a fluid undergoing solid-body rotation; $(a,b) = (y,-x)$, and the results are much improved.

A further improvement comes from noting that because the analysis is local to each triangle, it remains valid if the velocity $\vec{a}$ is replaced locally by (Figure 4)

$$\vec{a}^* = \vec{a} + \lambda \vec{a}_p \qquad (2.6)$$

where $\vec{a}_p$ is the component of $\vec{a}$ parallel to the level lines of u.

Since $\vec{a}_p \cdot \nabla u = 0$, the residual is unchanged. We generate a new set of coefficients $k_i^* = k_i + \frac{1}{2}\lambda \vec{a}_p \cdot \vec{n}_i$, but the scheme does not 'know' which set is correct, and remains monotone whichever is used. Intuitively, the 'best' choice is

$$\vec{a}^* = \vec{a}_m = \frac{(\vec{a} \cdot \nabla u)}{(\nabla u \cdot \nabla u)} \nabla u \qquad (2.7)$$

which is the smallest of the equivalent advection speeds, and permits the largest timesteps. Conceptually, $\vec{a}_m$ may be regarded as a wavespeed arising from interaction between the data (∇u) and the p.d.e. ($\vec{a}$). At the scalar level, such a thought may seem fanciful, but we believe it holds the key to correct treatment of systems. In our scalar experiments, both resolution and convergence are usually enhanced by choosing $\vec{a}_m$, except that in regions where the solution is nearly constant, ∇u, and hence $\vec{a}_m$ are ill defined. We have used

$$\vec{a}^* = \vec{a}_m + \beta(|\nabla u|)\vec{a}_p \qquad (2.8)$$

where β tends to unity as $\nabla u \to 0$. This gives the results in Fig. 1c.

It is possible to define a second-order, Lax-Wendroff type scheme by taking

$$\alpha_i = \frac{1}{3} + \frac{1}{2}\nu_i \text{ with } \nu_i = -\frac{k_i \Delta t}{S_i} \qquad (2.8)$$

The results from this scheme appear in Fig. 1d. It is not strictly legitimate to use k_i^* in (2.8) because second-order accuracy requires $\vec{a}$ to be consistently defined from triangle to triangle. In the steady state, solutions obtained from (2.8) are identical [3] to those obtained from the SUPG finite-element method [4] if the free parameter τ appearing in that method is taken to be $\frac{1}{2}\Delta t$.

Figure 3 summarises the results of this section by plotting $u(x)$ along $y = 0$.

III NON-LINEAR SCALAR PROBLEMS

Since all the analysis is triangle-by-triangle, one can define equivalent linear problems in each element. For example, to solve

$$u_t + f(u)_x + g(u)_y = 0 \tag{3.1}$$

one defines $a = f_u$, $b = g_u$ in each triangle as

$$\tilde{a} = \frac{f_P - f_Q}{u_P - u_Q} \quad , \quad \tilde{b} = \frac{g_R - g_S}{u_R - u_S} \tag{3.2}$$

where f_P, g_S, u_P, u_S are obtained by linear interpolation along the edges (Figure 5). It is easily shown that the fluctuation φ is unaltered by this substitution; hence the scheme remains conservative. Results for a problem with $f(u) = \frac{1}{2}u^2$, $g(u) = u$ appear in Figure 6. They were obtained using scheme S1 with the definition (2.8).

IV SYSTEMS OF EQUATIONS

A system of conservation laws such as

$$\underline{u}_t + \underline{F}(\underline{u})_x + \underline{G}(\underline{u})_y = 0 \tag{4.1}$$

can be reduced to a set of scalar problems if and only the Jacobian matrices $A = F_u$ and $G = B_u$ commute, which is rare in practice. However, interaction of particular data with the p.d.e. always generates disturbances that can be represented by a finite number of simple waves [5,6]. Indeed, the representation is not unique, and the alternatives still have to be explored. However, once a local scalar decomposition is arrived at, we can make use of the above scalar schemes on each component. It has proved essential to identify the correct propagation direction for each wave. For example, entropy waves are propagated in the direction ∇S, not in the direction of the flow. Some preliminary results for the Euler equations are shown in Figure 7. The problem is shock reflection at $M_\infty = 2.9$. The wave decomposition used is a slight variation of [5], although [6] has also been used successfully.

V FUTURE WORK

Two major steps remain to be completed before a viable scheme for multi-dimensional conservation laws will exist. At the scalar level, a high-order positive scheme can be created by applying limiters, not to the element gradients, but to the distribution formulae. At the system level, the relative merits of different wave models need to be explored.

With all the ingredients in place to create a scheme that reflects the true multi-dimensional physics, we hope to see benefits not only in accuracy but also in convergence and robustness.

REFERENCES

1. P.L. Roe. Lecture Notes in Physics 141 p354, Springer 1981.
2. SP Speckreijse. Math. Comp. 49, p135, 1987.
3. H. Deconinck, R. Struijs, P.L. Roe. VKI Lecture Series Notes 1990 - 03.
4. C. Johnson, VKI Lecture Series Notes. 1990 - 04.

5. P.L. Roe, J. Comp. Phys. 63, p 458 1986
6. H. Deconinck, Ch. Hirsch, J. Peuteman, Lecture Notes in Physics 264, p 216, Springer, 1986.

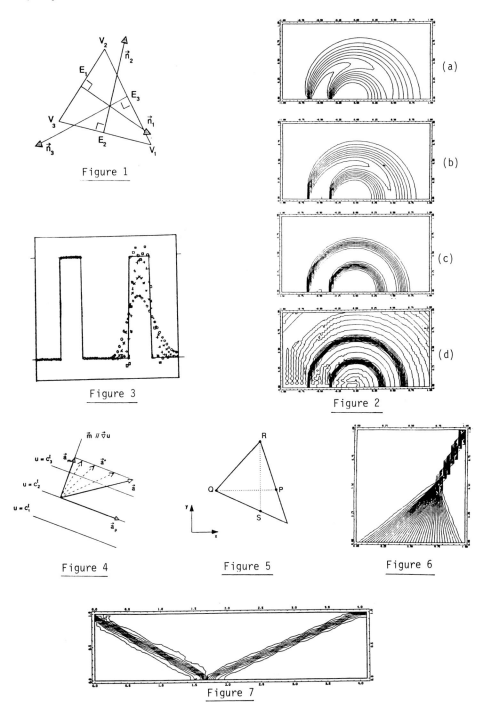

Figure 1

Figure 3

Figure 2

Figure 4

Figure 5

Figure 6

Figure 7

Nonlinear Galerkin Method with Finite Element Approximation

Jacques LAMINIE, Frédéric PASCAL and Roger TEMAM
Laboratoire d'Analyse Numérique, bât. 425
Université PARIS SUD
91405 ORSAY CEDEX, FRANCE

In numerical simulations of incompressible flows represented by the Navier–Stokes equations :

$$\begin{aligned}\frac{\partial u}{\partial t} - \nu \Delta u + u.\nabla u + \nabla p &= f & \text{in } Q &= \Omega \times [0,t] \\ \nabla.u &= 0 & \text{in } Q & \\ u(x,t)_{|\partial \Omega} &= 0 & \forall t &\in [0,T] \\ u(x,0) &= u_0(x) & & \end{aligned} \quad (1)$$

one of the main difficulties is to integrate the equations on large intervals of time with values of the parameters which are physically relevant.

When we reach a turbulent regime, flows become time dependent, and an essential aspect of these flows is the interactions between small and large eddies. The Nonlinear Galerkin Method that stems from recent developments in dynamical systems theory proposes a new method for taking into account these interactions.

It was shown that the permanent regime is represented by a global attractor $\mathcal{A}$ which attracts all the orbits. Since the attractor can be a complicated set, the idea of an inertial manifold which contains $\mathcal{A}$ and which attracts all the orbits at an exponential rate was introduced [1]. But the existence of inertial manifolds for the Navier–Stokes equations is not proved, and the idea of the Nonlinear Galerkin Method is to look for the solution on some approximate inertial manifolds which attract all the orbits in a thin neighborhood.

Until now, theory [6] and numerical tests [3] have been made with spectral methods ; we present here this method in the case of finite elements discretizations. A natural way for introducing small and large structures in finite elements is by using multilevel discretizations and hierarchical bases.

1 Hierarchical basis for P1–elements in the two–dimensional case.

Let Ω be an open bounded set of $\mathcal{R}^2$. We are given an admissible triangulation of Ω denoted by T_{2h} and a finer triangulation T_h, generated by subdiving a triangle τ of T_{2h} in d^2 ($d \geq 2$) congruent subtriangles (fig. 1).

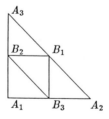

Figure 1: Subdivision of the triangle τ into 4 subtriangles

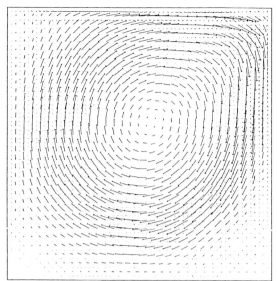

Figure 2: Driven cavity, $Re = 1000$: velocity decomposed on $V_{2h} \oplus W_h$.

Associated with the triangulation T_h (resp. T_{2h}), we have the finite element space V_h (resp. V_{2h}) which is the space of continuous piecewise linear functions from Ω into $\mathcal{R}$, which are linear on each $\tau \in T_h$ (resp. $\tau \in T_{2h}$). Let us denote by $V_{0,h}$ (resp. $V_{0,2h}$), the space of functions v of V_h (resp. V_{2h}) such that $v_{|\partial \Omega} = 0$. Of course we have :

$$V_{0,2h} \subset V_{0,h} \subset H_0^1(\Omega)$$

Let S_h (resp. S_{2h}) and n_h (resp. n_{2h}) denote the set and the number of interior nodes of T_h (resp. T_{2h}). Nodes created by the refinement are numbered from $n_{2h} + 1$ to n_h. Let $\phi_{h,i}$ (resp. $\phi_{2h,i}$) denote the canonical basis function of $V_{0,h}$ (resp. $V_{0,2h}$) associated to the ith nodes. We define the hierarchical basis of $V_{0,h}$ by :

- the canonical basis functions $\phi_{2h,i}$ of $V_{0,2h}$ for the first n_{2h} nodes
- the canonical basis functions $\phi_{h,i}$ of $V_{0,h}$ for the last $n_h - n_{2h}$ nodes.

Let $W_{0,h}$ be the space spanned by the functions of the canonical basis of $V_{0,h}$ associated to the $n_h - n_{2h}$ nodes created by the refinement, then $V_{0,h} = V_{0,2h} \oplus W_{0,h}$. If $u_h \in V_h$ then :

$$u_h = \sum_{i=1}^{n_{2h}} y_i \phi_{2h,i} + \sum_{i=n_{2h}+1}^{n_h} z_i \phi_{h,i}$$

where y_i and z_i are uniquely defined by (fig. 2) :

$$\begin{aligned} y_i &= u_h(A_i) & A_i \in S_{2h} \\ z_i &= u_h(B_i) - \frac{u_h(A_{i,1}) + u_h(A_{i,2})}{2} & B_i &= \mathrm{mid}[A_{i,1}, A_{i,2}] \in S_h \setminus S_{2h} \end{aligned} \qquad (2)$$

2 Numerical tests with the Dirichlet problem

Some preliminary studies have been made for solving the Dirichlet problem in a cavity to test the behaviour and the implementation of the hierarchical finite elements on several grids. We consider the classical Dirichlet problem : find u solution of

$$-\Delta u = f \text{ in } \Omega \; ; \; u_{|\partial\Omega} = 0 . \tag{3}$$

By projecting the equations on $V_{0,2h}$ and $W_{0,h}$, (3) is discretized in :

$$\begin{array}{ll}(\nabla(y_h + z_h), \nabla \bar{y}_h) = (f, \bar{y}_h), & \forall \bar{y}_h \text{ in } V_{0,2h} \\ (\nabla(y_h + z_h), \nabla \bar{z}_h) = (f, \bar{z}_h), & \forall \bar{z}_h \text{ in } W_{0,h}\end{array} \tag{4}$$

where $u_h = y_h + z_h$. This is equivalent to the following linear system :

$$A\begin{pmatrix} y \\ z \end{pmatrix} = \begin{pmatrix} A_{cc} & A_{cf} \\ A_{fc} & A_{ff} \end{pmatrix}\begin{pmatrix} y \\ z \end{pmatrix} = \begin{pmatrix} b_c \\ b_f \end{pmatrix}$$

where $A_{cc}, A_{cf} = {}^tA_{fc}, A_{ff}$ are matrices, and y is the column of coordinates of y_h and z is the column of coordinates of z_h in the hierarchical basis.

Solvers of this linear system are based on block iterations of a symmetric, block, Gauss-Seidel method. The iterative procedure is initialized by solving $A_{cc}y^{(0)} = b_c$. The 2 linear systems which have to be solved are smaller than the original one. The equation in A_{cc} can be solved :

- inductively (using coarser grids)
- directly (if the grid is coarse enough)
- with a conjugate gradient which is preconditionned by an incomplete Cholesky method.

Since the values of z are small, only a few iterations of a conjugate gradient or a symmetric Gauss-Seidel method are sufficient to solve the equation in A_{ff}. These multigrid algorithms are compared with :

- a conjugate gradient of A which is preconditionned by the first block iteration of the above described procedure in the case where $V_{0,h}$ is provided with the hierarchical basis.

- a conjugate gradient of A which is preconditionned by an incomplete Cholesky method in the case where $V_{0,h}$ is provided with the nodal basis.

We observed that the block iterative method based on the hierarchical basis is very performant (as it is observed with the usual multigrid method). The difference is greater as the number of nodes and the number of grids increase (fig. 3) .

3 Implementation of the hierarchical basis in the Navier-Stokes problem

The approximation of the Navier-Stokes problem is based on the mixed finite element $IsoP1 - P1$ [2]. The 2 finite elements spaces are :

- $(V_{0,h})^2$ for velocity
- V_{2h} for pressure.

When $V_{0,h}$ and V_{2h} are provided with their nodal basis, the approximate variational problem gives the usual Galerkin method, namely : find $u_h \in V_{0,h}$ such that :

$$\begin{array}{ll}\frac{\partial(u_h, \bar{u}_h)}{\partial t} + a(u_h, \bar{u}_h) + (u_h.\nabla u_h, \bar{u}_h) + (\nabla p_h, \bar{u}_h) = (f, \bar{u}_h) & \forall \bar{u}_h \text{ in } V_{0,h} \\ (\nabla.u_h, \bar{p}_h) = 0 & \forall \bar{p}_h \text{ in } V_{2h}\end{array} \tag{5}$$

where $a(u, v) = \nu(\nabla u, \nabla v)$

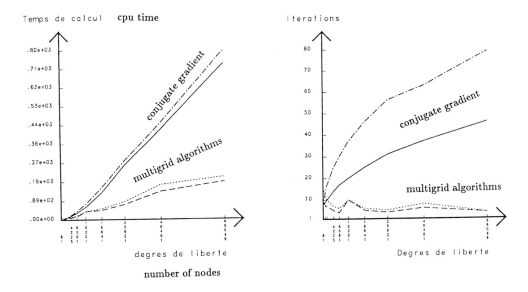

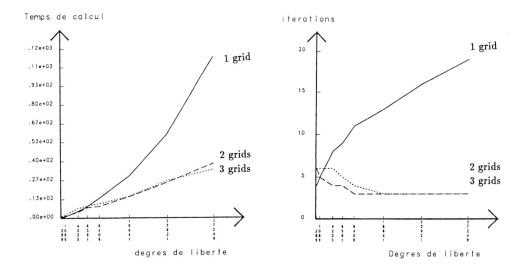

Figure 3: Comparison of the different agorithms for the Dirichlet problem

The discretization in time is done with a 3 steps splitting method [4]. 2 steps consist in solving a linear Stokes problem (by the Uzawa method) and 1 step consists in solving a nonlinear problem (by a least square method).

When $V_{0,h}$ and V_{2h} are provided with their hierarchical basis [5], then the approximated variational form of the Navier-Stokes equations is :

$$u_h = uy_h + uz_h \quad uy_h \in V_{0,2h}, \quad uz_h \in W_{0,h}$$

$$p_h = py_h + pz_h \quad py_h \in V_{4h}, \quad pz_h \in W_{2h}$$

$$\begin{aligned}
\frac{\partial(uy_h + uz_h, vy_h)}{\partial t} &+ a(uy_h + uz_h, vy_h) + (\nabla py_h, vy_h) \\
+ \ ((uy_h + uz_h).\nabla(uy_h + uz_h), vy_h) &= (f, vy_h) \quad \forall vy_h \in V_{0,2h} \\
(\nabla.uy_h, qy_h) &= 0 \quad \forall qy_h \in V_{4h}
\end{aligned} \quad (6)$$

$$\begin{aligned}
\frac{\partial(uy_h + uz_h, vz_h)}{\partial t} &+ a(uy_h + uz_h, vz_h) + (\nabla py_h, vz_h) \\
+ \ ((uy_h + uz_h).\nabla(uy_h + uz_h), vz_h) &= (f, vz_h) \quad \forall vz_h \in W_{0,h} \\
(\nabla.uz_h, qz_h) &= 0 \quad \forall qz_h \in W_{2h}
\end{aligned} \quad (7)$$

where :

$$V_{0,h} = V_{0,2h} \oplus W_{0,h}$$

$$V_{2h} = V_{4h} \oplus W_{2h}$$

At each time step, y_h is determinated on the coarse grid by solving the 3 steps of a splitting method, then the correction z_h is computed on the fine grid by solving 1 step of an Euler method with taking into account the new value of y_h. Tests are in the process of being realized.

References

[1] C. Foias, O. Manley and R. Temam, *On the interaction of small and large eddies in two-dimensional turbulent flows*, Math. Mod. and Num. Anal. (M2AN), 22, 1988, 93-114.

[2] M. Bercovier and O. Pironneau, *Error estimates for Finite Element Method of the stokes problem in the primitive variables*, Numer. Math. 33, p211-224, (79).

[3] F. Jauberteau, C. Rosier and R. Temam, *A Nonlinear Galerkin Method for Navier Stokes Equations*, to appear in Proc. Conf. on Spectral and High Order Methods for P.D.E. ICOSAHOM'89, Italie, 1989.

[4] R. Glowinski, B. Mantel and J. Periaux, *Numerical solution of the time dependent Navier-Stokes equations for incompressible viscous fluids by finite element and alternating direction methods*, Numer. Meth. in Aero. Fluid Dyn., ed Roe, Reading 1981.

[5] M. Marion and R. Temam, *Nonlinear Galerkin Methods ; the finite elements case*, Numerische Mathematik, 57, 1990, p1-22.

[6] M. Marion and R. Temam, *Nonlinear Galerkin Methods*, SIAM J. Num. Anal., 26, 1989, p1139-1157.

[7] R. Temam, *Sur l' approximation de la solution des équations de Navier-Stokes par la méthode des pas fractionnaires(I)*, Arch. Rational Mech. Anal., 32, 1969, p135-153

[8] R. Temam, *Sur l' approximation de la solution des équations de Navier-Stokes par la méthode des pas fractionnaires(II)*, Arch. Rational Mech. Anal., 33, 1969, p377-385

THREE-DIMENSIONAL HIGHLY ACCURATE MmB SCHEMES FOR VISCOUS, COMPRESSIBLE FLOW PROBLEMS

Huamo WU[†] and Koichi OSHIMA[‡]

[†] Computing Center, Academia Sinica; Beijing 100080 CHINA
[‡] The Institute of Space and Astronautical Science, Sagamihara 229, JAPAN

Abstract

In this paper the so-called MmB schemes [1,2,3] with high accuracy and high resolution are developed and applied to three-dimensionsl viscous flow at various Mach numbers. Typical flow fields around cone-like bodies are calculated.

1 INTRODUCTION

The so-called total variation diminishing (TVD) schemes are widely applied in CFD [4], especially for complex flows with strong discountinuities. These methods show high accuracy and high resolution abilities. Unfortunately, there is no second order accurate TVD scheme for scalar conservation laws in two-dimensions [5], although the numerical experiments show that the space splitting seems to work well in two-dimenensions. Hence, it may be of interest to creat new concepts beyond TVD in multi-dimensional case.

Given the natural requirement that the numerical solution u of the Cauchy problem for the equation

$$\frac{\partial u}{\partial t} + \frac{\partial}{\partial x} f(u) = 0$$

at any point P is bounded by the local maximum and minimum bounds of the initial data in the smallest union of mesh element on the previous time step containing the domain of dependence of u on the mesh element with center at point P, we derived a new class of second order accurate high resolution schemes - local Maxima and $m$inima Bounds preserving (MmB) schemes [1,2,3]. This MmB scheme in one-dimension is almost identical to the TVD schemes. In the multi-dimensional cases, we have two classes of second-order accurate MmB schemes, one of which consists of space-dimensionally splitting schemes using one-dimensional second-order MmB (also TVD) schemes. Another class of the schemes consists of unsplitting modifications of the well known Lax-Wendroff scheme [1,2,3].

2 MmB SCHEMES FOR CFD

There are a variety of possible extentions of the MmB schemes to the nonlinear systems of conservation laws. In [3], we suggested to modify the MacCormack scheme using the Jacobian matrix splitting A into $A+$ and $A-$ technique. Here, for the steady flow calculations, we present a physically-splitting approach.

Consider, for example, the Euler equation in one-dimension

$$\frac{\partial Q}{\partial t} + \frac{\partial Qu}{\partial x} + \frac{\partial P}{\partial x} = 0, \qquad (1)$$

where $Q = (\rho, \rho u, e)^T$, $P = (0, p, pu)^T$, and ρ, u, e, p are the density, the velocity, the total energy and the pressure of the fluid. We split (1) into two sets of equations

$$\frac{\partial Q}{\partial t} + \frac{\partial F}{\partial x} = 0, \tag{2}$$

$$F = Qu, \tag{3}$$

$$F = P. \tag{4}$$

Firstly, the equations (2) are discretized according to their characteristic form to obtain

$$\hat{Q} = Q - \frac{\Delta t}{\Delta x} \Delta_0 F + \frac{1}{2} \frac{\Delta t}{\Delta x} \Delta_-(|A|Q\Delta_+ Q), \tag{5}$$

where

$$\hat{Q} = Q(t_{n+1}), \qquad Q = Q(t_n),$$

$$|A| = R |\wedge| R^{-1}, \qquad |\wedge| = \text{diag}(|\lambda_1|, |\lambda_2|, |\lambda_3|),$$

and λ_1, λ_2, λ_3 are the eigenvalues of $A = \partial F/\partial Q$, R_1, R_2, R_3 are the right eigenvectors of A. And R is a matrix whose three columes consist of R_1, R_2 and R_3, the symbols Δ_0, Δ_+ and Δ_- denote the central, forward and backward differences, respectively. The final difference scheme is unsplitted and has form

$$\hat{Q} = Q + C(Q) + D(Q), \tag{6}$$

where $C(.)$ and $D(.)$ are obtained from (3) and (4), repectively.

The obtained scheme (6) is of first-order accuracy. It is modified further by using flux limiters [6] to get higher resolution of discountnuities and second-order accurate for steady flow. In the case of the Navier-Stokes equations, the viscous terms V(Q) should be introduced to the right-hand side of (6):

$$\hat{Q} = Q + C(Q) + D(Q) + V(Q). \tag{7}$$

These viscous terms are obtained by using central differences.

3 NUMERICAL EXAMPLE

Present method has been used to calculate the viscous flows around bodies. The body under consideration has a plan of symmetry XOZ. It is a combination of a sphere (or a flatted sphere) with a cone-like configuration. The three curves of the intersection of this body surface with the lower-half plane XOZ ($\phi = 0$), upper-half plane YOZ ($\phi = \pi/2$) and upper-half XOZ ($\phi = \pi$) are depicted in Figures 1, 2 and 3 by solid lines with shadow.

The Navier-Stokes equations

$$\frac{\partial Q}{\partial t} + \frac{\partial F}{\partial x} + \frac{\partial G}{\partial y} + \frac{\partial H}{\partial Z} = V \tag{8}$$

are transformed into a strong consrvation form in the general body-fitting coordinates t, ξ, η, ζ:

$$\frac{\partial \tilde{Q}}{\partial t} + \frac{\partial \tilde{F}}{\partial \xi} + \frac{\partial \tilde{G}}{\partial \eta} + \frac{\partial \tilde{H}}{\partial \zeta} = \tilde{V}, \tag{9}$$

where $\tilde{V}$ is the viscous term, LHS is the Euler part of the equation.

The Boundary Conditions
The plane $\eta = 0$ represents the axis of the body which is treared as an interior line using equation (8), and $\eta = 1$ is the outflow boundary where linear extrapolation is performed. The planes $\zeta = \pm 1$ are the planes of symmetry. On the body ($\xi = 1$), the temperature T_W is specified as $T_W = \alpha T_\infty$ ($\alpha > 0$) and $(u, v, w)|_{\xi=1} = 0$. At the inflow boundary $\xi = 0$, two kinds of boundary treatment are

considered depending upon the inflow state. Namely, if the inflow is hypersonic, then all of the inflow parameters (ρ, p, u, v, w) are given at $\xi = 0$. In the case of subsonic inflow then if the Jacobian matrix of $\tilde{F}$ has m negative eigenvalues, m relative Riemann invariants are extrapolated from the interior mesh points and $5 - m$ inflow parameters are specified. The convergence occures after about 2000 iterations with residuals of oder 10^{-3}. The parameters used in the caluculations are as follows: $M_\infty = 5$, 0.1; $Re_L = 10^5$; L- the radius of the sphere; $\alpha = 1$; the numbers of the mesh points in ξ, η, ζ are $31 \times 41 \times 17$.

4 RESULTS

The problem we were given is to find out the optimum shape of a projectile which is ought to make sustained flight over a wide range of speed; from supersonic to very low speed. Particular interests are in the pressure distribution around the nose, so as to find out a position, if any, where the pressure coefficient is kept almost constant throughout all these speed range.

Because we want to use single algorithm for these various Mach numbers, and also to find out any singular or steep change of flow patterns without being obscured by viscous effect (real or artificial), the above-discussed MmB method was extended to the low speed range.

Figures 1, 2 and 3 show the distributions of pressure $\tilde{p} = p/p_{stag}$ along three generating curves of the bodies for sphere-cone-like (S-C) and flatted sphere-cone-like (FS-C) configurations. It is noted that very steep change of pressure distribution at the corner of flatted sphere is observed.

References

(1) H.M.Wu (1989) A new class of accureate high resolution schemes in two dimensions – Analysis and applications; Proc. 3rd ISCFD-Nagoya.
(2) H.M.Wu and S.L.Yang (1989) MmB – A new class of accurate high resolution schemes for conservation laws in two dimentions; Proc. 8th GAMM conference on numerical methods in fluid mechanics, pp.582-591.
(3) H.M.Wu and S.L.Yang (1989) MmB—A new class of acuurate high resolution schemes for conservation laws in two dimensions; Impact of computing in science and engineering, 1, pp.217-259.
(4) H.C.Yee (1987) Construction of explicit and implicit symmetric TVD scheme and their applications; J. Comput. Phys. vol.68 pp.151-179.
(5) J.B.Goodman and R.J.Le Veque (1989) On the accuracy of stable schemes for 2D conservation laws; Math. Comp. vol.45 no.171 pp.15-21.
(6) P.K.Sweby (1984) High resolution schemes using flux limiters for hyperbolic conservation laws; SIAM J. Numer. Anal. vol.21 pp.995-1011.

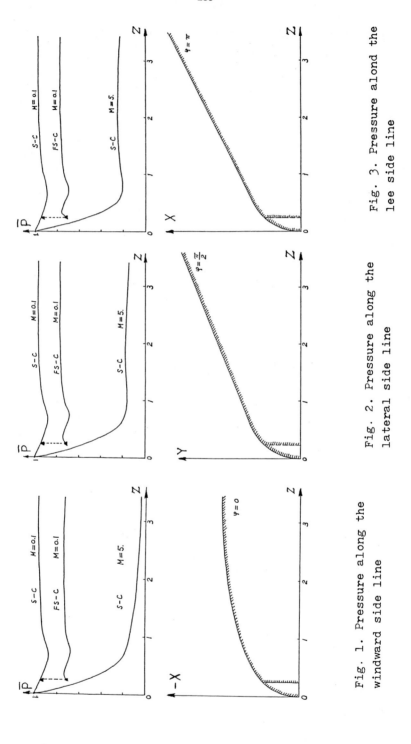

Fig. 1. Pressure along the windward side line

Fig. 2. Pressure along the lateral side line

Fig. 3. Pressure along the lee side line

A COMPARISON OF THE EFFECTS OF GRID DISTORTION ON FINITE-VOLUME METHODS FOR SOLVING THE EULER EQUATIONS

P. O'Brien and M. G. Hall
Aerodynamics Department, RAE Farnborough, Hants, GU14 6TD, UK

Summary

The accuracy of typical cell-centred and cell-vertex finite-volume methods on ideal and distorted two-dimensional grids is compared by performing test calculations for the flow past the RAE 2822 aerofoil. Both methods show appreciable sensitivity to grid distortion. Use of a modified cell-vertex method with an improved smoothing algorithm is shown to yield much lower sensitivity as well as lower error levels.

1 Introduction

In practical computations geometric and other constraints frequently force the production of distorted grids in spite of their known disadvantages. Such grids may be highly stretched, with large changes in cell size from one cell to the next, or may consist of elongated or skewed cells. Any appreciable departure from uniformity and orthogonality tends to lead to a loss of accuracy. Theoretical analysis of the performance of a numerical algorithm is difficult in these circumstances. In this paper, we describe a numerical experiment consisting of calculations of the compressible flow about the RAE 2822 aerofoil. High accuracy around the leading edge is especially desirable, because the overall flow is strongly influenced by the details there. However, grid distortions are generally unavoidable in this region because of the large surface curvature and, moreover, the solution gradients are high, so that the achievement of high accuracy is especially difficult.

The computations were performed on two grids with different degrees of distortion at the leading edge using three different finite-volume methods (A, B and C) for solving the Euler equations in two dimensions. The first of these is derived from Ref 1 and is of the cell-centred type; the second (Ref 2) is a cell-vertex code while the third method differs from the second only in having an improved smoothing algorithm. Two flow conditions have been treated, with subcritical and supercritical flow respectively in the region of the leading edge, but only the results for the subcritical flow are presented here because of limitations of space, but the conclusions for supercritical flow are essentially the same.

For all three methods we solve the system of equations

$$\frac{d}{dt}(hU) + QU - DU = 0, \qquad (1)$$

where U is the solution vector, h is a cell area, Q is a spatial discretisation operator, and D is the smoothing (or artificial dissipation) operator. The methods differ in the details of the system and in particular in the form adopted for the smoothing operator D.

2 Outline of methods

In method A the dependent variables are specified at the cell centres and the solution is advanced using a Runge-Kutta time-marching algorithm. Smoothing terms of second and fourth order are adopted to minimise odd-even decoupling and oscillations near regions of high solution gradient such as stagnation points and shock waves. Thus the smoothing operator D in (1) takes the form

$$DU = D_\xi U + D_\eta U, \qquad (2)$$

where, for example

$$D_\xi U = \nabla_\xi \left(\frac{h}{\Delta t} \left\{ \varepsilon^{(2)} \Delta_\xi U - \varepsilon^{(4)} \Delta_\xi \nabla_\xi \Delta_\xi U \right\} \right). \qquad (3)$$

Here, Δ_ξ and ∇_ξ respectively represent forward and backward undivided difference operators in the increasing ξ direction, and $\varepsilon^{(2)}$ and $\varepsilon^{(4)}$ are empirical coefficients based on second pressure differences. This smoothing scheme is conservative.

In method B the solution is advanced in time by means of a Lax-Wendroff time-stepping algorithm. The dependent variables are specified at the cell vertices, so that second-order accuracy can in principle be obtained for the flux across a cell face even for non-smooth grids. Numerical smoothing is required here also and, for cells A, B, C and D surrounding a grid point P, the smoothing change DU in equation (1) is given by

$$DU = \frac{\mu h}{\Delta t} \left(\bar{U}_A + \bar{U}_B + \bar{U}_C + \bar{U}_D - 4 U_P \right), \qquad (4)$$

where, for example, $\bar{U}_C$ is an average value for cell C given by the arithmetic mean of the values of U at the vertices of this cell. The coefficient μ is given by

$$\mu = \mu_0 \Delta S + f(\rho, \bar{R}), \qquad (5)$$

where μ_0 is an empirical constant, ΔS is an average cell dimension and $\bar{R}$ is a residual for point P. The term $\mu_0 \Delta S$ is a background smoothing term whereas $f(\rho, \bar{R})$ provides strong shock smoothing where required. This smoothing scheme is not conservative.

Method C is essentially the same as method B with the exception of the smoothing algorithm. The smoothing operator D in equation (1) is here given the form

$$D_\xi U = \nabla_\xi \left(h \varepsilon^{(2)} \Delta_\xi U - \frac{h}{\Delta t} \varepsilon^{(4)} \Delta_\xi (\nabla_\xi \Delta_\xi' U) \right) . \tag{6}$$

This is conservative but it differs from method A. Firstly, the background smoothing term contains a divided difference $\Delta_\xi' U$. Then, $\varepsilon^{(2)}$ is constructed from third differences of density rather than second pressure differences, and $\varepsilon^{(4)}$ is an empirical constant.

3 Grid and calculation details

Two grids were used in the course of the investigation. Both grids are of C-type with 128×16 quadrilateral cells. As shown inset in Fig 1, the ideal grid has a cell aspect-ratio of about unity at the leading edge. This aspect ratio is maintained over much of the surface. The distorted grid has a cell aspect-ratio up to about four in the leading edge region, but is similar to the ideal grid elsewhere.

Calculations were performed for the RAE 2822 aerofoil at the subcritical flow condition $M_\infty = 0.676$, $\alpha = 1.0°$. Each of the three methods was applied to obtain fully converged results on both the ideal and distorted grids. Fig 1 shows distributions of the total pressure function $1 - P/P_\infty$ on the aerofoil surface. This function gives a measure of the accuracy of a particular method since its value should be zero upstream of a shock wave and be equal to that given by the Rankine-Hugoniot relations downstream. Converged values of lift and drag coefficient are included.

4 Results and conclusions

The results show that, while the cell-centred method (Fig 1(A)) has by far the largest errors on the ideal grid, the original cell-vertex method (Fig 1(B)) shows a greater sensitivity to the grid distortion at the leading edge. The lower sensitivity of the cell-centred method is also reflected in the lift and drag coefficients. In comparison, the improved cell-vertex method (Fig 1(C)) has a markedly lower sensitivity to grid distortion than either of the previous methods, as well as lower error levels. Note that the lift and drag coefficients for this method also show much less variation between the two grids. Thus we have found that a great improvement in accuracy can be obtained by use of an improved smoothing algorithm alone.

References

1   Jameson, A., Schmidt, W., Turkel, E., (1981). Numerical solutions of the Euler equations by finite-volume methods using Runge-Kutta time-marching schemes. AIAA paper 81-1259.
2   Hall, M.G., (1986). Cell-vertex multigrid schemes for solution of the Euler equations, in Numerical Methods for Fluid Dynamics II, Eds K.W. Morton, M.J. Baines, Oxford, pp 303-345.

© *Controller, Her Majesty's Stationery Office, London 1990*

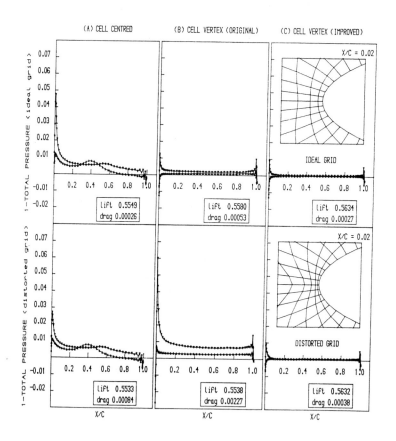

Fig 1 Subcritical flow results. RAE 2822, $M_\infty = 0.676$, $\alpha = 1.0°$

VORTICITY ERRORS IN MULTIDIMENSIONAL LAGRANGIAN CODES

John K. Dukowicz
Los Alamos National Laboratory, Los Alamos, New Mexico 87545
and
Bertrand J. A. Meltz
Centre d'Etudes de Limeil-Valenton,
B. P. 27, 94195 Villeneuve Saint-Georges Cedex, France

Introduction

It is well known that most multidimensional lagrangian fluid dynamics computations can only be continued for a short time before the mesh is destroyed by "tangling", or crossing of mesh lines. This is ultimately to be expected because the presence of shear and rotation causes a lagrangian fluid element to become so stretched and distorted that it can no longer be adequately represented by a discrete computational cell. Most frequently, however, a computation experiences premature tangling failure because of numerical errors associated with the presence of a distorted mesh, a type of error that is relatively poorly understood and whose remedy is frequently attempted by ad-hoc artificial viscosity methods. It is clear that a reduction in the mesh-distortion error would extend the scope and usefulness of lagrangian calculations.

A longstanding paradox in lagrangian hydrodynamics concerns the computation of one-dimensional irrotational flows which should be free from mesh tangling but which, computed using a nonuniform two-dimensional mesh, will usually develop increasing mesh distortion, and will eventually fail due to mesh tangling. This is best demonstrated by a frequently used and widely known test problem, called the Saltzman problem. This problem concerns a strong one-dimensional shock wave, propagating in an initially specified, nonuniform two-dimensional plane mesh. Most lagrangian codes experience severe difficulty with this test problem.

By analyzing the Saltzman problem we have demonstrated that the difficulty in computing accurate lagrangian vertex velocities is caused by a spurious vorticity generated in the presence of an irregular mesh. We further argue that there exists an acoustic mechanism for "healing" errors in the divergence of velocity, while errors in vorticity are merely transported and persist in time, thereby accounting for the above-mentioned mesh distortion. Based on this understanding, we have developed two methods for constructing lagrangian vertex velocities that eliminate this spurious vorticity, while preserving the true or intrinsic vorticity.

Role of Divergence and Vorticity

Fig. 1 illustrates the Saltzman problem. In the upper part of the figure we show a strong piston-driven shock wave calculated using an initially uniform lagrangian mesh. In the lower part is the paradoxical result of the Saltzman problem, obtained for the same conditions but using an initially nonuniform mesh. To investigate this further, we introduced two closely associated quantities: the lagrangian cell volume V, and the cell skewness Ω, defined by

$$\frac{dV}{dt} = \int \nabla \cdot \mathbf{u} \, d\tau, \quad \frac{d\Omega}{dt} = \int \nabla \times \mathbf{u} \, d\tau,$$

where the integration is over the volume of a computational cell. Changes in the lagrangian cell volume are a consequence of the effects of divergence, and similarly changes in the skewness are a consequence and a measure of the effects of the vorticity. These quantities are pictured in Fig. 2 for the Saltzman problem case at the time corresponding to Fig. 1. We note that the cell volume change relative to the initial volume behaves in the manner expected for this problem, while the skewness, which we would expect to be zero because of the presumed absence of vorticity, correlates closely with the mesh distortion observed in Fig. 1. We therefore conclude that the mesh distortion is caused by the existence of a spurious vorticity, while the divergence behaves substantially as expected.

The equation for vorticity $w = \nabla \times \mathbf{u}$ may be written as

$$\frac{d\vec{\omega}}{dt} = -\vec{\omega} D + (\vec{\omega} \cdot \vec{\nabla}) \mathbf{u} - \vec{\nabla}\frac{1}{\rho} \times \vec{\nabla} p ,$$

where $D = \nabla \cdot \mathbf{u}$, the divergence. Because the above problem is one-dimensional, the gradients of pressure and density are expected to be parallel and there should be no physical source of vorticity; any sources of vorticity must be numerical. The numerical generation of vorticity may be traced to the failure of the curl-grad identity for discretization on an irregular mesh. The above equation indicates that, aside from the terms that stretch and compress or expand the vorticity, the vorticity that is generated is merely transported by the flow. On the other hand, the divergence D satisfies the equation

$$\frac{dD}{dt} = S_d - \frac{1}{\rho} \vec{\nabla} \cdot \vec{\nabla} p ,$$

where

$$S_d = -\vec{\nabla} \mathbf{u} : \vec{\nabla} \mathbf{u} - \vec{\nabla}\frac{1}{\rho} \cdot \vec{\nabla} p$$

may be considered a source term. Consider the equation for isentropic pressure changes

$$\frac{dp}{dt} = -\rho c^2 D ;$$

combining this with the above divergence equation we obtain

$$\frac{d}{dt}\left[\frac{1}{\rho c^2}\frac{dp}{dt}\right] = -S_d + \frac{1}{\rho}\vec{\nabla} \cdot \vec{\nabla} p ,$$

a nonlinear acoustic wave equation. This implies that there exists an acoustic mechanism which disperses divergence errors, in contrast to the behavior of vorticity errors which are not subject to a similar mechanism. In principle, therefore, we should be able to correct the mesh distortion problem caused by the spurious vorticity by subtracting out the effect of the vorticity errors provided the spurious vorticity may be distinguished from the legitimate vorticity.

Reconstructing the Velocity Field

Theoretically, we can reconstruct a velocity field provided we know its true divergence and vorticity. Because of the existence of the above-mentioned mechanism for acoustically dispersing divergence errors, we may assume that the provisional vertex velocity field contains the correct divergence The true vorticity, however, must be determined from its sources. We thus solve the following two-dimensional inviscid vorticity transport equation, which for numerical reasons is best put in the form

$$\frac{d}{dt}\int \omega \, d\tau = \int \frac{1}{\rho^2} \frac{\partial p}{\partial e} \mathbf{k} \cdot \left(\vec{\nabla}\rho \times \vec{\nabla}e\right) d\tau ,$$

where the variables have their usual meaning, except that $\mathbf{k}$ is the unit vector normal to the two-dimensional plane, and $d\tau$ is a differential lagrangian volume element.

Assuming we have a provisional velocity field $\mathbf{v}'$, obtained as a result of a lagrangian calculation, its divergence D' and vorticity ω', and the true vorticity ω from the above equation, we have developed two methods for obtaining the correct velocity $\mathbf{v}$: (1)

$$\vec{\nabla}\cdot\vec{\nabla}\varphi = D' \quad ; \quad \mathbf{n}\cdot\vec{\nabla}\varphi = w , \quad w=\text{specified normal velocity on the boundary},$$

$$\vec{\nabla}\cdot\vec{\nabla}\mathbf{A} = -\vec{\omega} \quad ; \quad \mathbf{A} = 0 \quad \text{on the boundary}.$$

This determines a unique velocity $\mathbf{v}$ given by

$$\mathbf{v} = \vec{\nabla}\varphi + \vec{\nabla}\times\mathbf{A} ,$$

whose divergence is D' (=D) and vorticity is ω. In two-dimensions the problem involves solving two Poisson equations, one with Neumann and the other with Dirichlet boundary conditions.

(2)

$$\vec{\nabla}\cdot\vec{\nabla}\mathbf{A} = \vec{\omega} - \vec{\omega}' \quad ; \quad \mathbf{A} = 0 \quad \text{on the boundary}.$$

The velocity is given by

$$\mathbf{v} = \mathbf{v}' - \vec{\nabla}\times\mathbf{A} ,$$

which leaves the divergence undisturbed and merely corrects the vorticity. In two-dimensions the problem consists of solving a single Poisson equation with Dirichlet boundary conditions. We have found that both methods give equally good results; the second method, however, is obviously much simpler and more efficient.

Results

Fig. 3 shows the significant improvement obtained as compared to Fig. 1 when method (2) is applied to the Saltzman problem. We have also computed a Mach 2 shock refraction problem involving a shock impinging on an inclined interface with a density ratio of 1.5. This problem is of interest because, unlike the Saltzman problem, it possesses a nonzero intrinsic vorticity: a vortex sheet along the interface generated by the passage of the shock. The results are shown in Fig.4; in (a) we show the mesh computed using a standard lagrangian code, and in (b) we show the mesh at the corresponding time computed using method (2). The improvement is readily apparent.

Method (2) is still very costly when applied at every time step. In the future we plan to investigate the effect of partial convergence of an iterative method for the Poisson equation.

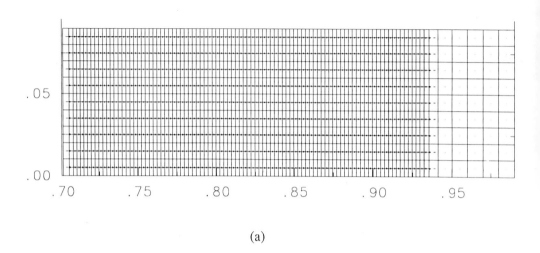

(a)

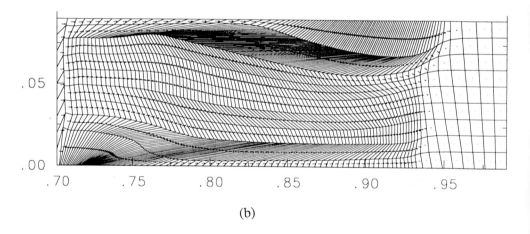

(b)

Figure 1: Saltzman Problem; a lagrangian calculation of a strong shock driven a piston. (a) initially uniform mesh, (b) initially nonuniform mesh.

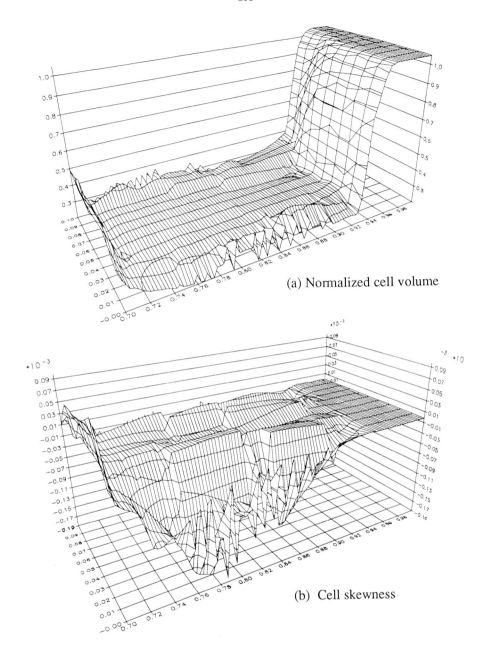

Figure 2: Normalized cell volume (a) and cell skewness (b) for the Saltzman problem calculation of Fig. 1(b).

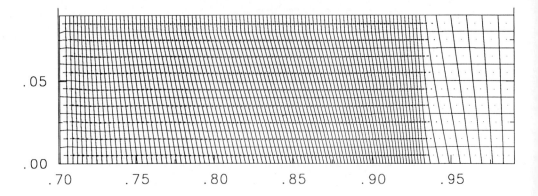

Figure 3: Saltzman problem of Fig. 1(b) calculated using Method (2).

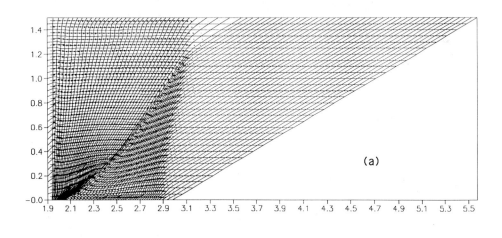

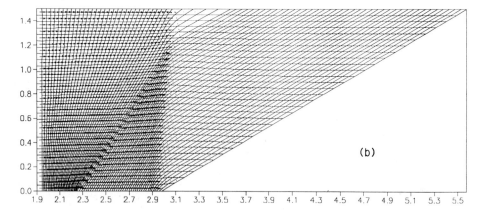

Figure 4: Mach 2 shock refraction by a 1.5 density ratio interface
(a) computed using a standard lagrangian method,
(b) computed using Method (2).

THREE DIMENSIONAL LARGE EDDY SIMULATIONS WITH REALISTIC BOUNDARY CONDITIONS PERFORMED ON A CONNECTION MACHINE

Jay Boris, Elaine Oran, Fernando Grinstein, Eugene Brown[*],
Chiping Li and Ronald Kolbe[**], and Robert Whaley[***]
Naval Research Laboratory
Washington, DC, 20375

We describe recent three-dimensional Large Eddy Simulations (LES) of unstable jets and turbulent flows performed using the highly parallel Connection Machine. The goals of the research are to study and predict complex fluid dynamic phenomena and turbulence in situations of scientific interest and engineering concern. We describe the large eddy aspects of the model and algorithms, the results of Connection Machine (CM) timing comparisons with similar models run on the Cray YMP, and briefly discuss the hardware and software available. We conclude with results of compressible three-dimensional turbulence and jet computations designed to calibrate the subgrid turbulence models intrinsic to our Flux-Corrected Transport (FCT) algorithms.

The Connection Machine runs an order of magnitude faster than the Cray for the Flux-Corrected Transport (FCT) algorithms considered here, making definitive numerical experiments on LES subgrid models at least possible. Parallel processing and LES models are necessary because the range of scales available to Direct Numerical Simulation (DNS), while growing, is not very large and thus Reynolds numbers are low in fully resolved simulations. Our CM models contain efficient implementations of nonperiodic boundary conditions for two and three dimensions to allow solution of realistic problems with walls, inflow and outflow. This can be done with only a few percent penalty relative to completely periodic boundary conditions despite the SIMD Connection Machine architecture.

This work seeks to organize, quantify and substantiate the strong evidence suggesting that monotone convection algorithms, designed to satisfy the physical requirements of positivity and causality, in effect have a minimal LES filter and matching subgrid model already built in[2]. The positivity and causality properties of FCT, not built into other easily parallelized convection algorithms, seem to ensure efficient transfer of the residual subgrid motions, as they are generated by resolved fluid mechanisms, off the resolved grid with minimal contamination of the well-resolved scales by the nonlinear monotonicity-preserving flux limiter.

The CFD requirements for DNS and LES algorithms are in fact somewhat different. In direct simulations, the smallest resolved scales are continuously being smoothed by viscous diffusion. The motions at the smallest dynamical scales are quite slow and the highest harmonics of the corresponding field variables are quite small, so local numerical errors have little effect. Since spectral methods shine at intermediate and long wavelength where physical viscosity gives relatively little smoothing, they generally have been a good match for DNS problems. In LES, the Reynolds number of the flow is so large that viscosity isn't effective in removing numerically-difficult steep gradients, even on the smallest resolved scales. The spectral and energy content of gradients on these scales is thus correspondingly larger in LES problems than in DNS problems.

It appears to be impractical to separate the formulation of the LES problem from the numerical method used for its solution. The entire LES approach is aimed at dealing with the

---

[*] Virginia Polytechnic and State University, Blacksburg VA
[**] Berkeley Research Associates, Springfield VA
[***] Thinking Machine Corporation, Cambridge MA

solution errors arising at small scale due to the finite numerical resolution and its interaction with the algorithms chosen to solve the model. This separation is valuable in principle, but the "subgrid fields" have to be matched to the "resolved fields" at the smallest resolved scales in practice – just where the distinctions between various methods and algorithms are the very largest. Since this matching should be done with some representation of the fluid dynamics at all scales included once but not twice, the short-wavelength errors in the specific algorithms chosen should not be ignored.

Monotone nonlinear convection algorithms were designed to limit errors in the shortest resolved scales in a physically meaningful way where sensible connection to a subgrid model is also required. Thus they seem a better choice for use in LES models than linear convection. The price for satisfying positivity and causality and the enhanced accuracy at short wavelength of monotone methods is somewhat larger errors at long wavelengths than found in spectral methods. Since these errors in long wavelengths are small in any case, the comparative advantage shifts to the monotone methods when accurate treatment of the smallest resolved scales is of paramount importance.

It is our experience that nonlinear monotone CFD algorithms such as FCT really have a built-in LES filter and a corresponding built-in subgrid turbulence model [2]. These monotone "integrated" LES algorithms are derived from the fundamental physical laws of causality and positivity in convection and do minimal damage to the longer wavelengths while still incorporating, at least qualitatively, most of the local and global effects of the unresolved turbulence expected of a large eddy simulation. These convection algorithms, when properly formulated, accept and transform the unresolved variability in the fluid field variables that is pushed into the short wavelengths by numerically resolved nonlinear effects and instabilities. This variability is locally converted to the correct macroscopic variables, e.g., viscous dissipation of the unresolved scales appears as heat. Diffusion of the eddy transport type is automatically left in the flow as required but the fluctuating driving effects of random phase, unresolved eddies on the large scales is missing unless specifically included as a subgrid phenomenology. This deficiency, however, is common to all the subgrid models in current use. In the turbulence simulations reported here, a stochastic backscatter model[3] will be included to augment the minimal eddy viscosity provided by the monotonicity preserving flux limiter once the behavior of the limiter alone is quantified.

CONNECTION MACHINE PERFORMANCE AND TIMING COMPARISONS

The Connection Machine is an SIMD fine-grained parallel architecture with thousands of scalar processors communicating data on a hypercube[4,5]. NRL has two Connection Machines, one with 16,384 (16K) processors and one with 8,192 (8K) processors. The instructions to and control of these processors is through one of several front-end computers: a VAX, a Symbolics, or a Sun. Each individual processor can be software subdivided into a larger number of virtual processors (in powers of two) to simplify programming of CFD problems where the number of cells is much larger than the number of processors. This is ideal for two dimensions but in 3D the number of virtual processors is limited by the storage required by the problem being solved since each vitual processor uses a significant amount of its memory for scratch storage.

Floating-point arithmetic is carried out by Weitek chips, each of which does pipelined processing of the floating-point operations for 32 of the scalar processors. Online color graphics are available so that it is possible to watch the evolution and make video tapes of two-dimensional computations as the CM performs the calculations. The user programs in one of several language extensions adapted to parallel processing, *Lisp, C*, or Fortran 8X. Compiling programs written in these extended languages produces a series of instructions in Paris (PARallel Instruction Set), microcoded routines that control the Connection Machine processors and the front-end interaction with the Connection Machine. There is also a lower-level machine language, CMIS (Connection Machine Instruction Set), which is extremely complicated to program but does produce some increase in speed over PARIS.

Our approach to fluid-dynamic computations on the CM has been to convert the most recent version of the Flux-Corrected Transport (FCT) algorithm[1], LCPFCT, first to C*, then to PARIS[4], and now to CMIS. The kernel of the algorithm consists of about 25 lines of code which is ideally suited for parallel implementation. We have structured the programs so that each virtual processor represents one fluid element and optimized library routines are used to access values of variables in the spatially neighboring processors. This flexible generic module is used to solve the appropriate sets of coupled continuity equations. It is combined with direction and timestep splitting to construct the two-dimensional and three-dimensional simulation models.

Comparative timings on the CM are shown in Table 1 of Reference 5. We have found that it is possible and practical to achieve up to factors of five over the Cray YMP for highly resolved computations ($256 \times 128 \times 128$ grid). The relatively small amount of memory on each processor keeps the virtual processor ratio (the ratio of virtual to physical processors) at one or two for these simulations, somewhat imparing the computational efficiency. As discussed by Boris[4], the limitation reduces the length of vectors presented to the Weitek floating-point chips, which are thus used to approximately 10% effeciency in C* and perhaps 40% efficiency in CMIS.

REALISTIC BOUNDARY CONDITIONS

A fundamental concern with the Connection Machine is how to avoid paying a large time penalty (factors of two to four) to do computations that do not have periodic boundary conditions. On the CM periodic boundary conditions are the default and cost no time or programming to implement. The hardware automatically ensures periodic boundary conditions in any number of dimensions but system sizes must be a power of two. While some good physics can be done in multiply periodic geometry, much physics and almost all engineering can not. The FCT models implemented for this paper contain efficient implementations of nonperiodic boundary conditions for two and three dimensions to allow solution of realistic problems with walls, inflow and outflow. This can be done with only a few percent penalty relative to completely periodic boundary condtions despite the SIMD Connection Machine architecture. We have considered several ways to deal with this problem:

1. **Uniform Boundary Condition Algorithms:** Using generic difference formulas for each stage of the FCT algorithm, the boundary cells can be treated just as internal cells. These so-called "uniform" algorithms[6] require additional computation to be performed everywhere and limit the complexity of boundary conditions which can be implemented. However, for important classes of problems and, in particular, for many supersonic flows, this extra computation can be minimized while implementing ideal symmetry, antisymmetry, slope, and value boundary conditions for only a few extra operations at each grid point.

2. **Boundary Condition Overlays:** One can also overlay values computed with the periodic boundary conditions with especially recomputed boundary values. This is more general than the uniform boundary algorithms and uses less memory but the overlay approach is geometrically limiting. Furthermore, the cost can be high if many different boundary conditions must be applied in different places.

3. **Boundary Conditions with a Segmented Data Structure:** In 3D computations on distinct boundary planes allow greater generality (e.g. inflow-outflow conditions) without wasted computation or memory. This approach is primarily useful in three-dimensions where each processor integrates equations for many different cells segmented into planes or blocks. The computational domain must be segmented into regions near boundaries and regions far from them. Then different, fully optimized parallel algorithms can be used to

update the distinct segments which have to be computed sequentially in any case. We have found that there is minimal loss of efficiency with this approach as long as there are enough boundary cells for all the physical processors to be kept busy simultaneously.

4. **Unfolding Loops in One Spatial Direction:** By writing a 1D scalar algorithm which works with planar cross-sections in one of the spatial directions, extremely high efficiency can be obtained at the cost of considerable extra scratch space. However, using CMIS programming in this special direction will recoup most of the scratch space while being able to take advantage of the fact that all of the data references lie in the same processor. This allows implementation of the most general boundary conditions in the chosen direction at no extra cost and allows system sizes other than a power of two. The programming would initially be complicated but the scratch space is reduced by a factor of thirty two this way.

5. **Realistic Inflow-Outflow Boundary Conditions:** Only the last three options lend themselves to efficient implementation of our best inflow-outflow boundary conditions[7] for subsonic flows. In these models the inflowing entropy and mass flux are specified and a separate first-order upwind, characteristic-like interpolation is used to accumulate time averaged values to be used as the outflow guard plane values.

COMPUTATIONS OF THREE-DIMENSIONAL JET FLOWS

The simulation of 3D jets is being undertaken to study intrinsically 3D phenomena which enhance mixing, to calibrate the built-in LES subgrid model provided by FCT, and to prepare the way for reactive jet and detonation simulations using parallel models of combustion chemistry. The simulations reported here feature a mode 3 helical perturbation on a circular jet which is intrinsically three-dimensional and allows comparison of measured growth rates with linear analytic estimates and results. Mode 3 was chosen so that intermediate resolution simulations of the same physical problem would be possible by changing to mode 2 and reducing the cell sizes by 50% while still keeping the total number of cells in each direction a power of 2.

Convergence tests compared solutions at $64 \times 32 \times 32$ (low) resolution, $128 \times 64 \times 64$ (medium) resolution, and $256 \times 128 \times 128$ (high) resolution. Figure 1 below shows a set of contour plots of the relevant flow variables taken in the vertical midplane of the jet after 0.7 ms has elapsed in an intermediate resolution simulation. The shear transition from jet fluid to background air occurs over only two cells in this case so shorter-wavelength harmonics of the initial perturbation (here mode 9) appear (their growth rates are faster) though the fundamental mode 3 can still be seen to dominate. In all of these tests the results of the intermediate and high resolution LES simulations, as measured by the integrated entrainment volume of background air into the jet, were virtually identical and only 10 – 20% different from the low resolution simulations with identical initial conditions.

Using periodic boundary conditions in the flow direction and stretched-grid side wall conditions to simulate transition to turbulence in the jet, different resolution computations were performed to calibrate the large eddy properties of the three-dimensional model. In our next series of turbulence simulations, a stochastic backscatter model[3] will be included to augment the minimal eddy viscosity provided by the monotonicity-preserving flux limiter[2]. In the nonlinear phase of the Kelvin-Helmholtz instability short wavelength helical streamers of fluid form, as shown in Figure 2, even though the initial shear layer interface was resolved over eight cells for this case. Though the low and intermediate resolution LES could not well resolve the thin mixing streamers, the measured entrainment was again virtually independent of resolution. This rapid convergence upon resolution improvement seems to indicate that little additional transport will be necessary from an added subgrid phenomenology. The results of some of these calculations and others depicting shock generation of turbulence will be presented and discussed.

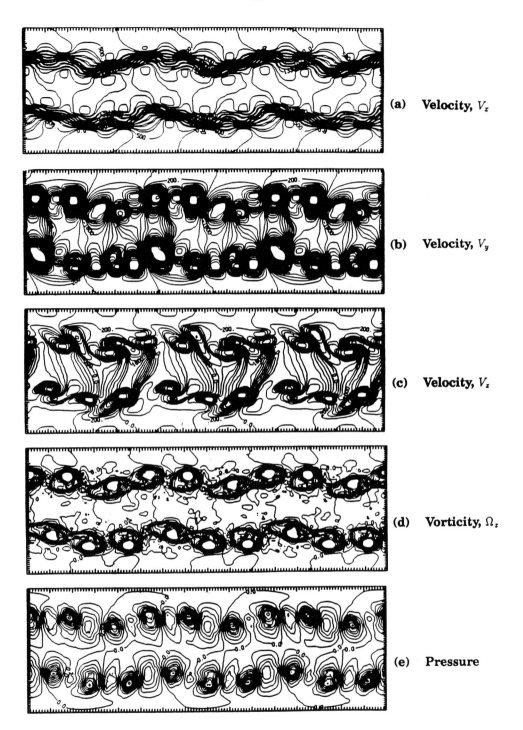

Figure 1. Contours of physical variables in the vertical midplane of the helically perturbed jet. Shorter wavelenth variations are evident than in Figure 1 because the jet had a sharper profile. Timestep 700, time 0.7 ms., is shown from the 128 × 64 × 64 simulation.

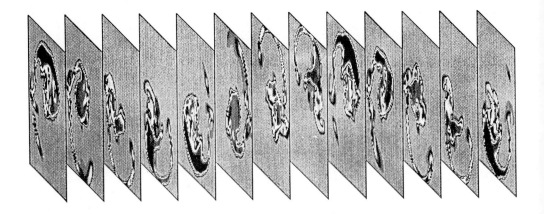

Figure 2. One and a half wavelengths of the fully developed helical Kelvin-Helmholtz Instability are depicted at Step 3001 (1.5 ms) through 13 cross-sections of the jet fluid density.

ACKNOWLEDGMENTS

This work was funded by the Naval Research Laboratory through the Office of Naval Research and by the Advanced Computation and Mathematics Program in the Defense Advanced Research Projects Agency. The authors would like to thank Fernando Grinstein, K. Kailasanath, John Gardner and Paul Boris for their help, and the staff of the NRL Connection Machine, particularly Henry Dardy, Eric Hoffman, and George Lam of NRL, for their advice and advanced support software.

References

1. J.P. Boris, and D.L. Book, *Solution of the Continuity Equation by the Method of Flux-Corrected Transport*, **Methods in Computational Physics**, 16, 85-129, 1976; and *LCPFCT – A Monotone Algorithm for Solving Continuity Equations*, J.P. Boris, J.H. Gardner, E.S. Oran, S. Zalesak, D.L. Book, R.H. Guirguis, to appear as a Naval Research Laboratory Memorandum Report, 1990.
2. J.P. Boris, *Comments on Large Eddy Simulation Using Subgrid Turbulence Models*, in **Whither Turbulence? Turbulence at the Crossroads**, Lecture Notes in Physics, No. 357, John Lumley (ed), presented at Cornell University, Ithaca NY, March 21 – 24 March 1989, (Springer-Verlag, New York, 1990).
3. C. Leith, *Stochastic Backscatter in a Subgrid-Scale Model: Plane Shear Mixing*, Physics of Fluids A, Fluid Dynamics, **2**, pp297–299, March 1990.
4. E.S. Oran, J.P. Boris, K. Kailasanath, C. Li, and E.F. Brown, *Reactive Flow Simulations on the Connection Machine*, Paper GI4, p. 2324, and J.P. Boris, E.S. Oran, F.F. Grinstein, C.P. Li, and E.S. Brown, *Boundary Conditions for Fluid Dynamics Simulations on the Connection Machine*, Paper GI3, p. 2323, 42nd Annual Meeting, American Physical Society Division of Fluid Dynamics, Palo Alto CA, B.A.P.S. Vol. 34, No. 10, 19–21 November 1989.
5. E.S. Oran, J.P. Boris, E.F. Brown, and R. Whaley, *Fluid Dynamics Computations on a Connection Machine – Preliminary Timings and Complex Boundary Conditions*, AIAA Paper No. 90-0335, American Institute of Aeronautics and Astronautics, Washington, DC, 1990.
6. C. Li, E.S. Oran, J.P. Boris, G. Patnaik, and J.L. Ellzey, *A Uniform Algorithm for Boundary and Interior Points for Computational Fluid Mechanics Applications on the Connection Machine*, NRL Memorandum Report being prepared for journal publication.
7. F.F. Grinstein, E.S. Oran, and J.P. Boris, J. Fluid Mech. **165**, p201 (1986).

A 3D NAVIER-STOKES SOLVER ON DISTRIBUTED MEMORY MULTIPROCESSOR

D. TROMEUR-DERVOUT (*), L. TA PHUOC (**) and L. MANE (*)

(*) ONERA, Div. Recherches Informatiques , BP 72, 92322 Chatillon Cedex, FRANCE
(**) LIMSI/CNRS and ONERA, BP 30, 91406 Orsay Cedex, FRANCE

In order to face the "large scale computing challenge" and to overcome limits of circuit technology, a new generation of computer based on parallelism has emerged. Two basic types of parallel computers are presently being investigated. The first type consists of a small group of very powerful processors sharing a large common memory (CRAY YMP).The second type, in attempt to overcome the memory bandwidth bottleneck, consists of a large number of processors, with their own local memory, connected through a communication network (hypercube for iPSC and Ncube, or grid topology for Transputer systems).
Since 1979 O.N.E.R.A. has conducted studies on the adequacy of parallel computers for numerical simulation. In this context a 3D Navier-Stokes code was implemented on a 32 processors INTEL iPSC2-SX and on an 8-processor CRAY YMP.

1 The 3D Navier-Stokes solver

The 3D Navier-Stokes equations are written following a velocity $\vec{V}$ and vorticity $\vec{\omega}$ formulation:

$$\frac{\partial \vec{\omega}}{\partial t} + (\vec{V}.\vec{\nabla})\vec{\omega} = (\vec{\omega}.\vec{\nabla})\vec{V} + \nu \nabla^2 \vec{\omega} \qquad (1)$$

$$\vec{\nabla}(\vec{\nabla}.\vec{V}) = \vec{\nabla} \times \vec{\omega} + \nabla^2 \vec{V} = 0 \qquad (2)$$

For boundary conditions we have Dirichlet condition for $\vec{V}$, and $\vec{rot}\vec{V}$ for $\vec{\omega}$.
The spatial discretization for each of the six scalar equations is performed by centred finite difference schemes. To solve these equations we present below different methods we investigated, noting that only the ADI solver was implemented on the INTEL hypercube.
We validated our code on the driven cavity problem and plan to extend it to more complex geometry such as a 3D-airfoil between two planes.

The ADI solver
We opted for a fractional step ADI method with stabilization corrections to solve the equations.
The ADI method to solve the equation: $\frac{\partial \mathbf{U}}{\partial t} = \mathbf{A}\mathbf{U} - \Phi$ is based on a splitting of $\mathbf{A}$:

$$\mathbf{A} = (L_x + L_y + L_z)$$

Then the equation is solved by the following fractional step algorithm:

$$(L_x - \frac{2}{\Delta t})\mathbf{U}^{k+\frac{1}{3}} = -(L_x + 2L_y + 2L_z + \frac{2}{\Delta t})\mathbf{U}^k + 2\Phi \qquad (3)$$

$$(L_y - \frac{2}{\Delta t})\mathbf{U}^{k+\frac{2}{3}} = L_y\mathbf{U}^k - \frac{2}{\Delta t}\mathbf{U}^{k+\frac{1}{3}} \qquad (4)$$

$$(L_z - \frac{2}{\Delta t})\mathbf{U}^{k+1} = L_z\mathbf{U}^n - \frac{2}{\Delta t}\mathbf{U}^{k+\frac{2}{3}} \qquad (5)$$

For no time depending equations, as Poisson's equations, we have an iterative fractional step ADI algorithm, replacing $\frac{2}{\Delta t}$ by accelerating parameters: $\omega_{k,x}, \omega_{k,y}, \omega_{k,z}$ at the kth iteration .
For a Poisson equation: $\mathbf{A} = \Delta$ and we take $L_x = \frac{\partial^2}{\partial x^2}, L_y = \frac{\partial^2}{\partial y^2}, L_z = \frac{\partial^2}{\partial z^2}$.
Considering data dependencies, each step of the method is then related to an implicite relation in one of the space directions.

Gradient's type and Multigrid solvers

Gradient type methods (GCR,DGCS) without preconditionning were tested because of their parallelizing facilities. Preconditioning was avoided because of its great cost in terms of computing and communication time. These methods, less accurate than ADI were dropped.

Multigrid methods were studied in order to improve code performance, owing to the fact that in general cases we might have difficulties in finding "good" ADI accelerating parameters. Multigrid algorithms present local parallelism benefits, particularly if the smoother is explicit. We plan to couple a Multigrid algorithms with ADI methods as smoothers.

Figure 1 shows the accuracy of these methods to solve a Poisson -like equation, where the analytical solution is $U = x^2 + y^2 + z^2$, on both a regular and an irregular mesh refined near boundaries.

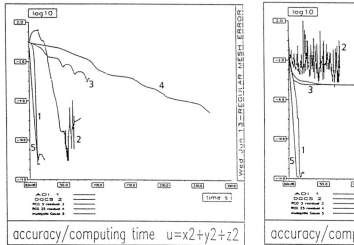
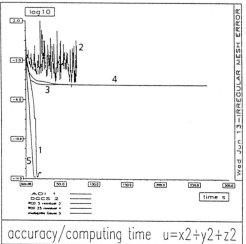

Figure 1: Accuracy comparison for: ADI - DGCS - RCG - Multigrid

2 The distributed memory multiprocessor iPSC2

The iPSC2-SX (Fig. 2) installed at ONERA is a 32-node multiprocessor. In this system each processor node is directly connected to 5 neighbouring nodes according to a hypercube topology. Processor activities are synchronized via message passing communication.

The peak performance of one individual processor (INTEL 80386 and Weitek 1167) is 1 MFlops (32 bits) and 0.6 MFlops (64 bits). On each node a "Direct-Connect Routing Module", supporting 8 two-directional channels, dynamically creates a hardware path for communication between nodes. The communication bandwidth is 2.8 MBytes/s and the start-up is 350μs. Each node has a 4 MBytes memory.

The programming languages are Fortran or C with extensions for sending or receiving messages.

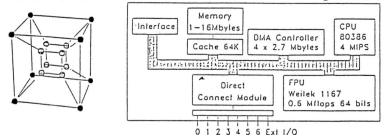

Figure 2: iPSC2-SX: 16-node hypercube and hardware node configuration

3 ADI implementation and performances on distributed memory multiprocessors

On distributed memory multiprocessors, algorithm implementation must take into account data dependencies and the fact that the computational domain must be spread over the processor's local memories. Efficient algorithms for these multiprocessors often result from minimizing data communication time between processors.

3D ADI solvers are therefore relatively complex to implement on these systems, compared to explicit methods or gradient type algorithms, because they induce data dependencies alternatively in the three spatial directions.

Different implementation strategies may be investigated for splitting the computational domain (planes, pencils, sub-cubes) and the corresponding mapping onto the processors depending on the network topology (ring, 2D or 3D-array, hypercube). We present two of these strategies below.

Implementation on a ring:
Considering a ring topology of P processors, the computational domain is split into P^2 pencils (Fig. 3). Assigning neighbouring subdomains in the data structure to neighbouring processors, each processor processes P Pencils. At each fractional ADI step, the linear systems are then solved by Gaussian elimination algorithm in P steps. A pencil boundary message is sent to the neigbouring processor after each step.

The ring algorithm has two major drawbacks: it generates a large number of inter-processor messages; and it limits the data parallelism to one dimension of the data structure.

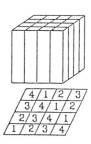

Fig.3: Data splitting for ring.

Implementation on a 3D grid:
Considering a 3D Grid topology of P^3 processors, the computational domain is split into P^3 sub-cubes (Fig. 4). At each fractional ADI step, the linear systems are solved by the substructuring algorithm.

The 3D grid algorithm requires fewer communications but more operations than the ring algorithm.

Fig.4: Data splitting for 3D grid.

Ring performances on iPSC2-SX:
Table 1 shows the performance obtained for one time step of the 3D Navier-Stokes code (16 ADI iterations for Poisson's equations) with a ring of 16 or 32 node iPSC2-SX. For a 66^3 domain, the elapsed time per mesh point and time step is $2.4\ 10^{-4}$s (8 Mflops) with 32 processors.

16-Processor iPSC2-SX			32-Processor iPSC2-SX		
Grid	1 Time step	Mflops	Grid	1 Time step	Mflops
50x50x62	57 s	4.6 Mflops	66x66x66	63 s	7.9 Mflops
66x66x66	102 s	4.9 Mflops	98x98x62	116 s	9.1 Mflops

Table 1: 3D Navier-Stokes Code on iPSC2-SX

Performance estimates:
Algorithm performance models can be constructed based on the following parameters :

(1) Architecture-characteristic parameters: start-up, communication bandwidth, computing speed and processor number P,

(2) Algorithm-characteristic parameters: number of operations, number and length of messages between processors, and mesh size N^3.

Figures 3 and 4 show, for the ring and 3D grid, the variation of the speed-up S as a function of the processor number, for different realistic values of the architectural parameters and mesh sizes.

$$S = \frac{time\ of\ the\ best\ sequentiel\ algorithm\ (Gauss)\ on\ one\ processor}{time\ of\ a\ parallel\ algorithm\ with\ P\ processors} \quad (6)$$

Figure 5 is representative of today's computers and shows that the substructuring algorithm is a prime choice when the machine is unbalanced due to the weakness of the communication network.

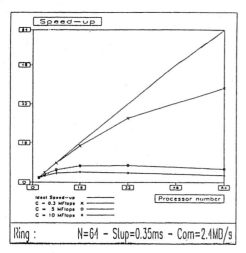 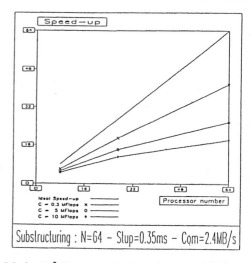

Figure 5: Ring and 3D grid Speed-up estimate: Mesh=64^3 Start-up=0.35ms Com=2.4MB/s

In Figure 6 we see that when the communication bandwidth of the network is more balanced with the computing speed of one processor, the ring algorithm completes successfully.

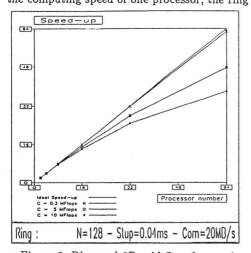 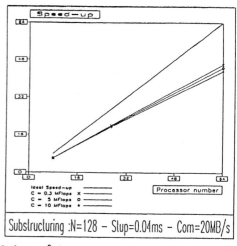

Figure 6: Ring and 3D grid Speed-up estimate: Mesh=128^3 Start-up=0.04ms Com=20MB/s

Those models lead us to estimate that for the 3D Navier-Stokes code with a 132^3 mesh, with a 3D grid of 125 processors, an elapsed time of $3.6\ 10^{-6}$s per point and time step (480 MFlops) may be reached with the hardware features of the iPSC2 communication network but with a processor computing speed of 10 MFlops, and $5.6\ 10^{-6}$s (310 MFlops) with 5 MFlops.

As mentioned before and for comparison, the 3D Navier-Stokes code achieved 110 MFlops on one processor of the CRAY2 and 1.2 GFlops on the 8-processor of the CRAY/YMP8 with autotasking.

4 Numerical results on a test case

Examples of numerical results for the driven cavity problem are presented for Reynolds number=100 (Fig. 7) and 3200 (Fig.8). The test case is a cubic cavity (1x1x1) with a regular 66^3 mesh for Reynolds 100 and a cubic cavity (3x1x1) with a regular 96x32x32 mesh for Reynolds 3200. We show the iso-vorticity lines for the middle plane.

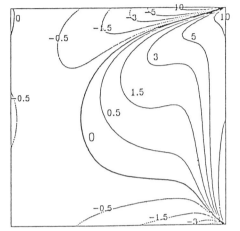
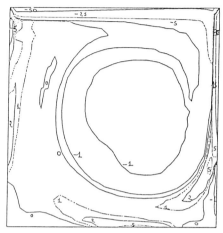

Figure 7: Reynolds=100 T=100   Figure 8: Reynolds=3200 T=100

5 Conclusion

The performance obtained for the 3D Navier-Stokes code on the iPSC2-SX shows the efficiency on this system of the parallelized ADI solver coupled with a Gaussian elimination mapped on ring topology. This performance and the time estimates for different architectural parameters show that algorithm efficiency on distributed memory multiprocessors is strongly dependent on start-up time and communication speed. Estimated times permit us to expect good performance from the ADI solver when coupled with substructuring algorithm and mapped on 3D grid topology network on massive parallel computers.

References

[1] P. LECA, L. MANE AND L. TA PHUOC, *Parallel algorithms for 2D and 3D Navier-Stokes equations on shared and distributed memory multiprocessors*. 7th Int. Conf. on Finite Element Methods in Flow Problems, U. of Alabama, Apr. 3-7 1989.

[2] NN. YANENKO, *Méthode à pas fractionnaires,résolution des problèmes polydimensionnels de physique mathématique*. Armand Colin Ed. 1968.

[3] J. DOUGLAS JR, *Alternating direction methods for three space variables*. Numerische Mathematik 4 p. 41-63, 1962.

[4] Y. SAAD, M.H. SCHULTZ, *Data communication in Hypercubes*. Research Report YALEU/DCS/RR-DRAFT, July 1985.

[5] Y. SAAD, *On the design of parallel numerical methods in message passing and shared memory environments*. Int. Sem. on Scientific Supercomputers, Paris, France Feb. 1987.

Parallel extrapolation methods for computational fluid dynamics

R. W. Leland and J. S. Rollett
Oxford University Computing Laboratory
11 Keble Road, Oxford, OX1 3QD, England

1 Introduction

Iterative solution methods for large, sparse linear systems of equations usually require significantly less time and memory than direct methods, and so are important in solving the systems which often arise in computational fluid dynamics. There are a number of *extrapolation* algorithms which have been closely studied in the past as methods of accelerating convergence of iterative processes [3]. Usually it has been stressed that these are very general, can achieve quadratic convergence for non-linear problems and require no knowledge of the way in which the sequence to be extrapolated is generated. But extrapolation techniques also have obvious application in parallel computing, and this has not been widely recognized in the literature. This paper explains and develops several parallel extrapolation methods and compares their performance using a test problem of Laplace type.

2 Ratio operated local extrapolation (ROLE)

In this method we record a number of the approximations to the value of each unknown generated by some iterative method. This state history is used to predict the converged value of the variable without reference to any other variable's value or history. In effect, linear algebra is used to fit the error decay curve associated with each variable. The communication between variables (and hence processors) which would have been required to reduce the error by the same amount using the underlying *base* iteration is avoided. Because the extrapolation phase of the algorithm requires no communication and can be load balanced simply by assigning the same number of unknowns to each processor, it can be implemented with perfect parallel efficiency.

To understand the method in more detail, consider solution of the $N \times N$ linear system $A\mathbf{x} = \mathbf{b}$ by an iterative method characterised by a constant iteration matrix M and homogeneous error equation $\mathbf{e_{k+1}} = M\mathbf{e_k}$ where $\mathbf{e_k} \equiv \mathbf{b} - A\mathbf{x_k}$ and $\mathbf{x_k}$ is the approximation at step k to the fixed point $\mathbf{x}$ of the iteration. A typical element, x_k, of $\mathbf{x_k}$ can be expanded in a subset the eigenvalues, λ_i, of M:

$$x_k = a_0 + a_1 \lambda_1^k + \ldots + a_p \lambda_p^k. \tag{1}$$

For a convergent iteration, $|\lambda_i| < 1$, hence $\lim_{k \to \infty} x_k = a_0$. The goal of the extrapolation is to find a_0 without explicit reference to the other a_i or to λ_i. Extrapolation is economical only if $p << N$, in which case a_0 is only an approximation to the converged value of an unknown since only the p lowest frequency error modes have been filtered.

We therefore apply the base iteration again to smooth the remaining higher frequency errors, and may also repeat the extrapolation.

An explicit formula for a_0 is arrived at as follows. We expand the shift vector $s_k = x_k - x_{k-1}$ in the $\lambda_i(M)$:

$$s_k = b_1 \lambda_1^k + \ldots + b_p \lambda_p^k \tag{2}$$

and note that s_k is the solution of a recurrence relation

$$c_p s_k + c_{p-1} s_{k-1} + \ldots + c_0 s_{k-p} = 0. \tag{3}$$

We write this for p subsequent steps in matrix form $Sc = 0$ where S is a *convolution matrix* of shifts with $S_{i,j} = s_{k-i-j}$. Because we have neglected some error modes we cannot solve this exactly and so solve instead the least squares problem

$$\min_{\|c\|_2 = 1} \left[\mathbf{c}^T S^T S \mathbf{c} \right]. \tag{4}$$

This is conveniently and accurately accomplished using a QR decomposition of S. Since $S^T S = R^T Q^T Q R = R^T R$, the $\mathbf{c}$ satisfying (4) will also minimise $\mathbf{c}^T R^T R \mathbf{c}$. We do just one step of inverse iteration on $R^T R$ using the triangularity of R and with $\mathbf{1} = (1, 1, \cdots, 1)^T$ as the starting vector. This avoids a potential division by zero in the calculation of a_0, but still gives an accurate $\mathbf{c}$ when R has one diagonal much smaller than the rest (as it will provided we do not extrapolate too early).

Next we write (1) for p steps in matrix form

$$\begin{bmatrix} 1 & 1 & \cdots & 1 \\ \vdots & \vdots & & \vdots \\ 1 & \lambda_1^p & \cdots & \lambda_p^p \end{bmatrix} \begin{bmatrix} a_0 \\ \vdots \\ a_p \end{bmatrix} = \begin{bmatrix} x_{k-p} \\ \vdots \\ x_k \end{bmatrix} \tag{5}$$

where $\lambda_1^{k-p} \ldots \lambda_p^{k-p}$ have been absorbed into the weights $a_1 \ldots a_p$. In matrix form this is $V\mathbf{a} = \check{\mathbf{x}}$, and has the counterpart $V^T \hat{\mathbf{c}} = \mathbf{d}$ based on the characteristic equation $c_p \lambda^p + \ldots + c_0 \lambda^0 = 0$ of (3). Here $\mathbf{d}$ is the column vector with first element $d \equiv \sum_{i=0}^{p} c_i$ and zero in all other positions, $\hat{\mathbf{c}}$ is $\mathbf{c}$ with element order reversed and $\check{\mathbf{x}}$ is the vector $(x_{k-p}, x_{k-p+1}, \ldots, x_k)^T$. Simple combination of these leads to the formula for the extrapolated value

$$a_0 = \frac{\mathbf{c}^T \hat{\mathbf{x}}}{\mathbf{c}^T \mathbf{1}} \tag{6}$$

where $\hat{\mathbf{x}}$ is the reversal of $\check{\mathbf{x}}$.

3 Minimal polynomial extrapolation (MPE)

It is advisable to defer extrapolation until the error is dominated by about p smooth modes. When the error is dominated by smooth functions of position on the mesh, calculations for nearby points are similar to one another, so most of the QR factorisations in ROLE use similar data and obtain similar results. We might therefore expect that there would be some reasonable way to combine the data and perform fewer minimisations. This is exactly what Minimum Polynomial Extrapolation does.

First we decide on which step to extrapolate and how many frequency terms to use: a sensible value for k is some small multiple of the the problem *gauge* G (the number of points on one side of the mesh), say $k = 1.5G$, since then the boundary conditions will have had time to propagate across the mesh; a convenient value for p is an integer near $\sqrt{G}$. Then we perform the initial base iterations as in ROLE, and at step $k - p - 1$ we begin recording the state history of each unknown. At step k we build the $N \times (p + 1)$ matrix U in which each row contains the last p shifts for a different mesh point. There is one row for each mesh point, so we write

$$U = [\mathbf{s_0}, \mathbf{s_1} \ldots \mathbf{s_p}]. \quad (7)$$

We then generate $S = U^T U$, a *single* $(p+1) \times (p+1)$ convolution matrix over the entire mesh. The minimising **c** and extrapolated values are found with the same algebra used in ROLE.

This specification of MPE is nearly that given by Mesina [2] in 1977, although he was working in a sequential computing context and so was unconcerned by the global communication implied in the formation of S. On a parallel machine, implementation of this product is a key issue since inefficiency arising from the communication overhead may outweigh the power of the extrapolation. On, for example, a distributed memory multiprocessor with mesh architecture, a block decomposition and ring shifting algorithm can be used to achieve efficiencies which go to unity as problem size increases. But for a given problem size or set of hardware parameters, the efficiency of the generation of S may be so poor relative to that of the base iteration or other less communication intense extrapolations that MPE is not worthwhile.

4 Ratio operated grain extrapolation (ROGE)

Numerical experiments suggest that MPE is the most potent extrapolation method tested in that it improves the solution most per unit of sequential work. This is not surprising as the method accounts for global interactions, but it is these interactions which make MPE a less naturally parallel algorithm than ROLE. We are led to consider some compromise between ROLE and MPE which uses shift information from many points but minimises interprocessor communication. An obvious candidate is to perform MPE on each *grain* of n mesh points independently, which is what ROGE does. Although the two become the same in the single processor limit, they have related but distinct numerical properties. Whereas the parallel version of MPE involves some complication if efficiency is to be kept high, ROGE is a naturally parallel algorithm with a straightforward implementation.

5 Results

Our test problem for comparing the methods was to solve the linear system generated by a five-point finite difference discretisation over a mesh of 36×36 interior points of

$$\frac{\partial^2 u}{\partial x^2} + \frac{\partial^2 u}{\partial y^2} \quad on \quad \Omega = [0, \pi] \times [0, \pi]$$

with $u = 0$ on $\partial \Omega$. We let $u_0(x, y) = N[\sin x + \cos x][\exp y + \exp -y]$, chose N so that $||u_0||_2 = 1$ and simulated on a SUN workstation the various extrapolation algorithms running on a mesh of 36 processors, each assigned a 6 by 6 sub-mesh. The base iteration was red-black SOR (RbSor) with $\omega = 1 + 3/4(\omega_{opt} - 1)$ chosen to simulate the typical situation in which we do not know the optimal acceleration parameter.

Our first observation was that the extrapolation methods were *more* robust than the base iteration in the sense that their convergence was much less sensitive to choice

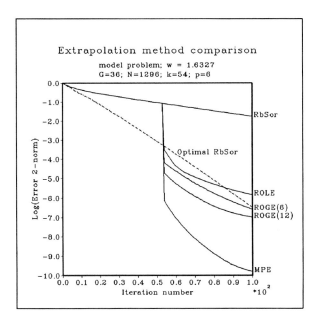

Figure 1: Comparison of extrapolation methods.

of ω. This is reported in more detail in [1] and can be explained in terms of standard theory locating the eigenvalues of the Sor iteration matrix.

Figure (1) compares the methods with respect to convergence with iteration number and figure (2) with respect to convergence in elapsed time. These confirm the prediction that MPE is the most potent extrapolation, but that the cost of global communication is so high that ROGE is as effective in real time. Since ROGE is simpler to implement, it would likely be preferred.

We next tested the methods using the conjugate gradient method (CG) as the base iteration and found that extrapolation was ineffective. This is because CG is in effect an extrapolation method itself.

We then compared the extrapolation methods with a multigrid algorithm (MG) applied to this problem using timing data provided by G.Shaw and A.Stewart[4], and found that MG took about $30/G$ as long as ROGE and so for very large problems is significantly better. Presumably this is because MG treats most error modes similarly well through a combination of smoothing iteration and grid alteration, whereas ROGE treats high frequency errors (through the base iteration) and low frequency errors (through the extrapolation) but does not filter middle frequency errors. ROGE is, however, much simpler to implement, and is actually *faster* on this test problem up to roughly 10^3 degrees of freedom.

Lastly, we considered using parallel extrapolation methods to solve the Navier-Stokes equations. These are difficult because the mass balance equations create zero diagonals for the pressure variables. There are various ways of overcoming this, including the use of block Sor methods in which (typically) all variables at a single node are adjusted together. Work continues to find an implementation of Sor which can be combined with extrapolation conveniently and effectively.

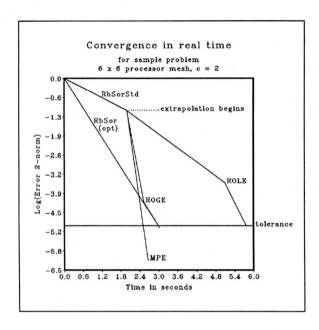

Figure 2: Performance of extrapolation methods on a 6 × 6 Sun workstation mesh.

6 Conclusions

Extrapolation methods enable us to use a standard parallel iterative method without the need to estimate the optimal acceleration parameter closely. ROGE is a extrapolation method which requires no inter-processor communication and therefore has a straightforward, efficient implementation on distributed memory multiprocessing machines. For the standard model problem, it has been shown to have performance competitive with that of multigrid for problems of up to roughly 10^3 degrees of freedom, yet is much easier to implement. With further tuning or multiple extrapolations, this limit may be increased.

Acknowledgement This material is based upon work supported under a National Science Foundation Graduate Fellowship held by Robert Leland.

References

[1] R. W. LELAND AND J. S. ROLLETT, *A comparison of parallel extrapolation methods*, in 6th GAMM Seminar, Kiel, W. Hackbush, ed., 1990.

[2] M. MESINA, *Convergence acceleration for the iterative solution of the equations $X = AX + f$*, Computer Methods in Applied Mechanics and Engineering, 10 (1977), pp. 165–173.

[3] D. A. SMITH, W. F. FORD, AND A. SIDI, *Extrapolation methods for vector sequences*, SIAM Review, 29 (1987), pp. 199–233.

[4] A. STEWART AND G. SHAW, *A parallel multigrid FAS scheme for transputer networks*, Parallel Computing (North-Holland), (To appear).

VARIABLE COEFFICIENT IMPLICIT RESIDUAL SMOOTHING

L. B. Wigton
Boeing Commercial Airplane Group
Seattle, Washington 98124, USA
and
R. C. Swanson
NASA Langley Research Center
Hampton, Virginia 23665, USA

The stability range of an explicit time marching scheme can be extended in a cost effective manner through application of implicit residual smoothing (IRS). Moreover IRS significantly enhances the smoothing characteristics of the time marching scheme for use with multigrid. In this paper we will derive the form of IRS introduced by Martinelli [1, 2]. An improved version of IRS is then constructed. Coupled with a careful choice of time step, the new version of IRS provides superior performance, particularly for very fine grid calculations.

Derivation of IRS Coefficients

Consider a semidiscrete representation of a 2-D scalar hyperbolic equation:

$$\frac{\partial w_{I,J}}{\partial t} + r_x \frac{w_{I+1,J} - w_{I-1,J}}{2} + r_y \frac{w_{I,J+1} - w_{I,J-1}}{2} = 0, \quad (1)$$

where we are using standard central differencing operators for the spatial derivatives. If we assume solutions of the form:

$$w_{I,J} = \hat{w}(t) e^{i\theta_x I} e^{i\theta_y J}, \quad (2)$$

then:

$$\frac{d\hat{w}}{dt} + f(r_x, r_y, \theta_x, \theta_y)\hat{w} = 0, \quad (3)$$

where:

$$f(r_x, r_y, \theta_x, \theta_y) = i(r_x \sin\theta_x + r_y \sin\theta_y), \quad (4)$$

is the Fourier symbol. If we apply IRS of the form:

$$(1 - \epsilon_x \delta_{xx})(1 - \epsilon_y \delta_{yy}), \quad (5)$$

then the Fourier symbol becomes:

$$f_s(r_x, r_y, \theta_x, \theta_y, \epsilon_x, \epsilon_y) = \frac{f(r_x, r_y, \theta_x, \theta_y)}{\beta_x \beta_y}, \quad (6)$$

where:
$$\beta_x = 1 + 2\epsilon_x(1 - \cos\theta_x), \quad (7)$$
$$\beta_y = 1 + 2\epsilon_y(1 - \cos\theta_y). \quad (8)$$

The coefficients ϵ_x and ϵ_y are positive and will be determined to ensure stability of the smoothed time marching scheme. We wish to maintain stability for time step $\Delta t = (\text{CFL})\Delta t_f$. CFL is user specified and is the same through out the flow field, while Δt_f is a fundamental time step parameter usually taken to be:

$$\Delta t_{f1} = 1/(r_x + r_y). \quad (9)$$

Δt_{f1} governs the numerical stability of the unsmoothed time marching scheme. A more physically meaningful fundamental time step parameter is given by:

$$\Delta t_{f2} = 1/\max(r_x, r_y), \quad (10)$$

which is the shortest time required for a signal to traverse the cell in any direction. With either choice of Δt_f we can form normalized versions of r_x and r_y given by:

$$\bar{r}_x = r_x \Delta t_f \quad (11)$$
$$\bar{r}_y = r_y \Delta t_f. \quad (12)$$

If the underlying time marching scheme is stable (along the imaginary axis) for Courant numbers up to CFL*, then the smoothed scheme will be stable if:

$$|Nf_s(\bar{r}_x, \bar{r}_y, \theta_x, \theta_y, \epsilon_x, \epsilon_y)| \leq 1, \quad (13)$$

where $N = \text{CFL}/\text{CFL}^*$ is essentially the factor by which we wish to increase the time step through application of IRS. A conservative (too large) estimate for ϵ_x and ϵ_y can be derived by first noting as in [3] that:

$$|f_s(\bar{r}_x, \bar{r}_y, \theta_x, \theta_y, \epsilon_x, \epsilon_y)| \leq \frac{\bar{r}_x \sin\theta_x}{\beta_x} + \frac{\bar{r}_y \sin\theta_y}{\beta_y}. \quad (14)$$

Then it is easily verified that:

$$g(\theta) = \frac{\sin\theta}{1 + 2\epsilon(1 - \cos\theta)}, \quad (15)$$

has a maximum value of $1/\sqrt{1+4\epsilon}$. Therefore the smoothed scheme will certainly be stable if:

$$\frac{\bar{r}_x}{\sqrt{1+4\epsilon_x}} + \frac{\bar{r}_y}{\sqrt{1+4\epsilon_y}} \leq 1/N. \quad (16)$$

It is desirable to find the smallest values of ϵ_x and ϵ_y which will ensure validity of this inequality. To this end we will minimize $\epsilon_x + \epsilon_y$ subject to equality holding in equation (16). Using the method of Lagrange multipliers, we consider:

$$\epsilon_x + \epsilon_y - \lambda(\frac{\bar{r}_x}{\sqrt{1+4\epsilon_x}} + \frac{\bar{r}_y}{\sqrt{1+4\epsilon_y}}). \quad (17)$$

If we equate the partial derivatives of this function with respect to ϵ_x and ϵ_y to 0, we find:
$$\frac{\overline{r}_x}{(1+4\epsilon_x)^{3/2}} = \frac{\overline{r}_y}{(1+4\epsilon_y)^{3/2}}. \tag{18}$$

Finally, substituting into equation (16) (with equality holding) we find:
$$\epsilon_x = \max\left\{\frac{1}{4}[N^2\overline{r}_x^{\,2}(1+(\frac{\overline{r}_y}{\overline{r}_x})^{2/3})^2 - 1], 0\right\}, \tag{19}$$
$$\epsilon_y = \max\left\{\frac{1}{4}[N^2\overline{r}_y^{\,2}(1+(\frac{\overline{r}_x}{\overline{r}_y})^{2/3})^2 - 1], 0\right\}. \tag{20}$$

These are precisely the IRS coefficients employed by Martinelli [1, 2]. The upper bound (14) used in the derivation of Martinelli's coefficients is sharp for a one dimensional problem. Indeed, in this case either $r_x = 0$ or $r_y = 0$ forcing the corresponding smoothing coefficient ϵ_x or ϵ_y to vanish and (14) reduces to an equality. However, if r_x and r_y are comparable to one another, so are ϵ_x and ϵ_y and the upper bound (14) is not sharp. In this case, corresponding to cell aspect ratios near one, formulas (19) and (20) considerably overestimate the appropriate IRS coefficients.

In order to find the "proper" values of smoothing coefficients a MACSYMA program was used to write a FORTRAN code which computed the minimum values of ϵ_x and ϵ_y (as measured by $\epsilon_x^2 + \epsilon_y^2$) which ensures validity of the correct inequality (13). Plots of ϵ_x and (ϵ_y/ϵ_x) versus (r_y/r_x) are shown in figures (1) and (2) respectively for various values of N. The fundamental time scales Δt_{f1} and Δt_{f2} have corresponding values of N related by:
$$\Delta t = N_1(\text{CFL}^*)\Delta t_{f1} = N_2(\text{CFL}^*)\Delta t_{f2}, \tag{21}$$

which implies:
$$N_2 = \frac{N_1 \max(r_x, r_y)}{r_x + r_y}. \tag{22}$$

In figures (1) and (2) it is assumed that N is N_2. For practical purposes, once we have determined the proper values of ϵ_x for (r_y/r_x) equal to 0 and 1, we can use a straight line to interpolate the rest of the curve (thus slightly overestimating ϵ_x). Also from figure (2) it is apparent that as soon as $N \geq 2$, we can safely assume that $(\epsilon_y/\epsilon_x) = (r_y/r_x)^{2/3}$. A FORTRAN code for computing the smoothing coefficients is now presented.

```
      SUBROUTINE CMPSMO(RXH,RYH,NX,NY,CFLFAC,EXH,EYH)
      REAL RXH(NX,NY),RYH(NX,NY),EXH(NX,NY),EYH(NX,NY)
C
C     INCOMING VALUES OF RX AND RY ARE STORED IN ARRAYS RXH AND RYH.
C     ARRAYS ARE OF SIZE NX BY NY.
C     CFLFAC IS N (ASSUMED TO BE .GE. 2).
C     SUBROUTINE PLACES APPROPRIATE SMOOTHING COEFFICIENTS INTO EXH AND EYH.
C
      EX1=.25*(CFLFAC**2-1.)
C
C     EX1 IS EX CORRESPONDING TO RY=0.
C     MARTINELLI'S FORMULA IS CORRECT FOR THIS CASE.
```

```
C
C  EX2 IS EX CORRESPONDING TO RY=RX.  THIS INVOLVES SOLVING A FIFTH DEGREE
C  POLYNOMIAL WHICH IS DONE USING A SIMPLE NEWTON ITERATION.
C
      S=0.
      DO 10 ITER=1,5
      F=(S**5+6*S**3+9*S)*CFLFAC-4*(1+S**2)
      FP=(5*S**4+18*S**2+9)*CFLFAC-8*S
      S=S-F/FP
   10 CONTINUE
      EX2=(1-S**2)*(1+S**2)/(4*S**2*(S**2+3))
C
C  NOW USE VECTORIZABLE LOOP TO COMPUTE SMOOTHING COEFFICIENTS.
C  NOTE CVMGT RETURNS FIRST ARGUMENT IF THIRD ARGUMENT IS TRUE
C  AND SECOND ARGUMENT IF THIRD ARGUMENT IS FALSE.
C  WE USE CVMGT TO INTERCHANGE ROLES OF RX AND RY IF RY .GT. RX.
C
      DO 100 J=1,NY
      DO 200 I=1,NX
      RX=RXH(I,J)
      RY=RYH(I,J)
      FACY=MIN(RX,RY)/MAX(RX,RY)
      EX=FACY*EX2+(1-FACY)*EX1
      EY=EX*FACY**(2./3.)
      EXH(I,J)=CVMGT(EX,EY,RX .GE. RY)
      EYH(I,J)=CVMGT(EY,EX,RX .GE. RY)
  200 CONTINUE
  100 CONTINUE
      RETURN
      END
```

When introducing this new version of IRS, care must be taken to modify the fundamental time step calculation in the code. Instead of using:

$$\Delta t_f = 1/(r_x + r_y + v_x + v_y), \tag{23}$$

where v_x and v_y refer to viscous terms when Navier-Stokes calculations are performed, one should use:

$$\Delta t_f = 1/\max(r_x, r_y, v_x + v_y). \tag{24}$$

With the new IRS and time step calculations in place, the code was found to be much more reliable, particularly when run on very fine grids (1024 by 256) for transonic airfoil flows.

References

[1] Martinelli, L.: "Calculation of Viscous Flows with a Multigrid Method," Ph.D. Thesis, Department of Mechanical and Aerospace Engineering, Princeton University, October 1987.

[2] Martinelli, L. and Jameson, A.: "Validation of a Multigrid Method for the Reynolds Averaged Equations," AIAA Paper 88-0414, January 1988.

[3] Radespiel, R.; Rossow, C.; and Swanson, R. C.: "An Efficient Cell-Vertex Multigrid Scheme for the Three-Dimensional Navier-Stokes Equations," AIAA Paper 89-1953, June 1989.

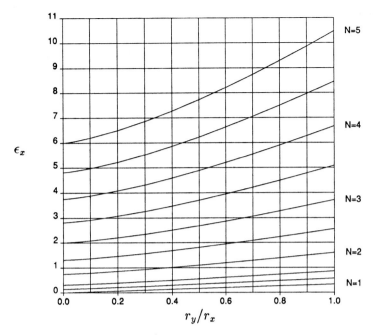

Figure 1: Plot of ϵ_x versus r_y/r_x for $N = N_2 = 1, 1.25, 1.5, 2, 2.5, 3, 3.5, 4, 5$.

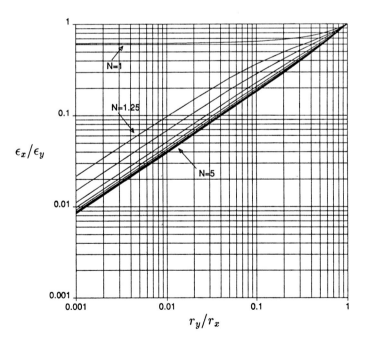

Figure 2: Plot of ϵ_x/ϵ_y versus r_y/r_x for $N = N_2 = 1, 1.25, 1.5, 2, 2.5, 3, 3.5, 4, 5$.

REACTIVE-FLOW COMPUTATIONS ON A CONNECTION MACHINE

E.S. Oran, J.P. Boris, and D A. Jones*
Laboratory for Computational Physics and Fluid Dynamics
Naval Research Laboratory
Washington, DC, 20375

This paper describes recent research and model development using the Connection Machine (CM) to perform multidimensional computations of highly compressible reacting flows. We briefly discuss computational timings compared to a Cray YMP speed, optimal use of the hardware and software available, treatment of boundary conditions, and parallel solution of terms representing chemical reactions. In addition, we show the practical use of the system for reactive-flow simulations by describing computations of multidimensional compressible reacting flows.

A CM consists of thousands of individual scalar processors connected by hypercube communications and all following a single instruction stream in a SIMD (single-instruction multiple-data) architecture. At NRL, there are two CMs, one with 16,384 (16K) processors and one with 8,192 (8K) processors. Communication to and control of these processors is through a front-end computer that may be either a VAX, a Symbolics, or a Sun. Each individual processor can be reconfigured by software to obtain powers of two increase in the number of virtual processors. The actual number of virtual processors is limited by the storage required by the problem solved. Floating-point arithmethic can be carried out by Weitek chips, each of which does pipelined processing of the floating-point operations for 32 of the scalar processors. Online color graphics are available to display the evolution of a calculation and to make video tapes. The user programs in one of several languages specially adapted to parallel processing: *Lisp, C*, or a special version of Fortran 8X. Compiling programs written in these extended languages produces a series of instructions in Paris (PARallel Instruction Set) microcoded routines that controls the CM processors and the front-end interaction with the CM. There is also a lower-level machine language, CMIS (CM Instruction Set), which is more complicated to program but produces some increase in speed over PARIS.

FLUID DYNAMICS ON THE CONNECTION MACHINE

We have converted the most recent version of the Flux-Corrected Transport (FCT) algorithm [1], LCPFCT, first to C*, then to PARIS, and now to CMIS. The kernel of the algorithm consists of about 38 lines of code well-suited for parallel instructions. Each virtual processor represents one fluid element and library routines are used to access values of variables in neighboring processors. This flexible generic module is used to solve appropriate sets of coupled continuity equations. It is combined with direction and timestep splitting to construct two-dimensional and three-dimensional computations.

A fundamental concern with the CM is how to avoid paying too severe a penalty (factors of two, four, or more in computational time) for computations that do not have periodic boundary conditions. Periodic boundary conditions are built into the CM hardware and cost no time or programming effort to implement for problems in any dimension, as long as the number of computational cells in each dimension is a power of two. We have tried or considered several approaches to programming more realistic boundary conditions: 1. *Boundary-Condition Overlays* – Recompute the boundary values after each computational timestep, and then overlay these solutions on the automatically computed periodic solutions (essentially ignore the problem and proceed serially); 2. *Uniform Boundary-Condition Algorithms* – Perform the same computations

---

* Materials Research Laboratory, Victoria, Australia

for all cells at all times, whether or not they are boundary cells, and choose the "correct" value for each cell; 3. *Segmented Data Structures* – Partition, or segment, the computational domain into regions near and far from the boundary, then use different optimized algorithms to update the distinct regions; and 4. *Unfolding Loops in One Spatial Direction* – Write a one-dimensional scalar algorithm that works with planar cross sections in one of the spatial directions (very efficient computationally, but requires a considerable amount of extra scratch space).

The choice of method for computing boundary conditions is closely tied to the problem being solved. The second approach, uniform computations everywhere [2], is useful for many supersonic flows and problems with relatively straightforward internal or external boundary conditions. Using generic difference formulas with a few additional constants in them at each stage of the FCT algorithm, the boundary cells can be treated the same as internal cells by choosing the proper values for the additional constants. The additional computation costs 5 to 25% extra. However, this approach is not the best for the most sophisticated subsonic inflow-outflow boundary conditions. For such computations in three-dimensions, we have implemented the segmented data-structure approach (see [3, 4]).

Comparative timings of test cases on the CM show that for a 128 × 128 grid problems on a 16K machine using 16K real processors (virtual-processor ratio of one), the speed of a computation is a factor of five faster than the Cray YMP using the CMIS version, and a factor of four using the simpler PARIS. For a three-dimensional computation with a 256 × 128 × 128 grid on a 16K machine (virtual-processor ratio of two) with optional boundary conditions, factors of more than four over the CRAY YMP are possible using PARIS and CMIS. These numbers essentially scale with the number of physical processors, so that the speed doubles if 32K processors are available. A more detailed description of the timing tests is given in [4].

MODEL FOR CHEMICAL REACTIONS AND ENERGY RELEASE

Chemical reaction waves, such as flames and detonations, are propagated in the fluid by physical processes such as convection, molecular diffusion, thermal condution, and radiation transport. In specific problems, certain of these processes dominate others. For example, in low-speed flames, diffusion and conduction cannot be neglected. However, for the types of high-speed, multidimensional computations we are now doing, it suffices to model just the chemical reactions and subsequent energy release or absorption, essentially local phenomena: physical diffusion effects are important on much longer timescales.

Here we replace a detailed set of ordinary differential equations representing the conversion of chemical species with a single variable that represents the progress of the reaction and then adjusts the individual densities, temperature, and pressure accordingly. The approach we are using here is based on the induction parameter [5], and a more detailed description of the current implementation is given in [6]. The input consists of three parameters which are tabulated as a function of temperature and pressure: the chemical induction time, the energy release time and form, and the final temperature, all of which could be derived from a full set of chemical kinetics equations, as we have done for hydrogen and oxygen, or from experimental data, as we have done for liquid nitromethane. Implementing algorithms for chemical reactions and energy release or absorption on the CM is relatively easy and fast because few computational steps are necessary and there is no interprocessor communication required because the processes are local.

MULTIDIMENSIONAL FLOW SIMULATIONS

Computations of imploding and exploding detonations and shocks are being used to study laboratory experiments of imploding and exploding detonations, astrophysical implosions and explosions such as star collapse and supernovas, and underwater explosions causing bubbles and then bubble collapse. The problem of initiation and propagation of an detonation from an initial high-energy "spark" is a element in all of these problems. When the initial energy put into a reactive medium is large enough, it can ignite a detonation either by first producing a shock that transitions to a detonation or by producing a flame that accelerates and becomes a detonation. In either case, whether or not a detonation is produced depends on the amount and time-history of the input of intial energy and on the inherent energy content of the material itself.

Figure 1 shows contours of several important physical variables in a computation of a steadily expanding explosion in a pure stoichiometric mixture of hydrogen and oxygen at 298 K, 1 atm. The initial conditions for this computation are determined by defining a detonation propagating in a spherically at the Chapman-Jouguet velocity at a given initial radius, r_o. The four outflow boundary conditions are extrapolation conditions and represent open boundaries. If r_o is too small, the detonation begins to propagate but eventually dies out. (Resolution tests have shown that this effect does not depend on the computational cell size.) This computation was done on a series of meshes ranging from 128×128 to 1024×1024. The induction parameter contours show that the reaction front is close to the shock front, here only several computational cells wide.The propagation is essentially spherical, more so where the curvature is small and therefore the expansion is the least. There is, however, considerable structure in the reacted gas due to several implosions and explosions that have happened in the course of the computation. When the pressure in the center becomes low enough compared to the surroundings, the fluid implodes. Although the surface of this explosion eventually goes unstable, the detonation front itself is so strong that it continues with little notice of the complex flow behind it.

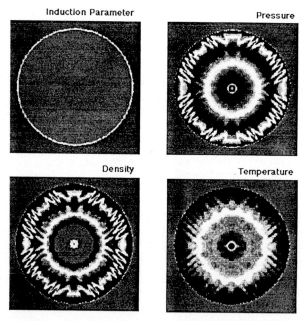

Figure 1. Spherical expansion of a detonation in a stoichiometric hydrogen-oxygen mixture at 298 K and 1 atm.

Another problem in which we are interested is the interaction of a shock with a vortex. This interaction is a fundamental interaction of compressible turbulence, and is important both in exothermic materials (combustion) and endothermic materials (reentry problems). Here we show preliminary results of a highly resolved computation. The interesting feature here is that the fluid velocity behind the Mach 1.5 shock is comparable to the maximum velocity in the compressible vortex, so that the interaction between the shock and vortex is strong. Figure 2 shows a sequence of frames from a simulation performed on a 512×256 grid. The inflow condition is maintained by setting the inflow variables to the shock condition. The outflow is an extrapolated conditions similar to that used in the hydrogen-spark problem described above. The upper and lower boundaries are refecting walls. Initially both the shock and the vortex are perturbed, but in time the shock straightens and is little affected by the vortex it has passed. The initially round vortex is flattened as it is compressed and becomes a rotating elliptical vortex. In addition, the interaction of the shock and the vortex generates a complex wave pattern whose origin is the vortex. An analysis of this interaction is being prepared by [7]. Future calculations with this model will include the effects of shock-ignition and the effects of this on the vortex and wave generation.

DISCUSSION

For a large class of explicit, chemically reacting problems typical of high-speed compressible flows, the current SIMD approach is excellent. When the data structure is properly chosen, SIMD parallel code is actually easier to write and shorter than the corresponding scalar code. However, there are important situations for which some MIMD features would be extremely useful. We have found a way around the boudary-condition limitation that is not too expensive, and for large enough three-dimensional problems using the segmented data-structure approach for boundaries, there is essentially no cost penalty after the complication of coding. Even though coding the uniform algorithm is much simpler generally, there are subtleties because there is more communication between cells in a fourth-order algorithm (arising from the half-step and whole-step operations) than in a simpler, less accurate algorithm.

In general, however, we have found that the CM is an excellent computer on which to carry out large, multidimensional, complex FCT computations. The timings are competitive and in fact can be a factor of four or five better than the Cray YMP. The CM is relatively user friendly and easy to program and the staff at the NRL has been particularly helpful and encouraging.

ACKNOWLEDGMENTS

This work was funded by the NRL through the ONR and by the Advanced Computation and Mathematics Program in DARPA. The authors would like to thank C. Li for sharing his insight into the boundary-condition problem, R. Kolbe for his help with the CM, J. Ellzey and J.M. Picone for their insights into the shock-vortex interaction, and R. Whaley from TMC for his solid advice and ready willingness to help.

REFERENCES

1. J.P. Boris, and D.L. Book, *Meth. Comput. Phys.* 16, 85–129, 1976; and J.P. Boris, J.H. Gardner, E.S. Oran, S. Zalesak, J. Ellzey, G. Patnaik, D.L. Book, R.H. Guirguis, *LCPFCT – A Monotone Algorithm for Solving Continuity Equations*, to appear as NRL Memorandum Report, 1990.
2. C. Li, E.S. Oran, and J.P. Boris, submitted to the SIAM Journal, 1990; to appear, NRL Memorandum Report, 1990.
3. J.P. Boris, E.S. Oran, E.F. Brown, C. Li, R.O. Whaley, and F.F. Grinstein, 12th International Conference on Numerical Methods in Fluid Dynamics, Oxford, England, July, 1990.
4. E.S. Oran, J.P. Boris, E. Brown, and R.O. Whaley, sl Fluid Dynamics Computations on a Connection Machine AIAA Paper No. 90-0335, AIAA, Washington, DC, 1990.
5. For a summary, see E.S. Oran, J.P. Boris, and K. Kailasanath, *Studies of Detonation Initiation, Propagation, and Quenching*, to appear in *Numerical Approaches to Combustion Modeling*, eds. E.S. Oran and J.P. Boris, to be published by AIAA, 1990.
6. E.S. Oran, D.A. Jones, and M. Sichel, submitted to the 29th Aerospace Sciences Symposium, Reno, NV, 1991.
7. J.L. Ellzey, E.S. Oran, and J.M. Picone, to appear, *Proc. 17th Int. Symp. Shock Waves and Shock Tubes*, AIP, New York, 1990; in preparation for *Phys. Fluids*, 1990.

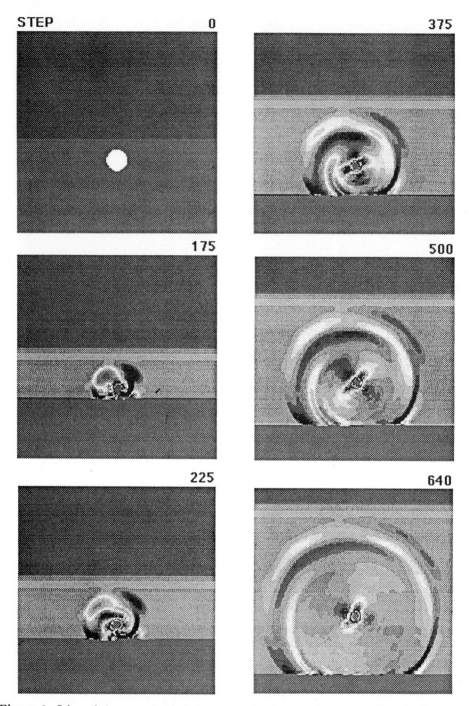

Figure 2. Selected timesteps in the interaction of a shock with a vortex. The shock comes from above the vortex and by step 175 has passed through.

NAVIER-STOKES SOLVERS USING LOWER-UPPER SYMMETRIC GAUSS-SEIDEL ALGORITHM

Seokkwan Yoon
MCAT Institute
Moffett Field, California, U.S.A.

Although notable successes have been reported with explicit schemes to solve the Euler equations for inviscid flows, implicit methods have emerged as useful tools for predicting viscous flows since the efficiency of explicit schemes to solve the Navier-Stokes equations is limited by a well-known stability criterion, the CFL condition. When the time step limit imposed by an explicit stability bound is significantly less than that imposed by the accuracy bound, implicit schemes are to be preferred. However, there is a trade-off between decreased number of iterations and increased operation count per iteration. Although the fastest convergence rate can be attained by the unfactored implicit scheme, direct inversion of a large block banded matrix is impractical in three-dimensions because of rapid increase of operation count with the number of mesh points and large storage requirement. The large factorization error associated with the alternating direction implicit method further limits the time step which is already restricted by the conditional stability. Conventional implicit methods which often achieve fast convergence rates suffer high cost per iteration.

A new implicit algorithm based on lower-upper(LU) factorization and symmetric Gauss-Seidel(SGS) relaxation offers very low cost per iteration as well as fast convergence. The LU-SGS implicit scheme, which shows unconditionally stability for a linear equation, is not only completely vectorizable but amenable to massively parallel processing. High efficiency is achieved by using two-dimensional arrays in three-dimensions on oblique planes of sweep. Despite its fast convergence characteristics, the LU-SGS scheme requires less computational work per iteration than most explicit schemes. In the framework of the LU-SGS algorithm, a variety of schemes can be developed by different choices of numerical dissipation models and Jacobian matrices of the flux vectors. The algorithm is especially efficient when the governing equations are coupled with source terms. A couple of new Navier-Stokes solvers are developed for the incompressible rotating(INS3D-LU) and the compressible reacting(CENS2D/3D) flows. Remarkable performance of the codes has been demonstrated on single processor of Cray YMP supercomputer by the computing times per grid point per iteration. The LU-SGS scheme not only achieved 150 MFLOPS but needs only 7 μsec for the full incompressible Navier-Stokes equations in three-dimensional generalized coordinates. Numerical examples include the incompressible internal flows through a rotating turbopump and the compressible external flows with strong shock waves and chemical reactions. Fig. 1 shows a pressure gradient across the inducer blades of the Space Shuttle Main Engine due to the action of the centrifugal force. Fig. 2 illustrates the interaction of the tip leakage and streamwise flows. Experimental validation in Figs. 3 and 4 (Mach 6 Nitrogen flow) not only points out the importance of real gas effects in high speed flows but also demonstrates the accuracy of the present hypersonic flow solver using multi-temperature finite-rate-chemistry.

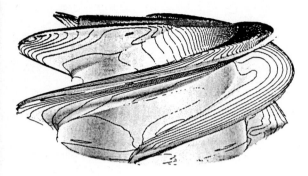

Fig. 1. Surface pressure contours for the Space Shuttle Main Engine turbopump inducer

Fig. 2. Particle traces for the suction side of the inducer

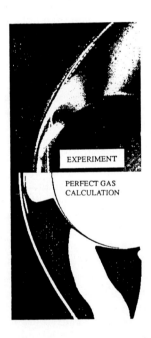

Fig. 3. Perfect gas solution for hypersonic blunt body flow compared to experimental data

Fig. 4. Thermochemical nonequilibrium solution using multiple temperatures

An Extension of Essentially Non-Oscillatory Shock-Capturing Schemes to Multi-Dimensional Systems of Conservation Laws

Jay Casper
Department of Mathematics, Old Dominion University
Norfolk, Virginia 23529

We wish to design high-order accurate essentially non–oscillatory (ENO) schemes for the numerical approximation of weak solutions of hyperbolic systems of conservation laws

$$u_t + f(u)_x + g(u)_y = 0 , \qquad (1)$$

subject to given initial and/or boundary conditions. With a semi–discrete formulation in mind, we note that for every every rectangle $(x_{i-1/2}, x_{i+1/2}) \times (y_{j-1/2}, y_{j+1/2})$, a weak solution of (1) satisfies

$$\frac{\partial}{\partial t} \bar{\bar{u}}_{ij}(t) = -\frac{1}{a_{ij}} \left[\hat{f}_{i+1/2,j}(t) - \hat{f}_{i-1/2,j}(t) + \hat{g}_{i,j+1/2}(t) - \hat{g}_{i,j-1/2}(t) \right] , \qquad (2a)$$

where a_{ij} is the area of the rectangle and $\bar{\bar{u}}_{ij}(t)$ is the cell average of u over that rectangle at time t. The fluxes $\hat{f}$ and $\hat{g}$ are given by

$$\hat{f}_{i+1/2}(t) = \int_{y_{j-1/2}}^{y_{j+1/2}} f(u(x_{i+1/2}, y, t)) \, dy \ ; \quad \hat{g}_{j+1/2}(t) = \int_{x_{i-1/2}}^{x_{i+1/2}} g(u(x, y_{j+1/2}, t)) \, dx \ . \qquad (2b)$$

We treat (2a) as a system of ordinary differential equations for the purpose of time discretization, using a "method–of–lines" approach. Along any $t =$ constant line, the right–hand side of (2a) is strictly a spatial operation, and we thus effectively "separate" the spatial and temporal operations for computing solutions of (1).

The multi–dimensional ENO schemes developed thus far in the literature are of finite–difference type, employing a split–flux approach. Due to their design, it is not clear, at present how to apply such schemes at boundaries, e.g. solid walls, in addition to the fact that curvature or discontinuity of the computational grid can pose fundamental problems for any finite–difference scheme in which stencils are not pre–determined. On the other hand, these issues are more readily handled using our finite–volume approach.

Let R^2 be a spatial operator which reconstructs $\{\bar{\bar{u}}_{ij}^n\}$, cell averages of a piecewise smooth solution $u(x, y, t^n)$ of (1), and yields a two–dimensional, piecewise polynomial function $R^2(x, y; \bar{\bar{u}}^n)$ which approximates $u(x, y, t^n)$ to r^{th} order, i.e.

$$R^2(x, y; \bar{\bar{u}}^n) = u(x, y, t^n) + O(h^r) , \qquad (3)$$

wherever $u(x, y, t^n)$ is sufficiently smooth, and $h = O(\Delta x) = O(\Delta y)$. In particular, the operator R^2 which we have developed is a "natural" two–dimensional extension of the one–dimensional "reconstruction by primitive" presented in [1]. The reconstruction involves polynomial interpolation in combination with an adaptive stencil algorithm which, in tandem, achieve high–order accuracy in smooth regions while avoiding oscillations near steep gradients.

The nature of the reconstruction operator can cause $R^2(x, y; \bar{u})$ to be discontinuous at cell interfaces. The relative size of these local "jumps" is on the level of the interpolation error in smooth regions and is $O(1)$ near discontinuities in u. The flux integrals (2b), then, are approximated by an appropriate quadrature, the flux contribution at each quadrature point being determined by the solution of the local Riemann problem whose initial states are determined by the reconstruction. The high–order accurate time discretization of (2a) is then achieved by the Runge–Kutta methods in [2]. The scheme is formally r^{th}–order accurate, in space and time, in the sense of local truncation error, wherever u is sufficiently smooth, and avoids $O(1)$ spurious oscillations near discontinuities.

Computational tests, using these finite–volume high–order ENO schemes to solve two–dimensional scalar equations, have yielded results that support both the accuracy as well as the non–oscillatory properties of these schemes. Furthermore, we have formally extended these schemes to general curvilinear co-ordinates and have performed numerical experiments which solve boundary–value problems governed by the two–dimensional Euler equations of gas dynamics. The reslts obtained represent the first successful attempt to employ high–order ENO schemes to two–dimensional problems with solid walls (*e.g.* Fig. 1).

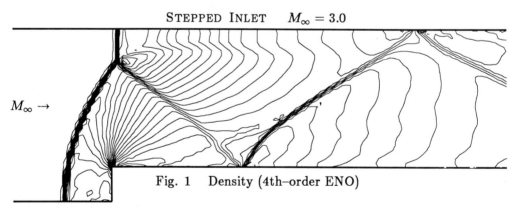

Fig. 1   Density (4th–order ENO)

REFERENCES

[1] A. Harten, B. Engquist, S. Osher, S. Chakravarthy, "Uniformly High–Order Accurate Essentially Non–Oscillatory Shock–Capturing Schemes III," *J. Comp. Phys.*, V. 71 (1987), pp. 231-323.

[2] C. Shu and S. Osher, "Efficient Implementation of Essentially Non–Oscillatory Shock–Capturing Schemes," *J. Comp. Phys.*, V. 77 (1988), pp. 439-471.

On a Modified ENO Scheme and its Application to Conservation Laws with Stiff Source Terms

Shih-Hung Chang
Department of Mathematics, Cleveland State University
Cleveland, Ohio 44115

In the study of numerical methods for reacting flow problems, LeVeque and Yee [J. Comput. Phys., 86(1990), pp.187-210] proposed the use of a certain model problem consisting of a one-dimensional scalar conservation law with parameter-dependent source term of the form $u_t + u_x = -\mu u \left(u - \frac{1}{2}\right)(u - 1)$, where μ is a parameter, for testing algorithms. This equation becomes stiff when the parameter μ is large. Their investigation showed that stable and second order schemes for reacting flows can be devised by using appropriate extensions of many current finite difference methods developed for non-reacting flows, including TVD schemes. However, in studying the ability of these extensions in dealing with propagating discontinuities on their model problem, it was reported that for a certain reasonably fixed mesh and for the very stiff case, all the methods produced solutions that look reasonable and yet are completely wrong, because the discontinuities are in the wrong locations. The difficulty is due to the smearing of the discontinuity in the spatial direction. This raises a fundamental question on extending numerical methods developed for non-reacting flows to reacting flows.

In our investigation we find that numerical methods can be devised to overcome the above mentioned difficulty. We construct a numerical scheme for solving conservation laws with stiff source terms, which, when applied to the model problem of LeVeque and Yee, results in stable solutions with excellent resolutions at the correct locations of discontinuities. This scheme is a modification of Harten's ENO scheme with subcell resolution, ENO/SR [J. Comput. Phys., 83(1989), pp.148-184], with which significant improvement in the resolution of contact discontinuities was achieved for conservation laws. We use the technique of ENO reconstruction with subcell resolution to locate the discontinuity within a cell and then accomplish the time evolution by solving the differential equation along characteristics locally and advancing in the characteristic direction. This scheme is denoted ENO/SRCD with CD standing for characteristic directions. We also extend the basic ENO and ENO/SR schemes to solve conservation laws with source terms. These are done by using Strang's time-splitting method.

All the schemes are tested on the model problem of LeVeque and Yee. Numerical results show that these schemes handle this intriguing problem very well. The ENO/SRCD scheme produces perfect resolution at the propagating discontinuity. The extensions of basic ENO and ENO/SR via time-splitting also perform very well, especially with ENO/SR showing almost perfect results, except for the very stiff case where some adjustment in the time step-size is needed.

In the following we describe briefly the construction of the scheme ENO/SRCD for the equation
$$u_t + a\, u_x = \psi(u), \qquad a > 0, \tag{1}$$

where the source term $\psi(u)$ arises from the chemistry of the reacting species. Observe that along the characteristic, $x = x_0 + at$, the solution to Eq.(1) evolves according to the following ODE

$$\frac{d}{dt}u(x_0 + at, t) = \psi(u(x_0 + at, t)), \qquad (2)$$

with initial data $u(x_0, 0)$. This equation will be solved approximately from the time step t_n to t_{n+1}. At t_n, suppose that we have obtained the numerical solution $v^n = \{v_j^n\}$, where v_j^n represents an approximation to $\bar{u}_j^n$, the cell average of u over $[x_{j-\frac{1}{2}}, x_{j+\frac{1}{2}}]$ at t_n. Then, to obtain v^{n+1}, we follow these steps:

Step 1. Obtain a reconstruction, $R(x; v^n)$, of the solution from the given values v^n.

Step 2. Locate the discontinuity, if any, within each cell $[x_{j-\frac{1}{2}}, x_{j+\frac{1}{2}}]$ using the subcell resolution technique and modify the reconstruction $R(x; v^n)$ to obtain $\hat{R}(x; v^n)$.

In both steps 1 and 2, we use the procedure of ENO reconstruction with subcell resolution of Harten. The reconstructed solution function $R(x; v^n)$ here is a piecewise quadratic polynomial obtained by using the primitive function approach.

Step 3. Advance $\hat{R}(x; v^n)$ via the ODE (2) along characteristics to t_{n+1} and then take cell averages to complete v^{n+1}.

If a discontinuity θ_j exists inside the cell $[x_{j-\frac{1}{2}}, x_{j+\frac{1}{2}}]$, then $R_{j-1}(x; v^n)$ is used as the solution to the left of θ_j and $R_{j+1}(x; v^n)$ to the right of θ_j at t_n. Since the solution to Eq.(1) evolves according to the ODE (2) along characteristics, we can obtain approximate solution values at t_{n+1} by solving (2) and advancing in the characteristic direction. If $w(x, t_{n+1})$ denotes the solution of (2) at t_{n+1} obtained by using the initial values $R_{j-1}(x; v^n)$ to the left of θ_j and $R_{j+1}(x; v^n)$ to the right of θ_j, the numerical solution v_j^{n+1} is then an approximation to

$$\frac{1}{\Delta x} \int_{x_{j-\frac{1}{2}}}^{x_{j+\frac{1}{2}}} w(x, t_{n+1}) \, dx.$$

Approximations to $w(x, t_{n+1})$ and hence v_j^{n+1} can be obtained easily.

The idea behind the construction of the scheme is that in order to obtain stable solutions with excellent resolutions at the correct locations of discontinuities, one needs a scheme that can track the discontinuity. When the information on the location of the discontinuity is used in treating the source term, significant improvement in numerical results to the present model problem was then achieved.

In conclusion, we have constructed a numerical scheme, ENO/SRCD, which has the potential to improve the resolution of contact discontinuities for non-reacting flow problems and also to produce excellent solutions to conservation laws with stiff source terms modeling reacting flows.

CONVENIENT ENTROPY-SATISFYING T.V.D. SCHEMES WITH APPLICATIONS

N.Clarke, D.M.Ingram, D.M.Causon, R.Saunders

Computational Fluid Dynamics Group,
Dept. Mathematics and Physics,
Manchester Polytechnic,
Chester Street
Manchester.

We have previously reported, to the 8th GAMM conference {1}, on the derivation of a "convenient" entropy satisfying TVD scheme for use with Lax-Wendroff type schemes {2} and have since incorporated this technique along with several others into a general purpose operator-splitting Euler solver. This approach involves adding a non-linear dissapative term to the corrector step of MacCormack's original two step explicit scheme, i.e.,

$$U_j^{n+1} = \mathcal{L}_{mac}(U_j^n) + \frac{1}{2}|\nu|(1-|\nu|)(G_{j+1/2}\Delta U_{j+1/2}^n - G_{j-1/2}\Delta U_{j-1/2}^n) \quad (1)$$

where
$$\Delta(\cdot)_{j+1/2} = (\cdot)_{j+1} - (\cdot)_j$$
and
G is a function of the flux limiter, $\Phi(r)$

The TVD term in (1) can be designed so as to implicitly apply artificial compression, in order to attempt to produce highly resolved discontinuities. However, in some cases, this can lead to non-physical *entropy-violating* expansion shocks. Constraining the scheme design so as to satisfy a discrete entropy condition as well as total variation diminution and second order accuracy (except in regions of extrema) simultaneously means that some of the most desirable limiters become forbidden fruit. Current theoretical work involves the design of schemes, of the type in (1), satisfying a *local* entropy condition which is rather less restictive than its global counterpart.

We also note that the Harten TVD constraints {3} are *sufficient* but not *necessary* conditions for the total variation diminution of the resulting scheme. We have obtained somewhat less restrictive conditions which still guarantee "TVDness" of the scheme but allow more freedom to possibly increase accuracy in regions of extrema.

It is also possible to design TVD schemes so as to enable a different TVD operator to be applied to each of the conserved variables in the Euler equations. This is done through the use of a Roe type approximate Riemann solver. This approach enables us to tailor the type of TVD term used to the features of the problem at hand and also to extract information, such as the characteristic directions. Clearly this requires some knowledge of the problem to be solved *apriori* but such information is normally available.

Our General purpose Euler code has been used to solve many problems ranging from external aerodynamics (flow past a finbody configuration) and hypersonics (flow past a double ellipse) to blast wave interactions (muzzle blast). The solver implements a variety of methods based upon the MacCormack scheme and is used as a work bench for the development of new schemes such as those discussed above. The solver is written in Fortran-77 and has been run successfully on a wide variety of computers.

We have obtained good results using the field-by-field decomposition, as described above, on the shock collision problem of Woodward and Colella {4} which are shown below together with results obtained using the Davis {5} symmetric TVD scheme on a mesh of 300 points for comparison. The results shown were obtained on a Silicon Graphics IRIS workstation.

References :

1. Clarke N, Causon D & Saunders R, A Convenient Entropy Satisfying Scheme for Computational Aerodynamics, NNMFM $\underline{29}$ Proc. 8th GAMM Conf., Wessling P. (Ed.) Vieweg, 1990.

2. Sweby P K, High Resolution Schemes using Flux Limiters for Hyperbolic Conservation Laws., SIAM J NM $\underline{21}$, No. 5, 1984.

3. Harten A, High Resolution Schemes for Hyperbolic Conservation Laws, JCP $\underline{49}$, 357 - 393, 1983.

4. Woodward P & Colella P, The Numerical Simulation of Two-Dimensional Fluid Flow with Strong Shocks, JCP $\underline{54}$, 115 - 173, 1984.

5. Davis S F, TVD Finite difference Schemes and Artificial Viscosity., ICASE Report 84-20, 1984.

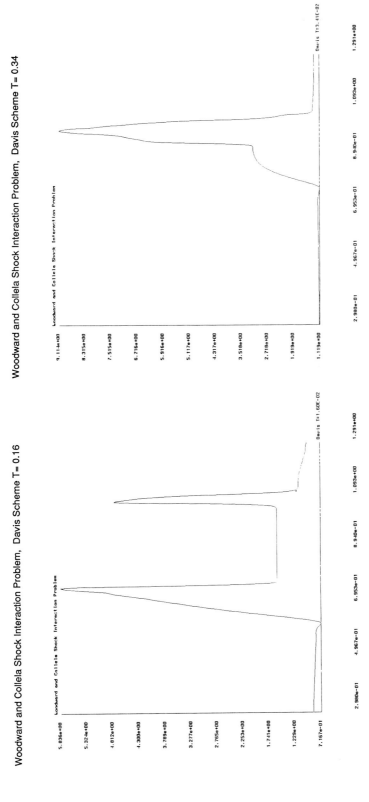

Woodward and Collela Shock Interaction Problem, Davis Scheme T= 0.16

Woodward and Collela Shock Interaction Problem, Davis Scheme T= 0.34

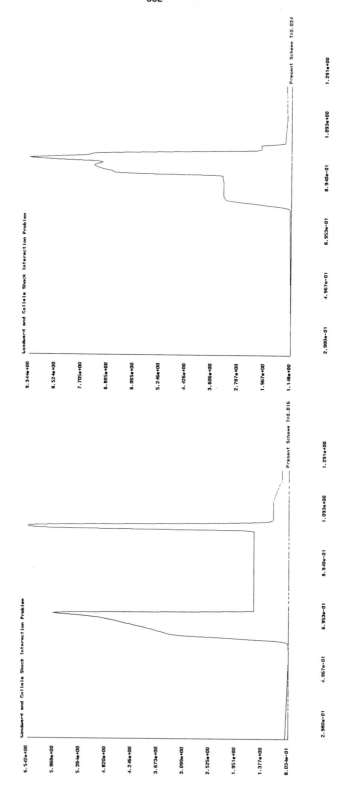

Woodward and Collela Shock Interaction Problem, Present Scheme T=0.016

Woodward and Collela Shock Interaction Problem, Present Scheme T= 0.34

SECOND ORDER DEFECT-CORRECTION MULTIGRID FORMULATION OF THE POLYNOMIAL FLUX-DIFFERENCE SPLITTING METHOD FOR STEADY EULER EQUATIONS

E. DICK

Department of machinery, State University of Ghent
Sint Pietersnieuwstraat 41, 9000 GENT, BELGIUM

The polynomial flux-difference splitting method and its multigrid formulation in first order form were presented in [1]. Here, we discuss the second order formulation and its implementation in multigrid form through the defect-correction approach.

The polynomial flux-difference splitting is based on the polynomial character of the flux-vectors. The splitting is a pure Roe-type splitting which however avoids square root evaluations [2]. The splitting is applied to the vertex-centered finite volume method, i.e. using control volumes centered around the vertices of the computational grid. In first order form, the discretisation leads to a set of equations which is positive in vector-sense, i.e. the matrix-coefficients involved have non-negative eigenvalues and the central matrix coefficient is the sum of the non-central ones. At boundaries, a formulation similar to the formulation in the flow field is possible. Due to the algebraically rigorous way in which the splitting is formulated, the exact number of independent equations, according to characteristic theory, is obtained. These equations are supplemented by boundary conditions, retaining the positivity of the discrete set of equations. As a consequence, the set of discrete equations can be solved by collective forms of classic relaxation methods. These relaxation methods have smoothing properties so that a multigrid formulation is possible.

A full multigrid formulation (multigrid with a starting cycle) using the non-linear equations (full approximation scheme) with W-cycles, full weighting as restriction for residuals, injection for projection of function values and bilinear interpolation as prolongation for corrections, is used. Gauss-Seidel relaxation in two forms is employed : symmetric successive relaxation and black-red relaxation (first even points, then odd points). The performance of both types of relaxation methods is comparable in terms of residual reduction per work unit (the number of operations equivalent to one relaxation for all grid points on the finest level). The black-red relaxation is fully vectorizable and parallelizable.

Second order accuracy is obtained by flux corrections according to the Chakravarthy-Osher flux-extrapolation technique. These corrections require flux-limiting. Here the MinMod-limiter is used [3]. The obtained discretisation is not positive anymore. Therefore a direct relaxation solution is not possible. To solve the second order problem, a defect-correction formulation with the first order discretisation as basis is employed. The second order correction is made only on the finest grid. This correction is frozen for all other grids and transferred by the usual restriction.

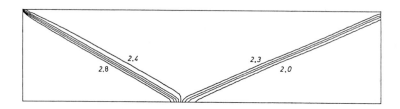

The figure above illustrates second order accuracy for the well known Harten-shock reflection problem, using a 96 × 32 cells grid on the finest level.

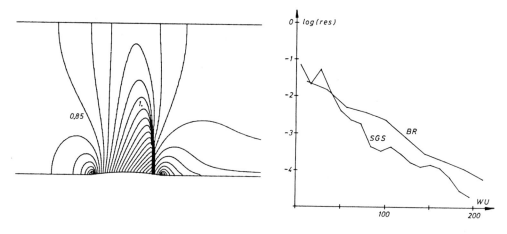

The figure above shows the second order result for the GAMM-transonic bump problem, also on a 96 × 32 cells grid. This result is only marginally different from the first order result. The figure shows the multigrid performance with symmetric Gauss-Seidel and black-red relaxation, using 4 grid levels.

References

1. DICK E., A multigrid method for steady Euler equations based on polynomial flux-difference splitting. Proc. 11th Int. Conf. Num. Meth. Fl. Dyn., Lecture Notes in Physics, 323, 1989, 225-229.

2. DICK E., A flux-vector splitting method for steady Euler equations. J. Comput. Phys., 76, 1988, 19-32.

3. DICK E., Multigrid formulation of polynomial flux-difference splitting for steady Euler equations. J. Comput. Phys., 1990, to appear.

Projection Shock Capturing Algorithms

Björn Engquist
Department of Mathematics
UCLA
405 Hilgard avenue
Los Angeles, CA90024

Björn Sjögreen
Department of Scientific Computing
Uppsala University
Sturegatan 4B
752 31 Uppsala
Sweden

We consider the system of nonlinear conservation laws

$$\mathbf{u}_t + \mathbf{f}(\mathbf{u})_x + \mathbf{g}(\mathbf{u})_y = 0$$

discretized in time and space with step sizes Δt, Δx and Δy, $t_n = n\Delta t$, $x_j = j\Delta x$, $y_k = k\Delta y$. The numerical solution in the point (t_n, x_j, y_k) is denoted by u_{jk}^n and the entire solution at time t_n is u^n and a function of j, k.

We consider the solution of the above problem with finite difference methods. The problem is difficult because of the occurrence of discontinuities in the solution.

One of the more widely used ways to derive a second order method with good shock capturing properties is to advance the solution with a first order method containing sufficient numerical damping

$$u^* = G(u^n)$$

and then subtract the viscosity where it is not needed, i.e. where the solution is monotone, giving

$$u^{n+1} = u^* - H(u^*)$$

this is the idea of the flux corrected transport algorithm (FCT) [1]. Alternatively, it is possible to make the correction on the old time level

$$u^{n+1} = u^* - H(u^n)$$

this is the idea of most of second order TVD schemes, see [2][3]. Here $G(u)$ has sufficient numerical viscosity to smooth out all shocks and $H(u)$ is a non-linear antidiffusive term, such that $G(u) - H(u)$ is the second order Lax-Wendroff scheme, whenever possible due to non-oscillation constraints.

In this poster we present results obtained with the following opposite approach. We first advance the solution with the Lax-Wendroff method

$$u^* = L(u^n)$$

and then make a correction step

$$u^{n+1} = u^* + M(u^*)$$

where $M(u)$ is a diffusive operator (filter, TVD projection), which removes the oscillations around shocks introduced by the scheme $L(u)$.

Some advantages of this opposite method are

- $M(u)$ is zero away from shocks, which makes the amount of work in the correction step small, even if we have to use complicated formulas, eigendecomposition, e.t.c. in the M operator.
- $M(u)$ is made independent of the scheme $L(u)$, and we can use any method we like in the first step to get u^*. This makes it possible to achieve higher order accuracy (>2) in a simple way and the algorithm can easily be incorporated into existing codes.

The formula for $M(u)$ used in the results presented here can be described in terms of moving grid points up and down due to certain constraints, see [4]. At most of the grid points the following formula describes the algorithm in the case of a one dimensional scalar problem

$$u^{n+1} = u_j^* + a_{j+1/2}(u_{j+1}^* - u_j^*) - a_{j-1/2}(u_j^* - u_{j-1}^*)$$

where with $r_j = (u_j^* - u_{j-1}^*)/(u_{j+1}^* - u_j^*)$,

$$a_{j+1/2} = \min(-r_j, 1/2)$$
$$a_{j-1/2} = 0$$

if $r_j < 0$ and $-r_j < 1$,

$$a_{j+1/2} = 0$$
$$a_{j-1/2} = \min(-1/r_j, 1/2)$$

if $r_j < 0$ and $-r_j > 1$. If $r_j > 0$ then $a_{j+1/2} = a_{j-1/2} = 0$.

For systems of equations, e.g. the Euler equations for compressible fluid flow, the algorithm has to be carried out field by field in the characteristic variables. Multidimensional problems are handled dimension by dimension.

On the poster we present results from solving the Euler equations in two space dimensions around aerofoils in the transonic regime, and around a disk at supersonic speeds. Methods of order of accuracy 2 and 4 are used together with the filter algorithm described above.

References:

[1] J.P.Boris, D.L.Book, "Flux-Corrected Transport. I.SHASTA, A Fluid Transport Algorithm That Works", J.Comp.Phys.,**11**, (1973), pp.38–69.

[2] A. Harten, "High Resolution Schemes for Hyperbolic Conservation Laws", J. Comput. Phys., **49** (1983), pp. 357–393.

[3] S.Osher, E.Tadmor, "On the Convergence of Difference Approximations to Scalar Conservation Laws", Math. Comp. 50 1988, pp.19–51.

[4] B.Engquist, P.Lötstedt, B.Sjögreen, "Nonlinear Filters for Efficient Shock Computation", Math.Comp.**52**, (1989), pp.509–537.

MULTIGRID TECHNIQUE APPLIED TO THE EULER EQUATIONS ON A MULTI BLOCK MESH

P.Å. Weinerfelt

CERFACS, 42 Av. G. Coriolis, 31057 Toulouse CEDEX, France.

and

Linköping University, Linköping, S-581 83, Sweden.

INTRODUCTION

The fast evolution of modern vector and parallel computers during the past decade has increased the need of modification of old algorithms and development of new ones, in order to utilize the full power of these machines. Especially, parallelization techniques have become important. For applications in CFD, domain decomposition leads naturally to a parallelization of the problem. The computational domain is split into sub domains or blocks in which the flow equations can be solved independently and in parallel. This approach also gives the possibility to compute the flow around complex geometries. For multi grid applications the multi block concept has the advantage that different strategies can be used in different blocks. Computing time can e.g be saved by only updating the solution every time step in blocks with large gradients. Domain decomposition also leads to algorithms which are well suited to parallel MIMD computers like Alliant FX/80 or iPSC/2 Hypercube. We will in the following paragraphs describe the application of this technique to the Euler equations and how to implement the solver on the iPSC/2 Hypercube. Finally some results are presented.

THE FLOW SOLVER

We are going to focuse on the solution to the steady Euler equations. The steady energy equation is replaced by the constant enthalpy equation along stream lines, which for uniform flow at the far field leads to constant enthalpy everywhere. Hence we obtain an algebraic equation for the pressure. The Euler equations are numerically solved by means of a cell vertex finite volume scheme. The computational domain is divided into quadrilaterals over which the equations are integrated numerically. As we are using a centered finite volume scheme, artificial viscosity needs to be added. We have followed the concept from Jameson using blended second and fourth order differences, formed as first and third order flux differences. The second order term, which is governed by a pressure sensor, are used locally to capture strong gradients. The fourth order term is added throughout the computational domain to eliminate spurious modes. It has to be switched off close to a shock in order to prevent oscillations. The resulting system of nonlinear ODEs is integrated explicitely in time by a Runge-Kutta scheme.

MULTIGRID TECHNIQUE

To speed up the convergence rate towards steady state a multi grid method has been used. An approximative solution is first obtained on the finest grid. The fine grid residuals are transferred to a coarser grid and used as a forcing term. A number of time step are then performed on the coarser mesh and the new residuals are transferred to the next coarser grid. When the coarsest grid is reached, the corrections are successively interpolated back to the finest grid and a new multi grid cycle can start. This procedure is repeated until the fine grid residuals are driven to zero.

MULTI BLOCK FORMULATION

The scheme above was first developed for a 2D single block O - mesh. In the next step towards a general multi block formulation the flow solver was rewritten to treat arbitrary boundary conditions on the four block sides. Finally, pointers describing the number of grid points, the type of boundary conditions, number of boundary conditions and connection to other blocks on each side, were introduced. The information is given in such a way that the updating at the boundaries can be vectorized and hence the overhead time minimized.

IMPLEMENTATION ON THE iPSC/2 HYPERCUBE

As mentioned before, the multi block approach has several advantages. The splitting of the computational domain into smaller subdomains suggests directly a way to make the computations in parallel. For each block and time step, the flow calculations are done independently followed by exchange of information between the blocks. This way of parallelizing the problem is well suited to a MIMD computer in which each node-program is executed independently. Since the nodes have local memories they can only communicate by sending and receiving messages. A 2D version of the flow solver has been implemented on a iPSC/2 Hypercube. Each block was associated with a node processor which was dealing with the local computations inside each block and sending/receiving information to other nodes or to the host processor. The principal tasks such as loading the node programs, i/o, checking of the convergence criteria and sending/receiving of the initial/final solution, were handled by the host processor.

CONCLUSION AND RESULTS

The single block flow solver was first tested and tuned for ETA 10 and Alliant FX/80. This code has proven to be robust, efficient and both vectorizable and parallelizable. In combination with the multi grid technique for steady state computations an almost grid independent convergence rate was obtained. The multi block flow solver has been used to compute the flow in different kind of geometries e.g valve cylinder assembly. The multi block mesh is better suited to this kind of complex geometry than a single block mesh. We also have the possibility to use different multi grid strategies in different blocks. Computing time can then be saved by e.g. updating the solution every time step only in blocks with large gradients. Concerning the implementation of the code on a MIMD computer we have found that even for rather small blocks the communication time on the iPSC/2 Hypercube is negligible compared to the computing time.

TREATMENT OF INERT AND REACTIVE FLOWS

BY A T.V.D. FORMULATION IN 2D SPACE
IN UNSTRUCTURED GEOMETRY

By A.J. FORESTIER
DEMT/SMTS
CEN SACLAY
91191 - GIF SUR YVETTE CEDEX - FRANCE -

I - ABSTRACT

This paper relates modelisation of Euler equation in Computational Fluid Dynamic (C.F.D.) for a two dimensional formulation with arbitrary mesh generated by quadrilaterals and triangles.

Treatment of boundary conditions is essential in global results and influence about the whole computation put modify all the solution : we distinct five types of boundary condition : total reflection, supersonic and subsonic inlet or outlet.

Different applications with this approach are also indicated.

II - T.V.D. SCHEME IN 1D FORMULATION

It is well know the Euler equation give the three conservation laws : mass, momentum and energy. We have :

$$\frac{\partial U}{\partial t} + \frac{\partial}{\partial x} F(U) = 0$$

where $U = (\rho, \rho u, \rho E)^T$ and $F(U) = (\rho u, \rho u^2 + P, (\rho E + P)u)^T$
with ρ : density, u : velocity, E : total mass energy,
P : pressure.

$$\begin{cases} \frac{\partial U}{\partial t} + \frac{\partial}{\partial x} F(U) = 0 \\ U(o,x) = \begin{cases} U_L & \text{if } x < 0 \\ U_R & \text{if } x > 0 \end{cases} \end{cases}$$

U_L and U_R two constant given states of the fluid. It is well know that the solution is :

$$U(t,x) = W_R\left(\frac{x}{t}, U_L, U_R\right)$$

when W_R is call the Riemann solver.

Let a simple uniform mesh Δx and note $x_j = j\Delta x$ with $U_i^n \simeq U(n\Delta t, x_i)$
A T.V.D. scheme gives in <u>each fundamental</u> variable a mean-value U_j^n in the mesh and a slope (here a vector three dimensionnal) Δ_i^n.

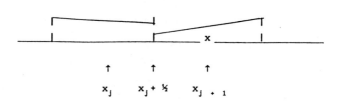

a) **First step**: Predictor of quantities $U_{j+\frac{1}{2},-}^n$ and $U_{j+\frac{1}{2},+}^n$ which depend only of local gradients in the mesh.

b) **Second step**: Riemann solver

$$V_{j+\frac{1}{2}}^n = W_R\left(0\,;\,U_{j+\frac{1}{2},-}^n,\,U_{j+\frac{1}{2},+}^n\right)$$

c) **Third step**: Conservation law

$$U_j^{n+1} = U_j^n - \frac{\Delta t}{\Delta x}\left[F\left(V_{j+\frac{1}{2}}^n\right) - F\left(V_{j-\frac{1}{2}}^n\right)\right]$$

d) **Fourth step**: Slope value by utilisation of monotone functions.

These ideas are initialy developped in [1] for U a scalar unknown.

III - BOUNDARY CONDITIONS IN 1D FORMULATION

In each mesh, three values are calculated through the vector (U_j^n).

So, for each boundary treatment, we associated a fictive state and then as the treatment of U_j^n in an internal mesh depends only on the adjacent values U_{j-1}^n and U_{j+1}^n. The algorithm is not modified.

The difference is founded about the definition of this fictive value :

. For <u>a reflection wall</u> if $U_j^n = \left(\rho_j^n,\,\rho_j^n u_j^n,\,\rho_j^n E_j^n\right)$ we take

$$\tilde{U} = \left(\rho_j^n,\,-\rho_j^n u_j^n,\,\rho_j^n E_j^n\right)$$

so for a state $(\rho,\,u,\,P)$ we take $(\rho,\,-u,\,P)$.

. For a <u>supersonic inlet</u> we give three different values as ρ, u, P or T, u, P (T for temperature).

. For a <u>subsonic inlet</u> we have P and ρ and u is calculated by the scheme. Giving at infinite amount H (enthalpy) and S (entropy) is identical.

. For a <u>supersonic outlet</u> nothing is given.

. For a <u>subsonic outlet</u> pressure P is prescribed and ρ and u are calculated by the scheme.

IV - 2D SCHEME FOR INERT AND REACTIVE FLOWS

The four step are conserved with a monodimensional Riemann solver associated to a state law (it is not an approximation solver but an exact Riemann solver) but here the Euler equation can be written as :

$$\frac{\partial U}{\partial t} + \frac{\partial}{\partial x} F(U) + \frac{\partial}{\partial y} G(U) = H(U)$$

where $U = (\rho, \rho u, \rho v, \rho E, \rho c_1, --, \rho c_N)^T$

$F(U) = (\rho u, \rho u^2+P, \rho uv, (\rho E+P)u, \rho uc_1, --, \rho uc_N)^T$

$G(U) = (\rho v, \rho uv, \rho v^2+P, (\rho E+P)v, \rho uc_1, --, \rho uc_N)^T$

$H(U) = (0, 0, 0, 0, \omega_1, -, \omega_N)^T$

In this source terme, the chemical cinetics are treated and c_i represents the concentration of i-component.

In step a), we predict the value of U at the interface of two cells with discontinuity at this interface.

In step b), we calculate the normal component of U (if $\vec{n}$ is the normal, it is $(\rho, un_x + vn_y, P)^T$) and after the value $U^{n+\frac{1}{2}}$ at the interface.

In step c), it is suffisant to apply conservation laws in the mesh.

Step d) we calculate the slopes in x and y so that we respect TVD formulation in x and y direction. The scheme in the 2D formulation is presented in [2].

For boundary conditions, the same treatment is developped for reflection wall, subsonic condition, and supersonic outlet. In a supersonic inlet we prescrive ρ, P and normal velocity.

V - APPLICATIONS

1 - FLOWS IN A NOZZLE IN INERT CONDITIONS

The first calculus corresponds to a convergent-divergent nozzle with subsonic inlet and outlet.

The second one is a double nozzle with a subsonic inlet and supersonic outlet.

2 - OXYGEN - HYDROGEN COMBUSTION IN A TUBE

We present a stock tube case where the bidimensionnal structures associated to the ZND theory.

VI - REFERENCES

[1]   Van Leer, B. : "Flux-vector splitting for the Euler equations". Lecture Notes in Physics. 170, Springer-Berlin, pp. 502-512.

[2]   Forestier, A., Gaudy, C. & Bung, H. : "Second order scheme for arbitrary mesh in Compressible Fluid Dynamic". I.C9MD, Williamsburg, July 1988.

UNSTRUCTURED UPWIND APPROACH FOR COMPLEX FLOW COMPUTATIONS

Kazuhiro Nakahashi and Kouichi Egami
Department of Aeronautical Engineering, University of Osaka Prefecture
4-804, Mozu-Umemachi, Sakai 591, Japan

In order to achieve an automation in the CFD, a grid generator using adaptive refinement techniques and an unstructured upwind flow solver are developed for two- and three-dimensional flow problems.

The unstructured grid is generated using two techniques; the geometry-adaptive refinement and the solution-adaptive refinement[1]. In the adaptation to the geometry, a coarse Cartesian grid which covers the entire flowfield is used as the initial grid. Then the grid refinement is applied to cells near the body surface. It is repeated until the minimum cell size becomes smaller than a specified value. The radius of curvature of the body surface is used to control the degree of the local refinement. This approach is basically similar to the quadtree/octree method[2], and easy to extend it to three-dimensional problems.

The grid generated by the geometry-adaptive refinement is improved by the solution-adaptive refinement. This refinement is applied to cell edges on which the absolute value of the gradient of the flow density is large. The geometry-adaptive refinement introduces information of the flowfield geometry into grid, and the solution-adaptive refinement introduces the fluid physics. A combination of these two techniques enables to generate a grid in a fully automatic manner.

Fig.1 shows a sequence of the grid generation procedure for a cascade problem. A three-dimensional grid generation procedure around a sphere is demonstrated in Fig.3.

In the present unstructured grid, several elemental shapes are used in a mixed manner for efficiency and flexibility. In two-dimensional field, both quadrilateral and triangular cells are used. Four types of elements, tetrahedron, pyramid, prism, and cubic, are used to discretize the three-dimensional flowfield.

The flowfield is solved using an unstructured upwind method. This is an extension of the modified flux-vector splitting method [3,4] to the use on the arbitrarily-shaped unstructured grid. The extension onto the unstructured grid is achieved using a finite volume concept. The MUSCL approach is used for the second-order scheme. The values outside of the element required for the MUSCL are evaluated using an extrapolation method to the neighboring cell.

The method has been applied to various flow problems and demonstrated its capability for complex flowfields including external and internal, single and multiple bodies. Fig.2 is the grid and the computed Mach contours for a compressor cascade. The computed Mach contours around two ellipsoid at Mach number 2 is shown in Fig.4.

References

[1] Nakahashi, K., "An Automatic Grid Generator for the Unstructured Upwind Method", AIAA Paper 89-1985-CP, 1989.

[2] Yerry, M.A. and Shephard, M.S, "Automatic Three-Dimensional Mesh Generation by the Modified-Octree Technique", Int. J. Num. Meth Eng., Vol.20, pp.1965-1990, 1984.

[3] Anderson, W.K., Thomas, J.L., and Van Leer, B., "Comparison of Finite Volume Flux Vector Splittings for the Euler Equations," AIAA Journal, Vol.24, No.9, pp.1453-1460, 1986.

[4] Hanel, D. and Schwane, R.,"An Implicit Flux-Vector Splitting Scheme for the Computation of Viscous Hypersonic Flow", AIAA 89-0274, 1989.

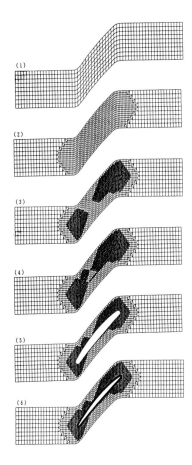

Fig.1 A sequence of grid generation for a compressor cascade.

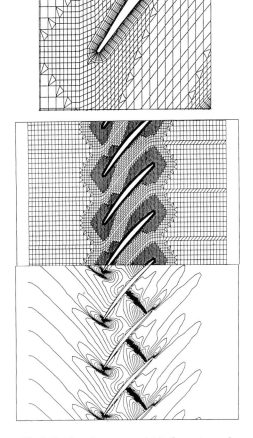

Fig.2 Grid and computed Mach contours for a compressor cascade.

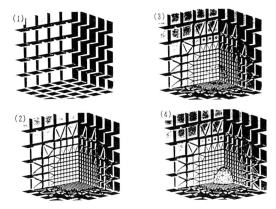

Fig.3 A sequence of three-dimensional grid generation for a sphere.

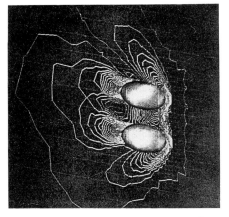

Fig.4 Computed Mach contours around two ellipsoid.

IMPLEMENTATION OF A HIGH RESOLUTION TVD SCHEME FOR COMPRESSIBLE INVISCID FLOW ON UNSTRUCTURED GRIDS

Luca Stolcis and Leslie J. Johnston
Department of Mechanical Engineering, UMIST
PO Box 88, Manchester M60 1QD, England

The use of total variation diminishing (TVD) schemes enables the development of high resolution shock-capturing methods which can be applied to a wide range of compressible flow problems. Such schemes are able to resolve the very complex shock structures that may be present in supersonic and hypersonic flows. However, as with more classical shock-capturing methods, limitations still remain in the application of these methods to complex geometries when using structured computational grids. More recently, increasing use has been made of unstructured or hybrid structured/unstructured grids in compressible flow calculations for complex geometries. The attraction of this approach is the ease with which grids can be generated for very complex geometries. Also, such an approach enables flow adaptive or grid enrichment techniques to be implemented, without major changes to the structure of the flow algorithm.

The objective of the present work is the implementation of a high resolution TVD scheme in an existing algorithm, which solves the two-dimensional Euler equations for compressible inviscid flows on unstructured grids. The existing flow solver is described in Ref.1, and uses a cell-centred finite-volume scheme to discretise the unsteady flow equations. These are marched in time to a steady-state solution using a multi-stage explicit time-stepping scheme. The basic algorithm uses artificial dissipation of the Jameson type both to stabilise the solution procedure and to ensure the clean capture of shock waves. The unstructured form of this flow algorithm, which is based on a consideration of cell edges rather than the cells themselves, enables the use of arbitrary-shaped polygonal computational cells.

The existing flow algorithm is to be extended to viscous flow, solving the Reynolds-averaged Navier-Stokes equations. Concern regarding the influence of the artificial dissipation on the viscous flow development has motivated the present study of a high-resolution TVD scheme, as an alternative approach to the Jameson-type artificial dissipation. The modified flow algorithm evaluates the numerical flux function by means of a non-MUSCL symmetric TVD scheme, based on the work of Yee-Roe-Davis (see Ref.2), modified for use with unstructured grids. The extension of the scheme to systems is performed via the local characteristic approach and the application to two-dimensional problems is performed by applying the one-dimensional scheme to each of the two coordinate directions.

The main implementation problems are related to the fact that for unstructured grids no curvilinear coordinates exist, so it is not a trivial task to define a suitable set of differences of characteristic variables. These problems have been overcome in the present work by using derivatives of the conserved variables to evaluate the differences. The same approach has been successfully applied for the evaluation of the flux limiters, which ensure second-order spatial accuracy, at least on regular grids.

The GAMM workshop channel test cases have been used for the initial validation of the modified flow algorithm. Figure 1 shows results for the inviscid supersonic case

($M_\infty = 1.4$, 4% thick bump). The calculations were performed on a relatively coarse grid containing 4800 cells and 7310 edges, obtained by directly triangulating a 81x31 structured quadrilateral grid. The solution obtained with the Jameson-type artificial dissipation is also shown because it was used as reference during the development of the present method. More recently, the method has been extended to laminar viscous flows and the influence of the artificial dissipation on the solution is under investigation.

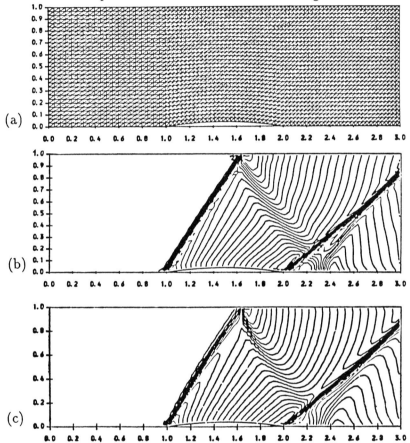

Figure 1: Supersonic channel flow ($M_\infty = 1.4$)
(a) grid
(b) iso-Mach contours for the Jameson-type artificial viscosity method
(c) iso-Mach contours for the present TVD method

References

[1] L.Stolcis and L.J. Johnston *Solution of the Euler equations on unstructured grids for two-dimensional compressible flows*, The Aeronautical Journal, Vol.94, No 936, June/July 1990, pp.181/195

[2] H.Yee *A class of high-resolution explicit and implicit shock-capturing methods*, Von Karman Institute Lecture Series LS 1989-04, March 1989

GENERALIZED FINITE VOLUME (GFV) SCHEMES ON UNSTRUCTURED MESHES: CELL AVERAGES, POINT VALUES, AND FINITE ELEMENTS

Steven T. Zalesak
Plasma Physics Division
Naval Research Laboratory
Washington D.C. 20375-5000

GFV Schemes Defined

Consider a general system of conservation laws, which takes the form

$$\frac{\partial w}{\partial t} + \nabla \cdot \mathbf{f}(w, \mathbf{x}) = 0 \tag{1}$$

where w and $\mathbf{f}$ are vector functions of the independent variables $\mathbf{x}$ and t, which we take to be space and time respectively. The classical finite volume (FV) method entails dividing the spatial domain into space-filling yet non-overlapping control volumes (cells), and applying the integral form of (1) to each cell. In the classical FV method the cells have sharp boundaries: a given infinitesimal fluid element may be in at most one cell.

Here we introduce the concept of a generalized finite volume (GFV) method: rather than have non-overlapping control volumes (cells) with sharp boundaries, we allow overlapping "fuzzy" control volumes with (usually positive) weight functions. These weight functions are constructed in such a way that at any given point in space, the sum of all cell weights is 1. Thus every infinitesimal fluid element is fully "counted" in a conservation sum, and the implementation of exact numerical conservation of physically conserved quantities becomes nearly as trivial as it is in the classical finite volume case. There is a drawback to the GFV method, however. Whereas an integration forward in time by a classical finite volume method requires only values of $\mathbf{f}$ at cell boundaries, with the GFV's more general definition of a control volume, we need in general to evalute $\nabla \cdot \mathbf{f}$ everywhere in space. In short, we need $\mathbf{f}$ defined globally.

Computing Pointwise Information

Just as the classical FV method must construct boundary-averaged values of $\mathbf{f}$ from cell averages of w, we must find a method by which we can extract pointwise information from our "cell averages" in the GFV method, since this is necessary for the algorithm's consistency, as mentioned above. If we constrain our conserved quantities to be expansions of basis functions, we find that we can recover globally defined point values from cell averages by merely solving a linear system of equations. We can show that this system need not be solved exactly either for accuracy or to enforce conservation.

The issue of conservation can be addressed in one of two ways. The first is to consider the cell averages to be the primary numerical quantities, with the evaluation of global pointwise information serving the same role as the evaluation of fluxes in the classical finite volume method: the pointwise information is used to update the cell averages in time, and then discarded. As long as the contribution of $\nabla \cdot \mathbf{f}$ to each cell is evaluated accurately, conservation is obtained trivially. The other way of addressing conservation is to force the integral of the pointwise construction of w to be exactly equal to that given by the cell averages, when the integral in taken over the whole domain. We have found that this latter constraint on w can be enforced easily within the context of solving the linear system approximately (iteratively).

The Finite Element Method

Our choice of a finite element terminology is intentional here, for what we have in fact described is a particular class of weighted residuals methods (we demand only that the sum of all weighting functions be 1 everywhere). Our basis functions obviously correspond to the finite element trial or shape functions, and our cell weighting functions correspond to the weighting functions used in the method of weighted residuals (MWR). The matrix in our system of equations for the pointwise construction of w given its cell averages corresponds to the finite element consistent mass matrix. Thus some GFV methods are not truly new, and have been around for quite some time. What is new is the interpretation we give here, and the uses to which that interpretation can be put. For example, we may regard the consistent mass matrix as an operator on an expansion in a specific set of basis functions which yields the cell averages of a function given its expansion coefficients; and we regard the normal finite element semi-discrete time evolution equation not as an implicit expression yielding time derivatives of expansion coefficients, but rather as an explicit equation for the time derivatives of cell averages.

Applications: Shock-Capturing Schemes, Conservative Remeshing

With the physical insight offered by a GFV interpretation of (many) finite element methods, it then becomes reasonably straightforward to implement some of the more modern shock-capturing schemes on a finite element mesh. In Fig. 1, we show results for the Sod shock tube problem, using a GFV method based on linear finite elements, characteristic decomposition in the sense of Roe, Runge-Kutta time advancement, and an enhanced version of the Roe-Sweby flux limiter. Clearly, the GFV/MWR/finite element method can be competitive with the best of the modern finite volume shock-capturing schemes, when playing by the same rules. In addition, the GFV method has the advantage of being well-defined on an unstructured mesh. Finally, we wish to point out that this GFV/shock-capturing formalism makes it possible to develop fully conservative remeshing algorithms for unstructured meshes, that are second-order accurate or higher, and free of spurious oscillations.

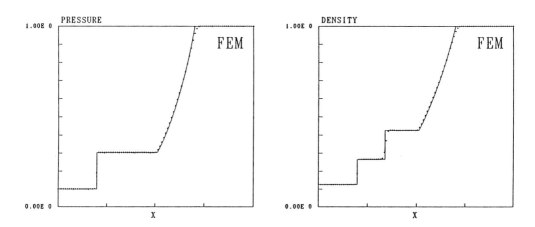

Fig. 1 Plots of the pressure and density for the Sod shock tube problem, computed using a GFV method based on linear finite element discretization.

SOLUTION OF STEADY-STATE CONSERVATION LAWS BY l_1 MINIMIZATION

JOHN LAVERY
Office of Naval Research, Arlington, Virginia

Capturing shock waves in two dimensions with the same high quality as in one dimension has been a particularly elusive goal of recent research in computational fluid dynamics. In the present paper, we create a truly two-dimensional algorithm based on l_1 minimization to solve the steady-state two-dimensional scalar Burgers' equation. One-dimensional Burgers' equations and Euler equations have been solved by l_1 minimization procedures in [1,2].

The inviscid Burgers' equation $(u^2)_x + \tau(u^2)_y = 0$ is discretized on each cell of a uniform m-by-n grid on the unit square by finite volumes. The boundary integrals in the finite-volume formulas are approximated using the values of the solution at the vertices of the cell by the trapezoidal rule, yielding the mn discrete equations

$$(-\Delta y - \tau \Delta x)u_{i-1,j-1} + (\Delta y - \tau \Delta x)u_{i,j-1} + (-\Delta y + \tau \Delta x)u_{i-1,j} + (\Delta y + \tau \Delta x)u_{i,j} = 0.$$

Let Dirichlet boundary conditions on all four sides of the unit square be given so as to create a zigzag shock in the interior (the zigzag line in Fig. 1). One then has $(m-1)(n-1)$ unknowns. The least-squares (l_2) solution of this overdetermined system of mn equations for $(m-1)(n-1)$ unknowns has a badly smeared-out transition layer instead of a discontinuity and not is a good approximation of the physically relevant discontinuous solution. Solving the system by least squares is therefore inappropriate. Instead, solving the system by l_1 minimization is proposed. The l_1 solution of this system is not unique but becomes unique if one adds a singular perturbation $-\epsilon u''$ or a nonsingular perturbation $+\epsilon u$ to the equation. Here $\epsilon > 0$ is, for mathematical purposes, arbitrarily small. (The nonsingular perturbation introduces artificial viscosity through a "back door": the error term for the trapezoidal integral of the nonsingular perturbation involves second-derivative terms and provides the slight amount of viscosity needed by the algorithm, but the fact that the second-derivative terms are not explicitly present means that the bandwidth of the matrix has not been increased.) A geometrically oriented l_1 algorithm that requires only $O(mn)$ operations is created to solve the l_1 minimization problem. (Standard l_1 algorithms require $O(m^2n^2)$ or more operations.) The l_1 procedure automatically concentrates the residuals of the overdetermined system in a small number of cells. These cells are the cells that contain the shock.

Shocks are captured in one cell with high accuracy and no oscillation at the nearest node points. A complete theory for the boundary-shock case is known. Computational results were obtained for internal shocks with $\tau = 0$, tan 15°, tan 30°, tan 45°, tan 60°, tan 75° and nonsingular perturbation parameter $\epsilon = 10^{-12}$. Square-cell grids of sizes 10x10, 25x25, 50x50, 75x75 and 100x100 and large-aspect-ratio grids of sizes 50x10 and 100x10 were used. The results for an interior-shock case with τ = tan 30° on a 50x50 grid, which are typical, are shown in Figure 1. Each symbol in the figure represents a cell. The cells denoted by "S" are the cells identified by the l_1 procedure as containing the shock. The position of the computed shock, which forms a band of cells only one cell wide, is seen to be an excellent approximation of the position of the physical shock (the zigzag line). The "0"'s and the "1"'s represent cells to the left and to the right of the shock, respectively. The maximum error of the values of the solution at all nodes is $\leq 1.00252*10^{-12}$ (physically relevant solution = +1 to the left of the shock and -1 to the right).

The l_1 procedure presented here is a truly two-dimensional algorithm that captures oblique and even zigzag shocks in bands of cells that are only one cell wide. The solutions are highly accurate and nonoscillatory.

Refererences:

[1] John E. Lavery, Calculation of Shocked Flows by Mathematical Programming, in *Proceedings 11th ICNMFD*, Lecture Notes in Physics 323, 360--363.
[2] John E. Lavery, Calculation of Shocked One-Dimensional Flows on Abruptly Changing Grids by Mathematical Programming, *SIAM J. Numer. Anal.*, January 1990.

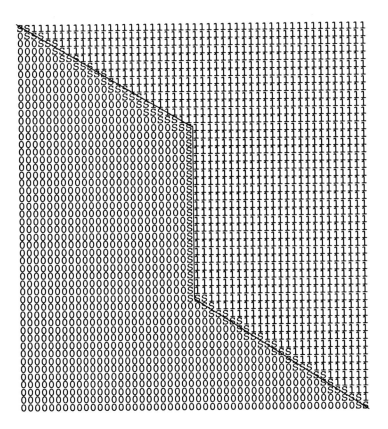

Figure 1. Position of numerical shock ("S" cells) vs. physical shock (zigzag line)

FLUX BALANCE SPLITTING - A NEW APPROACH FOR A CELL VERTEX UPWIND SCHEME

C.-C. Rossow
DLR Institute for Design Aerodynamics
Flughafen, D-3300 Braunschweig, F. R. Germany

Introduction For the schemes presently in use in computational fluid dynamics a significant classification can be made in upwind schemes and central differencing schemes using artificial dissipation. The latter methods use a symmetric discretization stencil an therefore need some artifice to provide the necessary amount of numerical dissipation to damp out spatial and temporal oscillations [1]. In contrast to that upwind schemes exploit the higher inherent numerical dissipation of one-sided difference formulas, and to determine the upwinding direction diagonalization of the flux jacobians is employed. Usually, upwinding is achieved by using spatial extrapolations of variables, flux vector components, or differences of characteristic variables [2], [3], [4]. In the scheme presented here a splitting of the conservative flux balance is achieved, and the splitted flux balances are than distributed to the mesh nodes to give an upwind-weighted cell vertex discretization scheme.
Solution Scheme The scheme deals directly with the conservative flux balances. Consider the one-dimensional Euler equations:

$$\frac{\partial \vec{u}}{\partial t} + \frac{\partial \vec{F}}{\partial x} = 0 \quad , \quad \text{where} \quad \frac{\partial \vec{F}}{\partial x} = A \frac{\partial \vec{u}}{\partial x} \quad , \quad A = \frac{\partial \vec{F}}{\partial \vec{u}}$$

Here $\vec{u}$ denotes the vector of conservative variables, $\vec{F}$ is the vector of flux density, and A is the corresponding flux jacobian. By the use of the matrix M^{-1} containing the left eigenvectors of A this equation can be transformed:

$$M^{-1} \frac{\partial \vec{u}}{\partial t} + M^{-1} \frac{\partial \vec{F}}{\partial x} = 0 = \frac{\partial \vec{w}}{\partial t} + \Lambda \frac{\partial \vec{w}}{\partial x}$$

where $\vec{w}$ are the characteristic variables and Λ is the diagonal matrix of the eigenvalues. The sign of the eigenvalues can than be used to obtain the upwinding direction. The present scheme consists of four steps:
1. Calculate the conservative flux balances in all cells
2. Calculate the characteristic flux balances by multiplication of the conservative balances with M^{-1} in each cell
3. Each cell sends its characteristic flux balance to either the right or left node, depending on the sign of the eigenvalues
4. Multiplication of the weighted characteristic fluxes bei the matrix M of the right eigenvectors to obtain the conservative flux balances in all cells

After this procedure time stepping can be performed by using the weighted conservative flux balances. As two-point difference formulas are used the scheme is first order accurate during the transient phase, and forward Euler time integration can be used. However, at steady state it is second order accurate.
In two dimensions the scheme is applied by transforming the equations with respect to both the x- and the y-direction. Using the left eigenvectors of the flux jacobian of the x-direction the 2-D equations may be transformed with respect to the x-direction. Note that the transformation is applied to the full equation, i.e. the total flux balance is transformed. Corresponding to the eigenvalues the transformed flux balances are distributed to the nodes, and than backtransformed. The same procedure is carried out for the y-direction, using the left eigenvectors of the flux jacobian of the y-direction. For the density and the energy equation the contributions of x- and y-direction are than weighted by the components of the velocity vector, and for the momentum equations only the contribution of the corresponding coordinate direction is used. For integration to steady state a 5-stage Runge-Kutta scheme is used, and it was found necessary to add fourth differences artificial dissipation to prevent odd- even decoupling. At shocks the artificial viscosity is switched off.
Numerical Results In Figure 1 results are shown for a one-dimensional flow in a Laval-nozzle. The inflow Mach number is 0.45. Figure 1a gives the Mach number distribution along the axis for the present scheme, a TVD-scheme and a central differencing scheme according to [1]. The present scheme captures the shock essentially within one cell. In Figure 1b total pressure losses are plotted. The present scheme does not exhibit any wiggles in the shock region, although no limiters had to be used. All schemes investigated needed about 1700 time steps to drop the residual of the density equation by 7 orders of magnitude.
Figure 2 shows iso-mach lines for a transonic duct flow over a circular arc. No oscillations have occured and the shock is captured within one cell. Figure 3a shows the results for the transonic flow around the NACA 0012 airfoil. The symbols mark the results of the upwind scheme on a grid with 80 by 16 cells. The solid lines are results of a central differencing scheme according to [5] on a 160 by 32 mesh. The shocks on the upper and lower surface are captured by the upwind scheme within one cell, the remaining pressure distribution is of the same quality as that delivered by the central differencing scheme on the finer mesh. Figure 3b gives the convergence history of the upwind scheme.
Conclusions The flux balance splitting provides a means for sharp resolution of shocks, and the overall accuracy is enhanced compared to a central differencing scheme. The upwinding formulation makes the scheme independent of the cell geometry, therefore it is appropriate for structured and unstructured meshes. The splitting in higher dimensions needs further validation, and future work will concentrate on a cell-by-cell decomposition along streamlines.

[1] Jameson, A., Schmidt, W., Turkel, E. *Numerical Solution of the Euler Equations by Finite Volume Methods Using Runge-Kutta Time Stepping Schemes.* AIAA 81-1259, 1981.

[2] Roe, P.L. *Approximate Riemann Solvers, Parameter Vectors, and Difference Schemes.* Journal of Computaional Physics 43, pp.357-372, 1981

[3] Van Leer, B. *Flux Vector Splitting for the Euler Equations.* Lecture notes in Physics, Vol 170, pp.507-512, 1982

[4] Steger, J.L., Warming, R.F. *Flux Vector Splitting of the Inviscid Gasdynamic Equations with Application to Finite Difference Methods.* Journal of Computational Physics, Vol 40, No 2, pp.507-512, 1981

[5] Rossow, C., Kroll, N., Radespiel, R., Scherr, S. *Investigation of the Accuracy of Finite Volume Methods for 2- and 3-Dimensional Flows.* AGARD-CPP-437, P17.1-11, 1988.

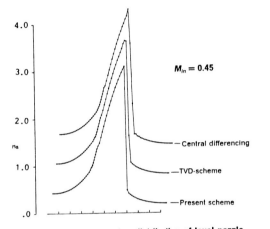

Figure 1a: Mach number distribution of laval nozzle

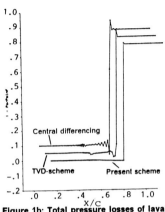

Figure 1b: Total pressure losses of laval nozzle

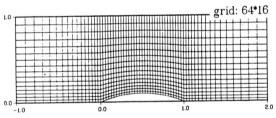

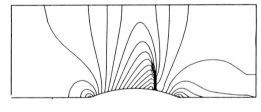

Figure 2: Transonic duct flow over circular arc for $M_{in} = 0.675$

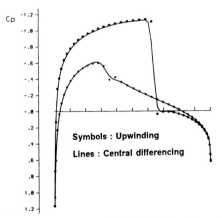

Figure 3a: Pressure distribution of NACA 0012 $M_\infty = 0.8$, $\alpha = 1.25°$

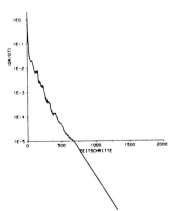

Figure 3b: Convergence history of upwind scheme

Multidimensional upwind schemes using fluctuation splitting and different wave models for the Euler equations.

R. Struijs, H. Deconinck and P. De Palma,
Von Karman Institute for Fluid Dynamics,
Steenweg op Waterloo 72, 1640 Sint-Genesius-Rode, Belgium

The subject of the present paper is the development of a multidimensional upwind method for the Euler equations. Considering first the scalar linear advection equation

$$u_t + \vec{a} \cdot \vec{\nabla} u = 0 \quad , \tag{1}$$

Roe [1],[2] developed a new class of schemes based on the fluctuation splitting concept, see also the companion paper in this conference [7]. This allows the construction of truly multidimensional conservative upwind schemes with reduced grid dependency. The fluctuation splitting method uses linear triangular elements T, on which the integral of u_t is calculated :

$$\phi^T = \iint_T u_t \, ds = - \iint_T \vec{a} \cdot \vec{\nabla} u \, ds = - \sum_{i=1}^{3} k_i u_i \tag{2}$$

Here, u is the value of the unknown at vertex i of element T, $k_i = \frac{1}{2} \vec{a} \cdot \vec{n_i}$, and $\vec{n_i}$ is the scaled inward normal of the edge opposite to node i. To update the solution at the vertices, the fluctuation of an element is distributed over its vertices with weights α_i, defining an explicit scheme which is conservative if $\sum_{i=1}^{3} \alpha_i^T = 1$. Classical finite volume schemes, as well as a Lax-Wendroff fluctuation splitting scheme can be expressed within the fluctuation splitting formulation. E.g. for a central scheme, $\alpha_i = \frac{1}{3}$. Upwind schemes are generated when the α_i depend on the unknowns and on a locally defined convection speed. It can be shown that the plane wave solution which optimally fits to the solution at timelevel n will travel in the direction of the gradient of the unknowns. Therefore the direction $\vec{m} = \vec{\nabla} u / |\vec{\nabla} u|$ becomes crucial for the fluctuation splitting method. Since the convection speed in this direction is always smaller than $|\vec{a}|$, larger timesteps are permitted, thus increasing convergence rate.

The method has been applied to systems of equations. Therefore, the system is decoupled into a set of characteristics, or simple waves. The characteristic speed of each simple wave is used to determine the upwind distribution coefficients α_i, which are used to split the part of the residual which corresponds to the simple wave over the vertices.

For the 2D Euler equations, two decomposition methods exist in literature. One has been proposed by Roe [3], and matches six simple waves to the solution at a given moment. The decomposition given by Deconinck-Hirsch [4] decomposes the Euler equations into four characteristic equations, which by construction are optimally decoupled. Both wave decomposition methods are implemented.

Results on a standard oblique shock reflection problem with comparison to literature results [5] are given in the figures. The calculations were performed on a symmetric mesh consisting of triangles in order to avoid bias from the grid. The isolines for the

Machnumber obtained with a monotone distribution scheme are given in fig. 1 for the four wave model. To illustrate the shock capturing capabilities, the Mach number along horizontal lines at heights $Y = 0$ and $Y = 0.5$ are given in fig. 2a. In figs 2b and 2c the cuts for the results obtained with the six wave model with first order monotone distribution, and the four wave model with the second order accurate non-monotone Lax-Wendroff scheme are given respectively. In all four wave cases, the gradient direction $\vec{m} = \vec{\nabla}u/|\vec{\nabla}u|$ has been used for the distribution step. For the result of fig. 1, comparison is made for pressure ratio's along these line with literature results (figs. 3a-c). From these figures, and with results from extensive testing on scalar equations [6],[7] a strongly reduced grid dependency has been found. The first order fluctuation splitting schemes perform about as good as classical second order TVD schemes.

References

1. P.L. Roe ; Numerical Algorithms for the Linear Wave Equation ; Royal Aircraft Establishment Technical Report 81047, 1981
2. P.L. Roe ; Linear Advection Schemes on Triangular Meshes ; Cranfield Institute of Technology CoA Report No 8720, Cranfield, Bedford, U.K., November 1987
3. P.L. Roe Discrete Models for the Numerical analysis of Time-Dependent Multidimensional Gas Dynamics ; Journal of Computational Physics 63, 458-476 (1986)
4. H. Deconinck, C Hirsch, J. Peuteman ; Characteristic Decomposition Methods for the Multidimensional Euler Equations ; Lecture Notes in Physics 264, pp 216-221, Springer, 1986
5. S.-H. Chang, M.-S. Liou ; A Numerical Study of ENO and TVD Schemes for Shock Capturing ; NASA Technical Memorandum 101355, 1988
6. H. Deconinck, R. Struijs, P.L. Roe ; Fluctuation Splitting for Multidimensional Convection Problems : An Alternative to Finite Volume and Finite Element Methods ; VKI Lecture Series 1990-03 'Computational Fluid Dynamics', March 5-9, 1990
7. P.L. Roe, H. Deconinck, R. Struijs ; Recent Progress in Multidimensional Upwinding ; ICNMFD Conf., Oxford, July 1990

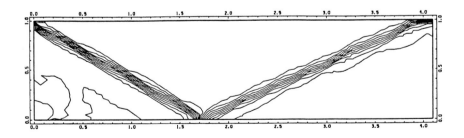

Figure 1. *Isolines of the Machnumber for the oblique shock reflection testcase. The Deconinck-Hirsch decomposition with a first order monotone distribution scheme is used, where the distribution is based on the gradient of the solution $\vec{\nabla}u$. The value of isolines starts from 1.9, with an increment of 0.05.*

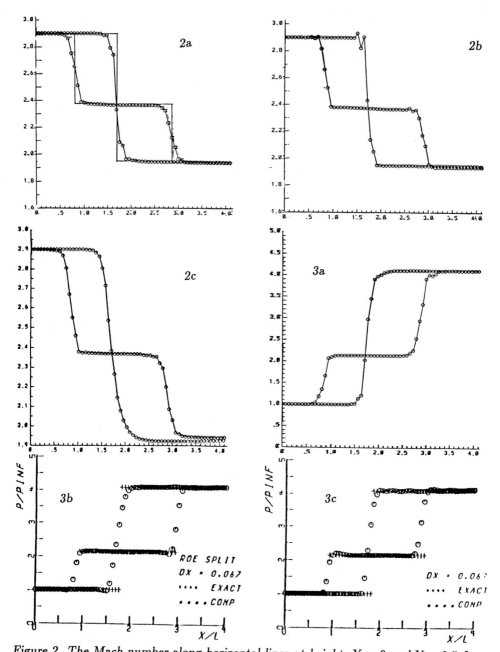

Figure 2. The Mach number along horizontal lines at heights $Y = 0$ and $Y = 0.5$ from the solution of figure 1 is given in figure 2a. Fig. 2b shows the cuts for case where Roe's six wave model (with a shear wave) and first order monotone distribution scheme is used. Fig. 2c shows the cuts for the non-monotone second order accurate Lax-Wendroff Fluctuation Splitting scheme, using the gradient of the solution $\vec{\nabla} u$ for the distribution.

Figure 3. Comparison of pressure ratio's along horizontal lines at heights $Y = 0$ and $Y = 0.5$. a) from the solution of figure 1. b) second order TVD result [5]. c) second order ENO result [5].

Interactions of a viscous fluid with an elastic solid: eigenmode analysis

R.M.S.M. Schulkes
Department of Mathematics, Delft University of Technology
P.O. Box 356, 2600 AJ Delft, The Netherlands

In this paper the interaction of a viscous fluid with a perfectly elastic solid is considered. The system to be investigated is as depicted in figure 1: a viscous fluid with a free boundary S occupies the region Ω_1, an elastic solid occupies the region Ω_2 and the common boundary of the fluid and the solid is denoted by G. Assuming that all time-dependent quantities exhibit a temporal behaviour of the form $e^{\lambda t}$, we find that the linearized equations governing the motion of the coupled fluid-structure system are as follows. In dimensionless form we have for the fluid

$$\lambda \mathbf{u} + \nabla p = \frac{1}{Re}\nabla^2 \mathbf{u}, \quad \nabla \cdot \mathbf{u} = 0 \quad \text{in } \Omega_1,$$

$$\sigma_n = -\frac{1}{\lambda}u_n, \quad \sigma_\tau = 0 \quad \text{on } S, \tag{1}$$

$$\mathbf{u} = 0 \quad \text{on } W_f,$$

for the solid

$$\lambda^2 \mathbf{d} = \tilde{E}\nabla \cdot \mathbf{T} \quad \text{in } \Omega_2,$$

$$T_n = 0, \quad T_\tau = 0 \quad \text{on } F, \tag{2}$$

$$\mathbf{d} = 0, \quad \text{on } W_s,$$

and continuity of stresses on the interface G gives

$$\sigma_n = \tilde{E}r_\rho T_n - \mathbf{k} \cdot \mathbf{d}_G$$

$$\sigma_\tau = \tilde{E}r_\rho T_\tau \tag{3}$$

$$\lambda \mathbf{d}_G = \mathbf{u}.$$

In the above equations $\mathbf{u}$ denotes the fluid velocity, p the pressure, $\mathbf{d}$ the displacement in the elastic solid and $\mathbf{T}$ the stress tensor of the elastic solid. We have introduced the dimensionless numbers $Re = UL/\nu$ denoting the Reynolds number (L is a length scale and $U = \sqrt{gL}$ is a velocity scale), $r_\rho = \rho_s/\rho_f$ being the ratio of densities of the solid and the fluid and $\tilde{E} = E/\rho_s gL$ denoting the scaled Young's modulus.

Our aim is to solve equations (1)-(3) by means of a finite-element discretization procedure. To that end a variational formulation of (1)-(3) is required. The deformation of both the fluid and the solid continua can be obtained, separately, from a variational principle. However, the different nature of the two continua (velocity and pressure unknowns in Ω_1 and displacement unknowns in Ω_2) complicate the formulation of a unified variational principle. This problem can be overcome by introducing a new variable $\mathbf{w}$. We define the new variable by $\mathbf{w} = \mathbf{u}$ in Ω_1 and $\mathbf{w} = \lambda \mathbf{d}$ in Ω_2. Note that the dimension of $\mathbf{w}$ is that of velocity in both Ω_1 and Ω_2. It can be shown (see Schulkes, 1990) that the unified variational formulation of the fluid-structure interaction problem reads

find $\mathbf{w} \in \Omega_1 \cup \Omega_2$ and $p \in \Omega_1$ such that for all functions $\mathbf{v} \in \Omega_1 \cup \Omega_2$ and $q \in \Omega_1$ the following equations are satisfied:

$$\int_{\Omega_1} (\lambda \mathbf{w} \cdot \mathbf{v} - p\nabla \cdot \mathbf{v})dx + \frac{1}{Re} a(\mathbf{w},\mathbf{v}) + r_\rho \int_{\Omega_2} \lambda \mathbf{w} \cdot \mathbf{v} dx +$$

$$\frac{1}{\lambda} r_\rho \tilde{E}\, b(\mathbf{w},\mathbf{v}) + \frac{1}{\lambda} \int_S w_n v_n ds + \frac{1}{\lambda} \int_G \mathbf{k} \cdot \mathbf{w} v_n ds = 0,$$

$$\int_{\Omega_1} q\nabla \cdot \mathbf{u}\, dx = 0,$$

where the functionals $a(\mathbf{w},\mathbf{v})$ and $b(\mathbf{w},\mathbf{v})$ are measures of the viscous dissipation of energy in the fluid and elastic bending energy of the solid respectively.

Application of the standard finite-element discretization prcedure in conjunction with a penalty-function approach to eliminate the pressure from the equations (see Cuvelier et al. 1986) yields a discrete eigenvalue problem of the form

$$\left(\lambda^2 \mathbf{A} + \lambda \mathbf{B} + \mathbf{C}\right)\mathbf{x} = 0. \tag{4}$$

Here $\mathbf{A}$ denotes the mass-matrix of the fluid and the solid, $\mathbf{B}$ denotes the stiffness matrix of the fluid and the matrix $\mathbf{C}$ contains terms related to the elastic bending energy of the solid and terms related to the free-surface and interfacial integrals. It is found that eigenvalue problem (4) can be solved efficiently using an inverse iteration procedure.

References

Cuvelier, C., Segal, A. & van Steenhoven, A.A. 1986 *Finite element methods and Navier-Stokes equations.* Reidel Publishing Company.

Schulkes, R.M.S.M. 1990 Interactions of a viscous fluid with an elastic solid: eigenmode analysis. To appear.

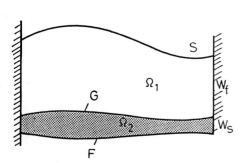

Figure 1: Schematic diagram of fluid-structure system.

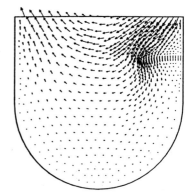

Figure 2: Eigenmode of a viscous fluid in a rigid container with an elastic baffle.

Parallel solution procedures for transonic flow computations

R.C.Hall[*] and D.J.Doorly[**]
* W.S.Atkins Engineering Sciences, Epsom, KT18 5BW
** Aeronautics Dept., Imperial College, London, SW7 2AZ

The application of parallel processing to the solution of compressible flow problems has been investigated, using a PC-hosted network of up to 8 transputers with the 3L Parallel Fortran compiler. Two-dimensional Euler and Navier-Stokes codes employing an explicit, finite volume, Runge-Kutta time stepping scheme have been developed, following the method of Jameson et al., and then parallelised using both the geometric (domain decomposition) and algebraic (processor farm) strategies.

Domain decomposition

A block-structured code has been developed with essentially two components: a Main task to perform all system input/output functions, prepare initial data for blocks, and collate the results from the various blocks, and a Solver task to carry out the calculations on a particular block. Parallel communications are employed both for the distribution/collection of initial/final data, and for the exchange of boundary information between Solver tasks.

Two alternative approaches for mapping an aerofoil C-type grid onto four processors are shown in Fig.1. The chain topology is the simplest to implement, while the grid arrangement minimises the number of boundary values to be interchanged, at the expense of additional complexity in defining neighbouring interfaces. The latter also allows better balancing of computational loads between processors.

The results shown in Fig.2. demonstrate that high parallel efficiencies may be obtained, particularly if an appropriate mapping of the solution blocks onto the processor topology is implemented.

Parallel Multiple Algorithm Solutions

A further advantage of the domain decomposition approach is that it may be extended to allow different processors to utilise different solvers in parallel. An example computation of a flat plate laminar boundary layer problem, involving the concurrent use of Euler and Navier-Stokes solvers in different blocks (corresponding to different regions of the flow) is illustrated in Fig.3. In order to balance the loads it is necessary to adjust the size of blocks to ensure that the computationally more expensive Navier-Stokes solver operates on appropriately smaller blocks than the Euler solver. The results demonstrate the advantages of this procedure.

Processor Farming

Finally, the task farming technique has been applied to the Runge-Kutta time-stepping portion of the solution algorithm. Special-purpose 3L software is used to control the routing of work packets to any 'free' processors and the sending back of results to the Master task, but this also results in a significant communications overhead. In conjunction with the lack of hardware-assisted through-routing, this may account for the disappointing performances obtained.

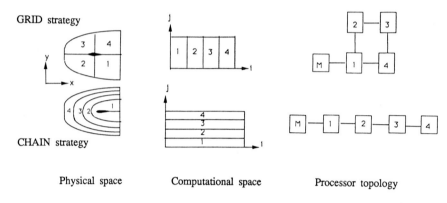

Fig.1. Partitioning strategies for aerofoil C-type mesh

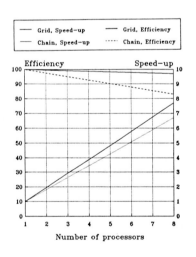

Fig.2. Computation speeds using the block-structured code

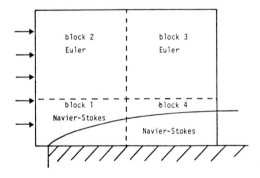

Fig.3. Parallel Euler and Navier-Stokes solution of laminar flat plate boundary-layer flow

COMPRESSIBILITY SCALING METHOD FOR MULTIDIMENSIONAL ARBITRARY MACH NUMBER NAVIER-STOKES CALCULATIONS OF REACTIVE FLOWS

L.A. Blinova, Yu.E. Egorov, A.E. Kuznetsov, M.A. Rotinijan,
M.Kh. Strelets, M.L. Shur
State Institute of Applied Chemistry
Dobrolyubov av, 14, 197198 Leningrad, USSR

New method for computation of multidimensional steady and unsteady flows of compressible gas and reactive gas mixtures (Compressibility Scaling Method - CSM) is presented. The approach used in CSM may be treated as generalization of the classical artificial compressibility method, widely spread in steady-state incompressible flow simulation, for the case of compressible flows. Unlike routine numerical methods for Navier-Stokes and Euler equations, which efficiency is decreasing sharply as flow Mach number M_0 is becoming much less than one, the efficiency of CSM is not affected by Mach number variation and in particularly does not diminish up to the limit value of the parameter $M_0=0$ ("incompressible flow" limit).

The main idea of CSM for steady flow calculation consists in removing the stiffness pertained to unsteady gas dynamic equations at $M_0 \ll 1$ by means of replacement of density ρ as a primary variable by nondimensional dynamic pressure $p'=(p - p_0)/\rho_0 U_0^2$ and subsequent regularization of the matrix of coefficients in the resulting system of equations ("compressibility scaling"). For numerical integration of thus obtained system an implicit finite-difference scheme of approximate factorization is formulated. The scheme is based on original form of physical processes splitting and can be solved by scalar tridiagonal matrix inversion algorithm.

For unsteady flows use of the compressibility scaling approach allows to construct efficient iterative implicit scheme of first or second order of accuracy in time. As it is in the steady-state case the scheme is implemented by scalar matrix inversion algorithm and is valid for arbitrary Mach number flow calculations.

Numerous computational experiments are carried out in order to get comprehensive information about CSM stability and convergence rate for various 2-D and 3-D flows in a wide range of Mach number variation. It is shown that for subsonic and mixed (subsonic and supersonic) flows with large subsonic zones CSM has great advantages over traditional implicit methods from the point of view of efficiency and robustness. As an illustration of the fact the number of iterations necessary for convergence of CSM (solid line) and well known implicit scheme by W.R.Briley and H.McDonald (dashed line) at optimum CFL value for each scheme is plotted on fig.1 as a function of Mach number for subsonic flow in plane channel with sudden expansion.

Capabilities of CSM are also illustrated by fig.2-4, on which the examples of it's application for numerical simulation of some mixed and supersonic flows are presented. In particularly fig.2 shows the flowfield in Laval nozzle with boundary layer separation at the exit and fig.3,4 demonstrate velocity vector fields and pressure contours in continuous wave supersonic HF-chemical laser cavities of two different constructions.

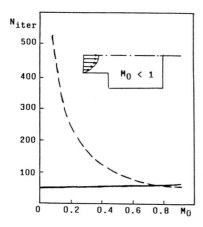

Fig. 1

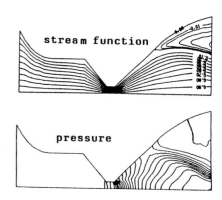

Fig. 2

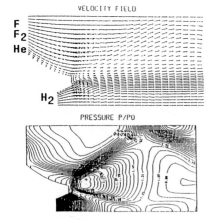

Fig.3

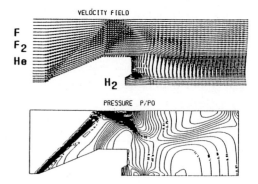

Fig. 4

Simple, Efficient Parallel Computing with Asynchronous Implicit CFD Algorithms on Multiple Grid Domain Decompositions

C.K. Lombard, S.K. Hong, J. Bardina, J. Oliger, S. Suhr,
J. Caporaletti, W.H. Codding and D. Wang

PEDA Corporation, Palo Alto, CA

An upwind scheme CSCM for the compressible Navier-Stokes equations in two and three space dimensions has been organized[1,2] in input addressable data structures of multiple and segmented grids that enable computation for arbitrary geometries without reprogramming. The method[3-5] has an extensible set of operationally explicit but unconditionally stable boundary point procedures of characteristics domain of dependent for both exterior[3] and interior[6] grid boundaries on the flow domain. Accordingly with natural multiple grid domain decompositions the approach is highly suitable for coarse grain parallel mapping onto concurrent computing architectures of moderate numbers of processors.

The pseudo time relaxation method based on DDADI diagonally dominant approximate factorizations[3,4] of the left hand side of the linearized implicit block matrix difference equations does not in principle depend for stability on synchronization between solution operations either among grids or among directions of sweep for the symmetric Gauss-Seidel space marching procedure[4,5] on a grid. Thus interprocess control simplifing asynchronous concurrent organizations offer intriguing possibilities.

The approach is equally appropriate for either global or distributed memory systems. Presently we are working toward low cost, comparatively high-performance microcomputer implementations, such as based around the Intel 860 chip. Distributed memory realizations that we are considering are sets of array processor boards in UNIX/VME workstations and integrated parallel mini-supercomputer workstations.

Here distributed memory applications are, organized on systems of multiple grids, each wholly contained within a single processor. With the diagonally dominant scheme, only boundary data obtained from either direct connection with or interpolation on overlapping adjacent grids [6] need be exchanged among memories. Flow computations are performed in local memory resident FORTRAN programs and data exchange is message-based through C programs, in some cases within the constructs of LINDA.

For dealing with geometric complexity, structured topologically quadrilateral block grid domain decompositions naturally support effective algebraic grid generation, efficient computation and coarse grain parallelism. Piecewise fitting body, domain boundary and major flow structures, blocks of mesh can always be directly connected into comparatively well conditioned systems generally at the price of globally singular coordinate topology. We make the latter problem readily manageable by introducing general directed graph data structures and a new, very flexible Graph-Object-Based programming style with high level language support (GRAPL) hosted in an interactive graphical environment U→S→E.

The present paper considers application of the asynchronous algorithm in coarse grain parallelism for multiple grid decompositions of the hypersonic jet interaction problem three space dimensions. The problem has otherwise been studied and reported in references 1,2 and 7. The chosen topical problem involves wide ranging geometric scales and topologies that can only be naturally and efficiently captured with appropriate good relative resolutions on a set of geometry/flow structure conforming grids.

As sketched in Figure 1 for the jet interaction problem with one jet issuing from a biconic test body, the minimal number of grids we find desireable with segmentation, i.e. steps, cavities or holes in computational coordinates, is three. One grid, the smallest, fits the exiting computed nozzle flow. A second, larger grid fits the local jet/external flow strong interaction region including the bow shock and primary separation surface over the jet flow. The third and largest scale grid, fits the body with jet-bow-shock induced upstream boundary layer separations and large external flow structures. These three grids can be further decomposed into additional grids to fill numbers of available processors and achieve rough balance of computational burden per processor. One fully connected organization involves 25 blocks, 19 with synthesis of grids 1 and 2. Smaller numbers of available processors, eg. four, can be very effectively employed by the algorithm by marching over sets of blocks successively.

A sample result, given in Figure 2, shows interaction of the vortical separations of the outer and jet flows in the symmetry plane on the windward side of the jet. The sharply captured recompression shock that is involved in the mechanism driving the jet flow separation is also evident near the top of the figure.

References

1. Lombard, C. K., Hong, S. K., Nystrom, G. A., and Bardina, J., "Asynchronous Concurrent Implicit CFD Algorithms", Fourth Conference on Hypercubes, Concurrent Computers, and Applications, March 6-8, 1989, Monterey, CA.
2. Lombard, C. K., Hong, S. K., Bardina, J., and Wang, D., "CSCM on Multiple Meshes with Application to High Resolution Flow Structure Capture in the Multiple Jet Interaction Problem," AIAA-90-2102, (Also AIAA-90-2103) July, 1990.
3. Lombard, C.K., Bardina, J., Venkatapathy, E. and Oliger, J.: " Multi-Dimensional Formulation of CSCM - An Upwind Flux Difference Eigenvector Split Method for the Compressible Navier-Stokes Equations," AIAA-83-1895, 1983.
4. Lombard, C.K., Venkatapathy, E. and Bardina, J.: "Universal Single Level Implicit Algorithm for Gasdynamics," AIAA-84-1433, 1984.
5. Bardina, J. and Lombard, C.K.: "Three-Dimensional Hypersonic Flow Simulations with the CSCM Implicit Upwind Navier-Stokes Method," AIAA-87-1114-CP, 1987.
6. Lombard, C.K. and Venkatapathy, Ethiraj: "Implicit Boundary Treatment for Joined and Disjoint Patched Mesh Systems," AIAA-85-1503, July, 1985.
7. Hong, S. K., Nystrom, G. A., Wang, D., Bardina, J., and Lombard, C. K., "Simulation of 3-D Jet-Interaction Flowfields with CSCM on Multiple Grids," AIAA-89-2552, July, 1989.

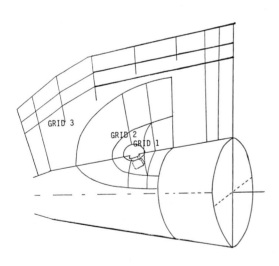

Fig.1 Schematic of multiple grid system for JI problem on blunted biconic test body.

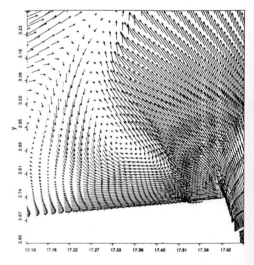

Fig.2 Plot showing resolution of small scale near field flow afforded by multiple grids.

VALIDATION OF AN UPWIND EULER SOLVER

David W. Foutch
Douglas R. McCarthy
Boeing Commercial Airplane Group
Seattle, Washington   USA

The goal of this validation is to determine how accurately the results of a numerical algorithm approximate the exact solution to the system of partial differential equations it is intended to solve. The validation technique could be applied to any PDE. Denoting the PDE operator by L and the boundary condition operator by B, the problem to be solved is:

$$L(Q) = 0 \quad \text{on D} \tag{1}$$
$$B(Q) = 0 \quad \text{on the boundary of D}$$

where Q is the exact solution to the continuum equations. In general, Q cannot be obtained in closed form, so it is impossible to know the exact solution to (1), complicating the job of determining the accuracy of the numerical result. It is possible, though, to pose a problem similar to (1) for which the exact solution is known. Substituting an assumed solution, $Q_A(x)$, $x \in D$, into (1) yields:

$$L(Q_A) = M \quad \text{on D} \tag{2}$$
$$B(Q_A) = C \quad \text{on the boundary of D}$$

where M and C are determined by the substitution. Since Q_A depends only on x, M and C are also functions only of x. The modified PDE, (2), is solved numerically and the result of the calculation is compared to the known solution, Q_A.

To illustrate the validation technique we apply it to the steady, two-dimensional Euler equations. McDonough [1] has applied this technique to the Navier–Stokes equations, developing a class of assumed solutions that approximate a physical flow. One contribution of the current work is to include the boundary conditions in the validation. We also argue that it is not necessary for the assumed solution to resemble a physical solution. It is sufficient that, when Q_A is substituted into operators L and B, none of the terms of L and B are zero. For this work, Q_A is based on the function:

$$f(x,y) = (1 + .125\cos(2\pi x))(1 + .125\cos(2\pi y))$$

We use the validation technique to confirm the order of accuracy of the numerical method, to investigate the effect of non–uniform grids on the solution, and to develop an estimate of the discretization error that does not depend on knowledge of the exact solution. The estimated error is calculated by the flow solver and is intended to be a measure of the accuracy of the numerical result when the exact solution is not known, for instance when M and C are zero.

The flow solver discretizes L using a conservative, finite volume scheme based on flux–difference splitting. The discretization is expected to be second order accurate for smooth flows on smooth grids. The discretized boundary conditions are also expected to be second order accurate. To determine the order of accuracy of the numerical algorithm, a series of calculations on successively finer grids are performed. The slope of the log–log plot of the maximum error vs. the grid spacing yields the order of accuracy. Figure 1 compares the order of accuracy of the solver on three types of grids. It also compares the known error to an error estimate produced by the flow solver. The slopes of the curves indicate that the solver is second order accurate on a Cartesian grid and on a smooth, curvilinear grid, but that it is first order accurate on a grid with a sharp kink. In the first two cases, the estimated error and the known error correlate very well. On the kinked grid the correlation is not as good, but the estimated error does show that the solver is first order accurate on that grid.

This validation technique provides valuable information about the properties of the flow solver. It shows that the solver treats both the Euler equations and boundary conditions with second order accuracy on smooth grids, but that the accuracy is reduced on non–smooth grids. It also demonstrates the usefulness of the error estimate provided by the solver. These conclusions are obtained without comparisons to experimental data or knowledge of the properties of the partial differential equation.

1. McDonough, J. M., "A Class of Model Problems for Testing Compressible Navier–Stokes Solvers," AIAA Paper 88-3646, 1988.

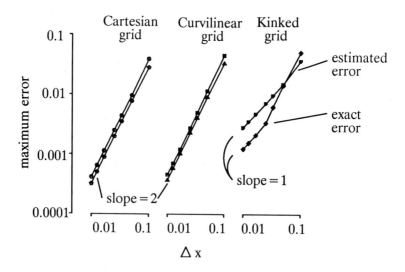

Figure 1. Exact and Estimated Error vs. Grid Spacing for Three Types of Grids

FINITE ELEMENT ANALYSIS OF AIRFOIL IMMERSED IN A SINUSOIDAL GUST

J. Shi and D. Hitchings
Dept. of Aeronautics, Imperial College of Science and Technology, London SW7, U.K.

1. The Sinusoidal Gust Problem and its FEM Formulation

For civil aircraft the critical loadings in structural strength design are often determined in gust encounter. To do analysis on this, one needs forces due to both harmonic airfoil motion and sinusoidal gust. Analytical solutions are usually limited to simple problem geometry like a plate. In this paper a general finite element formulation is proposed for airfoils of arbitrary shape.

The linear sinusoidal gust model is made of a uniform irrotational mean U_∞ flow and a rotational purtabation gust field $W_{g0} \sin\omega(x/U_\infty - t)$. Because of the special characteristic of this latter rotational flow field i.e. the curl of the velocity is only a function of x, we can replace the continuous rotational flow field by a number of discrete vortex lines of a 'consistent' strength. Here 'consistent' means that the vertical velocity difference between two points at some distance apart in x direction are the same in both discrete and continuous vortex system. In other words we have replace a smooth gust velocity profile with a series of step functions.

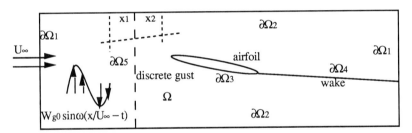

After the above simplification, we can define a velocity potential ϕ for which the governing equations and boundary conditions are:

$$\nabla^2 \phi = 0 \qquad \text{in } \Omega$$

$$\frac{\partial \phi}{\partial n} = u_\infty \qquad \text{on } \partial\Omega_1$$

$$\frac{\partial \phi}{\partial n} = W_{g0} \sin\omega(\frac{x}{U_\infty} - t) \qquad \text{on } \partial\Omega_2 \qquad (1)$$

$$\frac{\partial \phi}{\partial n} = 0 \qquad \text{on } \partial\Omega_3$$

$$\Delta\phi = f_1 \qquad \text{on }\partial\Omega_4$$

$$\Delta\phi = f_2 \qquad \text{on }\partial\Omega_5$$

where f_1 and f_2 are known functions and will be explained in more detail later. $\partial\Omega_5$ are the boundaries formed by the discrete vortex lines. Other variables are as defined in the above figure. Since finite element formulation the Laplacian equation in (1) is well established, here only finite element simulation of rotational thin layer or boundary conditions is discussed in detail. In the present formulation the same assumptions have been made as in [1]. Furthermore the continous gust field is replaced by a discrete one made of sharp edged gusts of consistent strength.(see [2] for more detail)

To solve the above problem, ϕ is resolved into:

$$\phi = (\phi_4 + \Gamma_0\phi_3) + (\phi_7 + \phi_6 + \Gamma_1\phi_1 + \Gamma_2\phi_2)\sin\omega t + (\phi_8 + \phi_5 - \Gamma_1\phi_1 + \Gamma_2\phi_2)\cos\omega t \quad (2)$$

where all the ϕ satisfies the governing equation but associated different boundary conditions:

	$\partial\phi/\partial n$ on $\partial\Omega_1$	$\partial\phi/\partial n$ on $\partial\Omega_2$	$\Delta\phi$ on $\partial\Omega_4$	$\Delta\phi$ on $\partial\Omega_5$
ϕ_1	0	0	$\sin\omega s/U_\infty$	0
ϕ_2	0	0	$\cos\omega s/U_\infty$	0
ϕ_3	0	0	1.0	0
ϕ_4	U_∞	0	0	0
ϕ_5	0	$-W_{g0}\cos(\omega x/U_\infty)$	0	0
ϕ_6	0	$W_{g0}\sin(\omega x/U_\infty)$	0	0
ϕ_7	0	0	0	$W_{g0}(\cos(\omega x_1/U_\infty) - \cos(\omega x_2/U_\infty))y$
ϕ_8	0	0	0	$W_{g0}(\sin(\omega x_2/U_\infty) - \sin(\omega x_1/U_\infty))y$

where x_1 and x_2 are the mid side node coordinates of two neigbouring columns of elements.

After ϕ is solved, we derive velocities from definition. Pressure can then be calculated according to Bernoulli's theorem. It is next integrated around the airfoil to get lift and moment.

2 Numerical example

The above theory has been tested on a number of cases. Finite element results for a plate are shown with analytical solutions [3] below:

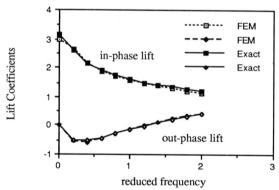

Aerodynamic response of a plate immeresed in a sinusoidal gust

References

[1] D.Hitchings and J.Shi, 'Calculation of Flutter Derivative and Speed by the FEM', in Int. Conf. on Hydro-Aeroelasticity, Prague, Dec.,1989.
[2] J.Shi, 'Finite Element Analysis of 2D Unsteady Incompressible Potential Flow', Ph.D Thesis, Univ. of London, 1990.
[3] B.C.Basu and G.J.Hancock, '2D Aerofoils and Control Surfaces in Simple Harmonic Motion in Incompressible Inviscid Flow', ARC CP1392, 1978.

Flame Capturing

S.A.E.G Falle,

Department of Applied Mathematical Studies, Leeds University,

Leeds LS2 9JT, U.K.

There are many circumstances in which premixed flames are very thin compared to typical length scales in the overall flow. This means that it would be computationally very expensive to treat the physics of the flame front correctly especially since the chemistry of flames is typically very complex and often poorly understood. However, in many combustible mixtures flames propagate at a definite velocity relative to the gas upstream and this velocity can be determined experimentally. So if one wants to study the gas dynamical effects of such flames all that is required is a model which ensures that the flame propagates at the correct speed and releases the correct amount of energy. The model must be chosen so that it gives stable flame speeds and little numerical noise even when the flame is only a few grid points wide.

A suitable model can be constructed by choosing a form of the reaction rate which depends only on a reaction progress variable. In this way the reaction can be spread over the whole flame rather than being concentrated in a zone at the rear of the flame as happens in real flames. This ensures that a steady flame speed can be achieved even with very low resolution (about 5 mesh points in the flame). The flame speed and width can be adjusted by varying the reaction rate and diffusion coefficient and these can be functions of upstream conditions. This makes it possible to use either experimental data or detailed theoretical models to ensure that the flame speed varies in the correct way as a function of the upstream state.

The numerical calculations are performed using a second order upwind conservative scheme. Since the sound speed varies considerably across the flame the scheme is made unconditionally stable by using an implicit first order step followed by an explicit second order step. The first order step uses the diagonalisation procedure described by Marx & Piquet (Int. J. Num. Methods in Fluids, 8, 1195, 1988) and so costs no more than an explicit step. The complete scheme can be shown to be linearly unconditionally stable and, unlike the usual diagonalised schemes, the second order step makes it conservative.

Figure 1 shows the temperature and reaction rate for a one dimensional flame and figure 2 the reaction rate and velocity vectors for a spherical flame calculated in axisymmetry. In both cases the numerical flame speed is within one percent of the theoretical value for a steady flame.

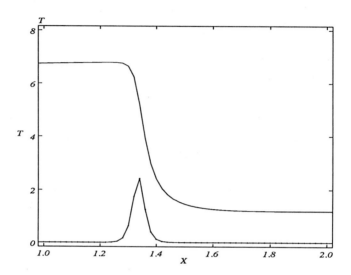

Figure 1. Temperature (no markers) and reaction rate (markers) for a one dimensional flame.

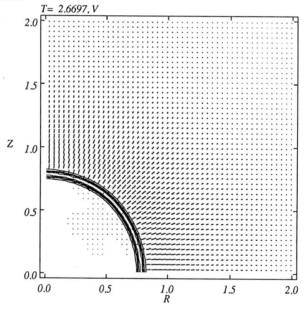

Figure 2. Velocity vectors and reaction rate for a spherical flame calculated in axisymmetry.

NUMERICAL RESOLUTION OF THE BURGERS EQUATION USING THE WAVELET TRANSFORM

C. Basdevant	*LMD, ENS, 24 rue Lhomond, 75005 Paris.*
M. Holschneider	*CPT, CNRS Case 907, 13288 Marseille Cedex 9*
J. Liandrat	*IMST, 12 av. Général Leclerc, 13003 Marseille*
V. Perrier	*ONERA, BP 72, 92322 Chatillon Cedex*
Ph. Tchamitchian	*CPT, CNRS Case 907, 13288 Marseille Cedex 9*

1. Introduction to the wavelet transform

Consider a fairly arbitrary function ψ, which is in general well localised and regular as for instance expressed by the following condition :

$$|\psi^{(k)}(x)| \leq C_{k,N} (1 + |x|^{-N}) \ , \ 0 \leq k \leq p.$$

In addition, we suppose that the first moments of ψ vanish:

$$\int_{\mathbb{R}} x^n \psi(x) \, dx = 0 \ , \ 0 \leq n \leq q.$$

Such a function will be called a wavelet since the vanishing of some moments forces ψ to have some oscillations. Dilating and translating the wavelet ψ we obtain a two parameter family of functions $\frac{1}{a} \psi \left(\frac{x-b}{a} \right)$. The parameter $b \in \mathbb{R}$ is a position parameter, whereas $a > 0$ may be interpreted as a scale parameter. The wavelet transform of an arbitrary function f with respect to the wavelet ψ is given by the following scalar products:

$$T(b,a) = \int_{\mathbb{R}} f(x) \frac{1}{a} \bar{\psi} \left(\frac{x-b}{a} \right) dx.$$

It is in general a smooth function over the position-scale half plane. Analysing a function with the help of the wavelet transform amounts to analyse it on different length scales around arbitrary positions. This transform is a sort of mathematical microscope, where $1/a$ is the enlargement and b is its position over the function to be analysed. The specific optic is determined by the wavelet itself.
This interpretation is useful in the analysis of fractal objects. More generally, such an analysis is particularly well adapted to the description of local, and even pointwise, singularities.
The wavelet transform is invertible, and preserves the energy (the L^2- norm) : it is in fact an isometry from $L^2(\mathbb{R}, dx)$ onto a subspace of $L^2 \left(\mathbb{R} \times \mathbb{R}^+, \frac{da\,db}{a} \right)$.

It has also been developped a discrete version of the wavelet transform. It can be defined, in an algorithmic way, special functions ψ such that the set of $\psi_{j,k}(x) = 2^{j/2} \psi(2^j x - k)$, $j,k \in \mathbb{Z}$, form an orthonormal basis of $L^2(\mathbb{R})$. Fast algorithms of decompositions and reconstructions are associated to those bases.

From a numerical point of view, the wavelet analysis may be thought as a compromise between Fourier analysis and finite element methods. It seemed us to be appropriate for a non linear problem, where generally a large amount of scales are involved, with the necessity to describe singularities in a localised way.

As a test problem, we considered the periodic Burgers equation, with a diffusion term or not. We present now two algorithms using the wavelet transform.

2. An adaptative wavelet pseudo-spectral scheme.

The first algorithm has been applied to the Burgers equation with small diffusion constant ($\nu = 10^{-3}$) on [0,1]. It is a kind of pseudo-spectral scheme, using bases of wavelets that are simultaneously spline functions of order 6. The main features are :
a) as in FFT algorithms, each time step requires $O(N \log N)$ calculations ;
b) the non linear term is treated classically by a collocation approximation ;
c) the very specificity of this algorithm lies in its non destructive adaptability. By this we mean that, in order to describe more adequately a strong gradient region, we add to the first rough version of the solution new wavelets at small scales that are localised near the shock. The criterion to flag the regions where adaptation is required is precise and cheap.

When we use a regular grid, the numerical results we obtained are of the same nature as for an usual spline method. When the adaptative form is used, the results are sharply improved.

3. A kind of particle method using wavelets

A second approach for numerically solving the Burgers equation (this time, it has been chosen in a conservative form) is inspired from particle methods : it consists in expanding the solution as a sum of a fixed number of wavelets :

$$u(x,t) = \sum_{i=1}^{N} c_i(t) \, \psi\left(\frac{x - b_i(t)}{a_i(t)}\right).$$

To each "atom" are associated the three parameters (c_i, a_i, b_i) of amplitude, scale and position, which evolve in time.

The PDE is then transformed in a linear system of ODEs governing the three parameters, by minimising the norm

$$\int_0^1 \left| \frac{\partial u}{\partial t} + u \frac{\partial u}{\partial x} \right|^2 dx.$$

This scheme leads to atom trajectories in the phase space (position-scale half plane), which characterises the behaviour of the solution. The advantage of this method in comparison with the previous one is that there is no necessity to define a mesh.

It must be added that, at the present time, the study is far from being complete : the first algorithm is not theoretically understood, and the second is still numerically in progress.

Bibliography

1. "Wavelets, time-frequency methods and phase space" Proceedings of a seminar held in Marseille, 14-18 Dec. 1987, J.M. Combes, A. Grossmann, Ph. Tchamitchian (eds.) Springer-Verlag, 1989.
2. M. Holschneider, On the wavelet transform of fractal objects, J. Stat. Phys. 5/6 (1988), 963-993.
3. V. Perrier, C. Basdevant, Periodical wavelet analysis, a tool for inhomogeneous field investigation. Theory and algorithms, La Recherche Aerospatiale, n° 1989-3.

APPLICATION OF BOUNDARY ELEMENT METHOD TO A SHALLOW WATER MODEL

BIN LI, MICHAEL G. SMITH,
ULF T. EHRENMARK, PETER S. WILLIAMS
Department of C.M.S.M.S., City of London Polytechnic,
London EC3N 1JY, United Kingdom.

In this research paper, the well known shallow water model is solved for the first time by the Boundary Element Method and applied to the Ramsgate sea area to investigate the velocity field. The Stokes fundamental solution is adopted to the shallow water model. The non-linear terms are treated as iteration terms. All the singular integration are performed analytically, so the accurate results of singular integration are obtained. The application of the model to Ramsgate sea area has allowed circulation patterns near the harbour to be resolved. The model results are found to agree well with the circulation observed in the field.

1. Shallow Water Model

The steady shallow water model has the form of

$$\frac{\partial q_k}{\partial x_k} = 0 \quad (k=1,2)$$

$$\frac{\partial}{\partial x_1}\left(\frac{q_1 q_1}{h}\right) + \frac{\partial}{\partial x_2}\left(\frac{q_1 q_2}{h}\right) + gh\frac{\partial \eta}{\partial x_1} + \left(\frac{g}{c^2 h^2}\right)|\mathbf{q}|q_1 = \varepsilon \Delta q_1$$

$$\frac{\partial}{\partial x_1}\left(\frac{q_1 q_2}{h}\right) + \frac{\partial}{\partial x_2}\left(\frac{q_2 q_2}{h}\right) + gh\frac{\partial \eta}{\partial x_2} + \left(\frac{g}{c^2 h^2}\right)|\mathbf{q}|q_2 = \varepsilon \Delta q_2$$

in which, is not included the Coriolis force and wind force, where q_1, q_2=discharges per unit width in x,y directions, g=gravitational acceleration, h=total depth of flow, η=water surface elevation above mean water level, c=Chezy bed roughness coefficient, ε=eddy viscosity.

If the term of $h\frac{\partial \eta}{\partial x_i}$ is rewritten as $h\frac{\partial \eta}{\partial x_i} = \frac{\partial h\eta}{\partial x_i} - \eta\frac{\partial h}{\partial x_i}$,

then the shallow water model can be written as

$$\frac{\partial q_k}{\partial x_k} = 0 \quad (k=1,2), \qquad \frac{\partial T_{jk}}{\partial x_k} = f_j$$

where $T_{jk} = -gh\eta\delta_{jk} + \varepsilon\left(\dfrac{\partial q_j}{\partial x_k} + \dfrac{\partial q_k}{\partial x_j}\right)$, $f_j = \dfrac{\partial\left(\dfrac{q_k q_j}{h}\right)}{\partial x_k} + t_j - g\eta\dfrac{\partial h}{\partial x_j}$, $t_j = \left(\dfrac{g}{c^2 h^2}\right)|\mathbf{q}|q_j$

2. Boundary Element Method for Shallow Water Model

The Boundary Element Method is based on the Green's function of the partial differencial equation. For two dimensional Stokes problem, the fundamental singular solution is

$$w_j^k(\mathbf{x}-\mathbf{x_0}) = -\dfrac{1}{4\pi\varepsilon}\left(\delta_{jk}\ln\dfrac{1}{|\mathbf{x}-\mathbf{x_0}|} + \dfrac{(x_j - x_{jo})(x_k - x_{ko})}{|\mathbf{x}-\mathbf{x_0}|^2}\right), \quad p^k(\mathbf{x}-\mathbf{x_0}) = \dfrac{-(x_k - x_{ko})}{2\pi|\mathbf{x}-\mathbf{x_0}|^2}$$

If this solution is applied to the shallow water model, the partial differential equation can be transformed to the integration equation as follows

$$\begin{pmatrix}1\\ \tfrac{1}{2}\\ 0\end{pmatrix} q_k(\mathbf{x}) = \int_s T_{ij}^{'} q_j n_i ds_o - \int_s w_i^k T_{ij} n_j ds_o + \int_\Omega f_i w_i^k d\Omega_o$$

$$h\eta g = \int_s -n_i p^j T_{ij} ds_o + \int_s n_i 2\varepsilon\dfrac{\partial p^j}{\partial x_{io}} q_j ds_o + \int_\Omega f_i p^i d\Omega_o$$

where $T_{ij}^{'} = T_{ij}(w^k) = \delta_{ij}p^k + \varepsilon\left(\dfrac{\partial w_i^k}{\partial x_{jo}} + \dfrac{\partial w_j^k}{\partial x_{io}}\right)$, $f_i = \dfrac{\partial\left(\dfrac{q_i q_j}{h}\right)}{\partial x_{jo}} + \left(\dfrac{g}{c^2 h^2}\right)|\mathbf{q}|q_i - g\eta\dfrac{\partial h}{\partial x_{io}}$

The integration equation mentioned above is used to calculate the velocity and water surface elevation.

3. Numerical Solution and Application

The calculating domain is discretised into limited triangular elements and the boundary is discretised into limited segments. The constant element method is adopted. The non-linear terms are treated as iteration terms. All the singular integration terms in the numerical integration are obtained by analytical method. Analysis shows that if the singular integration is performed by numerical method, a large error will occur. The model and method mentioned above has been applied to the Ramsgate sea area to investigate the velocity field. It has allowed circulation patterns near the harbour to be resolved. The model results are found to agree well with the circulation observed in the field.

Stabilization of the Navier-Stokes solutions via control techniques

M.O. Bristeau*, R.Glowinski**, J.Périaux†, M. Ravachol †, Y. Xiang ‡

========

Abstract

Control techniques have already played an interesting role in the construction of effective methods for solving nonlinear problems in Fluid Mechanics ([1], [2]). Indeed, some limiter techniques for compressible flow simulation can be seen as projectors on a set of physically admissible solutions defined by entropy inequalities [3].

In our presentation we would like to further discuss the projection treatment of entropy inequalities and the implementation of some concepts recently introduced by J.L. Lions [4] to stabilize via control formulations the spurious oscillations produced by centered schemes when applied to fluid flows at larger Reynold and/or Mach numbers.

We illustrate the above notions by considering an incompressible viscous flow model by the following Navier-Stokes equations

$$\frac{\partial u}{\partial t} - \nu \nabla^2 u + (u \cdot \nabla) u + \nabla p = f \quad \text{in} \quad \Omega, \tag{1}$$

$$\nabla \cdot u = 0 \quad \text{in} \quad \Omega,$$

$$u(x, 0) = u_o(x), \quad \text{with} \quad \nabla \cdot u_o = 0, \tag{2}$$

$$u = g \quad \text{on} \quad \Gamma, \quad \text{with} \quad \int_\Gamma gn d\Gamma = 0 \quad \text{and} \quad u_o \cdot n = g \cdot n \tag{3}$$

In order to control at large Reynold numbers the oscillations of the solution we introduce the following cost function

$$J(v) = \frac{1}{2} \int_0^T \|v\|^2 dt + \frac{\lambda}{s} \int_0^T dt \int_\Omega |\nabla \omega|^s dx \tag{4}$$

where in (4), $\|.\|$ is a norm to be chosen appropriately, (H_0^1 for instance) $\lambda > 0$ a regularization coefficient, s exponent ≥ 2 and $\{y, p\}$ the solution of the state equation of

$$\frac{\partial y}{\partial t} - \nu \nabla^2 y + (y \cdot \nabla) y + \nabla p = f - \nu \Delta v \quad \text{in} \quad \Omega, \tag{5}$$

$$\nabla \cdot y = 0 \quad \text{in} \quad \Omega,$$

---

* INRIA
** University of Houston and Paris 6, INRIA
† AMD/BA and INRIA
‡ Paris 6 and AMD/BA

$$y(x,0) = y_o(x), \quad \text{with} \quad \nabla \cdot y_o = 0, \tag{6}$$

$$y = g \quad \text{on} \quad \Gamma, \quad \text{with} \quad g \quad \text{satisfying (3)} \tag{7}$$

$$\omega = \nabla \times y \tag{8}$$

The influence of parameters λ and s will be studied.

Following J.L. Lions we should minimize $J(v)$ on an appropriate set of perturbation v. Indeed we can also use formulation which are more local in time and take advantage of operator splitting methods [5] to decouple the oscillation treatment of the convective term from the numerical treatment of incompressibility.

Several numerical realizations of these ideas will be discussed and compared and applied, combined to compatible finite element approximation [6] and first iterative solvers, to the practical solution of test problems originating from Aerospace Industry.

References

[1] R. GLOWINSKI, *Numerical Methods for Nonlinear Variational Problems*, Springer, New-York, 1985.

[2] M.O. BRISTEAU, R. GLOWINSKI, J. PERIAUX, P. PERRIER, O. PIRONNEAU, On the Numerical solution of nonlinear problems in fluid dynamics by least squares and finite element methods (1) Least squares formulations and conjugate gradient solution of the continuous problems. Comp. Methods. Appl. Mech. Eng. 17/18, 619-157 (1979).

[3] M.L. MERRIAM, Smoothing and the second law, Comp. Meth. in Applied Mech. and Engineering 64 (1987) pp. 177-193.

[4] J.L. LIONS, *Insensitive Controls*, Colloque Franco-Italo-Soviétique, Mathématiques Appliquées, Pavie, Octobre 1989.

[5] M.O. BRISTEAU, R. GLOWINSKI, J. PERIAUX, Numerical Methods for the Navier-Stoles Equations. Applications to the simulation of compressible and incompressible viscous flows. Computer Physics Report 6, (1987) North-Holland, Amsterdam, pp. 73-187.

[6] M.O. BRISTEAU, L. DUTTO, R. GLOWINSKI, J. PERIAUX, G. ROGE, Compresible viscous flow calculations using compatible finite element approximations, Proc. of FEMIF7 Conf. Huntsville, 1989, submitted to Special Issue of Int. Journal for Num. Methods in Fluids.

Approximation of the Stokes Problem with a Coupled Spectral and Finite Element Method

by Naima Débit and Yvon Maday

C.N.R.S. and Université Pierre et Marie Curie

4 place Jussieu, 75252 Paris Cedex 05, France

It is a common knowledge among numericists to state that finite element methods represent a powerfull tool for the numerical simulation of the Stokes and Navier-Stokes flows. There are many methods available that can fit many requirements of order (essentially 1 or 2), flexibility, structuration or unstructuration. On the other hand spectral type methods have begun to prove also their usefullness in the same range of problems and allow either to achieve orders of accuracy that are not available with the finite element methods or achieve a given accuracy at a lower price. Right now and as far as we know, there are only four types of spectral methods well suited for the resolution of the Stokes and Navier Stokes problem in primitive variables. They differ from the choice of compatible spaces that discretize the velocity and the pressure.

Our aim is to present and analyze a method of domain decomposition that allows for using a finite element method on some domains while using a spectral method on others. The idea is not new and has already been introduced in previous papers where the approximation of the Poisson problem with such a technique was proposed. The main difficulty for the Poisson equation is to use a good matching between the subdomains in order to be able to prove that the consistency errors and the approximation errors are optimal. We recall that this optimality has to be understood in the sense that these two errors (hence the global numerical errors) can be bounded by the sum of the local errors resulting from the chosen discretization on the subdomains, independantly of the other subdomains. The matching condition that has been introduced [BDM][BMS] has led to the notion of the *mortar element method* and we refer to [BMP] for a presentation of the general notions as well as the analysis in the case of the Poisson equation. We refer to [MMP] and [AMMP] for details on the implementation. In these papers, it is proven that the method is not only optimal from the theoretical point of view of the error bound, but also from the implementation point of view in terms of flexibility, locallity and parallelism.

In the case of the Stokes problem, another difficulty rises as regards the coupling of spectral and finite element method in that a compatibility condition has to be checked in the choice of discrete spaces for the velocity and the pressure. Moreover, the dependancy of the inf-sup condition that links the two discrete spaces with respect to the parameters of discretization has also to be evaluated. This is the aim of this presentation.

The coupling of the finite element and the spectral methods is achieved via the *mortar element method*. The interest of this approach relies in the flexibility in which the choice of the finite element and the spectral method so as the choice of the various parameters of discretizations is done. This was the conclusion of the analysis performed in the framework of the Poisson equation. The conclusion is proven here to pertain as long as it is based on the coupling of a compatible finite element and spectral discretization. The finite element method that is used can be any compatible one. The spectral method that is used can be any one of the four known ones but the notion of spurious modes that exists in two of them and the dependancy of the inf-sup condition that pollutes slightly each of them as regards the computation of the pressure implies a separate analysis of each of them. We analyze only two of the methods: one without any spurious modes (the $I\!\!P_N \times I\!\!P_{N-2}$ method) [MP] and one with spurious modes (the $I\!\!P_N \times I\!\!P_N$ method)[BMM].

Essentially, the conclusion is that, if the matching between the subdomains is done through the *mortar element method* for the velocity, any compatible method of finite or spectral element type can be used over each subdomain to discretize the Stokes problem and the resulting discretization will provide a numerical solution with an error that is bounded by the sum of the independant local errors.

The presentation will also include some numerical results that prove the ability of the method to approximate certain classes of problems in an optimal way.

References

[AMMP] G. ANAGNOSTOU, Y. MADAY, C. MAVRIPLIS & A. T. PATERA — On the Mortar Element Method: Generalizations and implementation. *in Proceedings of the 3rd Domain Decomposition Techniques*, SIAM, to appear.

[BDM] C. BERNARDI, N. DEBIT & Y. MADAY — Coupling Spectral and Finite Element Methods for the Laplace Equations, *Math. Comput.* (to appear).

[BMM] C. BERNARDI, Y. MADAY & B. METIVET — Calcul de la Pression dans la Résolution spectrale du Problème de Stokes, *La Recherche Aérospatiale* **1** (1987), 1–21.

[BMP] C. BERNARDI, Y. MADAY & A. T. PATERA — A New Nonconforming Approach to Domain Decomposition: The Mortar Element Method, *in Nonlinear Partialm Differential Equations and their Applications* Pitman (to appear).

[BMS] C. BERNARDI, Y. MADAY & G. SACCHI LANDRIANI— Nonconforming Matching Conditions for Coupling Spectral and Finite Element Methods, *J. of Applied Num. Math.* (to appear).

[MMP] Y. MADAY, C. MAVRIPLIS & A. T. PATERA — Nonconforming mortar element method: application to spectral discretizations, *Proceedings of the Second International Symposium on Domain Decomposition Methods*, T. F. CHAN, R. GLOWINSKI, J. PERIAUX and & O. B. WIDLUND eds, SIAM, Philadalphia (1989).

[MP] Y. MADAY & A. T. PATERA — *Spectral Element Methods for the Incompressible Navier Stokes Equations*, NOOR ed. ASME (1989).

STREAMWISE UPWIND ALGORITHM DEVELOPMENT FOR THE NAVIER-STOKES EQUATIONS

Peter M. Goorjian, Shigeru Obayashi,* and Guru P. Guruswamy
NASA Ames Research Center
Moffett Field, California 94035 USA

Introduction

Improvements have been made to a streamwise upwind algorithm so that it can be used for calculating flows with vortices. A calculation is shown of flow over a delta wing at a supersonic Mach number and at an angle of attack. Also the algorithm has been extended to an implicit algorithm that uses a moving grid system, for computations of unsteady flow fields over oscillating wings. A calculation is shown of flow over an oscillating rectangular wing at a transonic Mach number. The thin-layer, Navier-Stokes equations are used for the calculation. The results are compared with other computations and experimental data. The present method shows improvements in accuracy, convergence and robustness properties in the comparisons.

Numerical Algorithm

The streamwise, upwind algorithm is applied to the inviscid fluxes in the thin-layer, Navier-Stokes equations. The algorithm is described by the following formula for the inviscid, cell interface flux $\widehat{F}$ with a surface vector, $\mathbf{S} = (\eta_x, \eta_y, \eta_z)$,

$$\widehat{F}(Q_l, Q_r, \mathbf{S}_{j+\frac{1}{2}}) = \frac{1}{2}\frac{|\nabla \eta|}{J} \times \left\{ [\, F_l + F_r\,] + [\, F_l \text{sign}\,(U_l) + s_l\, \Delta^* F_l\,]\cos^2\theta_l \right. \tag{1}$$
$$\left. - [\, F_r \text{sign}\,(U_r) + s_r\, \Delta^* F_r\,]\cos^2\theta_r - |A|\,\Delta Q\ \sin^2\theta \right\}$$

where Q_l and Q_r are left and right states, respectively. For the first-order-accurate computations, $l = j$ and $r = j+1$. For third-order-accurate extensions, (used in the figures), the MUSCL approach is used. The rotation angle, θ, depends on the local stream direction, velocity and pressure gradient. The symbol * indicates local sonic values. The formulas for $\Delta^* F$ and the switches s_l and s_r use the speed q and the Mach number, q/c, rather than the normalized, contravarient velocity component U and the Mach number component U/c that many other upwind methods use. With the use of this rotated differencing, the switching of terms at transonic shock waves occurs independently of their alignment to grid lines. The first two terms in Eq. (1) correspond to central differencing and the remaining terms are formed from linear combinations of the eigenvectors derived from F.

To extend Eq. (1) from a fixed grid system to a moving grid system, the flow velocity relative to the fixed grid, $\mathbf{q} = (u, v, w)$, is redefined as the flow velocity measured relative to the moving grid, $\tilde{\mathbf{q}} = (u - x_t, v - y_t, w - z_t)$. Also the definition of the contravariant velocity from is modified from a fixed grid system, $\widehat{U}_{fixed} = \nabla \eta \cdot \mathbf{q}$, to a moving grid system, $\widehat{U}_{moving} = \nabla \eta \cdot \tilde{\mathbf{q}}$. The algorithm is described in detail in references 1-3 with the implicit formulas given in reference 3.

Results

Figure 1 shows conical flow calculations of the flow over a delta wing with a leading-edge sweep, $\Lambda = 75°$, at a Mach number, $M_\infty = 2.8$, an angle of attack, $\alpha = 16°$ and a Reynolds number, $Re = 3.565 \times 10^6$. Three-dimensional calculations are given in reference 2. The present results are compared with Roe's upwind method and a central-differencing method.

---

* MCAT Institute, San Jose, California

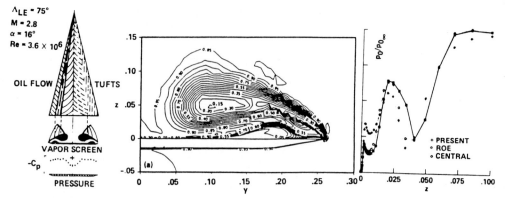

Fig. 1. a) Exp. Flow Field b) total pressure contours c) comparisons at y = 0.216.

Figure 2 shows results at the 50% span station for unsteady flow over a rectangular wing with a NACA 64A010 airfoil section and an aspect ratio of 4. The comparisons are made between experimental data, the present method, (UP-LU), and central differencing, (CD), with two values of dissipation, 0.01 and 0.02. The numerical instabilities in the CD results are shown by oscillations in the pressure profiles of CD(0.01).

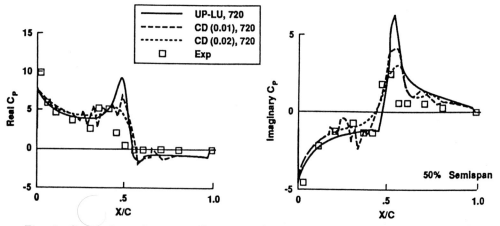

Fig. 2. Comparison of upper surface unsteady pressures with 720 times steps per cycle; Grid: $151 \times 25 \times 34$, $Re = 2 \times 10^6$, $M_\infty = 0.8$.

References

[1] Obayashi, S. and Goorjian, P. M., "Improvements and Applications of a Streamwise Upwind Algorithm," AIAA Paper 89-1957-CP, Proc. AIAA 9th Computational Fluid Dynamics Conference, pp 292-302, June 1989.

[2] Goorjian, P. M. and Obayashi, S., " A Streamwise Upwind Algorithm Applied to Vortical Flow over a Delta Wing," Eight GAMM Conference on Numerical Methods in Fluid Mechanics, Delft, the Netherlands, Sept. 27-29, 1989. Also, NASA TM 102225 Oct., 1989.

[3] Obayashi, S., Goorjian, P. M. and Guruswamy, G. P., " Extension of a Streamwise Upwind Algorithm to a Moving Grid System," NASA TM-102800, April, 1990.

A DOMAIN OF DEPENDENCE UPWIND SCHEME FOR THE EULER EQUATIONS

A. Dadone
Istituto di Macchine ed Energetica
Università di Bari
via Re David 200, 70125 Bari, Italy

B. Grossman
Dept. Aerospace and Ocean Eng.
Virginia Polytechnic Inst. & State Univ.
Blacksburg, Virginia 24061, USA

Among the numerical methods available for inviscid, compressible flows, upwind methods have the property of approximately mimicking physical wave-propagation phenomena. When dealing with one-dimensional flow problems, the choice of appropriate upwind directions is straightforward, since characteristic information may only be propagated forward or backward. For two or three-dimensional flow problems, the choice is more difficult, with most schemes using directions which are normal to the cell faces. Such grid dependent schemes do not always account for the domain of dependence of each grid point. The consequences are particularly evident when computing shocks oblique with respect to the computational mesh. In such a case, the domain of dependence is violated due to the fact that in the vicinity of the shock, the velocity component in one of the splitting directions may be subsonic, and accordingly, numerical information emanating downstream of the shock may actually reach the upstream region, causing excessive smearing of the oblique shock.

One of the existing efforts to develop grid independent schemes is the work of Hirsch *et al*, which involves a scheme with upwinding directions normal to particular characteristic waves. Even though they did not succeed in converging the angle of such directions, the results were encouraging. Levy *et al* have analyzed the influence of the choice of upwind differencing angles and have obtained interesting results. However, they were not able to impose angles normal to the oblique shock, which seems to be the most appropriate choice.

The objective of the present effort is to develop a grid-independent upwind scheme able to satisfy the domain of dependence of each mesh cell. The new scheme utilizes the approximate Riemann solver developed by Roe. The flux-difference splitting is applied in two orthogonal directions for each cell face. These directions are numerically computed based on pressure-gradient data, and the resulting angles are nearly normal and tangential to oblique shock waves. A filtering of the numerical pressure-gradient data is utilized in order to avoid convergence difficulties due to numerical noise.

Two different test problems have been considered: shock reflection on a flat plate and supersonic flow through a channel with a wedge on the lower wall. In the shock reflection problem, the incident shock angle was 29° and M_∞ was 2.9. A uniform 81 by 41 grid was employed. Numerical results computed with the first-order version of the new algorithm are shown in Figures 1a and 2a. Fig. 1a presents the resulting pressure contours. The pressure distributions in the axial direction are shown in Fig. 2a. The circles correspond to the pressure distribution along the flat plate and the squares indicate the pressures along a line $y = 0.125$. Figures 1b and 2b show the corresponding results using the classical (one-dimensional) Roe flux-difference splitting, with first-order spatial accuracy. These results indicate that the new scheme smears oblique shock waves much less than *classical* schemes with the same accuracy. In particular, the number of mesh intervals involved in the shock transition is roughly reduced by a factor of 2, while preserving the monotonicity of the results. Moreover, Fig. 3 indicates that the new scheme is capable of enforcing upwinding directions very close to the exact oblique shock angles. In Fig. 4 we see that the proposed scheme has a convergence rate very similar to the classical scheme, even with the use of upwinding angles which are continually computed (not frozen). Computed pressure contours on the $M_\infty = 3$ flow through a channel with a 5° compression ramp, shown in Figs. 5a and 5b exhibit similar features. Also, preliminary results with higher-order accuracy are encouraging.

1. Hirsch, C., Lacor, C. and Deconinck, H., "Convection Algorithms Based on a Diagonalization Procedure for the Multidimensional Euler Equations", AIAA 87-1163-CP, June 1987.
2. Levy, D. W., Powell, K. G. and Van Leer, B., "An Implementation of a Grid-Independent Upwind Scheme for the Euler Equations", AIAA 89-1931-CP, June 1989.

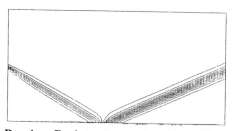
Fig. 1a. Shock impingement problem. New scheme, 1st-order

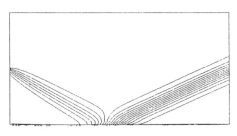
Fig. 1b. Classical scheme, 1st-order

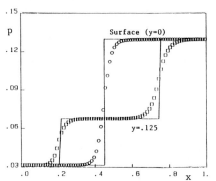
Fig. 2a. Pressure distribution parallel to the plate, new scheme.

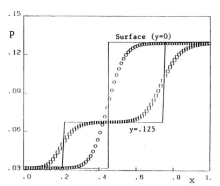
Fig. 2b. Classical calculation of distribution in fig. 2a.

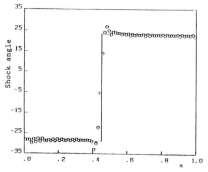
Fig. 3. Numerically computed shock angles.

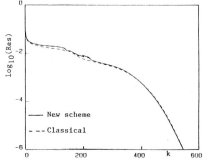
Fig. 4. Convergence rates.

Fig. 5a. Channel flow problem. New scheme, 1st-order

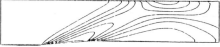
Fig. 5b. Classical scheme, 1st-order

NUMERICAL APPROXIMATION BY GODUNOV-TYPE SCHEMES OF SHOCKS AND OTHER WAVES

B. Favini, M. Di Giacinto

Dipartimento di Meccanica e Aeronautica
Università degli Studi di Roma "La Sapienza"
Via Eudossiana, 18 – 00184 – Roma

Godunov-type schemes are a class of methods designed for the numerical integration of the gasdynamic equations written in integral form in order to approximate compressible inviscid flows with shocks and contact discontinuities. The peculiarity of these schemes with respect to other techniques approximating the integral form, consists of explicitely modelling the non linear wave propagation phenomena. This fact ensures the consistency of the numerical solution with the physical model and, specifically, the increase of the entropy across shocks.

The first scheme exhibiting this property has been proposed by Godunov [1] who adopted, in its original formulation, an initial data reconstruction based on a piece-wise constant polynomial. Then the Riemann problems arising at the interfaces among the constant states are solved exactly by an iterative procedure and this allows to define analitically the gasdynamic state at a new time level. By integrating the solution inside each discretization cell, an averaged value may be defined in order to reconstruct the piece-wise constant distribution at the new time level.

In spite of the attempts of increasing the accuracy and the efficiency of Godunov's scheme have succeeded, a peculiar behaviour of this techniques when approximating slowly moving shocks has been shown. In particular, as probably pointed out first in [2], the post-shock state is spoiled by a low frequency oscillation that can be hardly smoothed out by the numerical diffusion inherent to the scheme. The interpretation proposed in [2], and reconfirmed in [3], connects the post-shock oscillation to the fact "... that the scheme will not provide sufficient dissipation to ensure that the correct amount of entropy production occurs. This situation arises when the speed of the characteristic of the family associated with the shock changes sign across the shock, i.e., when the shock is nearly stagnant" (quotation from [2] page 107). Afterwards a different interpretation of the error has been presented in [4]: "... the error can be explained in terms of the discrete shock structure. The requirement that no noise be generated behind the shock is equivalent to requiring that the internal zones of the discrete shock must move along the shock curve (...) passing through the post-shock state ... For nonlinear systems, this is not true, and it can be shown that this requirement implies that no unsteady shock can be represented with only one internal zone". In order to support his thesis, the author produces a comparison between the results obtained with Godunov's, Roe's and Osher's schemes, which shows that Osher's scheme strongly reduces the amplitude of the post-shock oscillations. As a matter of fact, Osher's scheme generates a wider shock-transition layer than Godunov's and Roe's schemes, i.e. including more than one internal zone.

In the present paper we propose a different answer to the question: these errors are basically originated by the presence of the averaging step in the Godunov-type schemes, and therefore it is, at the moment, an intrinsic limit of these techniques.

A Godunov-type schemes can be formulated in an abstract form as [5]:

$$\underline{w}_j^{n+1} = A(I_j) \cdot E(\tau^n) \cdot R(x^p; \underline{w}^n) \tag{1}$$

where $R(\cdot;\cdot), E(\cdot), A(\cdot)$ are three operators representing respectively: the data reconstruction obtained by interpolating the initial data $\underline{w}^n$ by means of polinomial of order p; the solution, exact or approximate, of the Riemann problem in the time interval τ^n; the evaluation of an average value of the solution inside the cell $I_j \equiv \left\{\forall x, x \in \left[x_j - \frac{\Delta x}{2}, x_j + \frac{\Delta x}{2}\right]\right\}$.

It can be shown that, in principle, the averaging step generates an error that can be possibly reduced, but not completely removed, by the successive steps. In particular, the outcome $\underline{w}_j^{n+1}$ of the averaging step usually do not correspond to a physical state connecting the gas states at the interfaces; in other words, different physical phenomena with respect to that obtained by the solution of the local Riemann problem are required in order to generate, in a whatever point inside the Δx interval, a gas state coincident with the average state. For instance, the average state belonging to a cell crossed by an isolated shock can be actually obtained as a physical gas state only by hypotizing a more complex system of waves (shock, rarefaction and/or contact discontinuity) connecting this state with the adjacent one.

In order to prove this statement we will resort to some theoretical considerations, on whose basis we can hypotize the behaviour of the Godunov-type schemes in dealing with the numerical simulation of some basic phenomena. Numerical experiences are also presented in order to confirm these expected behaviours. Then, we will show that numerical approximation of centered rarefaction fans are affected by the same kind of error, while the contact discontinuities are not.

REFERENCES

[1] S.K. Godunov, "A finite difference method for the numerical computation of discontinuous solutions of the equations of fluid dynamics", Mat. Sb., **47**, 1959, 357–93.

[2] P. Colella, "A direct eulerian MUSCL scheme for gasdynamics", SIAM J. Sci. Stat. Comput., **6,1**, 1985, 104–17.

[3] P. Colella, P. Wooward, "The piecewise parabolic method (PPM) for gasdynamical simulations", J. Comput. Phys., **54**, 1984, 174–201.

[4] T.W. Roberts, "The behaviour of flux-difference splitting schemes near slowly moving shock waves", Numerical Method for Fluid Dynamics III, ed. K.W. Morton and M.J. Baines, Clarendon Press, Oxford, 1988, 442–48.

[5] A. Harten, "Eno scheme with subcell resolution", J. Comput. Phys., **83**, 1989, 148–184.

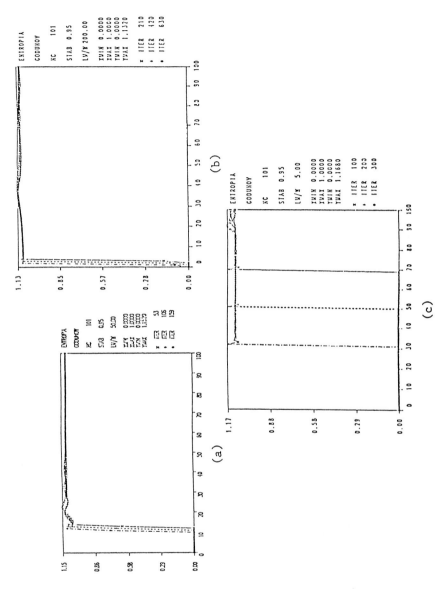

Fig. 1. Entropy plots for unsteady shocks with different speeds defined as the ratio between the maximum velocity of wave propagation (λ_{max}) and the shock speed (v_s): (1.a) $\lambda_{max}/v_s = 50$; (1.b) $\lambda_{max}/v_s = 200$; (1.c) $\lambda_{max}/v_s = 5$. The post-shock oscillations are related to an entropy wave prompted by the averaging operator. The wave lenght is a function of the ratio between the relative, with respect to the shock, fluid velocity behind it and the shock velocity. This fact sustains that the error exhibited by the numerical approximation of a fast moving shock (1.c) has the same origin.

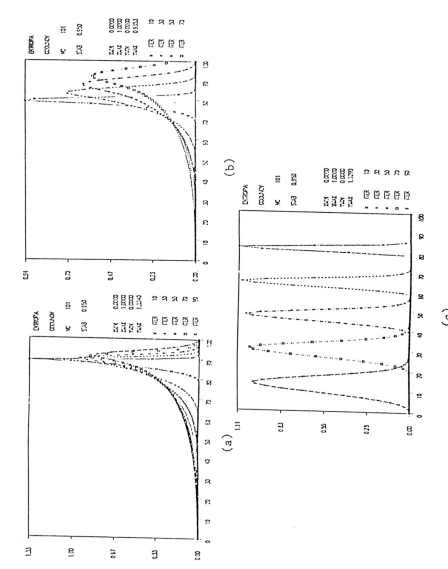

Fig. 2. Entropy plots for centered expansion waves with different velocity of propagation defined as the ratio between the speed of the head (v_h) and the speed of the tail (v_t): (2.a) $v_h/v_t = 5$; (2.b) $v_h/v_t = 20$; (2.c) $v_h/v_t = 1.1$. The average state computed on the cells inside the expansion fan does not represent physical states consistent with a rarefaction wave, and the flows is altered by the presence of contact discontinuities, whose intensitis and propagation velocities justify the form and the amplitude of the error in the numerical solution.

A MULTIBLOCK MULTIGRID THREE-DIMENSIONAL EULER EQUATION SOLVER

Frank E. Cannizzaro   Alaa Elmiligui

Old Dominion University, Norfolk, VA

N. Duane Melson   E. von Lavante
NASA Langley Research Center   University of Essen
Hampton, VA   Essen, West Germany

Current aerodynamic designs are often quite complex (geometrically). Flexible computational tools are needed for the analysis of a wide range of configurations with both internal and external flows. In the past, geometrically dissimilar configurations required different analysis codes with different grid topologies in each. The duplicity of codes can be avoided with the use of a general multiblock formulation which can handle any grid topology. Rather than 'hard wiring' the grid topology into the program, it is instead dictated by input to the program.

In the present work the compressible Euler equations, written in a body-fitted finite-volume formulation, are solved using a pseudo-time-marching approach. Two upwind methods (van Leer's flux-vector-splitting and Roe's flux-differencing) have been investigated. Two types of explicit solvers (a two-step predictor-corrector and a modified multistage Runge-Kutta) have been used with multigrid acceleration to enhance covergence.

A multiblock strategy is used to allow greater geometric flexibility. The solution domain is divided into multiple zones (blocks) and the grid for each block is then generated. If the blocks are chosen appropriately, the difficulty of generating a boundary-fitted grid is significantly reduced. Also, the placement of wall boundary conditions is limited only by the placement of the face of a block. The trade-off for this flexibility is the overhead required for the communication between the multiple blocks. In the present multiblock implementation, two simplifying assumptions have been made. First, the interfaces between blocks are assumed to have C^0 continuity. Second, the boundary condition on any face of a block is assumed to be homogeneous across the entire face (i.e., either completely a wall, an inflow/outflow boundary, or an interface with another block). On block faces that have either a wall or an inflow/outflow boundary condition, 'standard' boundary conditions are used. On faces that are interfaces, a special interface routine presets the values in two ghost cells (normal to the face) equal to the current values in the coincident interior cells in the adjacent block. The updates of the interface ghost cells are performed before each iteration in a given block. The iteration of each block can then proceed without the need for further information from adjacent blocks.

There are two possible strategies for the implementation of multigrid with a multiblock grid structure: (1) multigrid inside of multiblock and (2) multiblock inside of multigrid. With the first strategy, a complete multigrid cycle (or cycles) is (are) performed for a given block. Then work begins on the next block and so forth until all the blocks are complete. This strategy allows the flexibility of different numbers and/or types of multigrid cycles for different blocks, which are adjusted to speed convergence of slowly converging blocks (assuming only steady-state results are sought). Unfortunately, communication between the blocks is reduced. The rate of convergence is reduced. (The interface boundary conditions of the adjacent blocks have to remain fixed in time or a special interface condition has to be used for the blocks to communicate from different multigrid levels).

With the second strategy, multiblock inside of multigrid, all the points on the multigrid fine grid (grid h) in all the blocks are updated before the multigrid process continues to the coarse grid (grid 2h). There, all the points for all the blocks are updated before proceeding to the next coarser multigrid grid. This allows communication between the coarse grids in the multigrid cycle. This method can also identically reproduce the convergence history of a single block solution using an explicit algorithm– a useful debugging tool for the multiblock logic. This latter strategy was used in the present work to compute corner flow through a duct, flow from a jet exhaust mixing with freestream air, and transonic flow over an ONERA M6 wing, without changing the computer program.

ENO SCHEMES FOR VISCOUS HIGH SPEED FLOWS

F. Grasso
Dipartimento di Meccanica e Aeronautica
Università di Roma "La Sapienza"

F. Bassi
Istituto di Macchine
Università di Catania

A. Harten
School of Mathematical Sciences
Tel Aviv University

In this paper an essentially non oscillatory scheme is implemented for the solution of two-dimensional viscous high speed flows.

In the recent past total variation diminishing schemes have been developed for the solution of such flows, and their reliability has been demonstrated [1,3]. However the accuracy of TVD schemes reduces to order one at points of extrema of the solution. Recently essentially non oscillatory schemes have been designed to obtain arbitrarily high order of accuracy [4,6]. ENO schemes are interpolatory schemes that use an adaptive stencil. Their design involves: the reconstruction of the solution from its cell averages and time evolution through an approximate solution of the resulting initial value problem.

Results with ENO have primarily been presented for 1-D hyperbolic systems of conservation laws. Extensions to 2-D hyperbolic systems of conservation laws are being designed: Harten is extending the schemes to 2-D by using a two-dimensional polynomial reconstruction via deconvolution technique [6]; Chang and Liou have extended the method by using a Strang-type fractional-time splitting algorithm [7]. However Strang-type dimensional splitting is not suitable for higher order ENO schemes.

In the present work a polynomial reconstruction via primitives of the characteristic variables is used. The method is applied to the Navier Stokes equations in conservation form

$$\frac{\partial}{\partial t}\int^v W\,dV + \oint (f^e n_x + g^e n_y)\,ds = \oint (f^v n_x + g^v n_y)\,ds \qquad (1)$$

where

$$W = (\rho, \rho u, \rho v, \rho E)$$
$$f^e = \left(\rho u, \rho u^2 + p, \rho uv, \rho H\right)$$
$$g^e = (\rho v, \rho uv, \rho vv + p, \rho v H)$$
$$f^v = \left(0, \sigma_{xx}, \sigma_{yx}, u\sigma_{xx} + v\sigma_{yx} - q_x\right)$$
$$g^v = \left(0, \sigma_{xy}, \sigma_{yy}, u\sigma_{xy} + v\sigma_{yy} - q_y\right)$$

$$p = (\gamma - 1)\rho \left(E - \frac{u^2 + v^2}{2}\right)$$

$$E = e + \frac{u^2 + v^2}{2}$$

$$H = E + p/\rho$$

The semidiscrete form of eqn. (1) is

$$V_{ij}\frac{d}{dt}W_{ij} = -\sum_\alpha \hat{F}^e_\alpha \Delta s_\alpha + \sum_\alpha \hat{F}^v_\alpha \Delta s_\alpha \qquad (2)$$

where $\hat{F}^e_\alpha$ and $\hat{F}^v_\alpha$ are respectively the numerical Euler and viscous flux contributions. Central differencing is used for the viscous terms. A generalized Godunov type scheme is employed for the Euler terms:

$$F^e_\alpha = \frac{1}{2}\left[F^e_\alpha(W^L_\alpha) + F^e_\alpha(W^R_\alpha)\right] - \frac{1}{2}\left|\frac{\partial F^e}{\partial W}\right|_\alpha \left(W^R_\alpha - W^L_\alpha\right) \qquad (3)$$

The evaluation of the left (L) and right (R) states at cell face α is obtained by characteristic decomposition and ENO reconstruction as follows:

$$W^{L,R}_\alpha = R_\alpha(W^{L,R}_\alpha)C^{L,R}_\alpha = R_\alpha\left(W^{L,R}_\alpha\right)\mathcal{R}^{L,R}\left[x_\alpha; R^{-1}W\right]$$

where $\mathcal{R}^{L,R}$ are the reconstruction polynomials for cells adjacent to face α and R is the right eigenvector matrix of the Euler jacobian matrix, evaluated by Roe's average.

In the preliminary computations 3rd order polynomial reconstruction is used and Runge Kutta algorithm is employed for time integration of eqn. (2).

Results of flows over compression ramps show that the main flow features are well captured by the method. As expected we find that ENO reconstruction gives a sharp representation of shock waves and accurate resolution of the viscous layer. Detailed comparison of computed and experimental results will be reported.

References

1. H.C. Yee, NASA TM 101088, 1989.
2. S.R. Chakravarty, AIAA paper 86-0243.
3. F. Bassi, F. Grasso, M. Savini, Lecture Notes in Physics, vol. 323, 1988.
4. A. Harten et al., J Comp. Phys., vol. 71, no. 2, 1987.
5. S.R. Chakravarty, A. Harten and S. Osher, AIAA paper 86-0339.
6. A. Harten, to appear, 1989.
7. F. Grasso, ISCFD-Nagoya, 1989.
8. S.H. Chang, M.S. Liou, NASA TM 101355, 1988.

Design of Airfoils in Transonic Flows

M. Hafez and J. Ahmad
Department of Mechanical Engineering
University of California, Davis

Computational fluid dynamics has reached a point where with powerful computers and advanced algorithms, many practical problems can be analyzed using efficient numerical simulations. Recently, there is a great interest in design and optimization of aerodynamic configurations using the well developed analysis codes, available in industry. In this paper, the design problem of airfoils in transonic flows is studied. A new design procedure is proposed and tested and preliminary results are discussed.

There are mainly two approaches for design; the first is based on inverse methods pioneered by the work of Lighthill for incompressible flows. The extension to transonic flows was carried by Carlson, Volpe and Melnik and others. The second approach is based on iterated direct methods proposed by Garabedian and McFadden. Recently, Jameson designed successfully airfoils, with constraints, via control theory, using his analysis codes.

The present work has some features from both approaches. The full potential equation will be used for the algorithm development. The proposed procedure is not, however, restricted to irrotational flows and indeed the potential equation can be replaced by Euler equations for inviscid flows with vorticity. The design problem can be formulated as follows. The airfoil is mapped into a circle in the computational plane; the mapping is an unknown function. The continuity equation, in conservation form, is invariant under this transformation. The density is obtained from Bernoulli's equation, where the modulus of transformation explicitly appears in the velocity components. At the surface of the cylinder, two conditions are imposed: the normal velocity component must vanish, while the tangential velocity component must match a prescribed function consistent with the desired pressure distribution. In the far field, the potential is derived asymptotically for a flow over a rotating cylinder (uniform flow, a doublet and an irrotational vortex). Three constants appear in the governing equations and boundary conditions; the total enthalpy which is supplied as an input data and the speed at infinity which must be treated as unknown and it is obtained as a part of the solution of the problem. The circulation, or the strength of the irrotational vortex, is fixed, as usual, via the Kutta condition at the trailing edge of the airfoil.

Following Jameson, the above formulation can be replaced by a constraint optimization problem: Find the airfoil shape which minimizes the functional

$$I = \frac{1}{2} \int_C (q-q_d)^2 d\theta \qquad (1)$$

where q_d is the desired surface velocity distribution and q must satisfy the governing equations and boundary conditions.

The variation of (1) is

$$\delta I = \int_C (q-q_d)\, \delta q\, d\theta \qquad (2)$$

Hence, δq is chosen as

$$\delta q = -\lambda (q-q_d) \qquad (3)$$

where λ is a positive parameter.

The velocity components in the computational plane are given by

$$u = \frac{\phi_\theta}{rh} \quad \text{and} \quad v = \frac{\phi_r}{h} \qquad (4)$$

At the surface, the normal velocity component vanishes, and equation (3) reduces to:

$$\frac{\delta\phi_\theta}{rh} - u\frac{\delta h}{h} = -\lambda(u-u_d) \qquad (3')$$

For a conformal mapping, $\delta h/h$ satisfies Laplace equaiton and vanishes in the far field. In the proposed procedure, the flow equations and boundary conditions, in the computational plane, are discretized using finite differences. The resulting discrete nonlinear equations are solved simultaneously via Newton's method. At each iteration, the linear system of equations for the corrections are augmented by the discrete equations of the Laplacian for ($\delta h/h$) and the discrete version of equation (3'). The solution is obtained via a direct solver, based on banded Gaussian elimination (Linpack). Once the corrections are available, ϕ and h are updated and the process is repeated until convergence. Finally, based on h, the airfoil shape is constructed.

The parameter λ controls the reate of convergence of the iterative process. For $\lambda = 1$, quadratic convergence is obtained and the proposed procedure reduces to an inverse design method based on conformal mapping. The computational effort is comparable to that of the analysis mode except the calculation of the mapping is now coupled with the flow equations. Also, Gaussian elimination is not essential and can be replaced for example, by line relaxation, where ϕ and h are solved simultaneously only along lines and not in the whole field.

It should be mentioned, that the inverse method may not be well posed if the speed at infinity (q_∞) is fixed. The desired velocity distribution should determine the level of q_∞. The above formulation can be easily modified by normalizing the velocity by q_∞ in the continuity and Bernoulli's equations as well as in equation (3'). At each iteration q_∞ is chosen to minimize

$$\frac{1}{2} \int_C (\bar{u}^2 - (\frac{u_d}{q_\infty})^2)^2 d\theta \tag{5}$$

where $\bar{u}$ is the normalized tangential velocity at the surface.

Central differences are used everywhere. To capture shock waves in transonic flow calculations, artificial viscosity terms are added explicitly either to the continuity or to the energy (Bernonlli) equation. For example, the continuity equation is replaced by

$$\nabla \cdot \rho \vec{q} = \nu \nabla^2 \rho \tag{6}$$

Alternatively, the Bernoulli equation can be modified to read

$$\frac{\gamma}{\gamma-1} \frac{p}{\rho} + \frac{1}{2} q^2 - \nu \nabla^2 \phi = \text{constant} \tag{7}$$

The latter form can be obtained by integrating the momentum equations including the visocus terms assuming the flow to be irrotational.

These first order, constant viscosity models can be, of course, replaced by more sophisticated schemes if necessary provided that the nonuniqueness problem of the potential solution is avoided.

Some preliminary results have been obtained based on the present method where the surface pressure distribution is given and the corresponding shape of the airfoil is calculated. In figure (1), the history of the convergence of the iterative process is plotted for a test case. A grid of 90 x 30 points is used. The residuals of the three governing equations are reduced twelve orders of magnitude in 9 iterations. In figures (2) and (3) the correction of the circulation and the velocity at infinity (q_∞) are also plotted as functions of iteration. The Mach contours are shown in figures (4). As an initial guess for the design process, the flow over a circular arc airfoil at the same angle of attack (1°) and Mach number (0.85) is calculated first using an analysis code. The difference between the desired surface pressure distribution and the pressure distribution given by the initial guess, derives the design calculations until the solution converges to the final airfoil shape which is shown in figure (4).

More details will be given in a separate paper including comparison with other design methods to demonstrate the efficiency of the present algorithm with applications to practical problems.

References

Lighthill, M. J., "A New Method of Two Dimensional Aerodynamic Design" ARC, Rand M 2112, 1945.

Carlson, L. A., "Transonic Airfoil Analysis and Design Using Cartesian Coordinates" J. of Aircraft, Vol. 13, 1976, pp. 349-356.

Volpe, G. and Melnik, R. E., "The Design of Transonic Airfoils by a Well Posed Inverse Method", Int. J. Numerical Methods in Eng., Vol. 22, 1986, pp 341-361.

Garabedian, P. and McFadden, G., "Computational Fluid Dynamics of Airfoils and Wings", Proc. of Symposium on Transonic, Shock, and Multidimensional Flows, Madison, 1981, Meyer, R., (Ed.) Academic Press, New York, 1982, pp. 1-16.

Jameson, A., "Aerodynamic Design via Control Theory" Princeton MAE Report 1824, May 1988. Also, ICASE Report No. 88-64, Nov. 1988.

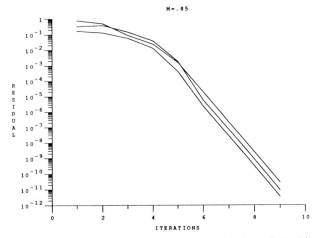

Figure (1) Convergence History of the Residuals of Continuity, Bernoulli and Mapping equations.

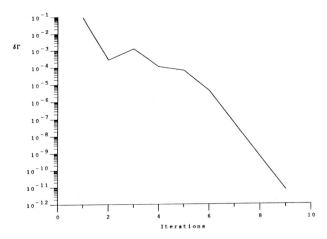

Figure (2) Correction of the Circulation versus Iteration.

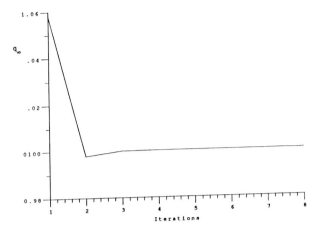

Figure (3)
The velocity at infinity (q_∞) versus Iteration

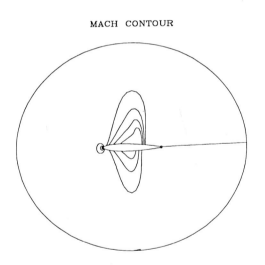

Figure (4)
A Plot of Mach Contours and the Designed Airfoil Shape.

SPLIT COEFFICIENT MATRIX (SCM) METHOD WITH FLOATING SHOCK FITTING FOR TRANSONIC AIRFOILS*

Peter M. Hartwich

ViGYAN, Inc., Hampton, Virginia 23666-1325, U.S.A.
and
NASA Langley Research Center, Hampton, Virginia 23665-5225, U.S.A.

I. Introduction

The Euler equations are a correct mathematical model for inviscid transonic flow. Unlike some potential formulation, they account for the entropy and vorticity generated across shocks. However, inspite of their theoretical shortcomings, potential methods are still widely in use for good reasons, among them computational speed and accuracy. Potential flow solvers have to solve for only one equation in both two and three dimensions. Euler methods need to solve for a nonlinear system of 4 (5) partial differential equations in computing 2-D (3-D) flow. Euler solvers could more than outweigh their economic handicap if they could fully exploit their conceptual advantage in computations of shocked flows. That is difficult to achieve with shock-capturing methods since they have to revert to first-order accuracy across shocks to ensure numerical stability [1]. Using techniques like upwinding and adaptive meshing helps to compensate for that loss in accuracy, but it also adds to the computational expenditure [2]. Alternately, shock fitting methods, which compute shocks as exact solutions using the Rankine-Hugoniot relations, have been reconsidered for Euler computations of transonic flow [3-6]. They are fast [4,6] and also becoming more general [6], a feature usually associated with shock capturing.

This paper briefly describes a recently developed general shock fitting procedure for the 2-D Euler equations and explores its efficiency and accuracy for applications to transonic flows over a NACA 0012 airfoil.

II. Governing Equations

Nonconservative schemes can incorporate upwinding very cheaply since the dependent variables can be chosen for convenience. Three, in a way "natural" choices are a (=speed of sound) and u, v (=Cartesian velocities) for they are required in any upwind scheme. The fourth choice is s(=entropy) to reduce the number and the complexity of the terms in the governing equations.

The two-dimensional, compressible, nonconservative Euler equations are written in general coordinates as

$$\mathbf{Q}_t + \mathbf{A}\mathbf{Q}_\xi + \mathbf{B}\mathbf{Q}_\eta = 0 \quad \text{with} \quad \mathbf{Q} = (a, u, v, s)^T. \tag{1}$$

All quantities are normalized with reference values for pressure, density, and length. Letting $\delta = (\gamma - 1)/2$, where γ is the ratio of the specific heats, a coefficient matrix $\mathbf{C}$ is defined as

$$\mathbf{C} = \begin{pmatrix} \lambda_5 & -\hat{\alpha}\delta\lambda_6 & \hat{\beta}\delta\lambda_6 & a\delta(\lambda_2 - \lambda_5) \\ -\hat{\alpha}\lambda_6/\delta & \hat{\alpha}^2\lambda_5 + \hat{\beta}^2\lambda_2 & \hat{\alpha}\hat{\beta}(\lambda_5 - \lambda_2) & a\hat{\alpha}\lambda_6 \\ -\hat{\beta}\lambda_6/\delta & \hat{\alpha}\hat{\beta}(\lambda_5 - \lambda_2) & \hat{\beta}^2\lambda_5 + \hat{\alpha}^2\lambda_2 & a\hat{\beta}\lambda_6 \\ 0 & 0 & 0 & \lambda_2 \end{pmatrix} \tag{2}$$

The eigenvalues of $\mathbf{C}$ are

$$\lambda_1 = \alpha u + \beta v - a\sqrt{\alpha^2 + \beta^2} \quad \lambda_2 = \lambda_3 = \alpha u + \beta v \quad \lambda_4 = \alpha u + \beta v + a\sqrt{\alpha^2 + \beta^2} \tag{3}$$

They are combined in Eq.(2) for reasons of compactness as $\lambda_5 = (\lambda_1 + \lambda_4)/2$ and $\lambda_6 = (\lambda_1 - \lambda_4)/2$. The metric coefficients in Eq.(2) are normalized as $\hat{\alpha} = \alpha/\sqrt{\alpha^2 + \beta^2}$ and $\hat{\beta} = \beta/\sqrt{\alpha^2 + \beta^2}$. For $\alpha = \xi_x$, and $\beta = \xi_y$, $\mathbf{C} = \mathbf{A}$, and for $\alpha = \eta_x$, and $\beta = \eta_y$, $\mathbf{C} = \mathbf{B}$.

III. Spatial Differencing

The coefficient matrix $\mathbf{C}$ can be split according to the sign of its eigenvalues [7]

$$\mathbf{C} = \mathbf{C}^+ - \mathbf{C}^- \tag{4}$$

with $\mathbf{C}^\pm = \mathbf{C}(\lambda_m^\pm)$, and $\lambda^\pm = (|\lambda_m| \pm \lambda_m)/2$, $(m = 1, 2, 3, 4)$. Upon inserting Eq.(4), Eq.(1) assumes

$$\mathbf{Q}_t - \mathbf{A}^-\mathbf{Q}_\xi + \mathbf{A}^+\mathbf{Q}_\xi - \mathbf{B}^-\mathbf{Q}_\eta + \mathbf{B}^+\mathbf{Q}_\eta = 0 \tag{5}$$

---

* NASA Langley Research Center has supported this work under Contract NAS1-18585.

A semi-discrete finite-difference approximation to Eq.(5) that is numerically stable in the sense of a linear stability analysis as in Steger and Warming [8] reads

$$\mathbf{Q}_t - \mathbf{A}^- \Delta^+_{i+1/2}\mathbf{Q} + \mathbf{A}^+ \Delta^-_{i-1/2}\mathbf{Q} - \mathbf{B}^- \Delta^+_{j+1/2}\mathbf{Q} + \mathbf{B}^+ \Delta^-_{j-1/2}\mathbf{Q} = 0 \qquad (6)$$

with

$$\begin{aligned}
\Delta^+_{i+1/2}\mathbf{Q} &= (3/2 - \phi_1)\Delta_{i+1/2}\mathbf{Q} - (1/2 - \phi_1)\Delta_{i+3/2}\mathbf{Q} \\
\Delta^-_{i-1/2}\mathbf{Q} &= (3/2 - \phi_2)\Delta_{i-1/2}\mathbf{Q} - (1/2 - \phi_2)\Delta_{i-3/2}\mathbf{Q} \\
\Delta^+_{j+1/2}\mathbf{Q} &= (3/2 - \phi_3)\Delta_{j+1/2}\mathbf{Q} - (1/2 - \phi_3)\Delta_{j+3/2}\mathbf{Q} \\
\Delta^-_{j-1/2}\mathbf{Q} &= (3/2 - \phi_4)\Delta_{j-1/2}\mathbf{Q} - (1/2 - \phi_4)\Delta_{j-3/2}\mathbf{Q}
\end{aligned} \qquad (7)$$

and

$$\Delta_{l-1/2}\mathbf{Q} = \mathbf{Q}_l - \mathbf{Q}_{l-1},\ l = i \text{ or } j$$

where $\mathbf{Q}_t$, $\mathbf{A}^{\pm}$, $\mathbf{B}^{\pm}$, and ϕ_k with $k = 1, 2, 3, 4$ are taken at the centroids (i, j). For $\phi_1 = \phi_2 = \phi_3 = \phi_4 = 0$, Eq.(6) gives the semi-discrete baseline scheme which uses second-order, one-sided differences. All mesh points, except those in the neighborhood of a shock wave, are computed with the same baseline scheme. The parameters ϕ_k are used to suppress differences which would otherwise be taken across shocks (for details, see Ref. 6).

IV. Time Differencing

A time-implicit operator containing block-tridiagonal matrices for the 2-D Euler equations is given by

$$[\mathbf{I} - \tau(\mathbf{A}^-\Delta_{i+1/2} - \mathbf{A}^+\Delta_{i-1/2} + \mathbf{B}^-\Delta_{j+1/2} - \mathbf{B}^+\Delta_{j-1/2})]^n \Delta\mathbf{Q}^n = -\tau \cdot RES(\mathbf{Q}^n) \qquad (8)$$

where $\mathbf{I}$=identity matrix, τ= time step, $\Delta\mathbf{Q}^n = \mathbf{Q}^{n+1} - \mathbf{Q}^n$, and n indicates the time level. As before, the Jacobians $\mathbf{A}$ and $\mathbf{B}$ are formed at the centroids (i, j). The algorithm is cast in delta form to compute steady-state solutions that are independent of the time step size. The residual $RES(\mathbf{Q}^n)$ comprises the spatial differences in Eq.(6) evaluated at the n-th time level with the provisos as mentioned in the preceding section.

Using a similarity transformation such as $\mathbf{C} = \mathbf{T}\mathbf{\Lambda}\mathbf{T}^{-1}$ with $\mathbf{\Lambda} = diag(\lambda_m)$ where $diag$ = diagonal matrix, a diagonalized approximate factorization (AF) algorithm [9] for Eq.(8) is written as

$$\{\mathbf{I} - \tau[(\mathbf{\Lambda}^-)^A \Delta_{i+1/2} - (\mathbf{\Lambda}^+)^A \Delta_{i-1/2}]\}^n \Delta\tilde{\tilde{\mathbf{Q}}} = -(\mathbf{T}^{-1})^A \tau \cdot RES(\mathbf{Q}^n) \qquad (9a)$$

$$\{\mathbf{I} - \tau[(\mathbf{\Lambda}^-)^B \Delta_{j+1/2} - (\mathbf{\Lambda}^+)^B \Delta_{j-1/2}]\}^n \Delta\tilde{\mathbf{Q}} = (\mathbf{T}^{-1})^B \mathbf{T}^A \Delta\tilde{\tilde{\mathbf{Q}}} \qquad (9b)$$

$$\mathbf{Q}^{n+1} = \mathbf{Q}^n + \mathbf{T}^B \Delta\tilde{\mathbf{Q}} \qquad (9c)$$

The time-implicit algorithm entailed in Eqs.(9) requires some minor modifications along shock fronts; they are discussed in Ref. 6.

V. Two-Dimensional Shocks

Let A and B be two points on a $\eta = const$ line, on either side of some shock. Furthermore, assume that $U = \hat{\xi}_x u + \hat{\xi}_y v \geq 0$ and that B belongs to the high pressure region. Using ideas from Davis [10] and Morton and Rudgyard [11], the direction cosines of the shock normal in a local orthogonal reference frame are evaluated as

$$\mathbf{N} = \Delta\mathbf{q}/|\Delta\mathbf{q}| \qquad (10)$$

where $\mathbf{N} = (N_1, N_2)^T$ and $\mathbf{q} = (V, V^*)^T$ with

$$\begin{aligned}
N_1 &= \Delta_{i-1/2}V/\sqrt{(\Delta_{i-1/2}V)^2 + (\Delta_{i-1/2}V^*)^2} \\
N_2 &= \Delta_{i-1/2}V^*/\sqrt{(\Delta_{i-1/2}V)^2 + (\Delta_{i-1/2}V^*)^2} \\
V &= \bar{\eta}_x u + \bar{\eta}_y v \quad \text{and} \quad V^* = \bar{\eta}_y u - \bar{\eta}_x v
\end{aligned} \qquad (11)$$

Thus, V is the scaled contravariant velocity component normal to $\eta = const$ lines, and V^* is perpendicular to V and in ξ-direction. The notation $\bar{\eta}_x$ and $\bar{\eta}_y$ indicates that the metric coefficients are computed as averages from the nodal values at i and $i-1$. Thus, the shock orientation as defined in Eq.(10) depends only on the change in the normal velocity component across a shock.

The velocity components normal and tangential to the shock, $\tilde{u}$ and $\tilde{v}$, are computed from

$$\tilde{u} = VN_1 + V^*N_2 \quad \text{and} \quad \tilde{v} = VN_2 - V^*N_1. \tag{12}$$

These quantities together with entropy and speed of sound are used to update the solution at point B using the Rankine-Hugoniot relations:

$$a_B = a_A \frac{\sqrt{(\gamma M^2 - \delta)(1 + \delta M^2)}}{(1+\delta)|M|} \qquad \tilde{v}_B = \tilde{v}_A$$

$$\tilde{u}_B = \tilde{u}_A - a_A \frac{M^2 - 1}{(1+\delta)|M|} \qquad s_B = s_A + \left[ln\frac{\gamma M^2 - \delta}{1+\delta} - \gamma \, ln\frac{(1+\delta)M^2}{1+\delta M^2}\right]/(2\delta\gamma) \tag{13}$$

The symbol M represents the relative shock Mach number:

$$sign(a_B - a_A) \, | \, M \, | = \frac{\tilde{u}_a - W}{a_A} \tag{14}$$

where W is the shock speed. Moretti [12] developed an implicit function relating the shock Mach number M to a shock parameter Σ:

$$\Sigma = \frac{\sqrt{(\gamma M^2 - \delta)(1 + \delta M^2)} + \delta(M^2 - 1)N_2}{(1+\delta)|M|} \tag{15}$$

To merge this approach to computing the shock Mach number M with the present approach to determine the shock orientation, the shock parameter Σ is here defined as

$$\Sigma = (a_B - \delta \Delta_{i-1/2} V^*)/a_A \tag{16}$$

The shock parameter Σ is evaluated at every grid point during the shock detection period. When it exceeds some tolerance Σ_o (typically $1.03 \leq \Sigma_o \leq 1.05$), then the pertaining grid point is accepted as seat of a shock. The corresponding shock Mach number is iteratively evaluated using Eqs. (15) and (16) [6,12].

The final values at B are obtained by assuming B lying right at the shock and applying a linear interpolation between shock front and the second node to the right of the shock. Finally, the shock is moved along the $\eta = const$ line by the increment $\Delta s = (W/N_2)\Delta t$. The calculation of the relative shock Mach number does not change at all for $V^* < 0$. If A belongs to the high pressure side, then the framework established by Eqs.(10) through (16) still holds true, provided the subscripts A and B in Eq.(16) are interchanged. The formulae for a shock crossing a $\xi = const$ line are obtained from interchanging the contravariant velocity components U and V and the definitions for N_1 and N_2 in Eq.(11).

VI. Boundary Conditions

The subsonic farfield boundary conditions are determined from characteristic-based formulations as readily found, for instance, in Ref. 2. The normal velocity along the airfoil contour, which here coincides with an $\eta = const$ line, is set to zero. This condition along with the definitions for V and V^* in Eq.(11) is used to recompute the Cartesian velocity components along the profile after they are updated via linear extrapolation. Entropy on the airfoil is computed from simple extrapolation. The speed of sound along the profile is iteratively calculated from the momentum equations by simplifying $\eta_x(\xi$-momentum$)+\eta_y(\eta$-momentum$)$:

$$-2\delta U(\eta_x u_\xi + \eta_y v_\xi) = (\xi_x \eta_x + \xi_y \eta_y)(a^2)_\xi + (\eta_x^2 + \eta_y^2)(a^2)_\eta. \tag{17}$$

VII. Results

Computed flow-field results are presented for three cases:

	Case 1	Case 2	Case 3
M_∞	0.63	0.80	0.80
α	2.00	0.00	1.25

These results were computed on a coarse and on a standard C-type mesh. The standard mesh with 161×33 grid points has been closely patterned after that used by Pulliam et al. [13] for computing subcritical shockless flow. A coarser grid has been derived by dropping every other grid point in each coordinate direction. The computational study by Pulliam et al. has been selected because, with its careful documentation, it allows for a well-defined comparison between (i) a quite accurate shock-capturing method whose accuracy has been further enhanced by using adapted meshes and (ii) the present method which is used only on grids for subcritical flow.

The comparison of the surface pressure distribution for the first test case (Fig. 1) indicates a good overall agreement between the present results on both grids and the results from Ref. 13 on the standard grid. Starting from freestream conditions, it takes 500 (1000) iterations to reduce the L_2-norm of all residuals by seven orders of magnitude on the coarse (fine) grid; asymptotic values for lift are established in less than 300 iterations (Fig. 2). Computing these results on a single processor of a Cray Y-MP requires about 4.8 μsecs/iteration/node.

In the second case, shocks develop at $x/c \approx 0.5$, which coincides with the coarsest chordwise resolution of the "subcritical" grids. The spatial step sixe $\Delta x/c$ in the neighborhood of the fitted shocks is about twice (161 × 33 grid) and four times (81 × 17 grid) as large as in the reference solution which was computed on an adapted 161 × 33 grid with 16 radial lines clustered around each shock with a regular spacing of $\Delta x/c = 0.01$. Even with that handicap, the surface pressure distributions and the shock locations as predicted with the present method on both grids agree quite well with the shock-capturing results on the adapted grid (Fig. 3). (Note that despite their crisp appearance, the captured shocks are resolved over four mesh intervals.) The surface entropy distributions in Fig. 4 demonstrate that (i) entropy is only generated across shocks, (ii) away from shocks entropy is merely convected and thus it remains at a constant level, and (iii) the fitted shocks are completely devoid of any spurious oscillations. As indicated by the convergence summary in Fig. 5, the L_2-norm of all residuals is reduced at a slightly faster rate than in the computations for the first case. The shock-fitting procedures add 0.5 μsecs/iteration/node to the aforementioned CPU time/node/iteration.

The reference solution for the third test case was computed on a 161 × 33 grid with 11 radial lines clustered around the upper (lower) shock with a constant step size of $\Delta x/c = 0.005 (= 0.01)$. The surface pressures as computed on the coarse and standard "subcritical" grid agree amazingly well with each other (Fig. 6). In line with other computations involving shock fitting [4,5], the present method predicts the location of the upper shock further downstream than the reference solution. Unlike previous applications of shock-fitting methods to this test case [4,5], the present procedure is capable even fitting the very weak lower shock. The surface entropy distribution demonstrates, as in the second test case, the accuracy of the present solutions. (Fig. 7). Asymptotic lift is established in about 200 (450) iterations on the coarse (fine) grid; the L_2-norm of all residuals is reduced by more than 6 orders of magnitude in about 500 (coarse grid) to roughly 800 (standard grid) iterations (Fig. 8). The computing time (about 5.1 μsecs/iteration/node) lies between the respective values for the first two test cases.

VIII. Concluding Remarks

The present general shock-fitting procedure for solving the 2-D Euler equations matches if not surpasses the computational efficiency of shock capturing schemes. Moreover, it produces solutions for both shocked and shockless flows which exhibit unusually little grid sensitivity. This brings down costs in two ways: savings in the total number of grid points required for a proper resolution of the flow translate into immediate shorter computing times, and reducing the need to generate adapted meshes improves turnaround.

IX. References

[1] Pulliam, T.H., *AIAA J.*, **24** (1986), 1931-1940.
[2] Pulliam, T.H., and Steger, J.L., AIAA Paper 85-0360 (1985).
[3] Albone, C.M.,*Num. Anal. Fl. Dyn. II*, Oxford Univ. Press (1986), Oxford, 427-437.
[4] Dadone, A. and Moretti, G.,*AIAA J.*, **26**, (1988), 409-424.
[5] Morton, K.W. and Paisley, M.F.,*J. Comp. Phys.*, **80**, (1989), 168-203.
[6] Hartwich, P.M., AIAA Paper 90-0108 (1990). (to appear in *AIAA Journal*)
[7] Chakravarthy, S.R., Anderson, D.A., and Salas, M.D., AIAA Paper 80-0268 (1980).
[8] Steger, J.L., and Warming, R.F., *J. Comp. Phys.*, **40**, (1981), 263-293.
[9] Pulliam, T.H. and Chaussee, D.S., *J. Comp. Phys.*, **39**, (1981), 347-363.
[10] Davis, S.F., *J. Comp. Phys.*, **56** (1984), 65-92.
[11] Morton, K.W., and Rudgyard, M.A.,*Lect. Notes Phys.*, **323**, Springer-Verlag (1989), New York, 424-428.
[12] Moretti, G., *Computers & Fluids*, **15** (1987), 59-75.
[13] Pulliam, T.H., Jespersen, D.C., and Childs, R.E., AIAA Paper 83-0344 (1983).

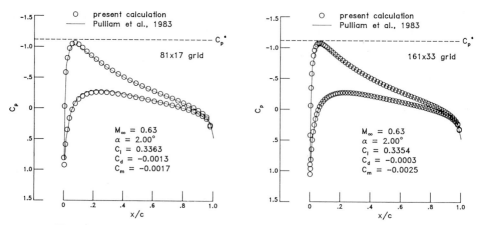

Figure 1. Surface pressure distributions for subcritical flow over a NACA 0012 airfoil.

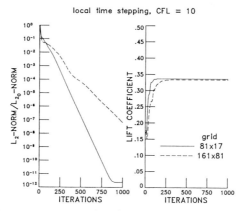

Figure 2. Convergence summary for subcritical flows past a NACA 0012 airfoil.

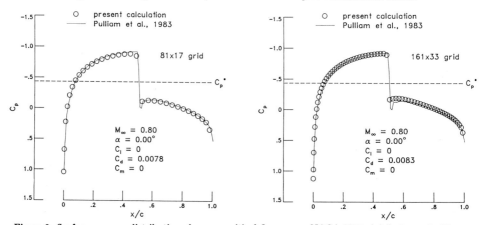

Figure 3. Surface pressure distributions for supercritical flow over a NACA 0012 airfoil at zero incidence.

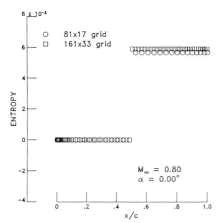

Figure 4. Surface entropy distributions for supercritical flow over a NACA 0012 airfoil at zero incidence.

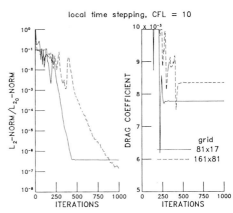

Figure 5. Convergence summary for supercritical flows past a NACA 0012 airfoil at zero incidence.

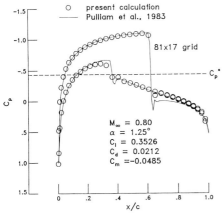
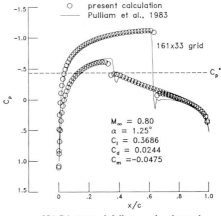

Figure 6. Surface pressure distributions for supercritical flow over a NACA 0012 airfoil at angle of attack.

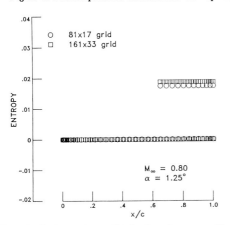

Figure 7. Surface entropy distributions for supercritical flow over a NACA 0012 airfoil at angle of attack.

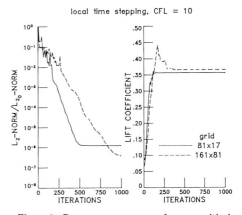

Figure 8. Convergence summary for supercritical flows past a NACA 0012 airfoil at angle of attack.

Supersonic Viscous Flow Calculations
For Axisymmetric Configurations

P.K. Khosla, T.E. Liang and S.G. Rubin
University of Cincinnati
Department of Aerospace Engineering
and Engineering Mechanics

Introduction

A pressure based flux-vector splitting technique has been applied for the computation of supersonic viscous interacting flows over axisymmetric configurations. The formulation is based on a reduced form of the Navier-Stokes equations, termed RNS, that accurately models the physics of large Reynolds number flows and leads to efficient computational procedures. The RNS development represents an enhancement of parabolized or PNS methodology, and is applicable to flows with strong viscous-inviscid interaction, including flow reversal and sharp shock wave/vortical discontinuity capturing.

A 'PNS' approximation was originally introduced for the computation of cold wall hypersonic flows, for which the effect of the axial pressure gradient, p_x, is negligible and initial value techniques are applicable, e.g. see refs. [1,2]. Subsequently [2,3], the PNS methodology was generalized to include p_x; however, it was shown, [2,4,6], that these initial value solvers are ill-posed and result in departure or exponentially growing solutions. In a series of papers, [7-10], the subsonic (elliptic) character of the full RNS system has been clarified and numerical methods that are free of the departure phenomena for all step-sizes have been developed. These procedures, which are considered for the calculations presented herein, appropriately split the acoustic fluxes into positive (initial value or 'PNS') and negative (boundary value or 'RNS') components, [10].

The present investigation considers the application of the RNS flux-split formulation for supersonic flow over several axisymmetric configurations. The effects of shock-boundary layer and shock-shock interaction, and the accurate resolution of regions of flow reversal are the primary results to be presented.

RNS Formulation and Discretization

The steady state form of the reduced Navier-Stokes or RNS equations, given here in cylindrical coordinates for the axisymmetric geometries to be considered, is as follows:

Continuity: $(\rho r u)_x + (\rho r v)_r = 0$

x momentum: $(\rho r u^2)_x + (\rho r u v)_r + r p_x = \frac{1}{Re}(r\mu u_r)_r$

r momentum: $(\rho r u v)_x + (\rho r v^2)_r + r p_r = 0$

energy: $\rho r(u H_x + v H_r) = \dfrac{1}{P_r Re}(r u H_r)_r + \dfrac{\gamma-1}{Re} M_\infty^2 (1 - \dfrac{1}{P_r})[u(r\mu u_r)_r + r\mu u_r^2]$

state: $p = \rho T / \gamma M_\infty^2$; $H = T + \dfrac{\gamma-1}{2} M_\infty^2 (u^2 + v^2)$

Where (u,v) are the normalized axial (x) and radial (r) velocities respectively, p the pressure, ρ the density, H the stagnation enthalpy and a the speed of sound. Re and M_∞ are the Reynolds number and Mach number, respectively. A coordinate transformation is applied such that the outer fitted shock and the body surface become the computational boundaries. A conical flow approximation is used to generate leading edge conditions; at the outflow, the (subsonic) negative acoustic flux is neglected in the boundary layer. All axial convective derivatives are discretized using two-point upwind differences. Radial convective and diffusion terms are approximated with two-point (trapezoidal) or three-point (central) differences, for details see refs. [8-13]. The flux vector splitting [10] leads to upwinding for all convective terms and the following representation for p_x:

$$p_x = \omega_{i-1/2}(p_i - p_{i-1})/\Delta x + (1 - \omega_{i+1/2})(p_{i+1} - p_i)/\Delta x = (\bar{p}_i - \bar{p}_{i-1})/\Delta x$$

where $\omega \leq \omega_M = \min[1,(u/a)^2]$ and the mid cell evaluation for $\omega = \omega_M$ leads to capturing of sharp shocks over three grid points. In terms of the grid point pressure value $\bar{p}_i = \omega_{i+1/2} p_i + (1 - \omega_{i+1/2}) p_{i+1}$ the momentum equations are in full conservation form, [12].

Although direct solvers have been applied to the flux split system [12], a marching procedure that is imbedded in a global pressure relaxation technique is considered to be more efficient for the class of supersonic flow problems considered here. A fully implicit coupled solver is applied at each axial location. With supersonic freestreams, the convergence rate of this relaxation method is a function of the degree of viscous-inviscid interaction, which includes flow reversal and shock formation, the radial dimension of the subsonic zone, and the axial mesh size, see refs. [7-10]. The application of multigrid (semi or full coarsening) methods [11] can accelerate the fine mesh calculations, so that the pressure relaxation procedure will converge in relatively few global sweeps, e.g., 5-15.

Solutions for several axisymmetric configurations, with shock boundary-layer interaction and regions of flow reversal, are computed. All bow shock waves are fitted, while imbedded shocks, arising from geometric and/or flow interaction, are captured. Artificial viscosity is not explicitly added; however, the numerical viscosity associated with the flux-split discretization is sufficient to capture smooth sharp shocks. This diffusion effect is minimized on fine grids.

Results

Three different axisymmetric configurations are considered in this paper. Flow past a cone-cylinder boattail, an aircraft forebody and a cone cylinder flare are investigated for supersonic Mach numbers, $M_\infty \geq 3$.

Figures (1a-c) depict typical laminar flow results for the supersonic flow over a cone-cylinder-boattail configuration. The effect of M_∞ and Re variation is shown. For $M_\infty = 3$, a region of flow recirculation occurs on the boattail afterbody. The flow reversal diminishes as M_∞ increases; for $M_\infty = 5$, 10, the flow is fully attached.

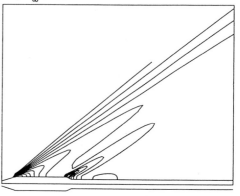

Fig. 1a. Cone-Cylinder-Boattail: Pressure Contours ; $M_\infty = 3$, Re = 7.5×10^4

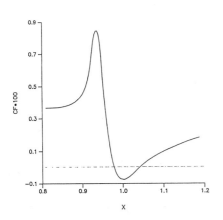

Fig. 1b. Cone-Cylinder-Boattail: Skin Friction; $M_\infty = 3$, Re = 7.5×10^4

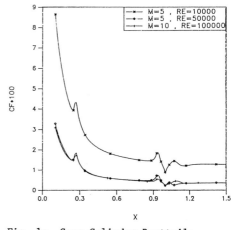

Fig. 1c. Cone-Cylinder-Boattail: Effect of Reynolds Number and Mach Number

Figures (2a-b) depict the results for flow over an axisymmetric aircraft forebody at several Mach numbers. Near the canopy an embedded shock which intersects the bow shock is captured. This result is typical of a number of shock-shock interactions that have been accurately computed with the RNS flux split formulation. From the pressure distribution, it is seen that there is a strong interaction over the entire geometry. A reversed flow region, due to the shock boundary layer interaction, occurs near the canopy. Once again, the size of the recirculation bubble decreases as the Mach number increases. For $M_\infty > 7$, the flow is fully attached.

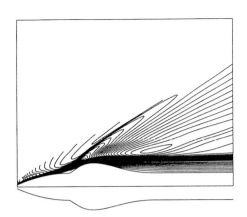 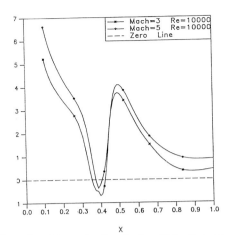

Fig. 2a. Aircraft Forebody: Isomach Contours; $M_\infty = 3$, $Re = 10^4$

Fig. 2b. Aircraft Forebody: Skin Friction

In Figures (3a-b), the solution for flow past a cone-cylinder-flare is depicted. Cones with 10° and 15° flare angles have been considered. A sharp oblique shock at the compression flare corner is captured. For Re = 10,000, the flow exhibits a small separation region only at the lower Mach number.

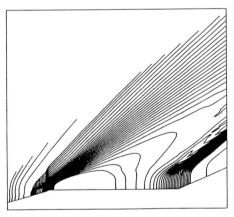 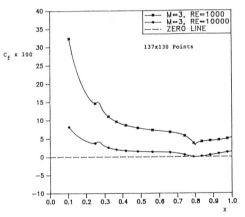

Fig. 3a. Cone-Cylinder-Flare: Pressure Contours; $M_\infty = 3$, $Re = 10^4$

Fig. 3b Cone-Cylinder-Flare: Effect of Reynolds Number on Skin Friction Coefficient

Turbulent flow computations have also been carried out for $Re=6\times10^6$. Cebeci-Smith and Baldwin-Lomax turbulence models have been considered in the present study. In Figures (4a-c), pressure and skin friction solutions for flow past a cone-cylinder boattail configuration are shown. Although the coefficient of pressure is almost identical for the two turbulence models, as is the case in many calculations, the skin friction is quite different, e.g., the large variation near the attachment point is evident. This difference, which appears for flows with recirculation, may be attributed to an inaccuracy in the estimate of the boundary-

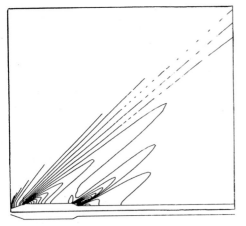

Fig. 4a. Cone-Cylinder-Boattail: Pressure Contours; Turbulent, $M_\infty = 3$, $Re = 6 \times 10^6$

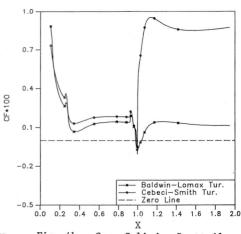

Fig. 4b. Cone-Cylinder-Boattail: Skin Friction; $M_\infty = 3$, $Re = 6 \times 10^6$

layer thickness required for the Cebeci-Smith model. These solutions are indicative of the sensitivity of the flow structure and skin friction, and the insensitivity of the pressure, to the turbulence model. Moreover, it is seen that small changes in pressure can lead to large variations in skin friction and flow structure. Such sensitivity has previously been observed near reattachment points where a form of laminar flow breakdown [11,13] indicative of incipient transition, occurs. This behavior has also been confirmed by full Navier-Stokes calculations.

For the calculations presented herein the computational grid ranges from 100x100 to 130x200 points in the axial and radial directions. Since the computer storage requirements are quite modest for this relaxation procedure, computations have generally been performed on a personal or a minicomputer. A typical rate of convergence plot is shown for the aircraft forebody in Figure 5. As expected, the convergence rate increases with mach number. For an engineering level of accuracy, e.g. maximum error of 10^{-4} - 10^{-5}, 15-40 global iterations are satisfactory. The results obtained herein reinforce the effectiveness of the RNS procedure as a tool for evaluating large Re complex flow physics at much lower costs than with full Navier-Stokes solvers.

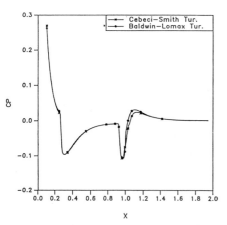

Fig. 4c. Cone-Cylinder-Boattail: Coefficient of Pressure, $M_\infty = 3$, $Re = 6 \times 10^6$

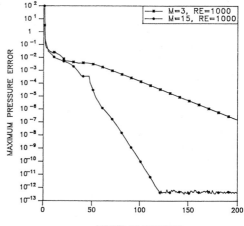

Fig. 5. Aircraft Forebody: Convergence History

Acknowledgement

This work has been supported by the Air Force Office of Scientific Research (L. Sakell, Technical Monitor) under Grant No. AFOSR-90-0096. The large scale computations were performed on the Cray-YMP at the Ohio Supercomputer Center.

References

1. Rudman, S. and Rubin, S.G., AIAA J., 6, 10, pp. 1883-1890, 1968; see also, Rubin, S.G. et al., AIAA J., 7, 9, pp. 1744-1751, 1969.

2. Lin, T.C. and Rubin, S.G., Computers and Fluids, 1, 1, pp. 37-58, 1973.

3. Cheng, H.K., Chen, S.Y., Mobley, R. and Huber, C.R., Rand Corp. RM 6193-PR, 1970.

4. Davis, R.T., Barnett, M. and Rakich, J.V., Computers and Fluids, 14, 3, pp. 197-224, 1986.

5. Lubard, S. and Helliwell, W.S., AIAA J., 12, 7, 1974.

6. Vigneron, Y.C., Rakich, J.V. and Tannehill, J.C., AIAA Paper No. 78-1137, and NASA TM 78500, 1978.

7. Rubin, S.G., Numerical Aspects of Physical Aspects of Aerodynamic Flows, T. Cebeci Ed., pp. 171-186, Springer-Verlag, Berlin, 1981.

8. Rubin, S.G. and Reddy, Computers and Fluids, vol. 11, pp. 281-306, 1983.

9. Khosla, P.K. and Lai, H.T., AIAA Paper No. 84-0458, 1984. Computers and Fluids, 16, 1988.

10. Rubin, S.G., Computers and Fluids, 16, 4, pp. 485-90, 1988.

11. Rubin, S.G. and Himansu, A., Int'l J. of Num. Methods in Fluid Dyn., 9, pp. 1395-1411, 1989.

12. Pordal, H.S. Khosla, P.K. and Rubin, S.G., AIAA Paper No. 90-0585, 1990.

13. Liang, T.E., Ph.D. Dissertation, University of Cincinnati, 1989.

Visualization Method for Computational Fluid Dynamics with Emphasis on the Comparison with Experiments

Yoshiaki Tamura, Kozo Fujii

The Institute of Space and Astronautical Science,
Yoshinodai 3-1-1, Sagamihara, Kanagawa, 229 Japan

1 Introduction

Postprocess of flow simulation is important as well as flow simulation itself and visualizing the computed result as a method of data processing is one of the useful postprocess. To help understanding the flow field, the capability to visualize various flow functions with various ways is required. Interactiveness and simplicity in use are also important[1]-[4].

Another aspect of the postprocess is to illuminate the flow physics. Streamline tracing is an example. There may be another methods and the choice of the methods to extract the feature of the flow fields depends on the problem. Bitz and Zabusky proposed a new concept of a postprocessor for this purpose[5].

Since the reliability of computed results is one of the recent topics in the CFD, validating the computer simulations can also be an aspect of the postprocess of the CFD. Digital data such as forces or pressure distributions on airfoils can be directly compared. When the entire flow field is to be compared, visualized images are used because there are few methods except visualizations to show entire flow field in experiments. The ordinary visualized images such as density contour plots are often compared with schlieren photographs and other kinds of pictures taken in experiments. Although they show similar images, this comparison is not fair and accurate because the process to generate images from computed results does not follow the principles of visualizations in experiments. To make the comparison fair and accurate, the methods to simulate visualization processes of experiments should be considered.

In the present paper, five visualization methods are simulated computationally and some of the obtained images are compared with experiments.

2 Methods to Simulate Experiments

2.1 Schlieren Photograph

One optical visualization method is schlieren photograph. In two-dimensional flow, the intensity of each point of schlieren photograph is proportional to the density gradient perpendicular to the knife edge because the deflection rate of ray is proportional to the density gradient[6]. When the knife edge is set to be perpendicular to the x axis, for instance, the intensity I is written as

$$I \propto \frac{\partial \rho}{\partial x} \qquad (1)$$

Figure 1 shows an example of the simulated color schlieren photograph. The flow field is two-dimensional supersonic intake($M_\infty = 3$). Figure 1 (a) is color schlieren photograph of the experiment and Figure 1 (b) shows the simulated schlieren photograph using the computed result. Figure 1 (c) is the density contour surface plots and the difference between (b) and (c) is obvious.

In three-dimensional flow, since the magnitude of density gradient varies along a ray of light, the intensity is proportional to the total deflection rate of the ray which is equal to the integration of the density gradient along the ray.

$$I \propto \int_{ray} \frac{\partial \rho}{\partial x} dl \qquad (2)$$

In this case, a numerical integration is necessary and it is done by the following manner. At first, the structured grid that we usually use is divided into tetrahedra to simplify the problem. Then tetrahedra that the traced ray of light goes through are searched.

Since the interpolation inside a tetrahedron can be linear, the integration within the tetrahedron I_t can simply be obtained as Eq. (3),

$$I_t = \frac{1}{2}\left(\left(\frac{\partial \rho}{\partial x}\right)_{in} + \left(\frac{\partial \rho}{\partial x}\right)_{out}\right) \times l \qquad (3)$$

where $()_{in}$ denotes the point where the ray comes in, $()_{out}$ denotes the point where the ray goes out and l is the length between these points. This integration is carried out in every tetrahedron that the ray goes through and each integrated value is summed up.

An example of a three-dimensional flow field is shown in Fig. 2. A hemisphere-cylinder is placed in a supersonic flow ($M_\infty = 1.96$) with 0 degree angle-of-attack. Figure 2 (a) is a schlieren photograph of experiment. The simulated three-dimensional schlieren photograph image is shown in Fig. 2 (b) and the simulated two-dimensional schlieren photograph image within a plane of symmetry is shown in Fig. 2 (c) for comparison. No density gradients are observed in the area between the shock wave and the hemisphere-cylinder in two-dimensional schlieren photograph. Since the rays of light through that area experience density gradients when crossing the shock waves, that area has a different color from the uniform flow region both in the experiment and in three-dimensional schlieren photograph. This effect may become important in the case of a more complex flow field.

2.2 Shadowgraph

Shadowgraphs in experiments are obtained in the similar manner as schlieren photographs, but without knife edges. The intensity is, now, proportional to the gradient of the deflection rate on the screen because the concentration and rarefaction of the rays occur according to the relative deflection rate of the rays. The screen does not become bright nor dark when the deflection rates are constant[6]. In two-dimensional flows, it can be written as,

$$I \propto \frac{\partial^2 \rho}{\partial x^2} + \frac{\partial^2 \rho}{\partial y^2} \qquad (4)$$

Figure 3 (a) shows the simulated shadowgraph pattern of the previous intake simulation.

In three-dimensional flow, it is not the integration of Eq. (4) but the gradient of the integration of $grad\,\rho$ in the directions perpendicular to the direction of the ray of the light. So Equations (1) and (2) are substituted by Eqs. (5)-(7).

$$I \propto grad\left(\int_{ray} grad\,\rho\, dl\right), \quad grad = \left(\frac{\partial}{\partial x}, \frac{\partial}{\partial z}\right) \qquad (5)$$

$$\vec{I_t} = \frac{1}{2}(grad\,\rho_{in} + grad\,\rho_{out}) \times l \qquad (6)$$

$$I = grad(\sum \vec{I_t}) \qquad (7)$$

Unlike schlieren photographs, the screen is not placed at the focal plane and the image is more or less distorted in actual experiments. This effect is not considered here.

2.3 Interferogram

Interferogram shows the light and shade pattern according to the difference of the light-path length that goes through the test section and the light-path length that goes through the reference area[6]. Since the light-path length depends on the density, the light and shade pattern corresponds to the difference of the density between the test section and the reference area. In two-dimensional flow, the numbers of drift of the light and shade stripe N is proportional to the difference of the density.

$$N \propto \Delta \rho = \rho - \rho_{ref} \qquad (8)$$

where ρ_{ref} is the density in the reference area. Figure 3 (b) shows the example of two-dimensional interferogram pattern of the intake.

In three-dimensional flow, N is proportional to the integration of the difference of the density that can be written as,

$$N \propto \int_{ray} \Delta \rho\, dl \qquad (9)$$

These three methods, especially two-dimensional simulations, have another advantage to ordinary visualization methods. Computed schlieren photograph, for example, can show compressions and expansions much more clearly than density contour surface plots and it makes understanding the flow fields easier.

2.4 Off-Surface Streamlines

Off-surface streamlines are often compared with particle path traces in experiments. Strictly speaking, off-surface streamlines are not exactly the same as particle path traces because particles have their own volumes and specific gravities and the particles and the air may not always be in an equilibrium state. However the difference between off-surface streamlines and particle path traces seems to be practically small and off-surface streamlines may be sufficient for qualitative comparison.

Three-dimensional streamlines are usually obtained by integration of the velocity vectors and integrations generally accumulate errors. For the visualizations, errors that are not recognized on the pictures are allowed. Errors of the streamlines, however, could exceed this allowance[7]. Figures 4 (a) and 4 (b) show the off-surface streamlines over a double-delta wing. This is the case of the vortex interaction ($\alpha = 12°$). The streamlines are plotted with the same computational result but with different integration method and they show so different pictures that we might deduce wrong interpretation of the flow characteristics.

Following three items are important for streamline tracing.
- Integration domain (physical/computational)
- Method to interpolate velocities and coordinates
- Integration method and integration step

Among them, the integration method and the integration step may be most influential to the accuracy. Two methods are examined here. The first integration method is sort of "Crank-Nicolson-implicit" method, that can be written as,

$$x(t + \Delta t) = x(t) + \tfrac{1}{2}\{u(t) + u(t + \Delta t)\}\Delta t \tag{10}$$

in physical domain. Figure 5 (a) shows the streamlines using this method. The other method is simple Euler-explicit-like that can be written as,

$$x(t + \Delta t) = x(t) + u(t)\Delta t \tag{11}$$

The streamlines in Fig. 5 (b) are plotted with this explicit method using the same integration step as Fig. 5 (a).

Because of the nature of numerical integrations, the integration becomes more accurate as the integration step becomes smaller. Figure 5 (c) shows the streamlines integrated by the Euler-explicit-like method with 1/20 integration step of the previous two examples and that gives much better results. This is only an example but the followings are generally concluded.
- Resultant streamline is (basically) not dependent on the way of integration if the integration step is small enough.
- The adequate integration step much differs depending on the integration method.

2.5 Oil-Flow Pattern

Oil-flow pattern is not a near-surface streamline because the oil has its own viscosity and thickness. It is not easy to estimate these effects but still necessary when we want to compare experimental photos with computed results, even though oil-flow may not give pure physical information than near-surface streamlines because of these effects.

Figure 5 (a) shows the near-surface streamlines on the double-delta wing and 5 (b) shows the oil-flow pattern in experiment. These pictures do not match, and as well as the positions of the separation lines, the pattern near the separation line is quite different. Although main reason for the discrepancy between the computed result and the experiment could be due to the transition to turbulent flow and the grid resolution, the effect of the oil viscosity may also contribute. Flow over the oil seems not to be able to turn so suddenly because of the oil viscosity or oil may have some effect of smoothing the velocities. Based on this idea, we correct the velocity by adding the second derivatives of the velocity to the velocity and draw near-surface streamline with it. Figure 6 (c) shows an example of near-surface streamline using the corrected velocity. The pattern becomes closer to the experiment. This is only a trial and we have no intention to say this is the true computed oil-flow pattern but at least we have to do something like this to compare computed results with the experiments of oil-flow.

3 Conclusions

The methods to simulate processes in experimental visualizations are important to compare the computed results with experimental pictures for the validation.

These methods, especially optical ones such as schlieren photograph, have another importance. In this paper, they were discussed only from an engineering point of view, that is a comparison of computations with experiments. Mathematically speaking, however, these methods are considered as ones that show certain functions of density, for instance second derivatives of density in the case of the shadowgraph. Since we have complete information about the flow field in computation, derivatives of any other flow functions can also be plotted. This kind of mathematical process may have possibility to illuminate features of flow fields.

Recent graphics workstations offer an environment to build up postprocessors of the CFD. There seems to be a tendency, however, to generate nice and beautiful pictures using the advanced computer graphics techniques without considering flow physics. The real postprocess is not only to visualize computed flow field but also to "cook" the computed flow field to illuminate the flow characteristics by means of mathematical approach such as integration and differentiation. To find out how to cook the computed results, we have to have good physical insight into the flow field.

References

1. P. G. Buning, J. L. Steger: AIAA Paper 85-1507-CP (1985)
2. T. Lasinski, et al.: AIAA Paper 87-1180 (1987)
3. D. L. Modiano, et al.: AIAA Paper 89-3138 (1989)
4. Y. Tamura, K. Fujii: AIAA Paper 90-3031 (1990)
5. F. J. Bitz, N. J. Zabusky: Computers in Phys. (1989)
6. H. W. Liepmann, A. Roshko: Element of Gas Dynamics, John Wiley (1956)
7. E. M. Murman, K. G. Powell: AIAA J. Vol. 27, No. 7, pp 982-984 (1989)

(a) experimental color schlieren photograph

(b) computer simulated schlieren photograph

(c) density contour plots

Figure 1   Two-dimensional supersonic intake

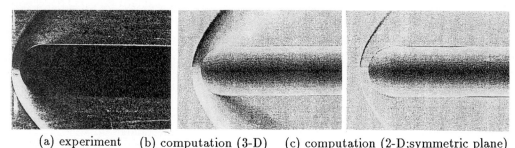

(a) experiment   (b) computation (3-D)   (c) computation (2-D;symmetric plane)

Figure 2   Schlieren photograph of hemisphere cylinder in supersonic flow

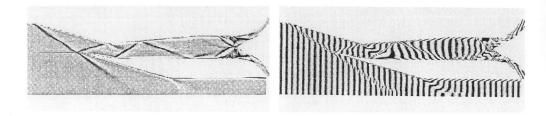

(a) shadowgraph   (b) interferogram

Figure 3   Computer simulated shadowgraph and interferogram pattern

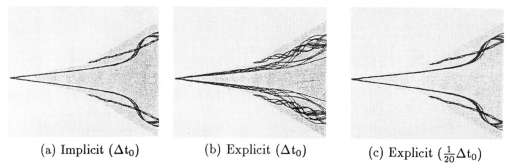

(a) Implicit (Δt_0)   (b) Explicit (Δt_0)   (c) Explicit ($\frac{1}{20}\Delta t_0$)

Figure 4   Off-surface streamline over double-delta wing

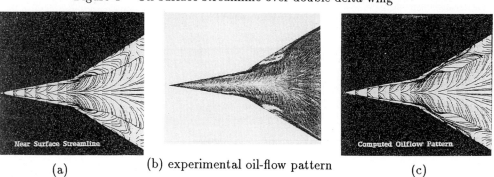

(a)   (b) experimental oil-flow pattern   (c)

Figure 5   Near-surface streamlines and oil-flow pattern

Navier-Stokes Calculations of Hypersonic Flow Configurations with Large Separation by an Implicit Non-Centered Method

by C. Marmignon, H. Hollanders† and F. Coquel.
ONERA, Chatillon, France.
† Presently, Aérospatiale, Les Mureaux, France.

Introduction

The recent space program HERMES has given a large impulse to European research in the computational aerothermodynamics of re-entry vehicles. The three-dimensional hypersonic flows over a space vehicle are governed by strong inviscid-viscous interacting phenomena (shock-boundary layer interaction with separation, horse-shoe vortical flow,...) which play an important role in the prediction of the thermal loading and of aerodynamic coefficients for flight mechanics. Such flows can be numerically predicted by solving the time-dependent Navier-Stokes equations. Moreover, in a first step, the perfect gas assumption can be done in order to predict the general behaviour of the flow, the laminar regime is assumed for an estimate of the heat flux at the wall.

The approach we have developed here is based on the solution of the complete time-dependent Navier-Stokes equations by an implicit non-centered finite-volume method first presented in [1]. In this method, the viscous fluxes are evaluated by using a space centered scheme similar to the Crank-Nicolson one. The inviscid part of the scheme is a finite volume formulation of the upwind scheme of Harten-Yee [2].

In [1], the validation of the method has been performed on 2D laminar hypersonic flows over ramp, blunt body and behind sphere-cones, and first 3D calculations of the flow over sphere-cone at incidence were also performed and compared with experimental data. Here, the method is applied to the calculation of the hypersonic laminar flows over a 2D ramp, over a double ellipse and a double ellipsoid at angles of attack of 0° and 30°. Comparison of results with experimental data are shown for the 3D flow over the double ellipsoid at 30° of incidence [3].

Survey of the method

The time-dependent Navier-Stokes equations for two and three-dimensional flows are solved in curvilinear meshes by means of an implicit method through a finite-volume approach. The method is first-order accurate in time, second-order accurate in space in regions of smoothness (only first-order elsewhere).

The 2D and 3D methods have been formulated in cartesian coordinates and are able to compute flows over general geometries. They are written in a delta formulation and a linearized conservative implicit form is used. Moreover, the method is of the cell-centered type since fluxes at the boundaries are used to prescribe boundary conditions.

Treatment of the inviscid flux by Harten-Yee's scheme. The present formulation of the inviscid scheme is deduced from the finite-difference discretization of the Euler equations written in arbitrary curvilinear coordinates. The metric based on the coordinate transformation is evaluated by means of control cell elements. The flux corrections of Yee's scheme are performed with limiters operating on characteristic variables in the normal direction to the cell side. The stability of the method is enhanced by using the entropy correction of Harten [2] with an entropy parameter δ depending on space and time[1].

Treatment of the viscous fluxes. The viscous fluxes are evaluated by using a space centered scheme similar to the Crank-Nicolson one where the velocity and energy space derivatives are calculated at the center of cell sides with formulae involving 6 points in 2D and 9 points in 3D.

At the implicit step of the method, the complete 2D (3D) operator is factorized by means of the ADI technique, and only 4x4 (respectively 5x5) block tridiagonal matrix systems are to be solved in each mesh direction. When the 3D mesh generation uses cylindrical coordinates, for computing the whole flowfield over a body, periodic boundary conditions in the azimuthal mesh direction appear. As a result, in this mesh direction, the block tridiagonal structure of the linear system is replaced by a cyclic one. An extended LU factorization technique is used for solving the corresponding linear system.

Results

The first application of the 2D method implemented on a CRAY-XMP, concerns the laminar flow over 15° wedge. Flow conditions for this case are:
- freestream Mach number M_0=5.0,
- freestream Reynolds number $R_{o,L}$=1500000 based on the distance L from the leading edge to the beginning of the ramp,
- freestream static temperature T_0=80K, wall temperature T_W=288K.

The computed Mach number is displayed in Fig.1. The flow exhibits a double recirculation (Fig.2).

The 2D and 3D method has been also implemented on a CRAY-2, in order to compute the flow over a double ellipse and a double ellipsoid for a free stream Mach number $M=8.15$, Reynolds number $Re=1.67E+7$ per meter and static temperature $T=56K$, and a wall temperature $Tw=288K$. The 2D and 3D flows have been investigated for two angles of attack $\alpha=0°$ and $\alpha=30°$. Results are shown in Fig. 3 to Fig. 8.

The 2D calculations have been performed with a 138x70 grid and convergence corresponds to a 6 order decrease of the maximal value of residues. Results presented in Fig.3 show the iso-Mach lines for $\alpha=0°$.

The 3D calculations have been performed with a grid made of 42 meridian planes, each of them using 64x64 mesh points for $\alpha=0°$, and a 79x61 mesh points for $\alpha=30°$ (extended domain in the streamwise direction). Calculations have been stopped after a decrease of the maximal residues of six orders of magnitude. Fig. 4 shows the field of isomach lines for $\alpha=0°$ in the symmetry plane. Fig. 5 gives the corresponding lines for $\alpha=30°$. In both configurations, the results predict a large separated region in front of the canopy, which induces a horse-shoe vortical flow structure (see Fig.6). C_p and Stanton wall distributions in the symmetry plane are compared to experimental data [3]. Fig.7 indicates a good agreement with experimental data for the C_p wall distribution. The numerical Stanton wall distributions plotted in Fig.8 correspond to two computations performed with two strategies of the entropy correction parameter: namely a fixed parameter set to $\delta=0.4$ and a varying in space parameter decreasing from $\delta=0.4$ to $\delta=0.01$ at the wall. The best agreement is surprisingly achieved with the fixed δ, while it was the inverse in previous calculations on simpler geometries [1]. So this phenomenon has not been yet explained and needs further studies. Let us underline that the wall C_p numbers were the same for the two strategies.

Conclusion

A 2D and 3D implicit finite volume Navier-Stokes solver has been developed for the prediction of hypersonic laminar flows over forebody shuttle-like geometries, and has shown its capability to capture strong shock waves and solve problems with large separation. Though it gives a rather good agreement with experimental data, some discrepancies remain especially with heat flux predictions. Attention must be paid on the numerical dissipation compared to the physical one.

References

[1] **H. Hollanders and C. Marmignon**: "Navier-Stokes high speed flow calculations by an implicit non-centered method", AIAA-89-0282, 1989.
[2] **H. C. Yee**: "Upwind and symmetric shock-capturing schemes", NASA TM 89464, 1987.
[3] **D. Aymer, T. Alziary, G. Carlomagno, L. de Luca**: "Experimental Study of Flow over Double Ellipsoid", Workshop on Hypersonic Flows for Reentry Problems, January 22-25, 1990 - Antibes (France).

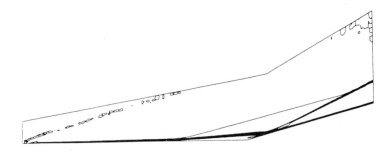

Fig.1 Mach contours

Fig.2 Visualization of the double recirculation by streamlines

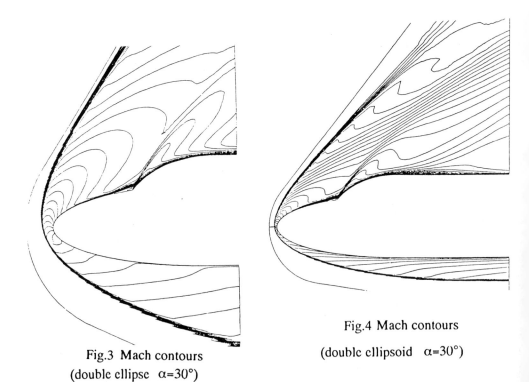

Fig.3 Mach contours
(double ellipse α=30°)

Fig.4 Mach contours
(double ellipsoid α=30°)

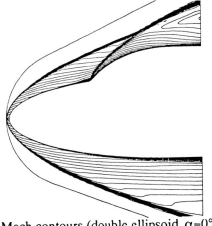

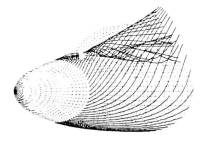

Fig.6 Visualization of the vortical structure by streamlines

Fig.5 Mach contours (double ellipsoid $\alpha=0°$)

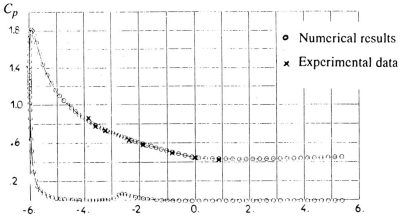

Fig.7 Wall C_p distribution in the symmetry plane

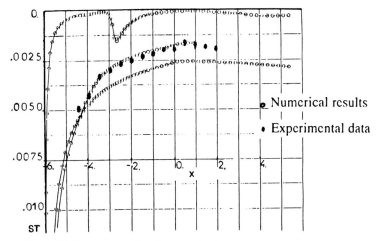

Fig.8 Wall Stanton distribution in the symmetry plane

AN UPWIND FORMULATION FOR THE NUMERICAL PREDICTION OF NON-EQUILIBRIUM HYPERSONIC FLOWS.

Salvatore Borrelli
C.I.R.A. Centro Italiano Ricerche Aerospaziali,Capua
Maurizio Pandolfi
Dipartimento di Ingegneria Aeronautica e Spaziale
Politecnico di Torino,Torino.

Introduction

We are interested on the numerical prediction of high speed flows, in the hypersonic regime. We assume a physical model based upon the Euler equations for the fluid dynamics and the equations of finite rate chemical reactions for the non-equilibrium phenomenology. Vibrational energy is assumed to be in equilibrium and ionization is not considered. Moreover, not only transport phenomena such as viscosity and thermal conductivity are neglected, but also the diffusion of the species.

The non-equilibrium chemical model we assume is the one proposed by Park in [1], that accounts for five species (O , N , NO , O_2 and N_2) and seventeen reactions.

In the following we review the basic equations and the finite volume approach. Then, the attention is focused on the **upwind** formulation used to evaluate properly the flux at the cell interfaces, as required by the finite volume approximation.

Equations and Discretization

Since we are interested in developing a procedure that allows for a correct numerical capturing of discontinuities, we write the governing equations as laws of conservations.

In the integral form, we have :

$$\frac{\partial}{\partial t} \int_{Vol} \rho_i \, dVol + \int_{\partial Vol} \rho_i \, \mathbf{V} \cdot d\mathbf{S} = \int_{Vol} \Omega_i \, dVol \qquad i=1,3$$

$$\frac{\partial}{\partial t} \int_{Vol} \rho \, dVol + \int_{\partial Vol} \rho \, \mathbf{V} \cdot d\mathbf{S} = 0 \qquad (1)$$

$$\frac{\partial}{\partial t} \int_{Vol} \rho \mathbf{V} \, dVol + \int_{\partial Vol} \rho \mathbf{V} (\mathbf{V} \cdot d\mathbf{S}) + \int_{\partial Vol} p \, d\mathbf{S} = 0$$

$$\frac{\partial}{\partial t} \int_{Vol} e \, dVol + \int_{\partial Vol} (p+e) \, \mathbf{V} \cdot d\mathbf{S} = 0$$

The first three equations of the system (1) refer to the production of the species O, N, NO, denoted in the following respectively by $i = 1, 2, 3$. The diffusion of the species is neglected and only convection and production are considered. The quantity Ω_i/ρ represents the rate of production the species and the partial density ρ_i comes from the mixture density ρ and the mass concentration Y_i :

$$\rho_i = \rho Y_i$$

The concentrations Y_4 of the molecular oxygen O_2 and Y_5 of the molecular nitrogen N_2 follow from the conservation of the atomic species.

The last three equations of the system (1) represent the conservation of mass, impulse and energy.

The finite volume approach is applied straightforward to the integral form of the governing equations. The computational domain is divided in quadrilater cells (volumes), denoted by the subscript m,n. Let us consider, for example, the first equation of the system (1). We have :

$$\frac{\partial}{\partial t} \int_{Vol} \rho_1 \, dVol + \int_{\partial Vol} \rho_1 \mathbf{V} \cdot d\mathbf{S} = \int_{Vol} \Omega_1 \, dVol \qquad (2)$$

Here, we assume that the dependent variables are known at the center of each cell m,n. Therefore, Eq. 2 takes the following approximated form :

$$\frac{d}{dt}(Vol_{m,n} \, \rho_{1m,n}) + Q_{1m,n} = \Omega_{1m,n} \, Vol_{m,n}$$

where $Vol_{m,n}$ is the cell volume, $Q_{1m,n}$ indicates the net flux out of the cell and $\Omega_{1m,n}/\rho$ is the rate of production of species 1. The net flux is given by :

$$Q_{1m,n} = \sum_{k=1}^{4} (|\Delta \mathbf{S}_k| \, f_{1k})_{m,n} \qquad (3)$$

The flux vector f_{1k}, on the k face of the cell, is evaluated according to the **Flux Difference Splitting** formulation, as shown hereafter.

The Flux-Difference Splitting Formulation

The basic steps of the **Flux Difference Splitting** formulation are the definition and solution of appropriate Riemann problems. We would like to address the reader not familiar with this formulation to [2] for a review on this matter.

At any interface between two cells (for example n,m and $n+1,m$), we define the directions normal (ξ) and tangent to this surface. Then we consider the flow properties at the center of these cells and define the Riemann problem on the base of these two sets of values by generating a discontinuity at the location $\xi_{n+\frac{1}{2},m}$. In doing this operation, we decompose the velocity vector into the normal u and tangential v components to the surface. The Riemann problem is now defined, as shown in Fig.1. The values at the center of the two neighboring cells correspond to those prescribed in the regions a and b, and the space coordinate ξ runs along the normal to the surface.

Now, we proceed to the approximate solution of this Riemann problem, according to the suggestions proposed in [3]. In addition, we make the approximation of considering frozen the concentrations of the species after the collapse of the discontinuity. Therefore, the initial concentrations in regions a and b, given by the initial data and generally different each other, remain unchanged through the *acoustic* waves (I,III), respectively in regions c and d. This approximation is added, somehow in agreement, to the original assumption of considering isentropic the evolution through the *acoustic* waves. Also, we assume the vibrational energy frozen to the levels of the initial regions a and b.

We describe the solution of the Riemann problem, that is the collapse of the initial discontinuity, on the basis of the following equations :

$$Y_{it} + u \, Y_{i\xi} = 0 \qquad (i = 1, 2, 3)$$

$$p_t + u\, p_\xi + \rho\, a_f^2\, u_\xi = 0$$

$$u_t + u\, u_\xi + \frac{p_\xi}{\rho} = 0$$

$$v_t + u\, v_\xi = 0$$

$$h_t - \frac{p_t}{\rho} + u\left(h_\xi - \frac{p_\xi}{\rho}\right) = 0$$

Note that the speed of sound a_f is defined as the frozen one.

A proper arrangement of these quasi-linear equations leads to the equations that express the advection of signals :

$$R_{jt} + \lambda_j\, R_{j\xi} = 0 \qquad (4)$$

Here λ_j represents the slope of the characteristic rays :

$$\begin{aligned}
\lambda_j &= u \qquad (j = 1, 2, 3, 6, 7) \\
\lambda_4 &= u - a_f \\
\lambda_5 &= u + a_f
\end{aligned}$$

and dR_j is the corresponding signal :

$$\begin{aligned}
dR_j &= dY_j \qquad (j = 1, 2, 3) \\
dR_4 &= dp - \rho\, a_f\, du \\
dR_5 &= dp + \rho\, a_f\, du \\
dR_6 &= dv \\
dR_7 &= dh - dp/\rho
\end{aligned}$$

On the basis of the advection equations, we can evaluate the flow properties in regions c and d. Always with reference to Fig.1, we consider the Eq. 4 for j=1,2,3,5,6,7 and we note that, through the wave I, we have (approximately for j=5,7) :

$$\begin{aligned}
Y_{ic} &= Y_{ia} \qquad (i = 1, 2, 3) \\
p_c + (\rho_a\, a_{fa})\, u_c &= p_a + (\rho_a\, a_{fa})\, u_a \\
v_c &= v_a \\
h_c - p_c/\rho_a &= h_a - p_a/\rho_a
\end{aligned}$$

On the contact surface (wave II), we impose the usual continuity of the pressure and velocity:

$$p_c = p_d \quad ; \quad u_c = u_d$$

Finally the advection equations for j=1,2,3,4,6,7 through the wave III, give (approximately for j=4,7):

$$\begin{aligned}
Y_{id} &= Y_{ib} \qquad (i = 1, 2, 3) \\
p_d - (\rho_b\, a_{fb})\, u_d &= p_b - (\rho_b\, a_{fb})\, u_b \\
v_d &= v_b \\
h_d - p_d/\rho_b &= h_b - p_b/\rho_b
\end{aligned}$$

The evaluation of the unknowns Y_1, Y_2, Y_3, p, h, u and v in regions c and d proceeds straightly from the above conditions. Then the density is obtained, being the pressure and enthalpy known and the total heat of formation h_{for} and the vibrational energy h_{vib} given by :

$$(h_{for})_c = (h_{for})_a \quad ; \quad (h_{vib})_c = (h_{vib})_a$$

$$(h_{for})_d = (h_{for})_b \quad ; \quad (h_{vib})_d = (h_{vib})_b$$

We look now at the direction of propagation of each wave and identify the region (one among a, b, c or d) that extends in time, at the location of the interface $\xi_{n+\frac{1}{2},m}$, after the collapse of the discontinuity. For instance, in the case of Fig.1, such a region is the region c. Then, we define the flux-vector at the interface (f_k of Eq. 3), on the basis of the flow properties that pertain to this region. Always in the particular case of Fig.1, we have :

$$f_k = [\rho_{1c} u_c, \rho_{2c} u_c, \rho_{3c} u_c, \rho_c u_c, (p_c + \rho_c u_c^2), \rho_c u_c v_c, u_c (p_c + e_c)]^T$$

The most interesting cases occur when a sonic transition shows up within one of the fans that describe the acoustic waves I, III. These are the cases of shocks or sonic expansions. We refer the reader to [4] for details on this matter.

Boundary Conditions

We assume the boundary coincident with one of the surfaces of the volume that lies on it. With reference to Fig.2, let C be the center of the cell where we predict the flow and B the point of the boundary, where we have to evaluate the flux f_k.

Since there is no volumes behind B, we cannot define a Riemann problem as at the interior volumes. Instead, we define a **half** Riemann problem with the flow properties evaluated at C and replace those values unavailable with the boundary conditions. Since the flow in C does not respect, in general, these boundary conditions, we expect a wave flowing backward from the boundary in the cell. In the new region generated behind this wave, we impose both the signals flowing on characteristics crossing the wave, and the boundary conditions.

In the particular case of the solid wall, the boundary condition requires the normal velocity u at the point B, of the interface, to be zero. Therefore, the only value needed at the interface is the pressure. According to the signal impinging on the wall (dR_4 or dR_5), the pressure p_k comes from :

$$p_k = p_B = p_c \mp (\rho_c a_{f_c}) u_c$$

the sign corresponding to the characteristic flowing towards the wall.

Integration

Before proceeding towards the integration, let us emphasize a important point. In solving the Riemann problem, we have introduced some approximations. They consist mainly in assuming isentropic the acoustic waves and frozen the chemistry and the vibrational energy through them. Therefore the splitting and the final evaluation of the flux at the interfaces can be somehow different from the results obtainable by the exact solution, that would not even be self-similar owing to the chemical relaxation. However, let us remind that the integration in time is carried out on the complete equations (Eqs. 1), with the generation of the proper dissipation through shock waves and the full description of the chemical reactions, and evaluation of the proper equilibrium vibrational energy. The Riemann problem and the related approximate solution represent just only tools for preparing the needed ingredients (f_k) for the final integration in time.

Because the momentum equation in system (1) is a vector one, we project it on two directions. In order to have strong conservation form, such a projection is done along two directions constant over the flowfield, namely the cartesian ones x, y along which the velocity presents the components U and V.

If we confine our attention to the first order upwind scheme, the updating from the step K to $K+1$ is obtained as it follows:

$$(w_i)_{m,n}^{K+1} = (w_i)_{m,n}^K - \frac{Dt}{Vol_{m,n}}(Q_{im,n}^K - \Omega_{im,n}\,Vol_{m,n}) \qquad (5)$$

where with $(w_i)_{m,n}$ we indicate one of the integration variables $(\rho_1,\rho_2,\rho_3,\rho,\rho U,\rho V,e)_{m,n}$. Once the new values of $(w_i)_{m,n}^{K+1}$ are known, we proceed to the decoding in order to find other flow properties.

At moderate values of the Damköholer number, that is for a non-equilibrium flow, far from equilibrium conditions and nearly close to the frozen one, the above fully explicit integration scheme provides stable numerical solutions. However, by increasing the Damköholer number and approaching the equilibrium conditions, the source terms in the chemical equations (Ω_i in the first three of Eqs. 1) induce severe numerical instabilities. The problem is known and expected. The remedy is easily found by carrying out an implicit evaluation of the source terms Ω_i, with reference to the dependence of them on the concentrations of the species. Therefore we proceed with a half-implicit procedure (implicit evaluation for the source term and explicit for the convection), adopted for the chemical equations, whilst we follow a fully explicit algorithm for the last three of Eqs. 1, that refer to the fluid-dynamics.

The reader interested on numerical results can find in [5] information not included in the present short contribution.

References

[1] C.Park, "On Convergence of Computation of Chemically Reacting Flows", AIAA Paper-85-0247, Jan. 1985.

[2] M.Pandolfi, "On the Flux Difference Splitting Formulation", Notes on Numerical Fluid Mechanics, Vol.24, Vieweg, 1989.

[3] M.Pandolfi, "A Contribution to the Numerical Prediction of Unsteady flows", AIAA Journal, Vol.22, N.5, 1984.

[4] M.Pandolfi and S.Borrelli, "An Upwind Formulation for Hypersonic Non-Equilibrium Flows", to appear in a book published by Springer Verlag, 1989.

[5] S.Borrelli and M.Pandolfi, "A Contribution to the Prediction of Hypersonic Non-Equilibrium Flows", Workshop on Hypersonic Flows for Reentry Problems, INRIA, Antibes, January 1990.

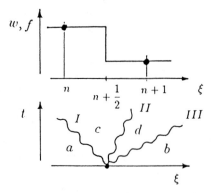

Fig.1 Interpretation of the initial data and collapse of the discontinuity.

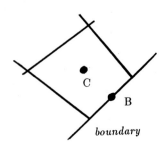

Fig.2 Cell on the boundary

NAVIER-STOKES RESULTS FOR THE HYPERSONIC BLUNT BODY PROBLEM WITH EQUILIBRIUM AIR PROPERTIES AND WITH SHOCK FITTING [*]

Frederick G. Blottner
Sandia National Laboratories
Albuquerque, New Mexico 87185-5800, USA

Introduction

There has been a significant improvement in the shock capturing technique to obtain accurate solutions in recent years. Although this technique can be used to capture shocks waves, the approach requires more grid points than a solution with shock fitting. Therefore there is still a strong interest in using and improving the shock fitting methods for production codes. In the present study, the unsteady thin-layer Navier-Stokes equations are solved for the hypersonic blunt body problem with a linearized block Alternating Direction Implicit solution procedure with central differences. In previous work [1] this solution procedure has been used to obtain accurate viscous flow solutions at low hypersonic Mach numbers for a perfect gas. After this code was modified to include real gas properties, steady-state solutions could not be obtained at $M_\infty > 15$ with the original solution procedure due to the numerical technique used at the shock wave. This paper is concerned with the changes made in the shock wave difference relations and the influence these changes have on the solutions. Numerical results are presented for flow over a spherical nose-tip to illustrate the properties of the modified shock fitting method.

Navier-Stokes Code

The governing conservation equations are first written in Cartesian coordinates x, y, and z and Cartesian velocity components u, v, and w; then the equations are transformed into curvilinear coordinates. The formulation uses the curvilinear coordinates $\xi = y^1$, $\eta = y^2$, and $\zeta = y^3$, which are defined as follows: ξ is the coordinate along the body surface starting at the nose and axis of the body; η is the coordinate around the body, which measures the distance along the surface from the windward to leeward plane of symmetry; and ζ is the coordinate from the body surface to the shock wave where the body surface is $\zeta = 0$. The grid points in the physical space across the shock layer are nonuniformly spaced to capture details of the flow.

This paper is concerned with the generation of blunt body solutions on axisymmetric bodies at zero angle of attack using the thin-layer Navier-Stokes equations. For laminar flow which is considered in this paper, the governing equations are completed with the equation of state and transport properties for air in chemical equilibrium which are expressed as $p = p(\rho, e_i)$, $T = T(\rho, e_i)$, $\mu = \mu(\rho, e_i)$, and $k_T = k_T(\rho, e_i)$. These properties

---

[*]This work performed at Sandia National Laboratories supported by the U. S. Department of Energy under Contract No. DE-AC04-76DP00789.

are obtained from analytical representation results of Liu and Vinokur [2] for the thermodynamic properties (pressure and temperature) while the viscosity and total thermal conductivity (constant pressure) use new curve fits of the Peng and Pindroh data [3]. The code has been set up to solve the flow between the body and the shock wave and from the stagnation streamline to some downstream outflow or exit plane where the maximum value of ξ occurs. The bow shock wave is treated as the outer boundary with shock fitting. Along the stagnation streamline boundary, the flow properties are extrapolated from the interior flow properties with a second-order accurate procedure developed in reference [4].

Shock Wave Equations

The properties behind the shock wave are determined from the Rankine-Hugoniot equations for a moving shock wave which are

$$\theta_{nsh} = \theta_{n\infty}(\rho_\infty/\rho_{sh}) \qquad \theta_{ni} = q_{ni} - b_n \qquad i = \infty \text{ or } sh \qquad (1)$$

$$p_{sh} = p_\infty + \rho_\infty \theta_{n\infty}^2 [1 - (\rho_\infty/\rho_{sh})] \qquad h_{sh} = h_\infty + \frac{1}{2}\theta_{n\infty}^2 \left[1 - (\rho_\infty/\rho_{sh})^2\right]$$

The velocity normal and relative to the shock wave is θ_{ni}; b_n is speed of shock wave normal to the wave and towards the body, and q_{ni} is the gas velocity normal to the shock wave. In the freestream this velocity is given as $q_{n\infty} = -(u_\infty \zeta_x + v_\infty \zeta_y + w_\infty \zeta_z)/\sqrt{g^{33}}$ and $\sqrt{g^{ii}} = \sqrt{(y_x^i)^2 + (y_y^i)^2 + (y_z^i)^2}$. The metric terms are evaluated at the shock and the subscripts indicate partial differentials. An additional relation at the shock wave must be obtained from the flow field properties between the body and the shock wave. Instead of a compatibility relation, the present approach extends a method developed by Thomas [5] and Kutler [6] where a "pressure equation" is used to determine the pressure behind the shock wave. For inviscid flow, air in chemical equilibrium, and in terms of curvilinear computational coordinates this equation becomes

$$\frac{\partial P}{\partial t} + U\frac{\partial P}{\partial \xi} + V\frac{\partial P}{\partial \eta} + W\frac{\partial P}{\partial \zeta} + \tilde{\gamma}PD = 0 \qquad (2)$$

where the new variables are defined as $P = \rho e_i = \rho h - p$ and $\tilde{\gamma} = h/e_i$. The specific internal energy of the gas is e_i. The parameter $D = u_x + v_y + w_z$ and $U, V,$ and W are the contravariant velocities. For a perfect gas, $\tilde{\gamma} = \gamma$ and $P = p/(\tilde{\gamma} - 1)$ and Eq (2) becomes the relation used by Kutler [6]. Note that the perfect gas equation does not give the real gas equation when γ is replaced with $\tilde{\gamma}$. After the shock velocity b_n is determined, the shock location is determined from the differential equations

$$\frac{dx_{sh}}{dt} = -b_n x_\zeta \sqrt{g^{33}} \qquad \frac{dy_{sh}}{dt} = -b_n y_\zeta \sqrt{g^{33}} \qquad \frac{dz_{sh}}{dt} = -b_n z_\zeta \sqrt{g^{33}} \qquad (3)$$

Solution of these equations gives the transient position of the shock wave as it evolves toward a steady-state location.

Numerical Solution of Shock Wave Conditions

<u>Original Shock Method</u>: After the interior flow between the body and the shock wave has been determined, the "pressure equation" (2) is solved to obtain P behind the shock wave. A explicit difference scheme is used which gives the difference relation

$$P_j^{n+1} = P_j^n - \Delta t_{sh}(A + \tilde{\gamma}DP_j)^n \qquad \text{where} \qquad A = UP_\xi + VP_\eta + WP_\zeta \qquad (4)$$

and the derivatives in the original scheme are evaluated, for example, as follows: $P_\xi = (P_{j+1} - P_{j-1})^n / 2\Delta\xi$ and $P_\zeta = (3P_L - 4P_{L-1} + P_{L-2})^n / 2\Delta\zeta$. The other derivatives in D and the contravariant velocities are evaluated with central differences in the $\xi-$ direction and one-sided differences are use for the $\zeta-$direction. The time step is obtained from $\Delta t_{sh} = 0.9/\sigma_{max}$ where the eigenvalue is the maximum of $\sigma = |U| + a\sqrt{g^{11}}$ or $\sigma = |W| + a\sqrt{g^{33}}$ and is evaluated at the shock wave. With P known behind the shock wave, the Rankine-Hugoniot relations are solved next to determine the flow properties behind the shock wave. The velocity $\theta_{n\infty}^2$ is eliminated from Eqs (1) to obtain the relation

$$\theta_{n\infty}^2 = (p_{sh} - p_\infty)/\rho_\infty [1 - (\rho_\infty/\rho_{sh})] = 2(h_{sh} - h_\infty)/\left[1 - (\rho_\infty/\rho_{sh})^2\right] \quad (5)$$

Since the enthalpy is $h = e_i + p/\rho$ and $P = (\rho e_i)_{sh}$, Eq (5) becomes

$$c\rho_{sh} + \rho_{sh} p_{sh} - \rho_\infty p_{sh} = d \quad c = 2\rho_\infty h_\infty - p_\infty \quad d = \rho_\infty (p_\infty + 2P) \quad (6)$$

This equation involves the unknows p_{sh} and ρ_{sh} since the freestream flow conditions are specified. It is solved with an iterative method along with $(e_i)_{sh} = P/\rho_{sh}$ and the equation of state $p = p(\rho, e_i)$. These equations are linearized with $\rho_{sh} = \bar\rho_{sh} + \Delta\rho$ and similiar relations for p_{sh} and $(e_i)_{sh}$. An iteration is used to determine the density ρ_{sh} and $\bar\rho_{sh}$ is the estimated density at each iteration. Once the density ρ_{sh} is determined, the internal energy $(e_i)_{sh}$, pressure p_{sh}, velocity $\theta_{n\infty}$, velocity $\theta_{n sh}$, and shock velocity b_n are obtained from the foregoing equations. The shock wave coordinates from Eqs. (3) are

$$x_{sh}^{n+1} = x_{sh}^n - Cx_\zeta \quad y_{sh}^{n+1} = y_{sh}^n - Cy_\zeta \quad z_{sh}^{n+1} = z_{sh}^n - Cz_\zeta \quad (7)$$

where $C = \sqrt{g^{33}} b_n \Delta t$ and Δt is the time step used in the interior flow solution.

Modified Shock Method: When the flow near the shock wave is supersonic, a marching solution technique could be used to solve this region of the flow. For this type of numerical scheme, the ξ derivatives in the shock equations should be evaluated with upwind difference relations rather than the central differences used in the original method with time marching. The modified method is identical to the original method except the ξ derivatives use the upwind difference $P_\xi = (3P_j - 4P_{j-1} + P_{j-2})^n / 2\Delta\xi$. In the pressure equation (2), P_ξ and the velocity gradients u_ξ, v_ξ, and w_ξ in D are evaluated with the upwind differences. The contravariant velocities involve the metric terms ζ_x and ζ_z which are a function of z_ξ and x_ξ. These metric terms are also upwind differenced. In the shock wave location equations (7), the second metric term g^{33} involves ζ_x and ζ_z which require z_ξ and x_ξ to be upwind differenced. The upwind differencing of the metric terms, velocity gradients, and P_ξ is required in the present method. In addition, a second-order upwind scheme is required in the nose region to obtain smooth results.

In the stagnation region of a blunt body the flow is subsonic and the central differencing of the ξ derivatives is appropriate. The switching from central to upwind differences is determined from the value of the eigenvalue $\sigma = U - a\sqrt{g^{11}}$ at the shock wave which changes sign when the flow goes from subsonic to supersonic in the ξ direction. If $j = j_s$ at the first shock point where σ has become positive, then central differences are used at $j = 1, 2, \cdots, j_s + 1$ while upwind differences are used at $j = j_s + 2, \cdots, j_{max}$. The numerical results do not appear to be sensitive to the location of the point of switching the differencing of the ξ derivatives.

Steady-state Solution: Assume an axisymmetric geometry with a polar coordinate system, the Cartesian coordinates are $x = x_0 - F\cos\theta$ and $z = r = F\sin\theta$ where F is the radial distance to the shock wave. The metric terms become $\zeta_x = -zJz_\xi$ and $\zeta_z = zJx_\xi$ while the Jacobian is $J^{-1} = z(x_\xi z_\zeta - z_\xi x_\zeta) = z\theta_\xi F F_\zeta$. With $v_\infty = w_\infty = 0$, the shock wave velocity $b_n = q_{n\infty} - \theta_{n\infty} = 0$ which gives $\zeta_x + \theta_{n\infty}\sqrt{g^{33}}/u_\infty = 0$ or $Sz_\xi = x_\xi$ where $S = \sqrt{(u_\infty/\theta_{n\infty})^2 - 1}$. After further development, this equation gives the shock location as the differential equation

$$F_\xi + \omega F = 0 \qquad \text{where} \qquad \omega = \omega(S, \theta) \tag{8}$$

For inviscid calculations, Lyubimov and Rusanov [7] previously developed an equation of this form and predicted the shock wave oscillations that occur with central differences. A difference technique with averaging was developed by these authors to remove the shock wave oscillations. The modified method for moving the shock wave of this paper also greatly reduces the shock wave oscillations as will be illustrated in the next section.

Numerical Results

The code has been used to obtain a real gas solution on a spherical nose-tip at an altitude of 30km and freestream Mach number of 15 to illustrate the behavior of the of the modified code. A time step is used such that the Courant number is 40 for the marching solution. For this case, the solution has been obtained with the original and modified shock difference relations. The variation of the L_2 residual (measure of how well the steady Navier-Stokes difference equations are satisfied) with time for the two methods used at the shock wave is given in Figure 1. The modified shock relations result in a more rapid decrease in the residual error. The wiggles in the shock wave properties for the two methods are illustrated in Figure 2 by the metric term $\partial\sqrt{g^{33}}/\partial\xi$.

Summary

The modified difference relations used at the shock wave have the following significant influences on obtaining accurate steady-state results:

1. High Mach number solutions are obtained that can not be obtained with the original shock method; modified shock method is more robust than original shock method.

2. Convergence of the solutions to a steady-state result is obtained more rapidly; approximately half the number of time steps required for the same L_2 residual.

3. Small wiggles in the solution at the shock wave observed in the present work and in previous investigations are significantly reduced.

References

1. F. G. Blottner, J. Spacecraft Rockets, Vol. 27, March-April 1990, pp. 113-122.
2. Y. Liu and M. Vinokur, AIAA Paper No. 89-1736, June 1989.
3. T-C Peng and A. L. Pindroh, Magnetohydrodynamics, Eds. A. B. Cambel, T. P. Anderson and M. M. Slawsky, Northwestern U. Press, Evanston, 1962.
4. F. G. Blottner, Proceedings 4th Symposium on Numerical and Physical Aspects of Aerodynamics Flows, Cal. State U., Long Beach, CA., Jan. 1989.

5. P. D. Thomas, M. Vinokur, R. A. Bastianon, R. J. Conti, AIAA J., Vol. 10, July 1972, pp. 887-894.
6. P. Kutler, J. A. Pedelty and T. H. Pulliam, AIAA Paper No. 80-0063, Jan. 1980.
7. A. N. Lyubimov and V. V. Rusanov, NASA TT F-714, Feb. 1973.

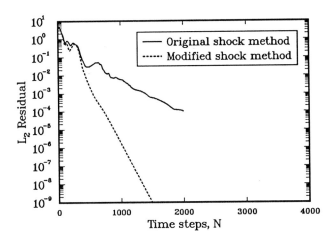

Figure 1: Influence of shock solution method on the convergence toward a steady-state solution for flow of air in chemical equilibrium over a sphere at $M_\infty = 15$ and an altitude of 30km ($J = 24$ and $L = 29$).

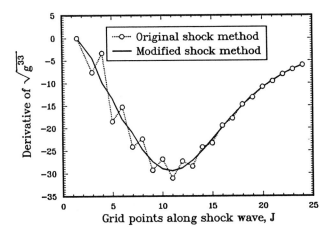

Figure 2: Influence of shock solution method on the metric term g^{33} at the shock wave for same case.

A POLYNOMIAL FLUX VECTOR SPLITTING APPLIED TO VISCOUS HYPERSONIC FLOW COMPUTATION

P. Lardy * & H. Deconinck
von Karman Institute, Rhode St Genèse, Belgium

1. Introduction

Upwind space discretization of the convective terms in the Navier-Stokes equations has been widely used over recent years, especially for the computation of high supersonic and hypersonic flows with strong shock waves. The crucial element in these methods is a Flux Vector Splitter (e.g. Steger-Warming, van Leer), or alternatively a Flux Difference Splitter (Roe, Osher) which allows to decompose the flux at a given cell face in contributions originating from the two sides by taking into account some physical propagation properties of the flow. The theory of Flux Difference Splitters (FDS) is based on the solution of the Riemann problem, while Flux Vector Splitter (FVS) follow from a particle model of the fluid. The particle model is inadequate when no particles are allowed to cross the cell face as occurs in a shear layer parallel with the mesh. Therefore, FVS are known to be more dissipative in resolving contact discontinuities when applied to inviscid flows and hence show a poor performance in resolving boundary layers and shear layers when applied to viscous flow.

On the other hand, Flux Vector Splitters are computationally very attractive because of their simplicity which in particular for implicit solvers allows for a straightforward analytical Newton linearization of the discrete fluxes. The purpose of the present paper is first to present a new and extremely simple class of Flux Vector Splitters, and to compare it with Roe's Flux Difference and van Leer's Flux Vector Splitter. Then, in a second step, the tangential velocity modification is introduced as proposed by Hänel in order to improve the performance of the new FVS in shear layers, and comparisons are made for viscous hypersonic flow.

2. Description of the Class of Schemes

In the following, the development of the new splitter is given for 1D flow. Consider the convective Euler flux F at a cell face given by

$$F = \begin{pmatrix} \rho u \\ \rho u^2 + p \\ \rho u H_t \end{pmatrix} = \begin{pmatrix} m \\ mu + p \\ m H_t \end{pmatrix}$$

where m denotes mass flux, p pressure, ρ density, u velocity, and H_t total enthalpy. A splitting of F is sought of the form $F = F^+ + F^-$ with

$$F^+ = \begin{pmatrix} m^+ \\ m^+ u + p^+ \\ m^+ H_t \end{pmatrix} \qquad F^- = \begin{pmatrix} m^- \\ m^- u + p^- \\ m^- H_t \end{pmatrix}$$

---

* presently at Aérospatiale, Les Mureaux, France

The objective is to decouple the treatment of the mass flux m^+, m^- from the pressure flux p^+, p^-. Hence these quantities are calculated as independent polynomials of Mach number M of the form

$$m^+ = \rho c \theta(M) \qquad m^- = \rho c (M - \theta(M)) \qquad p^+ = p\beta(M) \qquad p^- = p(1 - \beta(M))$$

The polynomials θ and β assure a smooth transition from $F^+ = F, F^- = 0$ at M=1 to $F^+ = 0, F^- = F$ at M=-1. Table 1 and figure 1 show the five first polynomials θ and β with increasing continuity of the derivatives at M=1 and M=-1. Note that the momentum flux splitting of van Leer [1] is obtained by the combination $\theta 2\beta 2$. For the energy flux, no splitting of the total enthalpy is introduced, as proposed by Hänel [2] and also studied by van Leer [3].

Extension to 3D is straightforward, using instead of M a Mach number based on the velocity normal to the cell face [4].

3. Properties of the Scheme

The degree of continuity of θ and β at M=1 and M=-1 assures the smoothness of the Jacobians of F^+ (resp. F^-) on the whole range of Mach numbers M. To allow for a crisp shock capturing, one eigenvalue of the cell face Jacobians should vanish. In fact, this is one of the design criteria of van Leer's original FVS [1]. In the present approach, it is approximately achieved by the definition of the split energy fluxes proportional to the split mass fluxes : the proportionality factor is total enthalpy, which remains constant in steady inviscid flow. Indeed, for

$$F^+_{energy} = \alpha F^+_{mass} \qquad \text{with} \qquad \alpha = H_t = c^t$$

the Jacobian of F^+ (resp. F^-) is singular and one eigenvalue vanishes.

The class of schemes is in principle infinite but if the degree of continuity n of θ and β at M=1 and M=-1 tends to ∞, the limits are :

$$\lim_{n \to \infty} \theta n = \begin{cases} 0 & \text{if } -1 \le M \le 0 \\ M & \text{if } 0 \le M \le 1 \end{cases} \quad \text{and} \quad \lim_{n \to \infty} \beta n = \begin{cases} 0 & \text{if } -1 \le M \le 0 \\ 1 & \text{if } 0 \le M \le 1 \end{cases}$$

which allows no upstream influence in subsonic region. Hence, there should be an optimal choice for combining θn and βm.

In order to determine the best combination of $\theta - \beta$, numerical experiments have been made based on a conical axisymmetric test case, namely the flow around a 10 degrees semi-angle circular cone at M_∞=8. As shown on figure 2, increasing the degree of θ gives a good improvement in shock capturing whereas increasing the degree of β does not produce significant results. The best choice was found to be $\theta 3\beta 2$ which fits very well with the original van Leer scheme taken as reference. $\theta 3\beta 2$ was successfully tested on a large range of Mach numbers (from 2 to 20) to prove its reliability in capturing weak to strong shocks (Fig. 3).

4. Viscous flow computations

The particular choice $\theta 3\beta 2$ has been compared with Roe's scheme, van Leer's FVS and Hänel's modification for a conical viscous hypersonic flow around a 10 degrees semi-angle circular cone at zero incidence and $M_\infty = 8$, $Re = 1.8 \cdot 10^5$. Without tangential velocity splitting, $\theta 3\beta 2$ gives a small improvement compared with van Leer's scheme (Fig. 4). But with the modification introduced by Hänel, the boundary layer is captured by $\theta 3\beta 2$ and Hänel's scheme in the same way as by Roe's FDS, although an oscillation appears on the pressure at the edge of the boundary layer for $\theta 3\beta 2$, and even more pronounced for Hänel's scheme (Fig. 5). If the same modification is applied to the original van Leer scheme, the temperature profile is completely wrong, giving a value of T/T_∞ of only 6.8, although the boundary layer thickness has become smaller. The dissipation in the shock is comparable for $\theta 3\beta 2$, Roe's and van Leer's schemes, but much higher for Hänel's scheme. This is due to the use of the critical sound speed as switching criterion in the flux function [2].

The second 2D test case is a conical viscous supersonic flow around a 9.09° semi-angle cone at 15° of incidence, $M_\infty = 2$ and $Re = 1.8 \cdot 10^5$. For this case, second order MUSCL extrapolations have been used. As was to be expected for 2nd order schemes, $\theta 3\beta 2$, Roe's and Hänel's schemes give similar pressure distributions around the cone (Fig. 6), quite different from van Leer's.

5. Conclusion

A new class of differentiable Flux Vector Splitters $\theta n \beta n$ has been studied. One of these schemes, $\theta 3\beta 2$, produces crisp shock profiles in a large range of Mach numbers (2 to 20). As opposed to the classical van Leer scheme, the new schemes perform well in shear layers after introducing the tangential velocity splitting proposed by Hänel; agreement of $\theta 3\beta 2$ with Roe's FDS is good for 1D and 2D conical viscous flows.

REFERENCES

[1] Van Leer, B.; Lecture Notes in Physics, Vol. 170, 1982, pp 507-512

[2] Hänel, D.; AIAA P 89-0274, 1989

[3] Van Leer, B.; Invited Lecture for the CFD Symp. on Aeropropulsion, NASA Lewis Research Center, April 1990, Cleveland, Ohio, USA

[4] Lardy, P. & Deconinck, H.; to be published in "L'Astronautique et l'Aéronautique", 1990.

$\theta1(x) = \frac{1}{2}x + \frac{1}{2}$

$\theta2(x) = \frac{1}{4}x^2 + \frac{1}{2}x + \frac{1}{4}$

$\theta3(x) = -\frac{1}{16}x^4 + \frac{3}{8}x^2 + \frac{1}{2}x + \frac{3}{16}$

$\theta4(x) = \frac{1}{32}x^6 - \frac{5}{32}x^4 + \frac{15}{32}x^2 + \frac{1}{2}x + \frac{5}{32}$

$\theta5(x) = -\frac{5}{256}x^8 + \frac{7}{64}x^6 - \frac{35}{128}x^4 + \frac{35}{64}x^2 + \frac{1}{2}x + \frac{35}{256}$

$\beta1(x) = \frac{1}{2}x + \frac{1}{2}$

$\beta2(x) = -\frac{1}{4}x^3 + \frac{3}{4}x + \frac{1}{2}$

$\beta3(x) = \frac{3}{16}x^5 - \frac{5}{8}x^3 + \frac{15}{16}x + \frac{1}{2}$

$\beta4(x) = -\frac{5}{32}x^7 + \frac{21}{32}x^5 - \frac{35}{32}x^3 + \frac{35}{32}x + \frac{1}{2}$

$\beta5(x) = \frac{35}{256}x^9 - \frac{45}{64}x^7 + \frac{189}{128}x^5 - \frac{105}{64}x^3 + \frac{315}{256}x + \frac{1}{2}$

Table 1 : Polynomials θ and β defining the class of schemes

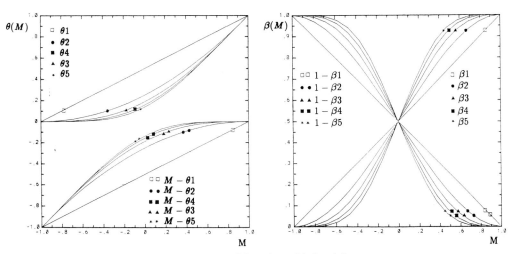

Figure 1 : Plot of the polynomials θ and β

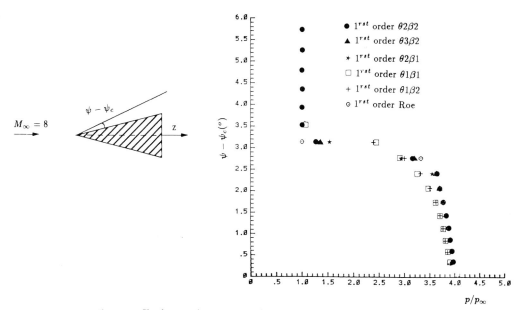

Figure 2 : Shock capturing properties for different choices of θ and β

$10°$ semi-angle cone flow at $M_\infty = 8$, $Re = \infty$

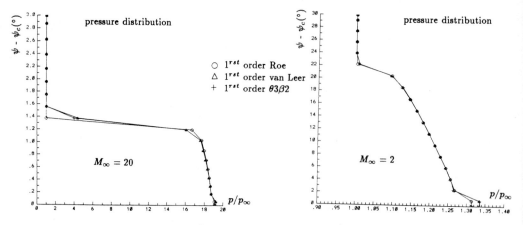

Figure 3 : Comparison between 1^{rst} order Roe, van Leer and $\theta 3\beta 2$ for 2 different shock strenghts 10° semi-angle cone flow at 0° incidence, $Re = \infty$

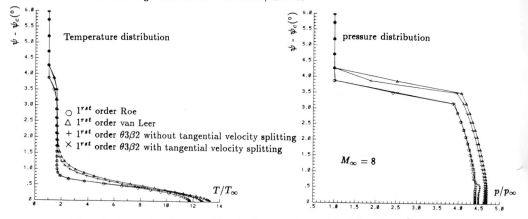

Figure 4 : Comparison between 1^{rst} order Roe, van Leer and $\theta 3\beta 2$ for viscous flow 10° semi-angle cone flow at 0° incidence, $M_\infty = 8$, $Re_\infty = 1.757\ 10^5$, $Pr = 0.72$

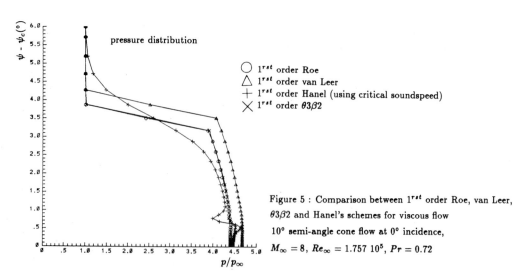

Figure 5 : Comparison between 1^{rst} order Roe, van Leer, $\theta 3\beta 2$ and Hanel's schemes for viscous flow 10° semi-angle cone flow at 0° incidence, $M_\infty = 8$, $Re_\infty = 1.757\ 10^5$, $Pr = 0.72$

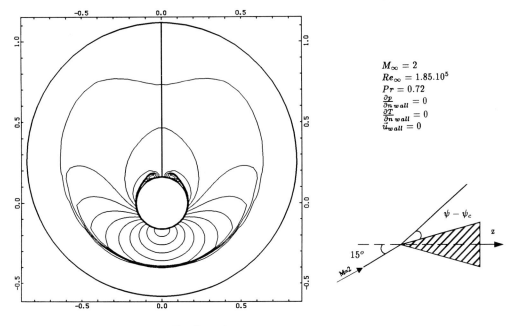

Density contours
9.09 degree semi-angle cone. Incidence 15 degrees
2nd order $\theta 3\beta 2$ with tangential velocity splitting

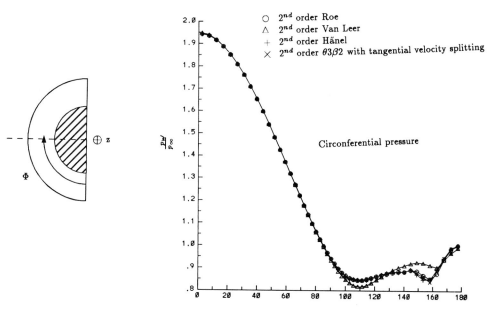

Figure 6: Comparison of Roe, Van Leer, Hänel and $\theta 3\beta 2$ schemes
9.09 degrees semi-angle cone. Incidence 15 degrees

SIMULATIONS OF INVISCID EQUILIBRIUM AND NONEQUILIBRIUM HYPERSONIC FLOWS

M. Pfitzner

Messerschmitt-Boelkow-Blohm GmbH
Dept. KT225, P.O. Box 80 11 69
D-8000 Munich 80, West Germany

Abstract

An algorithm for the simulation of 2-D and 3-D inviscid hypersonic flows using shock fitting is described. It has been implemented for the ideal, equilibrium and nonequilibrium real gas cases. The flow solver is based on a quasi-conservative split-matrix method with upwind-biased space discretisation coupled to Runge-Kutta time-stepping. The chemistry source terms are treated either explicitly or (point-) implicitly.

1. Introduction

The interest in the simulation of hypersonic flows has increased considerably in recent years due to the development of new space transportation systems around the world. In Europe, particularly the HERMES and SAENGER projects are a driving force behind the development of new technologies. There is an increasing demand for CFD predictions of the flow phenomena, since for hypersonic flows not all similiarity parameters can be reproduced simultaneneously in the wind tunnel and free flight experiments are expensive and extremely difficult to perform.

The high temperatures occuring in hypersonic flows cause the excitation of vibrational degrees of freedom of the molecules, dissociation of oxygen and nitrogen, the production of nitric oxide and (at very high temperatures) ionisation. If we are interested in the surface heating and aerodynamic loads of a shuttle-like vehicle at reentry, the modelization of ionization is not necessary.

Along the reentry trajectory, there are three flow regimes, which require a modelization of varying complexity. Below a freestream Mach number of $M_\infty < 4$ the ideal gas assumption holds. At flying heights below $H < 40$ km, and at Mach numbers above 4, the gas is approximately in local equilibrium. Above 40 km nonequilibrium effects come into play.

The shock-fitting Euler algorithm, which has been developed for the ideal gas[1] and equilibrium real gas[2] cases, and which has been optimized for the application to supersonic/hypersonic flows[3], is generalized to flows in chemical non-equilibrium. The chemical source terms are treated either explicitly or (point-)implicitly.

The code is applied to 2-D test cases and 3-D configurations in equilibrium and nonequilibrium flow and the results are discussed.

2. Flow Solver

The Euler equations of flow in chemical nonequilibrium are written in quasi-linear form

$$Q_\tau + AQ_\xi + BQ_\eta + CQ_\zeta = S \tag{1}$$

using the conservative variables

$$Q = (\rho, \rho u, \rho v, \rho w, \rho_1, \ldots, \rho_n, e)^T$$

The Jacobi matrices[4] are split according to the sign of the eigenvalues like, e.g.

$$A = T\Lambda T^{-1} = T\Lambda^+ T^{-1} + T\Lambda^- T^{-1} = A^+ + A^- \qquad (2)$$

$$\Lambda = \text{diag}(\lambda_1, \lambda_2, \lambda_3, \ldots, \lambda_3)$$

and the derivatives Q_ξ are split accordingly[5]:

$$AQ_\xi \longrightarrow A^+ Q_\xi^+ + A^- Q_\xi^- \qquad (3)$$

The derivatives $Q_\xi^\pm$ are calculated using a third order upwind-biased formula.

Since the eigenvalue λ_3 is strongly degenerated, the evaluation of the matrix products of the Jacobi matrices with an arbitrary vector $\vec{Y}$ is simplified

$$\begin{aligned}A\vec{Y} &= T\Lambda^\pm T^{-1}\vec{Y} \\ &= (\lambda_1^\pm - \lambda_3^\pm)(\vec{t_1^-} \cdot \vec{Y})\vec{t_1} + (\lambda_2^\pm - \lambda_3^\pm)(\vec{t_2^-} \cdot \vec{Y})\vec{t_2} + \lambda_3^\pm \vec{Y}\end{aligned} \qquad (4)$$

which requires only the nth row vector $\vec{t_n^-}$ of the right eigenvector matrix T and the nth column vector $\vec{t_n}$ of the left eigenvector matrix T^{-1}, $(n = 1, 2)$. Thus only the nondegenerate eigenvectors corresponding to λ_1, λ_2 have to be calculated and the nondegenerate ones do not appear. The amount of computational work increases only linearly with the species number.

The time relaxation algorithm is a second order accurate three-step Runge-Kutta scheme optimized for the use with the upwind-biased space discretization. For steady state caculations a local time step is applied using CFL numbers between 0.5 and 1.5.

The total enthalpy $h_o = (e + p)/\rho$ is a constant along streamlines in the steady state. This is also valid across the shock and for nonequilibrium flows. To aid convergence rate and robustness of the code, after each complete iteration step the energy e is corrected so that the total enthalpy of the free stream is reproduced:

$$e = \rho h_o^\infty - p \qquad (5)$$

The pressure is evaluated using the uncorrected energy and the equation is not iterated. In the supersonic part of the flow field, the pseudo space marching method is applied[3].

3. Boundary conditions

At the body the post correction technique is applied. In the first step at the boundary the normal flow equations are solved, omitting the derivatives towards the body. Then this predicted value Q^* is corrected with

$$Q = Q^* - \beta \vec{t_i} \qquad (6)$$

where the vector $\vec{t_i}$ is the left eigenvector corresponding to the characteristic coming from the body and the scalar β is calculated so that Q satisfies the boundary condition.

At the shock the five Rankine-Hugoniot equations, one characteristic equation for the influence from behind the shock, and the condition of unchanged concentrations across the shock are solved for the flow variables and the shock velocity[4,5]. The nonequilibrium shock fitting algorithm can be formulated very similar to the equilibrium one presented in[2].

4. Chemistry Terms

The reacting air was modeled using the five species (N_2, N, O_2, O, NO) and the 17 nonionizing reactions from[6]. The forward rates and equilibrium constants are also taken from [6]. Thermal equilibrium was assumed for the vibrational degrees of freedom. The chemical source term was treated either explicitly or (point-) implicitly. In the simulation of nonequilibrium reentry air flows at heights above 50 km explicit treatment of the sources was sufficient.

5. Results

The present capabilities of the 3-D shock fitting code are demonstrated in figs.1,2. Fig.1 shows Mach number isolines of the top of HERMES in equilibrium air at $M_\infty = 8, \alpha = 30°, \beta = 0°, H = 42.5$ km and $M_\infty = 25, \alpha = 30°, \beta = 5°, H = 75$ km. The wing shock moves from the winglets toward the body as one goes to higher freestream Mach numbers. Fig.2 displays Mach number isolines of flow about a double ellipsoid at $M_\infty = 8.15, \alpha = 0°$ (ideal gas), $M_\infty = 25, \alpha = 0°$ (equilibrium real gas). The captured canopy shock interferes nicely with the fitted bow shock, leaving behind a slip line, which can be seen particularly well in the $M_\infty = 8.15$ case. In the real gas case the fitted shock is very close to the body. Fig.3 shows a comparison of fringe patterns in flow of partially dissociated nitrogen[8] about a 2 inch diameter cylinder at ($p_\infty = 2910 Pa, T_\infty = 1833 K, u_\infty = 5590 m/s, c_\infty^N = 0.073, M_\infty = 6.14$). Whereas the shock position is reproduced well, the fringe pattern differ somewhat due to the one- temperature model used. Fig.4 shows a comparison of temperature contours of flow about a 1/4 inch sphere in air at ($p_\infty = 664 Pa, T_\infty = 293 K, u_\infty = 5280 m/s, M_\infty = 15.3$), for the ideal, equilibrium and nonequilibrium cases. Also shown is the comparison of the shock contour with an experiment of Lobb[7]. The experimental shock standoff distance is slightly larger than the calculated one owing to thermal nonequilibrium effects. Fig.5 displays the application of the 3-D nonequilibrium code to the HERMES nose at $M_\infty = 25, \alpha = 30°, H = 75$ km. The typical zones of strong temperature gradients near the shock and the body are seen well.

6. Conclusion

The split-matrix algorithm using shock-fitting and Runge-Kutta time-stepping has been generalized successfully to the nonequilibrium case. The code is in the process of validation and has been applied already to realistic 3-D configurations. In the near future the implementation of thermal nonequilibrium and the coupling to a 3-D nonequilibrium boundary layer code is planned.

7. Acknowledgement

The author thanks G. Hartmann for performing the calculations around the double ellipsoid.

References

[1] Weiland C., Pfitzner M.: Lecture Notes in Physics, Vol. **264**, pp. 654-659, Springer, 1986.
[2] Pfitzner M., Weiland C.: AGARD-CP 428, Paper No.22, Bristol, 1987.
[3] Pfitzner M.: Notes on Numerical Fluid Dynamics, Vol. **24**, pp. 489-498, Vieweg, 1988.
[4] Poeppe C.: HERMES research report H-DS-1-1059-AMD.
[5] M. Pfitzner, to be published.
[6] Park C.: AIAA paper **85-0247**.
[7] Lobb R.K.: in "The High Temperature Aspects of Hypersonic Flow", ed. W.C. Nelson, Pergamon, 1964.
[8] Hornung H.G.: J. Fluid Mechanics, Vol. **53**, pp.149-176, 1972.

8. Figures

Fig.1: HERMES-top, equilibrium air, $\alpha = 30°$, Mach number contours, $\Delta = 0.5$

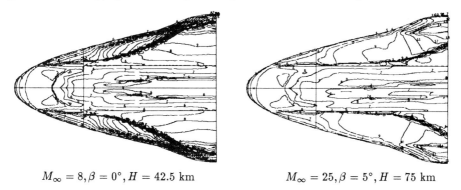

$M_\infty = 8, \beta = 0°, H = 42.5$ km $\qquad$ $M_\infty = 25, \beta = 5°, H = 75$ km

Fig.2: double ellipsoid, $\alpha = 0°$, Mach number contours, $\Delta = 0.25$

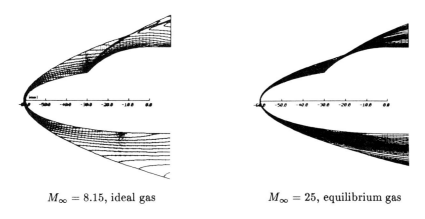

$M_\infty = 8.15$, ideal gas $\qquad\qquad$ $M_\infty = 25$, equilibrium gas

Fig.3: 2 inch cylinder in nitrogen, $p_\infty = 2910 Pa, T_\infty = 1833 K, u_\infty = 5590 m/s, c_\infty^N = 0.073$

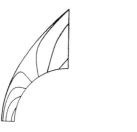 

fringe contours of $\rho/\rho_\infty(1 + 0.28 c_N)$ $\qquad\qquad$ Experiment[8]

Fig.4: 1/4 inch sphere in air, $p_\infty = 664 Pa, T_\infty = 293 K, u_\infty = 5280 m/s$

T/T_∞ contours, $\Delta = 1$

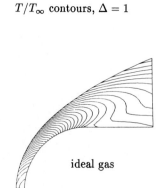

ideal gas

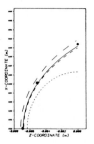

comparison of shock contour with experiment[7]

(- - ideal, - - equilibrium,

— nonequilibrium, • experiment)

equilibrium

nonequilibrium

Fig.5: HERMES nose, nonequilibrium air, $M_\infty = 25, \alpha = 30°, H = 75$ km

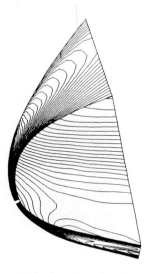

T/T_∞ contours, $\Delta = 1$

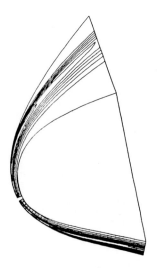

c_{NO} contours, $\Delta = 0.01$

Numerical Solutions of the Boltzmann Equation for Hypersonic Flows

Nobuyuki Satofuka, Koji Morinishi and Yoshito Sakaguchi

Kyoto Institute of Technology
Matsugasaki, Sakyo-ku, Kyoto 606 JAPAN

1 Introduction

The determination of the flow field around an aerospace vehicle such as NASP or HERMES is one of the most challenging problems in the field of computational fluid dynamics. Much of the efforts is focused on a better understanding of the physics and chemistry of high temperature air flows based on the continuum assumptions [1,2]. However at high altitudes where the density is low, the Navier-Stokes Equations become no longer valid. In such cases it is necessary to solve the Boltzmann equation. Needless to say, the direct numerical integration of the equation is computationally too costly due to the complexity of the collision term. So far most of the attempts to solve the Boltzmann equation has been carried out using the Direct Simulation Monte Carlo (DSMC) method proposed by Bird [3]. This method, however, needs a large number of sample particles as well as large computer capability in order to obtain a sufficiently accurate solution because of the inherent statistical scattering.

Recently we proposed a time dependent finite-difference method to solve a model Boltzmann equation, and showed some qualitative results at a Mach number of 4.0[4]. In this paper, we apply a method of lines approach to the solution of a model Boltzmann equation for hypersonic flows of a rarefied gas over two-dimensional bodies of arbitrary shape. Computations were made for the cases of right circular cylinder and a double ellipse for Mach number 4.0 and 8.15.

2 Kinetic Model Equation
2.1 BGK model

In order to avoid the complex Boltzmann collision integral, BGK model of the kinetic equation is employed in this paper. We may write the model equation in standard notations:

$$\frac{\partial f}{\partial t} + \mathbf{c} \cdot \frac{\partial f}{\partial \mathbf{X}} = \nu (f_0 - f), \tag{1}$$

where f is the distribution function, $\mathbf{c} = (c_x, c_y, c_z, c_{r_1}, c_{r_2})$ the molecular velocity, $\mathbf{X} = (x, y, z)$ the spatial coordinates, and ν the collision frequency. In this equation, the equilibrium distribution f_0 is given as follows:

$$f_0 = \frac{\rho}{(2\pi RT)^{5/2}} \exp\left[-\frac{c^2}{2RT}\right], \tag{2}$$

where $c = \sqrt{(c_x - u)^2 + (c_y - v) + (c_z - w)^2 + c_{r_1}^2 + c_{r_2}^2}$ is the peculiar velocity.

The moments ρ, $\mathbf{U}(u,v,w)$, and T are given by

$$\left. \begin{aligned} \rho &= \int f d\mathbf{c}, \\ \rho \mathbf{U} &= \int \mathbf{c} f d\mathbf{c}, \\ \frac{5}{2}\rho RT &= \int \frac{1}{2} c^2 f d\mathbf{c} \end{aligned} \right\} \tag{3}$$

where R denotes the gas constant, ρ the density, $\mathbf{U}$ the macroscopic velocity, and T the temperature.

2.2 Non-dimensionalization

With the characteristic length of a flow field L and the most probable speed c_∞ defined as $c_\infty = \sqrt{2RT_\infty}$, the following dimensionless variables are introduced:

$$\left.\begin{array}{ll} \bar{\rho} = \rho/\rho_\infty, & \bar{T} = T/T_\infty, \\ \bar{\mathbf{U}} = \mathbf{U}/c_\infty, & \bar{f} = fc_\infty^5/\rho_\infty, \\ \bar{\mathbf{X}} = \mathbf{X}/L, & \bar{\mathbf{c}} = \mathbf{c}/c_\infty, \quad \bar{t} = t/L/c_\infty. \end{array}\right\} \quad (4)$$

Without any confusion, using the same notations, the dimensionless form of Eq.(1) is written as the same along with

$$f_0 = \frac{\rho}{\pi T^{5/2}} \exp\left(-\frac{c^2}{T}\right). \quad (5)$$

A collision frequency ν usually is a function of moments and independent of molecular velocities. For Maxwell molecules, the dimensionless collision frequency is written as

$$\nu = \frac{8\rho}{5\sqrt{\pi}Kn}, \quad (6)$$

where Kn is the Knudsen number based on the free stream mean free path l_∞ given as

$$l_\infty = \frac{16\mu_\infty}{5\rho_\infty\sqrt{2\pi RT_\infty}}. \quad (7)$$

2.3 Reduced distribution function

To reduce the number of independent variables, the following reduced distribution functions are introduced for two-dimensional problems:

$$\left.\begin{array}{l} F(x,y,c_x,c_y,t) = \iiint f \, dc_z \, dc_{r_1} \, dc_{r_2}, \\ H(x,y,c_x,c_y,t) = \iiint (c_z^2 + c_{r_1}^2 + c_{r_2}^2) f \, dc_z \, dc_{r_1} \, dc_{r_2}. \end{array}\right\} \quad (8)$$

The corresponding equations for the reduced distribution functions with BGK model of collision integral are written in an arbitrary curvilinear coordinate system as follows:

$$\left.\begin{array}{l} \dfrac{\partial F}{\partial t} + c_\xi \dfrac{\partial F}{\partial \xi} + c_\eta \dfrac{\partial F}{\partial \eta} = \nu(F_0 - F), \\ \dfrac{\partial H}{\partial t} + c_\xi \dfrac{\partial H}{\partial \xi} + c_\eta \dfrac{\partial H}{\partial \eta} = \nu(H_0 - H), \end{array}\right\} \quad (9)$$

where c_ξ and c_η are the contravariant molecular velocity components defined as

$$\left.\begin{array}{l} c_\xi = (c_x y_\eta - c_y x_\eta)/J, \\ c_\eta = (-c_x y_\xi + c_y x_\xi)/J, \end{array}\right\} \quad (10)$$

in which x_ξ, x_η, y_ξ, and y_η are the metric coefficients and J is the Jacobian of the transformation,

$$J = x_\xi y_\eta - x_\eta y_\xi .\tag{11}$$

The equilibrium distributions F_0 and H_0 are given,

$$\left.\begin{array}{l} F_0 = \dfrac{\rho}{\pi T}\exp\left[-\dfrac{(c_x - u)^2 + (c_y - v)^2}{T}\right], \\ H_0 = \dfrac{3}{2}TF_0 . \end{array}\right\}\tag{12}$$

3 Computational Procedure
3.1 Spatial discretization

In our method of lines approach we first discretize the spatial derivatives in Eq.(9) by second order upwind differencing. For example ξ derivatives of F at grid point (i, j) are approximated as

$$\dfrac{\partial F}{\partial \xi}\bigg|_{i,j} = \begin{cases} \dfrac{-3F_{i,j} + 4F_{i+1,j} - F_{i+2,j}}{2\Delta\xi} & (c_\xi < 0) \\ \dfrac{3F_{i,j} - 4F_{i-1,j} + F_{i-2,j}}{2\Delta\xi} & (c_\xi > 0), \end{cases}\tag{13}$$

where $\Delta\xi$, $\Delta\eta$ are the grid spacing and i, j are grid induces in ξ, η directions, resectively.

3.2 Time integration scheme

Substitution of the spatial discretization into the BGK model kinetic equation (9) yields the system of ordinary differential equations in time,

$$\dfrac{d\vec{F}}{dt} = \vec{G}(\vec{F}) ,\tag{14}$$

where $\vec{F}$ denotes $(F_{i,j}, H_{i,j})^T$. The time integration scheme adopted in this paper is a rational Runge-Kutta scheme. As applied to Eq.(14), the RRK scheme can be written in the following two stage form,

$$\begin{aligned} \vec{g}_1 &= \Delta t \cdot \vec{G}(\vec{F}^n) , \\ \vec{g}_2 &= \Delta t \cdot \vec{G}(\vec{F}^n + C_2\vec{g}_1) , \\ \vec{F}^{n+1} &= \vec{F}^n + \dfrac{2\vec{g}_1(\vec{g}_1,\vec{g}_3) - \vec{g}_3(\vec{g}_1,\vec{g}_1)}{(\vec{g}_3,\vec{g}_3)} , \end{aligned}\tag{15}$$

where $\vec{g}_3 = b_1\vec{g}_1 + b_2\vec{g}_2$, $b_1 + b_2 = 1$ and $(\vec{g}_1, \vec{g}_3)$ denotes the scalar product of vectors $\vec{g}_1$ and $\vec{g}_3$. we chose the values of parameters as $b_1 = 2$, $b_2 = -1$ and $c_2 = 1/2$.

3.3 Boundary conditions

Along inflow boundary ahead of the shock wave, we specified the equilibrium distribution with prescribed free stream properties for molecules incoming from upstream. On the outflow boundary zero gradient along η direction for the reduced distribution functions is assumed as,

$$\frac{\partial F}{\partial \eta} = \frac{\partial H}{\partial \eta} = 0 . \tag{16}$$

For the interaction of molecules with the body surface, fully diffuse reflection is assumed. The density of molecules emitted from the body surface is determined from the mass balance on the surface,

$$\int_{C_n>0} C_n F dC_x dC_y = \int_{C_n<0} -C_n F dC_x dC_y . \tag{17}$$

4 Numerical Results

Numerical computations were performed on hypersonic flows past a circular cylinder and a double ellipse. The conditions for the numerical experiments are listed in the Table 1. Figure 1 shows the density contours for the flow past a circular cylinder under two different free stream conditions, i.e., $M_\infty = 4.0$, $Kn = 0.125$ and $M_\infty = 8.15$, $Kn = 0.04$. For the purpose of comparison the Navier-Stokes solutions are also shown in the lower part of the figure. In either case, the flow field feature ahead of the cylinder is similar to each other. For the case with $M_\infty = 4.0$ and $Kn = 0.125$, the locations of shock-like structure likely to apart further from the body than that predicted from the continuum equations. The density contours for the flow past a double ellipse at $M_\infty = 8.15$, $Kn = 0.04$ and $\alpha = 30°$ are shown in Fig.2, while the velocity distribution along the body surface is shown in Fig.3. Even for this small Knudsen number significant velocity slip is evident in the leeward side of the ellipse.

References
[1] J.A.Desideri, N.Glinsky and E.Hettena, *Hypersonic Reactive Flow Computations*, Computers & Fluids Vol.18, No2, pp.151-182, 1990
[2] H.C.Yee, G.H.Klopfer and J.L.Montagne, *High-Resolution Shock-Capturing Schemes for Inviscid and Viscous Hypersonic Flows*, Journal of Computational Physics 88, pp.31-61, 1990
[3] G.A.Bird, *Molecular Gas Dynamics*, Clarendon Press, Oxford, pp.113-115, 1976
[4] H.Oguchi, K.Morinishi and N.Satofuka, *Time-Dependent Approach to Kinetic Analyses of Two-Dimensional Rarefied Gas Flows*, Rarefied Gas Dynamics Vol.I Edited by O.M.Belotserkovskii, M.N.Kogan, S.S.Kutateladze and A.K.Rebrov, Plenum Publishing Corporation, 1985
[5] A.Wambecq, *Rational Runge-Kutta Methods for Solving System of Equations*, Computing Vol.20, pp.333-342, 1978

Table 1 Conditions for the numerical experiments.

	M_∞	K_n	α (deg)	c_x	c_y
Circular Cylinder	4.0	0.125	0.	-6 ~ 8	-7 ~ 7
	8.15	0.04	0.	-12 ~ 14	-13 ~ 13
Double Ellipse	4.0	0.1	0.	-6 ~ 8	-7 ~ 7
			30.		-6 ~ 8
	8.15	0.08	0.	-12 ~ 14	-12 ~ 14
			30.		-12 ~ 14
		0.04	30.		-12 ~ 14

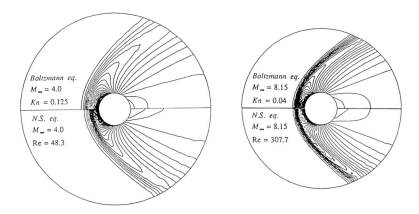

Figure 1 Density contours for hypersonic flows past a circular cylinder.

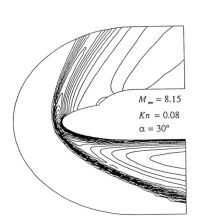

Figure 2 Density contours for hypersonic flows past a double ellipse.

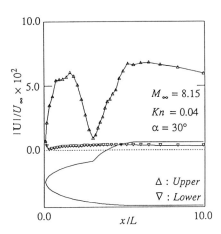

Figure 3 Velocity distribution along the surface of the ellipse.

A High Resolution Cell Vertex TVD Scheme for the Solution of the Two and Three Dimensional Euler Equations

N. Kroll and C. Rossow

DLR, Institute for Design Aerodynamics

Flughafen, D-3300 Braunschweig, West Germany

Overview

Recent interest in hypersonic vehicles has focused attention on the numerical simulation of high speed flows. The necessity of accurately resolving the strong shocks, expansion fans and shear layers associated with these flow fields has led to the development of a variety of upwind-biased computational algorithms. In contrast to central difference techniques, these methods use local wave propagation theory to evaluate flow properties near discontinuities without the need of additional artificial dissipative terms.

This paper presents a modification of the central difference cell vertex method described in [1] for high speed flows. The original scheme used a compact discretization operator for the approximation of the spatial derivatives at a grid node involving all neighboring points. The dissipation model according to Jameson [2] performed the required artificial damping. This scheme, implemented in a finite volume formulation, has proven successful in calculating sub-, trans- and low supersonic flow fields around complex configurations [1, 3, 4]. In a previous attempt to increase the shock capturing ability of the cell vertex algorithm [5], the artificial dissipation model was replaced by a flux limited dissipation model described by Yee [6]. In this preliminary effort, however, no significant improvement was demonstrated. This paper discusses the shortcomings and presents a modified cell vertex scheme following the TVD approach of Harten and Yee [7]. Inviscid calculations of the flow past a two dimensional blunt body with and without an interfering shock and the three dimensional flow field around the HERMES reentry vehicle demonstrate the greatly improved shock capturing capability of the current method in multiple dimensions.

Numerical Method

The current effort focuses solely on the improvement of the shock resolution of the cell vertex scheme described in [1, 4]. Since the shock capturing ability of a computational algorithm solving the fluid flow equations relies mainly on the discretization of the convective terms, this investigation considers only the inviscid equations. An efficient extension of the present cell vertex TVD scheme for viscous flows is given in [8]. For convenience, the numerical algorithm is presented here in two dimensional notation.

Taking the flow quantities $\vec{W}$ at cell vertices, a finite volume spatial discretization of the Euler equations in integral form for a perfect gas leads to

$$\left(\frac{\partial \vec{W}}{\partial t}\right)_{i,j} = -\frac{1}{V_{i,j}}\left(\vec{Q}_{i,j} + \vec{D}_{i,j}\right). \tag{1}$$

Here, $\vec{Q}_{i,j}$ represents a central approximation of the flux balance over the control volume $V_{i,j}$. $\vec{D}_{i,j}$ denotes an operator which either represents an artificial dissipation to suppress the tendency of central schemes to decouple, or it describes an antidiffusive flux function modifying the central flux stencil to that of an upwind biased scheme.

In the cell vertex method described in [1], $\vec{Q}_{i,j}$ may be interpreted as the flux through the boundary of a "supercell" $V_{i,j}$ (Figure 1a) involving all neighboring points of grid node i, j. The artificial dissipation, $\vec{D}_{i,j}$, is provided by the dissipation model of Jameson [2] using a blend of second and fourth differences of the conserved variables.

In order to improve the shock resolution of the original scheme, the artificial dissipation is replaced by a flux limited term following Yee [6] or Harten and Yee [7]. Since this model, in contrast to Jameson's scalar dissipation, is based on the local characteristic approach, the central discretization operator, $\vec{Q}_{i,j}$, for the inviscid flux can now be converted to an upwinded operator. As usual, the upwind operator for multiple dimensions is formed by successive application of the one dimensional operator in each direction of the curvilinear coordinate systems. Using this approach, the update of the state vector at node i, j does not take information from the diagonal nodes into account. Therefore, the original central operator $\vec{Q}_{i,j}$ is modified from a "nine point" to a "five point" stencil. Connecting the centers of the cells surrounding a node forms an auxiliary cell as sketched in Figure 1b. The resulting flux stencil is similar to a cell centered scheme for the auxiliary cell, but it still offers the advantages of cell vertex methods in the treatment of boundaries and singularities. The inviscid flux through the cell face $i + \frac{1}{2}$ is now evaluated as follows

$$\vec{Q}_{i+\frac{1}{2},j} + \vec{D}_{i+\frac{1}{2},j} = \frac{1}{2}\left(\bar{\bar{F}}_{i,j} + \bar{\bar{F}}_{i+1,j}\right) \cdot \vec{S}_{i+\frac{1}{2},j} + \frac{1}{2} R_{i+\frac{1}{2},j}\, \vec{\Phi}_{i+\frac{1}{2},j} \qquad (2)$$

where $\bar{\bar{F}}$ denotes the tensor of the flux density, $\vec{S}_{i+\frac{1}{2},j}$ is the surface vector of face $i+\frac{1}{2}$ and R is the right eigenvector matrix of the flux jacobian in transformed space. The flux function $\vec{\Phi}$ is a function of differences of characteristic variables and modifies the symmetric approximation $\vec{Q}_{i+\frac{1}{2},j}$ of the inviscid flux in accordance with local wave propagation. Limiter functions bring this upwind scheme to second order. Two flux functions suggested by Yee [6] and Harten and Yee [7] have been implemented which both lead to a scheme with TVD properties. Only results obtained with the dissipative flux function given in [7] are presented. The spatial discretization results in a system of ordinary differential equations in time which is solved by an explicit multistage Runge Kutta scheme.

Results

The first set of figures demonstrates the improved shock resolution of the modified cell vertex TVD scheme for the flow past a 2D cylinder. In Figures 2 and 3 results obtained with the dissipative flux function following [7] in combination with both the central five and nine point stencil are shown for a freestream Mach number of 10. Figure 2 displays iso-Mach lines with an increment of 0.25 and Figure 3 contains the Mach number distribution along the stagnation streamline. With the nine point stencil the shock smears over 3-4 cells, while the five point stencil resolves the shock within 2 cells. In this case, the crisp shock resolution obtained in one dimension could be retained in two dimensions. For comparison, Figures 2 and 3 also contain the solutions with Jameson's 2nd and 4th order dissipative terms. Using this dissipation model the shock smears over 4 cells for both stencils and the iso-Mach lines exhibit oscillations in the vicinity of the shock. Figure 4 displays line plots of Mach number obtained with the modified cell vertex scheme for freestream Mach numbers 5, 10 and 20. The Mach number distributions are plotted along the stagnation streamline and along a ray inclined 36 degrees from the vertical. This investigation emphasizes the sharp shock resolution afforded by the current method. The resolution is independent of freestream Mach number. The standoff distance for the various Mach numbers agree well with the values published in classical literature.

The next figure demonstrates the applicability of the present scheme for a complex shock-shock interaction problem. Figure 5 contains the solution for a 2D cylinder with an interfering shock at freestream Mach number 8.03. This flow field, identified in the literature as Type IV, is characterized by multiple length scales which arise due to the formation of intricate features within the interaction region. Figure 5 displays the 60×40 grid and iso-Mach lines with an increment of 0.25 showing crisply resolved bow shock, interfering shock and slip lines. Of particular interest is the interaction region itself. Here, the TVD scheme resolves the formation of a supersonic jet bounded by upper and lower slip lines and a terminating strong normal shock. Figure 5 also includes a comparison of the calculated and measured surface pressure.

The final set of figures shows results for the method in three dimensions. The left frame of Figure 6 exhibits the solution for the HERMES reentry vehicle at a freestream Mach number of 8 and an angle of attack of 10°. The C-type grid given in [5] has been used for this calculation. Figure 6 shows iso-Mach lines in the symmetry plane and three selected cross planes. Throughout the flow field, the shock retains the resolution obtained in two dimensions. Figure 6 also contains an enlarged view of the nose region in the symmetry plane. The solution displays no oscillations, even near the singular line occurring in the C-grid structure. The Mach number distributions along specific grid lines in the symmetry plane emphasizes the crisp shock resolution of the current method. This set of figures demonstrates that this shock resolution ability remains unchanged in multiple dimensions.

Summary

A modification of a central difference cell vertex scheme for high speed flow has been presented. In this scheme the discretization operator for the inviscid fluxes has been changed to a cell centered type operator which is applied to a set of auxiliary cells but still provides the update of the state vector at the nodes of the grid cells. The artificial dissipative terms have been replaced by a dissipative flux function following Harten and Yee which results in a scheme with TVD properties. Two and three dimensional calculations of high Mach number flows demonstrate the greatly improved shock resolution of the current method in multiple dimensions.

Acknowledgements

This work was accomplished within a scientific exchange program between DLR and U.S. Air Force during the first author's stay at Wright Patterson AFB. The authors would like to thank M. Aftosmis and Dr. R. Radespiel for many helpful discussions.

References

[1] C. Rossow, N. Kroll, R. Radespiel, and S. Scherr. "Investigation of the Accuracy of

Finite Volume Methods for 2- and 3-Dimensional Flows". *AGARD-CPP-437*, 1988.

[2] A. Jameson, W. Schmidt, and E. Turkel. "Numerical Solutions of the Euler Equations by a Finite Volume Method Using Runge-Kutta Time Stepping Schemes". *AIAA Paper 81-1259*, 1981.

[3] N. Kroll, C. Rossow, S. Scherr, J. Schöne, and G. Wichmann. "Analysis of 3-D Aerospace Configurations Using the Euler Equations". *AIAA Paper 89-0268*, 1989.

[4] R. Radespiel, C. Rossow, and R.C. Swanson. "An Efficient Cell-Vertex Multigrid Scheme for the Three-Dimensional Navier-Stokes Equations". *AIAA Paper 89-1953*, 1989.

[5] J. Schöne, N. Kroll, C. Rossow, H. Li, and Th. Sonar. "A Central Finite Volume TVD Scheme for the Calculation of Supersonic and Hypersonic Flow Fields Around Complex Configurations". *AIAA Paper 89-1975*, 1989.

[6] H.C. Yee. "Construction of Implicit and Explicit Symmetric TVD Schemes and Their Applications". *Journal of Computational Physics*, 68:151–179, 1987.

[7] H.C. Yee and A. Harten. "Implicit TVD Schemes for Hyperbolic Conservation Laws in Curvilinear Coordinates". *AIAA Journal*, 25:266–274, 1987.

[8] R. Radespiel and N. Kroll. "A Multigrid Scheme with Semicoarsening for Accurate Computations of Viscous Flows". *DLR-IB 129-90/19*, 1990.

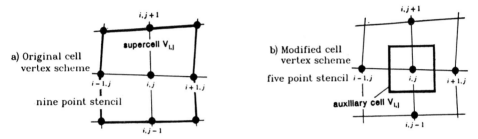

Figure 1: Nodes involved in the central approximation of the inviscid flux.

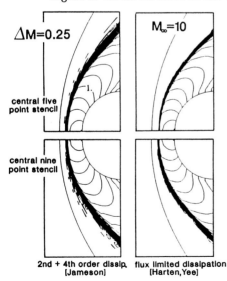

Figure 2: Mach contours for a 2D cylinder on a 40 × 40 grid.

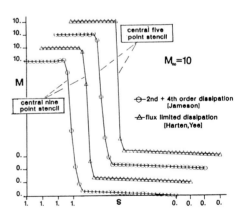

Figure 3: Mach number distribution for a 2D cylinder at $M_\infty = 10$ along the stagnation streamline.

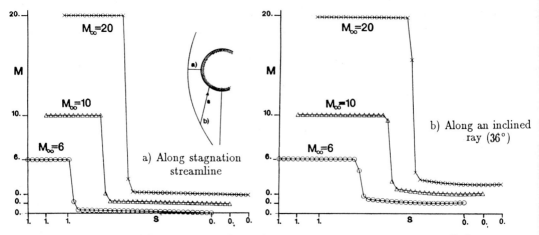

Figure 4: Effect of freestream Mach number on shock resolution for a 2D cylinder.

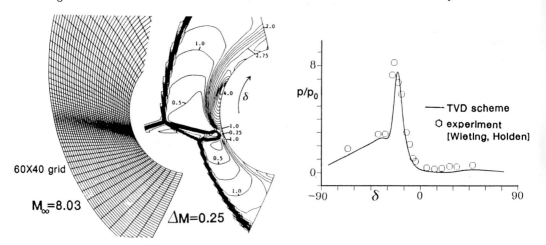

Figure 5: Mach contours and surface pressure for Type IV interaction.

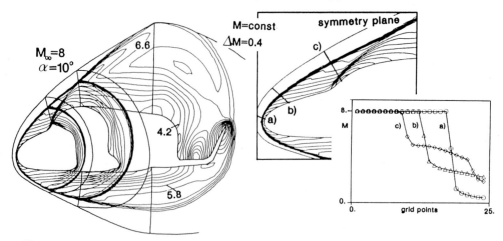

Figure 6: Mach contours and local Mach number distribution for the HERMES configuration.

Computation of hypersonic flows around blunt bodies and wings

Charles-Henri BRUNEAU, Jacques LAMINIE
Laboratoire d'Analyse Numérique, Bâtiment 425
Université Paris-Sud, 91405 ORSAY (France)

The method presented in former conferences is applied to blunt bodies and delta wings with rounded leading edge in order to show its ability to capture strong bow shocks.

The steady Euler equations are written in conservative form for the density and the components of the velocity in $2D$ and $3D$. The pressure is replaced in the equations of conservation of momentum using Bernoulli's equation. Then the solution is obtained by means of Newton linearization and least-squares embedding.

The method is first applied in $2D$ to compute hypersonic flow around a circular cylinder or an ellipse at $M_\infty = 3$ and 8. Depending on the initialization, an entropy corrector and a mesh adaptation procedure are required to capture the shock and to fit the mesh to the shock respectively. Then the flow over a sphere is computed and the location of the shock is very close to the theory. In both cases the shock is captured within a cell.

At last, the method is used to compute the flow around a delta wing with smooth leading edge. It is not clear yet, whether vortical phenomena can occur or not around smooth surfaces using Euler model.

References.
Ch.-H. Bruneau, J. Laminie, J.J. Chattot, Computation of hypersonic vortex flows with an Euler model, 11th ICNMFD, Williamsburg 1988, *Lecture Notes in Physics 323*, 1989.
Ch.-H. Bruneau, Computation of hypersonic flows round a blunt body in $2D$ with Euler model, to appear in Computers and Fluids.

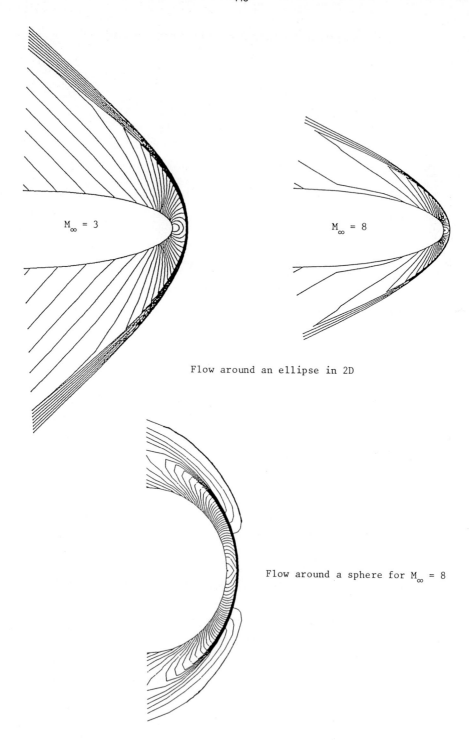

Flow around an ellipse in 2D

Flow around a sphere for $M_\infty = 8$

HYPERSONIC FLOWS ABOUT WAVERIDERS WITH SHARP LEADING EDGES

Kevin D. Jones, GRA, F. Carroll Dougherty, Asst. Prof.,
and Helmut Sobieczky, DLR, Visiting Prof.
Aerospace Engineering Sciences, University of Colorado
Box 429, Boulder, Colorado 80309-0429, USA

Hypersonic waverider design has recently become an important research topic in the aerospace industry. An inverse design effort has been initiated where the waverider body is determined from a specified shock shape (ref. 1). The ultimate goal of this research is to develop computational methods where the supersonic inlet shape and inlet flow conditions are specified and an optimized forebody that provides those conditions is found. One author has been working on the inverse design code, while the other two have been working with Euler and Navier-Stokes codes to verify the inverse design results. This presentation addresses the application of the Euler codes to waverider shapes with sharp leading edges. Previously, calculations were done on bodies with rounded leading edges or with conical solutions for the nose initial conditions. Solutions about waverider shapes have been simulated without resorting to either shortcut.

The approach taken has been to use the F3D code developed at NASA/Ames Research Center (ref. 2). The F3D code utilizes a two-factored approximate-factorization algorithm with flux-splitting in the streamwise direction. The flux-splitting provides an "upwinding" that adds inherent stability for supersonic and hypersonic regimes. The code is well-documented, robust, and solves both the Euler and Navier-Stokes equations in generalized coordinates for time-dependent or steady-state simulations. It has been used to solve hypersonic flows up to Mach 25.

Well-constructed grids are crucial to the accuracy and stability of solutions to finite-difference problems. Grid clustering, the treatment of sharp edges on the body, the outer boundary shape, and orthogonality and volume variation of grid cells all play a part in the effectiveness and accuracy of the flow field simulation. In the hypersonic regimes of this study, the flow field solutions are very sensitive to these grid effects. After experience with existing grid generators, it became obvious that a more specialized grid generator was necessary. A new code, HYGRID, was developed specifically for the waverider type-configurations. Details about the new grid generator and the necessary changes to the flow solver can be found in ref. 3.

Several test cases were used to debug and validate the flow solver for the waverider configurations. Limitations to the code's performance were evaluated with respect to the control parameters and the grid quality. Some of the results are discussed below.

The planar shock waverider is a special case of conical waveriders where the radius of the conical shock is driven to infinity, resulting in a planar shock across the lower surface of the configuration. A test case at Mach 5.5 with a Prandtl-Meyer expansion upper surface was run. The shock is visibly planar, and the pressure jump across the shock is within 1% of the analytic solution. The surface and pressure isolines at a typical cross-section are shown for this simulation in fig. 1.

Several conical shock waverider test cases corresponding to wind tunnel models under investigation at NASA/Langley were run. A Mach 6 waverider with a Prandtl-Meyer expansion upper surface and pressure isolines at a typical cross-section are shown in fig. 2. The computed shock is visibly conical and the pressure jump across the shock is within 2% of the analytic solution.

Long term goals of this project include the simulation of shock/inlet interactions and the analysis of the inlet flow. The computation of these features presents several difficulties. In order to fit grid boundaries to both the forebody and inlet surfaces, the Chimera multiple grid scheme is used (ref. 4). A cone with an axisymmetric wedge inlet is being run as a test case to validate the multiple grid flow solver. The symmetry plane grids and pressure isolines for a preliminary solution are shown in fig. 3. The results reveal the grid sensitivities of the Chimera scheme and illustrate the need for grid refinements, but nevertheless, the basic flow features are captured.

References:
1. Sobieczky, H., Dougherty, F. C., and Jones, K. D., "Hypersonic Waverider Design for Given Shock Waves," to be presented at the *1st International Waverider Symposium*, University of Maryland, Oct. 17-19, 1990.
2. Steger, J. L., Ying, S. X., and Schiff, L. B., "A Partially Flux Split Algorithm for Numerical Simulation of Compressible Inviscid and Viscous Flow," Proceedings of the Workshop held by the Institute of Nonlinear Sciences, University of California, Davis.
3. Jones, K. D., and Dougherty, F. C., "Computational Simulation of Flows About Hypersonic Geometries with Sharp Leading Edges," to be presented at the *8th Applied Aerodynamics Conference*, Portland, AIAA Paper No. 90-3065, Aug. 20-22, 1990.
4. Dougherty, F. C. and Kuan, J. H., "Transonic Store Seperation Using a Three-Dimensional Chimera Grid Scheme," AIAA Paper No. 89-0637, Jan., 1989.

Acknowledgements:
This work was supported by NASA/Langley through contract number NAG-1-880.

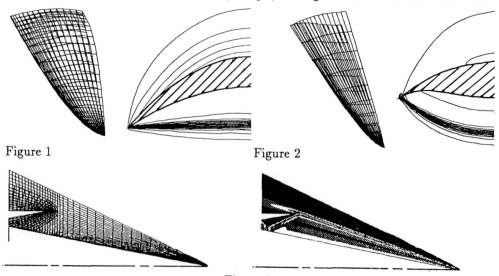

Figure 1   Figure 2

Figure 3

UPWIND RELAXATION METHOD FOR HYPERSONIC FLOW SIMULATION

B. Müller
DLR, Institute for Theoretical Fluid Mechanics,
Bunsenstr. 10, D-3400 Göttingen, F.R. Germany

The 3D Euler and thin-layer Navier-Stokes equations are solved by an upwind relaxation method for hypersonic flow over blunt and pointed bodies.

The inviscid fluxes are discretized by the second-order (first-order at extrema) upwind total variation diminishing (TVD) method of Harten and Yee. Special care in enforcing the entropy condition is necessary. For inviscid hypersonic flows over round leading edges, unphysical attached shocks can be avoided by reducing the spatial accuracy to first-order everywhere when starting from freestream initial conditions. For laminar hypersonic flows over blunt bodies, numerical stability in the stagnation point region requires Harten's entropy fix to be applied to the linear waves in the wall tangential directions as well. The 'entropy parameter' is anisotropicly scaled by the spectral radii of the inviscid Jacobian matrices. The viscous fluxes are second-order central-differenced.

The time derivative is approximated by the first-order Euler implicit formula. The linearization of the inviscid fluxes is first-order in space only and non-conservative to ensure stability and the TVD property, respectively. The resulting linear system is solved by a relaxation method. Alternating Gauss-Seidel relaxation is applied in the streamwise direction. In the crossflow planes, Jacobi line relaxation is employed in the circumferential direction, followed by simultaneous Gauss-Seidel line relaxation in the normal direction. The solution of the mutually independent block-tridiagonal linear systems is vectorized on the CRAY Y-MP. For supersonic and hypersonic flow over pointed bodies with attached shocks, the time-dependent difference equations in crossflow planes are iterated to the steady state by marching downstream from a conical solution near the apex. Local time stepping is employed to accelerate the convergence to the steady-state.

The method is validated for inviscid hypersonic flow of perfect gas over a waverider [1]. Here, an application to the Saenger forebody is considered at $M_\infty = 4.5$ and $\alpha = 6°$ [1]. The convergence history of the conical solution in the crossflow plane $i=2$ next to the apex shows a reduction of the L_2-norm of $\Delta\rho^n = \rho^{n+1} - \rho^n$ by 5 orders of magnitude within 200 time levels (Fig. 1). Fig. 2 indicates that the pressure maximum in the outflow plane occurs on the lower side close to the round leading edge. The compression near the concave part on the lower side is caused by the body shape changing from almost elliptical convex cross sections near the apex to wing-body cross sections downstream. The strong expansion around the leading edge is closely followed by a compression due to the concave upper part of the Saenger forebody in the outflow plane. Using grid adaptation, the outer shock can be better resolved even on coarser meshes [1].

In spite of the coarse 25x15x25 (streamwise x circumferential x normal) mesh, the surface pressure distribution determined by the present shock capturing method is close to a shock fitting result [2] and experimental data [3] for laminar hypersonic flow over a sphere-15° cone at 15° angle of attack (Fig. 3). It is noted that the Stanton number distribution is in good agreement with [2] as well.

For laminar hypersonic flow over a circular cone with 10° half angle at 24° incidence, the skin friction lines exhibit a conical behaviour, as their projection on a plane normal to the axis indicates (Fig. 4). In particular, the separation line is straight apparently starting from the singular apex.

References

[1] Müller, B., Niederdrenk, P., Sobieczky, H.: Simulation of Hypersonic Waverider Flow, Proc. of First International Hypersonic Waverider Symposium, University of Maryland, Oct. 17-19, 1990.

[2] Riedelbauch, S., Müller, B., Rues, D.: 3D Viscous Hypersonic Flow over Blunt Bodies, DFVLR-FB 89-09, 1989.

[3] Cleary, J.W.: An Experimental and Theoretical Investigation of the Pressure Distribution and Flow Fields of Blunted Cones at Hypersonic Mach Numbers, NASA TN D-2929, 1965.

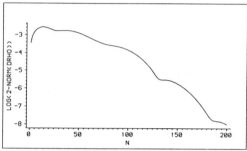

Fig. 1 Convergence history of $\|\Delta\rho^n\|_2$ for Saenger forebody, $M_\infty = 4.5$, $\alpha = 6°$, Euler equations, adapted 16x43x16 mesh.

Fig. 3 Surface pressure coefficient for sphere-15° cone, $M_\infty = 10.6$, $\alpha = 15°$, $Re_{\infty,R} = 1.1 \times 10^5$, $T_\infty = 47.333\ K$, $T_w/T_\infty = 6.3381$, 25x15x25 mesh.

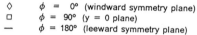

◇ $\phi = 0°$ (windward symmetry plane)
□ $\phi = 90°$ (y = 0 plane)
— $\phi = 180°$ (leeward symmetry plane)

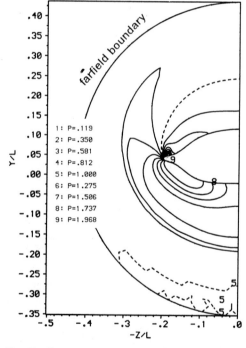

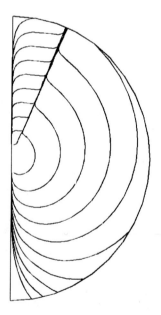

Fig. 2 Pressure contours ($p/p_\infty = const$) for Saenger forebody, $M_\infty = 4.5$, $\alpha = 6°$, Euler equations, unadapted 31x83x31 mesh.

Fig. 4 Skin friction lines for 10° cone, $M_\infty = 7.95$, $\alpha = 24°$, $Re_{\infty,L} = 4.2 \times 10^5$, $T_\infty = 55.39\ K$, $T_w/T_\infty = 5.456$, 11x23x21 mesh.

PNS SOLUTION USING SPARSE QUASI-NEWTON METHOD FOR FAST CONVERGENCE

N. Qin and B.E. Richards
Department of Aerospace Engineering
University of Glasgow, Glasgow G12 8QQ, Scotland

For hypersonic viscous flows around spaceplane shapes, the parabolized Navier-Stokes (PNS) solution appears to be a reasonable choice due to the fact that (1) the inviscid flow outside the boundary layer is most likely to be supersonic in the streamwise direction; (2) streamwise separation is unlikely to happen under the design condition and (3) it considerably reduces the computing time and the storage requirement, which are the most serious factors to limit full 3D NS hypersonic viscous flow simulation. Therefore, the PNS approach can provide a useful tool for the design of hypersonic vehicle using currently available supercomputers, such as IBM 3090 or CRAY XMP.

In the discretization of the PNS equations, the streamwise direction is treated as the marching direction and an implicit scheme is usually employed to avoid the strict limit on the step size for stability with an explicit scheme. In the crossflow plane, a high resolution scheme (in the sense of both shock wave and shear layer resolution) is preferred for an accurate prediction of flowfield and aerothermodynamic parameters such as heat transfer rate. After the discretization, we obtain a large sparse nonlinear system for Q_i

$$R(Q_{i-2}, Q_{i-1}, Q_i) = 0$$

where Q_{i-2}, Q_{i-1} are known solutions of the two upstream stations. The currently widely used noniterative implicit procedures attempt to linearize the system in various simplified ways.

In solving steady state problems in CFD, implicit time dependent schemes are generally used, where time accuracy is not of interest. Therefore, approximations and simplifications in the implicit operator may be made. Although these approximations and simplifications can degrade the convergence of an implicit scheme as pointed out in [2,3], but the spatial accuracy is not influenced as long as a steady state solution is achieved. However, in PNS solution, inconsistent simplifications might cause accumulated errors, which will propagate further downstream. Hence, PNS approaches have to focus on the consistency and accuracy of the linearization.

After an exact linearization is implemented on the nonlinear system, a Newton method is obtained written as

$$\left(\frac{\partial R}{\partial Q_i}\right)^k (Q_i^{k+1} - Q_i^k) = -R(Q_{i-2}, Q_{i-1}, Q_i^k)$$

The problem is that the Jacobian of the nonlinear system arising from discretization of the NS equations is very difficult to obtain if at all possible. This is due to (1) the complex physical modelling involving shock waves, viscous effects, turbulence and chemical reacting gas and, therefore, (2) the complicated formulation of modern high resolution spatial discretization schemes, such as TVD and flux difference splitting schemes.

In order to obtain fast convergence without calculating the Jacobian analytically, we propose a sparse quasi-Newton method [1] for the PNS solution, which updates an approximation to the Jacobian at each iteration extracting information of the system from the solution obtained. In this way the system is treated as a black box and the procedure is general and robust. Previous efforts of the present authors have demonstrated the theoretically proved superlinear convergence of the sparse quasi-Newton method for Euler [2] and NS [3] solutions using high resolution schemes. The general procedure can be written as

$$A^k S^k = - R(Q_i^k)$$
$$Q_i^{k+1} = Q_i^k + S^k$$
$$A^{k+1} = A^k + P_j [D^+ R(Q_i^{k+1})(S^k)^T]$$

More details on the formulation of the SQN method for the nonlinear block structured sparse system arising from CFD can be found in [2,3]. Strategies for initialization are also given there. In using the sparse quasi-Newton method for the PNS solution, a good initial value at station i, Q_i^0 can be obtained from the converged solution of the previous station.

The hypersonic flows around a sharp cone (Tracy's cone) or blunted cones (RAE case) have been simulated using the PNS code, where spatial discretization was carried out by Osher's flux difference splitting scheme. The solutions on the starting surface for the spatial marching were obtained by a locally conical NS solver for the sharp nose cases and a full 3D NS solver for the blunt nose cases. Only two or three iterations at each station were required for convergence.

References:

1. Schubert, L.K., Modification of a Quasi-Newton Method for Nonlinear Equations with a Sparse Jacobian, Math Comp vol. 24 , 1970, pp27-30.

2. Qin, N. and Richards, B.E., Sparse Quasi-Newton Method for High Resolution Schemes, Proc 7th GAMM Conf on Num Meth in Fluid Mech 1987.

3. Qin, N. and Richards, B.E., Sparse Quasi-Newton Method for Navier-Stokes solution, Proc 8th GAMM Conf on Num Meth in Fluid Mech, 1989.

NUMERICAL SIMULATION OF NON-EQUILIBRIUM HYPERSONIC FLOW USING THE SECOND-ORDER BOUNDARY LAYER EQUATIONS

M.L. Sawley and S. Wüthrich

Institut de Machines Hydrauliques et de Mécanique des Fluides,
Ecole Polytechnique Fédérale de Lausanne,
ME - Ecublens, CH-1015 Lausanne, Switzerland

Within the framework of the Hermes European space shuttle project, the viscous flow of a chemically-reacting and radiating gas over a body at hypersonic velocity has been studied. The boundary layer equations have been solved numerically for either two-dimensional planar or axisymmetric laminar flow.

The gas is assumed to be a mixture of perfect gases, with species N_2, O_2, N, O and NO considered, corresponding to air at temperatures up to 8000 K. The enthalpy of each species is given as a polynomial function of temperature obtained by fitting statistical thermodynamic results. The model developed by Straub is used to calculate the transport coefficients (viscosity, thermal conductivity and multi-component diffusion) as a function of the composition and thermodynamic state of the mixture. Both equilibrium and non-equilibrium chemistry models have been considered. To simulate non-equilibrium flows, a set of seventeen chemical reactions is used.

For first-order (in $Re^{-1/2}$) boundary layer calculations, there exists no pressure gradients normal to the surface of the body. The pressure throughout the boundary layer is then set equal to the inviscid surface pressure determined using an Euler code into which the same chemical models have been incorporated. For second-order calculations, the effects of the longitudinal surface curvature with the associated normal pressure gradients are accounted for by the inclusion of the appropriate terms in the governing equations. In addition, for high Mach number flows the bow shock is strongly curved, leading to an entropy gradient at the edge of the boundary layer. The effect of "entropy layer swallowing" is taken into account by using the pressure, and normal gradients of pressure and tangential velocity at the body surface calculated by the Euler code to determine the appropriate values at the edge of the boundary layer.

The boundary layer equations are discretized using finite differences, and solved in a decoupled manner with global iteration accounting for coupling between the equations. Special attention is taken in the resolution of the equations along the stagnation line in order to obtain accurate profiles, which are used in the initiation of the space-marching procedure. The chemical source term is treated in a semi-implicit fashion, which provides for an efficient and robust resolution of the species continuity equations over the entire Mach range considered.

The influence of the various physical and chemical models has been examined by calculating the flow over a hyperboloid (the equivalent axisymmetric body for the American space shuttle) using the first-order equations. Calculations have been undertaken, using different chemistry models, for the range $9 \leq M_\infty \leq 28$. Both fully catalytic and non-catalytic wall surfaces have been considered, with either a fixed temperature or radiation adiabatic wall boundary condition. For the case of a fully catalytic wall, only a slight difference is observed between the non-radiative heat flux at the wall calculated assuming non-reactive flow, chemical equilibrium or non-equilibrium chemistry. However, major differences in the profiles of temperature and species concentrations are observed. The heat flux calculated for non-equilibrium flow over a non-catalytic wall is significantly lower.

An study of second-order effects has been undertaken by calculating the non-equilibrium flow over the windward surface of an ellipse (a = 60 cm, b = 15 cm) with extension flat plate (of length 100 cm) at an angle of attack of 30°. Freestream conditions of $M_\infty = 20$, $p_\infty = 10$ Pa, $T_\infty = 250$ K and $Re_\infty = 6 \times 10^4$ m^{-1} were chosen. A non-catalytic wall surface, with $T_W = 1500$ K, was assumed. The results show that the second-order effects result in an increase along the entire length of the body of the wall heat flux and skin friction coefficient. The figure below show that there is significant vorticity (associated with the entropy gradient) at the edge of the boundary layer, which is accounted for by the second-order calculations. Also, the longitudinal surface curvature along the ellipse section produces a substantial pressure gradient in the boundary layer region.

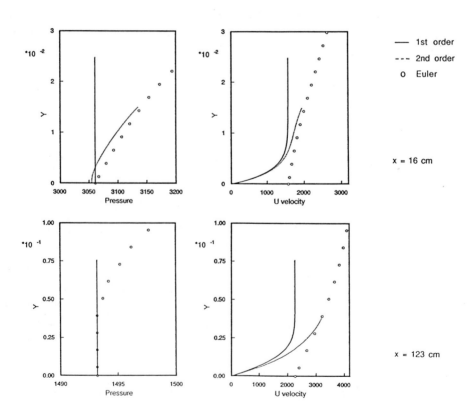

CHEMICAL EQUILIBRIUM AND NON-EQUILIBRIUM INVISCID FLOW COMPUTATIONS USING A CENTERED SCHEME

J.B. Vos[†] and C.M. Bergman[‡]

[†] Hydraulic Machines and Fluid Mechanics Institute (IMHEF)
Swiss Federal Institute of Technology - Lausanne, CH-1015 Lausanne, Switzerland
[‡] European Centre for Research and Advanced Training in Scientific Computing
(CERFACS), 42 Avenue Gustave Coriolis, F-31057 Toulouse, France

ABSTRACT

Within the framework of the collaboration between IMHEF and CERFACS a 2D Inviscid Flow solver for Hypersonic flows has been developed. The Euler equations are discretized in space on a structured mesh using the Finite Volume method with centered differences. The resulting system of ordinary differential equations is integrated in time using the explicit Runge Kutta scheme. Artificial dissipation terms are added to damp odd/even oscillations allowed for by centered space differences, and to damp spurious oscillations near discontinuities. External shock waves in the flow field are treated by a shock fitting procedure, while (weaker) internal shock waves are captured by the numerical scheme. A complete description of the numerical method can be found in [1].
The strong shock waves present in hypersonic flows give rise to high temperatures directly behind the shock wave, which may result into the dissociation of air. This is a process which costs energy, hence temperatures in the flow field will be reduced. Air dissociation can be modelled on different levels, which depend on the ratio of the characteristic time scales of the flow and the chemistry. If the characteristic time scale of the chemistry is much smaller than that of the flow, it can be assumed that the flow is in chemical equilibrium, i.e. chemical reactions are taking place, but the production of a chemical species is balanced by its destruction. The other limit is that the chemistry time scale is much smaller than that of the flow, hence no chemical reactions are taking place. The chemistry is frozen, and the air is treated as a thermally perfect gas. If the time scales are of the same order of magnitude the flow is in chemical non-equilibrium. These three levels of modelling have been included in the Euler solver. Incorporation of equilibrium and frozen chemistry is straightforward for the centered scheme described

above, since only the relation which connects the pressure to the density and total energy had to be changed. This has been done using the effective $\tilde{\gamma}$ approach, where $\tilde{\gamma}$ is the effective ratio of specific heats. For explicit schemes, this $\tilde{\gamma}$ needs to be calculated only once per 100 time steps. Non-equilibrium chemistry has been incorporated by solving three partial differential equations for the partial densities of the species N, O and NO using the Runge Kutta scheme described above, together with two algebraic equations for the species N2 and O2. The time step used in the Runge Kutta time stepping is in this case the minimum of the chemical time step and the fluid dynamic time step. Calculations showed that only in the initial phase of the time integration process the chemical time step was the smallest of the two. Figure 1 shows the calculated temperatures for the flow around a sphere. Owing to the shock fitting procedure the external bow shock is sharp and oscillation free. The highest temperatures are found when the gas is treated as frozen since in this case no dissociation is taking place. The lowest temperatures along the stagnation line are obtained when it is assumed that the flow is in equilibrium. For the non-equilibrium calculation, the flow is frozen across the shock wave, and is in chemical equilibrium at the stagnation point. This explains the strong temperature gradient between shock and body.

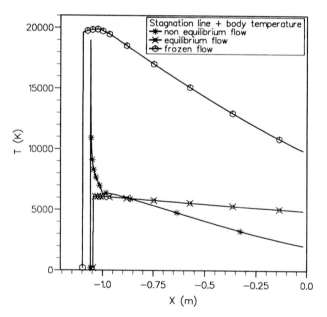

Figure 1. Flow over a sphere, p_∞= 10 Pa, T_∞=200 K, M_∞ = 25.

[1] Vos, J.B. and Bergman, C.M., "A 2D Euler Solver for Hypersonic Flows Including Equilibrium and Non-Equilibrium Chemistry". Report IMHEF T-90.

NUMERICAL SIMULATIONS OF SEPARATED FLOWS AROUND BLUFF BODIES AND WINGS BY A DISCRETE VORTEX METHOD COMBINED WITH PANEL METHOD

Shigeru ASO

Department of Aeronautical Engineering, Kyushu University, Fukuoka 812, JAPAN

Masanori HAYASHI

Nishinippon Institute of Technology, Fukuoka 800-03, JAPAN

1. Introduction

A combination of a discrete vortex method and a panel method[1] is one of the useful methods for simulating separated flows around bluff bodies and wings. In the method the procedures to determine the circulations and the locations of shedding vortices are investigated carefully. The authors have proposed a vortex shedding model in order to express a separated shear layer as a weak and fine row of vortices for the calculations of separated flows around rectangular and concave cylinders. A new method for the calculation of separated flows around a wing section at specified angle of attack or pitching wing section are also proposed.

2. Numerical Results and Discussions

Calculated results of separated flows around rectangular cylinders are shown in Fig. 1. The results show excellent agreements with experiments. Also calculated C_D(Fig.2) show sharp peak at $d/h=0.66$ and characteristics of critical geometry is predicted at excellent agreement with experiments[2]. Also separated flows around trapezoidal and concave cylinders are simulated by the same method[2]. The results are not shown here due to the limitation of space, however calculated flow patterns and aerodynamic characteristics show good agreements with experiments.

Separated flows around airfoils of NACA0012 and NACA4412 at various attack angles are

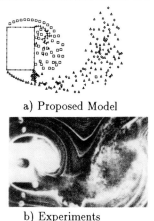

a) Proposed Model

b) Experiments
Fig.1 Flow pattern of square cylinder($d/h=0.667$)

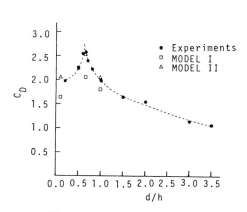

Fig.2 Comparisons of C_D with experiments

simulated. In the calculation separation points are estimated by solving boundary layer equations and the circulations of shedding vortices are estimated using local velocities near separation points[3]. Fig. 3 shows an example of comparison of calculated pressure coefficients with experimental data. The results show excellent agreement with experiments.

Flow patterns of separated flows around a pitching airfoil is shown in Fig. 4. Figures show that the separation point, flow pattern and the extent of the separated region in increasing or decreasing process of attack angle are quite different even at the same angle of attack in a cycle. Fig. 5 shows the behavior of C_L versus. And the maximum lift is greater than that for steady case. Also those features express the characteristics of the deep stall regime qualitatively. The results show excellent agreements and a hysteresis of lift for pitching airfoil at dynamic stall is also simulated properly.

References
[1] Sakata,H. et. al.:J. Japan Soc. of Mech. Eng.,49B(1983),801- 808.
[2] Aso,S. and Hayashi,M.: Memoirs Faculty. Eng. Kyushu Univ., 50-3 (1990).(to be published.)
[3] Aso, S. et. al.: Proc. The Sym. Mech. for Space Flight 1988 of ISAS (1988), pp.11-20.

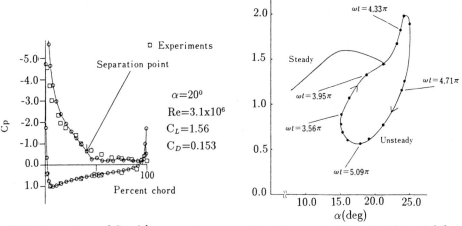

Fig.3 Comparison of C_p with experiments(NACA4412,$\alpha=20°$)

Fig.5 Hysteresis of C_L of pitching airfoil (NACA4412,$\alpha=20° + 5°$ sinωt, k=0.2)

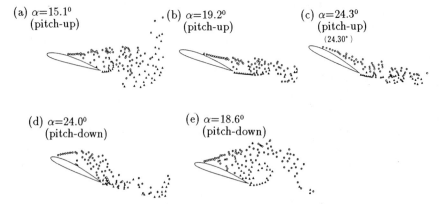

Fig.4 Flow patterns around pitching airfoil (NACA4412,$\alpha=20°+ 5°$sinωt, k=0.2)

NUMERICAL STUDIES OF TRANSVERSE CURVATURE EFFECTS ON COMPRESSIBLE FLOW STABILITY

Michèle G. Macaraeg
NASA Langley Research Center, Hampton, Virginia

Q. Isa Daudpota
Old Dominion University Research Foundation, Norfolk, Virginia

A spectral cylindrical linear stability code (SPECON) is presented for temporal stability analysis of compressible flow with transverse curvature. The mean flows studied include a boundary layer profile over a cylinder and a parabolized Navier Stokes solution on a blunted cone. The effect of curvature is contrasted for a range of Mach numbers extending from incompressible to Mach 6.8. For low Reynolds numbers, increasing curvature tends to stabilize the "viscous" first mode; however, at high Reynolds number this mode is destabilized with increasing curvature (see Fig. 1). In the planar limit, growth rate decreases with Mach number for a fixed Re, in agreement with previous Cartesian results.

Unlike the first modes, supersonic modes in the planar limit do not have growth rates that decrease monotonically with Mach number. Another feature distinguishing of supersonic modes is that they tend to be destabilized with curvature as Mach number is lowered at constant Re.

The second mode, which is dominant at higher Mach numbers, is found to be stabilized with increasing curvature. However, at equivalent streamwise wavenumber and very high Reynolds number an inviscid mode similar in structure is found to be destabilized with curvature. In addition, previously unreported essentially inviscid, unstable modes with a fine scaled structure, are found at very high Reynolds number at streamwise wavenumbers above the second mode. A possible relation with experimentally discovered modes at frequencies above the second mode is conjectured. Experiments [1] have shown the presence of unstable modes at wavelengths higher than those predicted by past linear stability results. Such modes are found to be unstable in the present study but at higher Re than those in the experiments. These modes have a highly peaked behavior in the critical layer, and their structure near the wall suggests an inviscid character.

Stability of flow on a blunt body has been conducted to see the effect of imposing different far field boundary conditions on a flow bounded by a shock. For the case studied it is found that Neuman conditions give rise to unstable disturbances, which decay to a nonzero constant (see Fig. 2). Dirichlet conditions are unsuccessful at producing this unstable mode. Further study is merited for adequate understanding of disturbance boundary conditions at a shock.

1. Stetson, K.F., Thompson, E.R., Donaldson, J.C, and Siler, L.G., AIAA 83-1761, 1983.

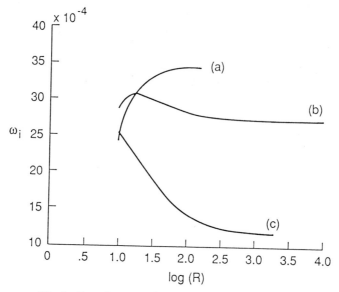

Fig. 1 Growth rates vs log(radius) for $M_\infty=1$, $\alpha = 0.2$, $\beta = 0$.
  a) Re = 2200
  b) Re = 3500
  c) Re = 5000.

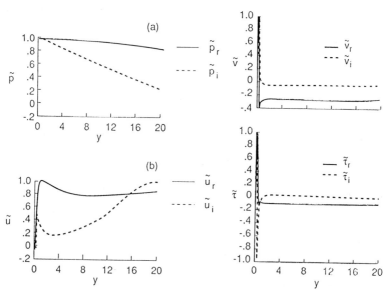

Fig. 2 Disturbance eigenfunctions associated with PNS blunted cone solution. Me = 3.367, Re = 34438.7, Te = 103.7K, radius = 7.3, $\alpha = 0.1$.

CONSERVATIVE SCHEMES FOR SUPERSONIC FLOWS WITH INTERNAL SHOCKS

L.I.Turchak

U.S.S.R. Academy of Sciences Computing Centre
Vavilov Str.40, 117967 Moscow

The author and his colleagues from the Computing Centre considered a number of problems of supersonic flows with discontinuities (shocks, contact discontinuities, etc.) past blunt bodies. Numerical simulation was based on the non-conservative shock-fitting Grid-Characteristic method /1/. In spite of the scheme being non-conservative the results obtained correspond well to the main physical features of such flows. In many cases there was a good agreement with the experimental data.

But, for more accurate numerical simulation of gasdynamic flows with interaction of some discontinuities, it is necessary to use conservative algorithms. Besides, for increasing the accuracy of blunt body problems numerical solutions it is desirable to fit the main discontinuities. To achieve the shock fitting it is possible to transform the problem from the Cartesian coordinate system to a curvilinear one. Constructing a conservative algorithm in a curvilinear coordinate system one meets certain difficulties connected with the metric coefficients approximations /2/.

In the report a method of the metric coefficients approximations is proposed. Its main idea is to use the Cartesian coordinates of the corner points of the numerical cells, and a hybrid scheme of a conservative characteristic method is developed. The scheme approximation order depends on the solution smoothness. Another numerical algorithm is constructed on the basis of Osher's ENO-scheme /3/.

Using the algorithms outlined above, some supersonic flows with internal shocks are studied: a plane shock diffraction on a blunt body; a body moving in a explosion domain; internal flow in a channel with a forward facing step.

REFERENCES: /1/ O.M.Belotserkovskii, A.S.Kholodov, L.I.Turchak. In: Current Problems in Computational Fluid Dynamics (1986), Mir Publishers, Moscow. /2/ V.F.Kamenetsky, L.I.Turchak. Soviet Union - Japan Symposium on CFD (1989). /3/ A.Harten, S. Osher, B. Engquist and S. Chakravarthy. J. Appl. Numer. Math. 2, 347 (1986).

The Flow Computation of a Liquid Rocket Engine Combustor of Complex Geometry

By

Tsung Leo Jiang
Institute of Aeronautics and Astronautics
National Cheng Kung University
Tainan, Taiwan 70101

Abstract

A computational model for the gasphase flow in arbitrarily-shaped combustors is formulated and applied to the flow computation of a liquid rocket engine combustor with an acoustic cavity. The SIMPLER algorithm with staggered grids is revised for a general non-orthogonal curvilinear coordinate system. The contravariant velocity components are selected as the dependent variables of the transformed momentum equations. However, the usual nine-point formulation for the pressure equation is avoided in a two-dimensional flow field by some manipulations of the transformed momentum equations of Cartesian velocity components. This computational model has been tested satisfactorily by the comparison of the numerical solutions with the experimental results of a laminar flow through a tube with an axisymmetric contriction. The flow computation of a liquid rocket engine comubstor with a hypothetic acoustic cavity is then conducted to study the detailed flow pattern around the acoustic cavity. The strongly recirculating flow in and out of the acoustic cavity has been predicted, and the effects of combustion on the flow pattern has been examined.

Introduction

The modern design of aerospace propulsion system calls for the sophisticated computer code to aid in the analysis of combustion performance of combustors of complex geometry such as the variable thrust engine (VTE) in the orbit maneuvering vehicle (OMV) [1]. However, before all the physical models can be implemented in the computer code to describe the complex heterogeneous combustion phenomena, the accurate flow computation is considered as the basis and should be assured beforehand. Among all the computational models installed in the computer code for the flow computation, the computational algorithm to compute the gasphase flow in the general body-fitted non-orthogonal curvilinear coordinates is obviously the most essential part of the whole computer code. It deserves more attention than others, and its accuracy and efficiency should be examined before the code development can be further pursued.

The central issue to solve the flow field in a non-orthogonal curvilinear coordinates is the proper selection of the dependent variables of the transformed momentum equations. The possible candidates for these dependent variables include Cartesian velocity components [2], covariant velocity components [3] and contravariant velocity components [4]. The selection of Cartesian velocity components has the merit of simpler transformed equations. However, it is needed to solve all the Cartesian velocity components at each control-volume face to obtain the contravariant velocity components which are used to interpret the convective flow through the control-volume faces. This drawback implies that more computational efforts are needed if the staggered grid system is used. The selection of covariant velocity components as the dependent variables of the transformed momentum equations warrants the five-point formulation in the pressure and pressure correction equations in a two-dimensional flow situation. This is because of the alignment of covariant velocity components with the grid lines such that only the pressure gradient's component along the grid line exists in the transformed momentum equations. However, in order to obtain the contravariant velocity components to describe the convective flow through the control-volume faces, interpolation is usually performed to acquire the other covariant velocity components so that the contravariant velocity components at the control-volume faces can be estimated. Furthermore, the continuity equation in terms of the covariant velocity components is generally nonhomogenous due to the non-orthogonal grids. On the other hand, the selection of contravariant velocity components as the dependent variables of the transformed momentum equations has the following advantages. First, the transformed continuity equation, which is used to formulate both pressure equation and pressure correction equation in the SIMPLER algorithm [5], preserves the same equation form exactly as that in the orthogonal coordinate system, and does not possess a nonhomogeneous source term for the non-orthogonal grids. Therefore, the same solution procedures for the pressure equation and the pressure correction equation as those in the orthogonal coordinate system can be used. Secondly, since the contravariant velocity components are computed directly from the transformed momentum equations, it is not necessary to exercise the interpolation or to compute all the velocity components to obtain the contravariant velocity components at every control-volume face.

The difficulty associated with the selection of contravariant velocity components as the dependent variables of the transformed momentum equations is that the direction of pressure gradient is not aligned with the contravariant velocity components in the non-orthogonal grids in general. Therefore, a nine-point formulation for the pressure equation in a two-dimensional flow computation was usually proposed to overcome this difficulty. The present paper, however, treats this problem in a different way. The transformed continuity equation and the transformed momentum equations are arranged to precisely preserve the same forms as those in the orthogonal curvilinear coordinates. Therefore, a five-point formulation can be proposed, and the revision of SIMPLER algorithm for the non-orthogonal curvilinear coordinates is straightforward and much simpler than others.

Governing Equations

The steady conservation equations for a general dependent variable ϕ can be given in the following general form in terms of the non-orthogonal curvilinear coordinates (ξ, η):

$$\frac{1}{\sqrt{g}}\frac{\partial}{\partial \xi}(\sqrt{g}\rho U\phi - \sqrt{g}g^{11}\Gamma_\phi \frac{\partial \phi}{\partial \xi}) + \frac{1}{\sqrt{g}}\frac{\partial}{\partial \eta}(\sqrt{g}\rho V\phi - \sqrt{g}g^{22}\Gamma_\phi \frac{\partial \phi}{\partial \eta}) = S \tag{1}$$

This general form can also represent the continuity equation and the momentum equations as ϕ is taken to be 1 and the Cartesian velocity components respectively. For the turbulent flow, K-ε two-equation turbulence model is used in which wall function method is adopted in the near wall region. The combustion process is assumed to be controlled by both chemical kinetics and turbulent mixing, therefore, the reaction rate is dominated by the slower one. The one-step Arrhenius rate law is adopted for the process controlled by the chemical kinetics, while the EBU (Eddy-Break-Up) model [6] is used to estimate the rate of turbulent mixing. Namely, the reaction rate is given by

$$\omega = -\text{Min}(\|R_{Arr}\|, \|R_{EBU}\|). \tag{2}$$

Computational Algorithm

From the transformed momentum equations with Cartesian velocity components being the dependent variables, the transformed momentum equation associated with contravariant velocity component U can be derived from u-equation as:

$$\frac{1}{\sqrt{g}}\frac{\partial}{\partial \xi}(\sqrt{g}\rho U x_\xi U - \sqrt{g}g^{11}\mu_{\text{eff}}\frac{\partial x_\xi U}{\partial \xi}) + \frac{1}{\sqrt{g}}\frac{\partial}{\partial \eta}(\sqrt{g}\rho V x_\xi U - \sqrt{g}g^{22}\mu_{\text{eff}}\frac{\partial x_\xi U}{\partial \eta})$$
$$= -\frac{1}{\sqrt{g}}(y_\eta \frac{\partial p}{\partial \xi} - y_\xi \frac{\partial p}{\partial \eta}) + Su \tag{3}$$

or derived from v-equation as:

$$\frac{1}{\sqrt{g}}\frac{\partial}{\partial \xi}(\sqrt{g}\rho U y_\xi U - \sqrt{g}g^{11}\mu_{\text{eff}}\frac{\partial y_\xi U}{\partial \xi}) + \frac{1}{\sqrt{g}}\frac{\partial}{\partial \eta}(\sqrt{g}\rho V y_\xi U - \sqrt{g}g^{22}\mu_{\text{eff}}\frac{\partial y_\xi U}{\partial \eta})$$
$$= -\frac{1}{\sqrt{g}}(x_\xi \frac{\partial p}{\partial \eta} - x_\eta \frac{\partial p}{\partial \xi}) + Sv \tag{4}$$

In general, there are two components of pressure gradient along both coordinate directions present in the equations. Therefore, a nine-point formulation is needed to properly account for the inter-influence between the pressure difference and the velocity components. However, if equation (3) is multiplied by x_ξ and equation (4) is multiplied by y_ξ, we can obtain a new equation by summing both equations, and only one component of pressure gradient is present in the new U-equation. This equation is the transformed momentum equation associated with the covariant velocity components in nature. However, we can have the contravariant velocity components as the dependent variables by some further manipulations. Thus, we obtain the transformed equation associated with contravariant velocity component U as:

$$\frac{1}{\sqrt{g}}\frac{\partial}{\partial \xi}(\sqrt{g}\rho UU - \sqrt{g}g^{11}\mu_{\text{eff}}\frac{\partial U}{\partial \xi}) + \frac{1}{\sqrt{g}}\frac{\partial}{\partial \eta}(\sqrt{g}\rho VU - \sqrt{g}g^{22}\mu_{\text{eff}}\frac{\partial U}{\partial \eta}) = -\frac{1}{E}\frac{\partial p}{\partial \xi}$$
$$+ \frac{1}{E}(x_\xi S u + y_\xi S_v + S u^*) \quad (5)$$

where $E = x_\xi^2 + y_\xi^2$. In the same way, we can derive the equation for V as:

$$\frac{1}{\sqrt{g}}\frac{\partial}{\partial \xi}(\sqrt{g}\rho UV - \sqrt{g}g^{11}\mu_{\text{eff}}\frac{\partial V}{\partial \xi}) + \frac{1}{\sqrt{g}}\frac{\partial}{\partial \eta}(\sqrt{g}\rho VV - \sqrt{g}g^{22}\mu_{\text{eff}}\frac{\partial V}{\partial \eta}) = -\frac{1}{F}\frac{\partial p}{\partial \eta}$$
$$+ \frac{1}{F}(y_\eta \tilde{S}v + x_\eta \tilde{S}_u + S v^*) \quad (6)$$

where $F = x_\eta^2 + y_\eta^2$. Also, the continuity equation in terms of the contravariant velocity components is given by

$$\frac{1}{\sqrt{g}}\frac{\partial}{\partial \xi}(\sqrt{g}\rho U) + \frac{1}{\sqrt{g}}\frac{\partial}{\partial \eta}(\sqrt{g}\rho V) = 0 \quad (7)$$

Both continuity equation and momentum equations shown above have preserved the same forms as those in the orthogonal coordinates. Therefore, the same procedure of SIMPLER algorithm can be applied and the revision from orthogonal coordinates to non-orthogonal coordinates is much simpler.

Results And Discussions

This computational model has been used to study the laminar flow through a tube with an axisymmetric contriction to check the numerics and the implementation of the present model. This problem has been studied experimentally by Young and Tsai [7], numerically by Rastogi [8] using orthogonal grids, and by Karki and Patankar [9] using non-orthogonal grids. The computed flow pattern is shown in Fig. 1, and the comparison of the present results with experiments and previous computational results is given in Table 1. The predicted location of separation point is in good agreement with the experimental result. However, the location of reattachment point is predicted slightly upstream of measured one. This discrepancy has also been reported by Rastogi [8] and Karki et al. [9]. But the difference is not significant. This comparison indicates that the present model has been implemented correctly to solve the flow field using non-orthogonal grids.

Table 1

Comparison of Predicted Separation and Reattachment Points with Experiments [7] and Other Computed Results [9] for the M-2 Case of Reference 7.

	Separation Point (x_s/x_o)	Reattachment Point (x_r/x_o)
Present Model	0.38	2.0
Experiment	0.37	2.2
Reference 9	0.40	2.1

This computation model is then applied to the flow analysis of a liquid rocket engine combustor installed with two annular-ring injectors. The propellant-combination selected in the computation is Nitrogen Tetroxide (N_2O_4) and Monomethylhydrazine (MMH). The flow conditions and combustor configurations are given as follows: $u_o = 40 m/sec$, $P_c = 100 psi$, $m_o = 1.36 kg/sec$, $\dot{m}_f = 0.865 kg/sec$, $M_R = 1.57$, $L_c = 2.22 m$, $D_c = 1.26 m$, and $d = 0.03 m$. The computed flow pattern of a liquid rocket combustor without an acoustic cavity is shown in Fig.2 for the cold flow and Fig.3 for the combusting flow. A strong recirculation zone is predicted in the outer corner of the combustor for both cases because of the arrangement of injectors. For the cold flow, the recirculation zone is longer and the velocity is accelerated to the nozzle exit purely due to the nozzle expansion. Down to the nozzle exit, the combustion flow velocity is accelerated due to both effects of combustion and nozzle expansion, and the maximum flow velocity is observed near the outer corner of the nozzle exit.

In order to study the effects of an acoustic cavity on the flow field, an acoustic cavity is then installed in the outer edge of the injector face. The predictions of the flow field are shown in Figs.4-5 for different configurations of acoustic cavity. It is observed that as the acoustic cavity is installed, the reattachment point is moved upstream. For the flow in the combustor without an acoustic cavity, the recirculation flow would encounter a dead end when it flows back to the injector face. Thus, more space is needed to accommodate the recirculating flow. On the other hand, the acoustic cavity is located just at where the recirculation flow is present and can provide additional space for the flow to recirculate. Therefore, the flow recirculation zone is shifted into the acoustic cavity and the reattachment point is moved more upstream than that of the flow in a combustor with no acoustic cavity.

The acoustic cavity used as a damping device in the liquid rocket engine combustor is mainly to suppress the combustion instability. The present predictions, on the aspects of flow field, indicate that the acoustic cavity can accommodate part of the recirculation flow. However, the understanding of the coupling between the flow field and the pressure oscillation suppression mechanism needs further study on the flow unsteadyness inside the acoustic cavity.

Concluding Remarks

A new computational model is presented for the flow computation of complex geometry. The key issue of this model is to use the contravariant velocity components as the dependent variables of the transformed momentum equations but waive the difficulty associated with the usual nine-point formulation for the pressure equation. A two-dimensional computer code based on this model has been developed for laminar/turbulent, non-reacting/reacting flows. The comparison of the numerical results with experiments for a test case has validated the present computational model. As it is applied to the analysis of a liquid rocket engine combustor with/without an acoustic cavity, the observations indicate that the strong recirculation flow can be absorbed into the acoustic cavity such that the flow recirculation zone is smaller in the combustor with an acoustic cavity.

REFERENCE

1. Chiu, H.H., T.L. Jiang, A.N. Krebsbach, and K.W. Gross, "Numerical Analysis of Bipropellant Combustion on Orbit Maneuvering Vehicle Thrust Chamber," *AIAA Paper 90-0045*.
2. Correa, S.M. and W. Shyy, "Computational Models and Methods for Continuous Gaseous Turbulent Combustion," *Prog. Energy Combust. Sci.*, Vol. **13**, pp. 249–292, 1987.
3. Karki, K.C. and S.V. Patankar, "Calculation Procedure for Viscous Incompressible Flow in Complex Geometries," *Numerical Heat Transfer*, Vol. **14**, pp. 295–307, 1988.
4. Demirdzic, I., A.D. Gosman, and R.I. Issa, "A Finite-Volume Method for the Prediction of Turbulent Flow in Arbitrary Geometries," *Lecture Notes in Physics*, Vol. **141**, pp. 144–150, 1980.
5. Patankar, S.V., Numerical Heat Transfer and Fluid Flow, Hemisphere, 1980.
6. Spalding, D.B., "Mathematical Models of Turbulent Flames, A Review," *Combustion Science and Technology*, Special Issue on Turbulent reacting Flow, Vol. **13**, p. 13 (1976).

7. Young, D.F., and Tsai, F.Y., "Flow Characteristics in Models of Arterial stenoses — I. Steady Flow," *J. Biomech.*, Vol. **6**, pp. 395–410, 1973.
8. A.K. Rastogi, "Hydrodynamics in Tubes Perturbed by Curvilinear Obstructions," *ASME J. Fluids Eng.*, Vol. **106**, pp. 262–269, 1984.
9. Karki, K.C. and S.V. Patankar, "Solution of Some Two-dimensional Incompressible Flow Problems Using A Curvilinear Coordinate System Based Calculation Procedure," Numerical Heat Transfer, Vol. **14**, pp. 309–321, 1988.

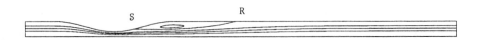

Fig.1 Predicted streamlines for laminar flow through a tube with an axisymmetric contriction ($\psi/\psi_{\text{in}} = 0.25, 0.50, 0.75, 1.0, 1.1$)

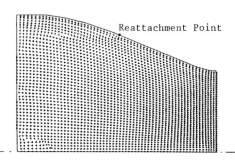

Fig.2 Predicted cold flow velocity vectors in a liquid rocket engine combustor without an acoustic cavity

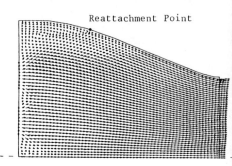

Fig.3 Predicted combustion flow velocity vectors in a liquid rocket engine combustor without an acoustic cavity

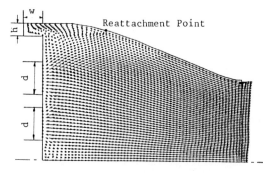

Fig.4 Predicted combustion flow velocity vectors in a liquid rocket engine combustor with an acoustic cavity (h/d =0.4, w/d=0.6)

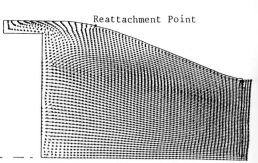

Fig.5 Predicted combustion flow velocity vectors in a liquid rocket engine combustor with an acoustic cavity (h/d =0.4, w/d=1.2)

Computation of the Transition to Turbulence and Flame Propagation in a Piston Engine

Ken Naitoh*, Kunio Kuwahara**, and Egon Krause***

* NISSAN Motor Co., Ltd. 1 Natsushima-cho, Yokosuka, Kanagawa, 237, Japan
** Institute of Space and Astronautical Science, Kanagawa, Japan
*** Aerodynamisches Institut, RWTH Aachen, Germany

I. Introduction

The transition to turbulence during the compression process and the flame propagation in the internal combustion engine are simulated without using explicit turbulence models. The compressible Navier-Stokes equations are solved with a reaction model of octane, which obeys the Arrhenius Kinetics Law. In order to compute the unsteady flow with density variations, a finite-difference method is developed with a scheme useful for the high-Reynolds number flows.

II. Mathematical formulation

II.I Basic equations for spatial variations

The basic equations to evaluate the spatial-varied scalar quantities and velocity, in non-conservative form, are as follows:

$$\rho \frac{Du_i}{Dt} + p_{2,i} = \sigma_{ij,j} \tag{1}$$

$$\frac{DT_2}{Dt} = -\frac{dT_1}{dt} + \frac{1}{\rho C_p'}[\frac{Dp_2}{Dt} + \frac{dp_1}{dt}] + \frac{1}{\rho C_p'}[-q_{i,i} + \sigma_{ij} u_{i,j} + \dot{Q}] \tag{2}$$

$$p_{2,ii} = -\rho \overline{D}_t - \rho[u_i u_{j,i}]_{,j} + \frac{p_{2,i}}{\rho} p_{2,i} + \frac{\sigma_{ij,ij}}{\rho} \tag{3}$$

$$\overline{D} = -\frac{C_p'-R}{C_p'} \frac{1}{p}[\frac{dp_1}{dt} + \frac{Dp_2}{Dt}] + \frac{1}{\rho C_p'T}[-q_{i,i} + \sigma_{ij} u_{i,j} + \dot{Q}] \tag{4}$$

$$p = \rho R T \tag{5}$$

$$\frac{DY_m}{Dt} = \dot{Y}_m - \frac{1}{\rho}[\rho J_{m\ i}]_{,i} \tag{6}$$

where the subscripts '1' and '2' denote 'space-averaged' and 'spatial variation', respectively. These are defined as follows.

$$f = f_1(t) + f_2(t,x), \quad \int f_2(t,x) dV = 0, \quad [\text{for } f = p, T, \rho\ ] \tag{7}$$

Equations (1)-(7) are obtained by a simple transformation from the Navier-Stokes equations. In the right hand side of Eq. (4), three basic types of density variations are explicitly stated, which are connected with the homogeneous compression, compressibility, and heat transfer.

II.II Basic equations for space-averaged quantities and the boundary conditions

In order to derive correct boundary conditions and space-averaged scalar quantities, the integral forms of equations for mass, momentum, and energy are used. These can be written as:

$$\frac{d}{dt}[\rho_1 V_{all}] = \int \rho (U_{s_i} - u_i) dS_i \tag{8}$$

$$\int p_{,i} dS_i = -\int \rho \overline{D}_t dV + \int [u_j u_{i,j}]_{,i} dV + \int [\frac{p_{2,i}}{\rho} p_{2,i} + \frac{\sigma_{ij,ij}}{\rho}] dV \tag{9}$$

$$\frac{d}{dt}[p_1 V_{all}] + \int \frac{R}{(C_p'-R)} p \overline{D} dV + \int p(U_{s_i} - u_i) dS_i = \int \frac{R}{C_p'-R}[-q_{i,i} + \sigma_{ij} u_{i,j} + \dot{Q}] dV \tag{10}$$

The above equations can further be manipulated to suit the topology of the computational domain. In the case of the low Mach number flows, which constitute the principal subject of this study, the following approximations can be made.
(The case of open region) Under the assumption that the physical quantities are constant at the inlet in the intakeprocess, Eqs.(8) and (9) can be reduced to Eqs. (11) and (12).

$$u_{in} S_{in} = u_{pis} S_{pis} + V_{all} \frac{1}{\rho_1} \frac{d\rho_1}{dt} \tag{11}$$

$$(p_2)_{in,1} S_{in} = (p_2)_{pis,1} S_{pis} - \rho_1 \frac{d^2}{dt^2}[\ln\rho_1] V_{all} \tag{12}$$

By the above relations, the velocity and pressure gradient at the inlet can be estimated.
(The case of closed region) In the case of the compression process, Eqs. (8) and (10) can be reduced to Eqs. (13) and (14).

$$\frac{d}{dt}[\rho_1 V_{all}] = 0 \tag{13}$$

$$\frac{dp_1}{dt}[\int (1-\frac{R}{Cp})dV] + \rho_1 \frac{d}{dt}V_{all} = \int \frac{R}{Cp}(-q_{i,i}+\sigma_{ij}u_{i,j}+\dot{Q})dV \tag{14}$$

The above equations evaluate the space-averaged density and pressure correctly. Furthermore, the boundary condition for pressure gradient at the walls can be found by using Eq. (9) [1]. Details on the boundary conditions are described in reference [2]. As shown above, Eqs. (1)-(10) express the basic governing equations including the boundary conditions.

II.III Chemical reaction model and physical constants

In this report, the one-step chemical reaction is considered only for octane. The progress rate is modeled by the well-known formula [3]. The dependence of the viscosity coefficient and thermal conductivity on temperature is assumed to be given by the Sutherland law. The Prandtl number and Schmidt number are set to be 0.7 and 1.0, respectively.

III. Numerical Procedure

The computational algorithm is the extended version of the ICE method[4]. The discretized equations are obtained:

$$\rho^n [\frac{Du_i}{Dt}]^{n+1} + p_2^{n+1}{}_{,i} = \sigma_{ij,j}{}^{n+1} \tag{15}$$

$$\frac{DT_2{}^{n+1}}{Dt} + \frac{dT_1{}^{n+1}}{dt} - \frac{1}{[\rho Cp']^n}[\frac{Dp_2}{Dt} + \frac{dp_1}{dt}]^{n+1} = \frac{1}{(\rho Cp')^n}[-q_{i,i}{}^{n+1} + \dot{Q}] \tag{16}$$

$$p_2^{n+1}{}_{,ii} = -\rho^n \frac{1}{\delta t}[\overline{D}^{n+1} - u_{i,i}{}^n] - \rho^n[u_i u_{j,i}]_{,j}{}^n + (\frac{p_{2,i}}{\rho})^n p_2^{n+1}{}_{,i} \tag{17}$$

$$\overline{D}^{n+1} = -(\frac{Cp'-R}{Cp'} \frac{1}{p})^n [\frac{dp_1}{dt} + \frac{Dp_2}{Dt}]^{n+1} + \frac{1}{(\rho Cp'T)^n}[-q_{i,i} + \dot{Q}]^{n+1} \tag{18}$$

where $[Df/Dt]^{n+1} = 1/\delta t [f^{n+1}-f^n] + u^n{}_i f_{,i}^{n+1}$

Equations (15)-(18) with the equation of state and the equations of species are solved so that Eqs. (7)-(10) are filled by using a iteration procedure in this report. The Euler backward scheme is used for the time-differencing and the third-order upwind scheme [7] is employed for the convective terms. The second-order central differencing is used for the other terms.
In this report, the moving generalized coordinate system is adopted. Details on the coordinate system are described in reference [8].

IV. Results and Discussion

IV.I Transition to turbulence in the compression stage of an model engine with a square cylinder

Figure 1 shows the density contours of the computed results and the corresponding experimental observation during the intake and compression processes in a model engine with a square cylinder. The density-contour interval is about 0.012kg/m^3. As the working fluid, freon 12 is used; the speed of sound is very low and the engine speed is set at 998 rpm. Thus, the maximum Mach number is over 0.5. The compression rate is 3.9. The number of the grid points is 100x100x10. In the intake process, small vortices are generated around the inlet due to the shear instability. In the compression process, a large vortex spins up and it breaks up suddenly into small ones near the end of the compression process. It is due to the shrinking chamber height and the increasing compression rate.

IV.II Wrinkled flame propagation in a realistic engine

The flow near the end of the intake process in a realistic engine using air as the working fluid is displayed in Fig. 2. This engine has a curved intake-port and flat head. The compression process and flame propagation are calculated after this initial flow. The computational particulars are shown in Table 1. In Fig.3, small vortices before ignition are generated by the same physical procedure as in Fig. 1. The grid system is depicted in Fig. 4. The number of the grid points is 70x70x20. Figure 5 illustrates that the wiggle of the flame in the engine is produced by these small vortices due to the piston motion.

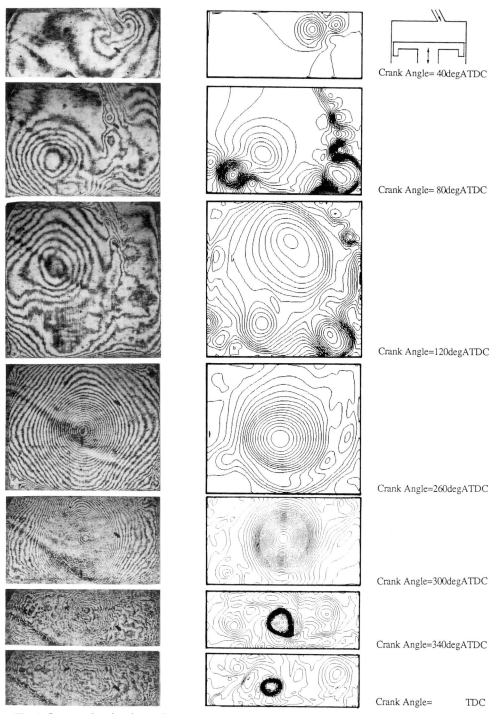

Fig. 1 Computational and experimental density contours in the intake and compression processes of an engine with a square cylinder : Experimental data by the Mach-Zehendar interferometry[10]

In Fig. 5, the small vortices and the expansion flow by the heat release of chemical reaction are captured over the grid-size. In this case, the size of grid is about 1 cu mm. In order to determine the flame speed quantitatively, much more grid points are needed because the Kolmogolov scale and the flame thickness are about the order of 0.01mm - 0.1mm.

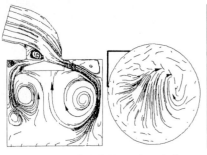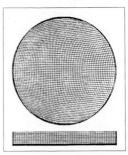

Fig. 2 The flow field near the end of the intakeprocess of an engine with a curved intakeport

Fig. 3 The flow field before ignition in an engine with a curved intake port

Fig. 4 The grid system for the compression process of an engine with a curved intakeport

IV.III Two-dimensional direct simulation of the effect of the flame thickness on turbulent flame speed

In the low and middle engine speed of usual engines, the turbulent flame speed increases according to the increase of the engine speed. However, in case of the higher engine-speed condition, the turbulent flame speed is saturated in spite of the turbulence increase. It is because the ratio of flame thickness to the vortex scale gives much influence to the turbulent flame speed. In order to know the physical essence behind this phenomenon, detailed computations are carried out in a two-dimensional closed region. The grid size and time increment are 0.05 mm and $2\mu s$. The number of grid points is 500x150. Computations are performed in six cases and the conditions are shown in Table 2. Instead of the real turbulence, artificial disturbance is given initially by using the equations $u'=u_0 \sin y$ and $v'=v_0 \sin x$. In Cases A-II and A-III, the vortex scale is about ten times as thick as the flame thickness. In Cases B-II and B-III, the vortex scale is about three times as thick as the flame thickness. The computed laminar burning velocity is 70 cm/s and 95 cm/s for CASE A-I and CASE B-I, respectively. By these results, it is evident in Fig. 6 and 7 that the flame speed in CASE B is smaller than in CASE A. Thus, small scale turbulence gives less influence on the flame speed. Also, cusps are observed in CASE A-III, though not in CASE B-III. This phenomenon is known in an experimental study [12].

V. Conclusion

Turbulence and the characteristics of the wrinkled flame in piston engines are computed by using the above-mensioned numerical method.

Nomenclature

C_p	constant-pressure specific heat	u_i	velocity vector
C_p'	$C_p' = \partial C_p/\partial T$	u', v'	initial disturbance
		U_{si}	the moving speed of the boundary
$\overline{D}/Dt$	divergence of the velocity ($\equiv u_{i,i}$) substantial derivative	U_l	laminar burning velocity
		U_t	turbulent burning velocity
		V	volume
h	entalphy ($\equiv C_p T$)	V_{all}	control volume in which the flow exsists
		Y_m	mass fraction of species m
J_{mi}	molecular diffusion flux of the m th species ($\equiv -D_m Y_{m,i}$)		
k	thermal conductivity	σ_{ij}	viscous stress tensor ($\equiv \mu u_{i,jj}+1/3\mu(u_{j,j})_{,i}$)
m	species	δt	time increment
n	time step		
p	pressure	subscripts	
$\dot{Q}$	heat release by chemical reaction	1	Space-averaged
q_i	heat flux ($\equiv -kT_{,i}$)	2	Space-varied
R	gas constant	pis	on piston surface
R_m	gas constant of secies m	in	on inlet
S_i	the vector normal to wall	t	temporal difference
T	temperature		

Table 1 Computational particulars for 3-d calculation in an engine with a curved intakeport

Engine Speed	1400 rpm	Bore x Stroke	85 mm x 86 mm
Compression Rate	12.0	Ignition timing	30degBTDC

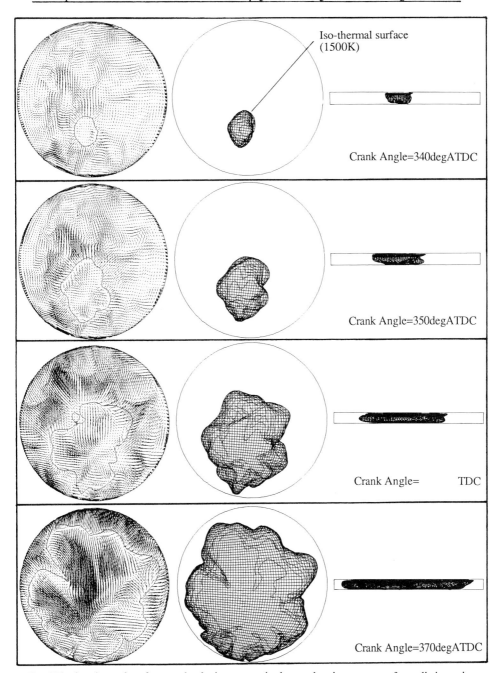

Fig. 5 The iso-thermal surfaces and velocity vectors in the combusting process of a realistic engine

Table 2 Computational particulars for 2-d calculation in a closed region with a constant volume

CASE	A-I	A-II	A-III	B-I	B-II	B-III
Initial Temperature			680 K			
Initial Pressure		5 atm			1 atm	
Initial disturbance $(u_0\ v_0)$	0m/s	1.5m/s	3.0m/s	0m/s	1.5m/s	3.0m/s

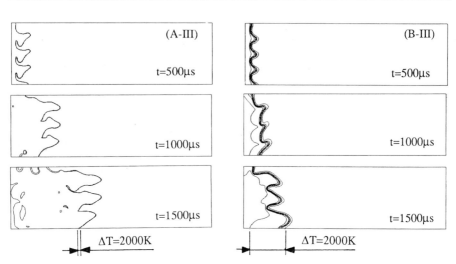

Fig. 6 Computational iso-thermal contours during the flame propagation process in a closed region
CASE A : thin flame case, CASE B : thick flame case

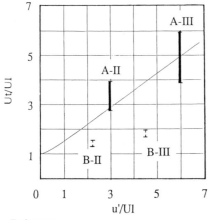

Fig. 7 The influence of the turbulence intensity and flame thickness on the flame speed

⊢━━⊣ A-II,III : Initial pressure 5atm
⊢─⊣ B-II,III : Initial pressure 1atm
────── : Yakhot - Orszag theory [11]

References
[1] P.J. Roache : Computational Fluid Mechanics, Hermosa Publishers Inc. 1976.
[2] K. Naitoh, H. Fujii, Y. Takagi and K. Kuwahara ; SAE paper 900256.
[3] C. K. Westbook and F. L. Dryer : Combst. Sci. and Tech., Vol. 27, 1981.
[4] F. H. Harlow and A. A. Amsden : J. Comp. Physics, Vol.8, 1971.
[5] A.D. Gosman, Y.Y. Tsui, and C. Vafidis : SAE paper No. 850498.
[6] A. Majda and J. Setian : Combst. Sci. and Tech., Vol. 42, 1985.
[7] T. Kawamura and K. Kuwahara : AIAA paper No. 84--340.
[8] K. Naitoh, Y.Takagi, K.Kuwahara, E.Krause, and K. Ishii : AIAA paper 89-1886.
[9] K. Naitoh and K. Kuwahara : to be published as the Proceedings of COMODIA '90,1990.
[10] B. Binninger, M. Jeschke, D. Hanel, W. Limberg, E. Krause ; Proceedings of the 2nd Int. Conf. on Supercomputing Applications in Automotive Industry, October 1988.
[11] V. Yakhot and S. Orszag : Comb. Sci. and Tech., 1989.
[12] A. O. zur Loye and F. V. Bracco : SAEpaper 870454.

COUPLED NAVIER-STOKES MAXWELL SOLUTIONS FOR MICROWAVE PROPULSION

S. Venkateswaran and Charles L. Merkle
Department of Mechanical Engineering
The Pennsylvania State University, PA 16802

INTRODUCTION

Microwave propulsion is a promising new concept for space applications. In this concept, microwave energy is used to heat a working fluid, which is then expanded through a supersonic nozzle to produce thrust. Several experimental programs [1] are currently studying microwave absorption in gases for propulsion applications but the problem has not been studied analytically so far. A theoretical analysis of the problem demands a detailed investigation of the physics and the dynamics of the coupled microwave-gasdynamic interaction. We have developed a comprehensive analytical model for this purpose. The approach features a coupled numerical solution of the Navier-Stokes equations for the fluid dynamics and the Maxwell equations for the electromagnetics. Time-dependent techniques using finite differences are employed in the model.

The analytical study is closely integrated with a companion experimental program. In these experiments, nitrogen and helium discharges generated in a microwave resonant cavity are being studied. The geometry employed in the calculations is similar to this experimental setup and is shown in Fig. 1. The gas flow is confined in a sphere-cylindrical tube while the microwave energy is fed into a cylindrical cavity surrounding the tube. The length of the microwave cavity may be adjusted to obtain the required standing wave mode in the cavity (TM_{012} has been used so far). Figure 1 also shows a typical grid system and an undistorted TM_{012} standing wave. This particular mode generates three points of maximum intensity along the centerline – two at the ends and one at the middle – each of which may serve as a site for the plasma. The key issues center around the characteristics of the discharge (such as plasma location, size and shape) and the coupling efficiency. The model will be used to examine these issues alongside the experiments and, in addition, to extend experimental results to wider regimes (such as scale-up of size and power).

THEORETICAL MODEL

The fluid dynamics are governed by the full Navier-Stokes equations in axisymmetric co-ordinates. The flow in the sphere-cylinder is typically of very low Mach number but the flow is still compressible because of the large temperature gradients. The contribution from the microwave field are the $\sigma|\mathbf{E}^2|$ source term in the energy equation and the Lorentz force terms in the momentum equations. Time-iterative algorithms [2] that march the unsteady N-S equations in time until a steady state is reached have become the prefered method of solving compressible flow problems. At low Mach numbers, however, the convergence of these algorithms is seriously impaired. Following Merkle and Choi [3], we have used a perturbation expansion to derive a low Mach number version of the N-S equations that has well-conditioned eigenvalues at arbitrarily low Mach numbers. Finally, the time-discretized equations are solved by an implicit approximate factorization method [2].

For the microwave field, the Maxwell's equations need to be solved. For the TM mode, the axial and radial magnetic fields and the tangential component of the electric field are zero. The Maxwell's equations then reduce to the following form (in cylindrical co-ordinates),

$$\frac{\partial \tilde{Q}}{\partial t} + \frac{\partial \tilde{E}}{\partial x} + \frac{\partial \tilde{F}}{\partial y} = \tilde{H}, \tag{1}$$

where $\tilde{Q} = (H_\theta y, E_x y, E_y y)^T$ and

$$\tilde{E} = \begin{pmatrix} (1/\mu)E_y y \\ 0 \\ (1/\epsilon)H_\theta y \end{pmatrix}, \quad \tilde{F} = \begin{pmatrix} -(1/\mu)E_x y \\ -(1/\epsilon)H_\theta y \\ 0 \end{pmatrix}, \quad \tilde{H} = \begin{pmatrix} 0 \\ -(\sigma/\epsilon)E_x y \\ -(\sigma/\epsilon)E_y y \end{pmatrix}.$$

Here E_x is the axial electric field intensity, E_y is the radial field intensity and H_θ is the tangential component of the magnetic field. Also, ϵ is the permittivity coefficient, μ is the permeability coefficient and σ is the electrical conductivity.

The standard method for solving the time-dependent Maxwell's equations is described by Yee [4]. The algorithm is a central difference based explicit scheme that is second order accurate in both time and space. The explicit formulation allows for easy coding and minimum storage According to this scheme, the flux vectors $\tilde{Q}$, $\tilde{E}$ and $\tilde{F}$ may be split up as $\tilde{Q} = \tilde{Q}_1 + \tilde{Q}_2$, $\tilde{E} = \tilde{E}_1 + \tilde{E}_2$ and $\tilde{F} = \tilde{F}_1 + \tilde{F}_2$. The vectors are defined so that those with subscript 1 contain only magnetic field components and those with subscript 2 contain only electric field components. The algorithm may then be represented by the following two step procedure.

$$\tilde{Q}_1^{n+\frac{1}{2}} - \tilde{Q}_1^{n-\frac{1}{2}} = -2\Delta t \left(\frac{\partial \tilde{E}_2}{\partial x} + \frac{\partial \tilde{F}_2}{\partial y} \right)^n \tag{2a}$$

$$\left(1 + \frac{\sigma}{\epsilon}\Delta t\right)\left(\tilde{Q}_2^{n+1} - \tilde{Q}_2^n\right) = -2\Delta t \left[\left(\frac{\partial \tilde{E}_1}{\partial x} + \frac{\partial \tilde{F}_1}{\partial y} \right)^{n+\frac{1}{2}} - \tilde{H}^n \right] \tag{2b}$$

Here, we note that $\tilde{H} = -\frac{\sigma}{\epsilon}\tilde{Q}_2$. Since the Maxwell equations are hyperbolic in time, a method of characteristics procedure is employed at the boundaries. Here, the incoming characteristic represents the power that is input into the cavity (or the incident power) and has to be specified as a boundary condition. The outgoing characteristic represents the power that is reflected back by the load to the source. The value of this characteristic depends on the solution of the field in the cavity and the appropriate characteristic equation is used instead of a boundary condition.

RESULTS

The coupled electromagnetic-gasdynamic problem has been solved for the sphere-cylinder combination shown in Fig. 1. A fluid dynamic grid of 101 X 51 grid and a microwave grid of 121 X 121 were used. A typical run (about 1000 time steps) took around 30 minutes of CPU time on a single processor of the CRAY-YMP. In this section we present some basic computational results and comparison with experiments.

A typical result for the sphere-cylinder is shown in Fig. 2. These calculations are for a pressure of 1 atm, an inlet temperature of 1000 K and a mass flow rate of 4×10^{-5} kg/s. The gas here, as in all calculations, is helium. The input power was 3 KW and the absorbed power was 2 KW giving a coupling efficiency of 67 %. Figure 2(a) shows the temperature and velocity contours. The temperature contours show a peak temperature of 9450 K with the plasma forming in the center of the sphere. Experimental results consistently show the plasma forming at the middle node as well. The velocity contours show the gas decelerating as it approaches the plasma, indicating the traditional *blockage* effect. Figure 2(b) shows the corresponding axial and radial electric fields for this case. The electric fields are strongly distorted from their zero loss shape and are seen to go to a minimum at the surface of the plasma. Of particular interest are the very steep gradients in the electric field in this region.

Various geometries in addition to the experimental configuration were studied to assess their effect on the discharge characteristics. Figure 3 shows the temperature contours for the experimental sphere-cylinder and a straight duct. In both cases, the pressure was 1 atm and the mass flow rate was 1.0×10^{-5} kg/s. The major difference in the straight duct solution is that the plasma now forms in the lowest end node instead of the middle node. This trend was also observed in experiments with a straight duct. The peak temperature and plasma size and shape are, on the other hand, quite comparable for the two cases.

Next, we present comparisons of the computational results to experimental data. We look at some global parameters such as peak temperature and coupling efficiency since measurements of temperature and velocity profiles are not available yet. Figure 4 shows measured electron temperatures and computed peak gas temperatures for different mass flow rates. Since local thermodynamic equilibrium (LTE) has been assumed in the calculations the gas temperature is the same as the electron temperature. Both experiments and calculations show that peak temperature is quite independent of the flow conditions. Overall, the agreement is very good with the calculations underpredicting temperature by about 5 to 10 %. This slight discrepancy may be due to the assumption of LTE in the plasma.

Coupling efficiencies from measurements and computations were also compared and good general agreement was observed. Efficiency ranged from 70 to 80 % and was not a strong function of mass flow rate or power. The one major experimental result that cannot be duplicated is the threshold power. The computed threshold powers method are consistently larger than those observed experimentally. The most probable reason appears to be that the LTE assumption is incorrect; however, this needs to be investigated further. Future calculations then will include extending experimental measurements to larger power levels and to more practical configurations. In addition, calculations will be extended to include the rocket nozzle as well in order to estimate thrust and overall efficiencies.

REFERENCES

[1] Balaam, P. and Micci, M. M., "Investigation of Free-Floating Nitrogen and Helium Plasmas Generated in a Microwave Resonant Cavity", AIAA Paper No. 89-2380, AIAA/ASME/SAE/ASEE 25th Joint Propulsion Conference, July 1989.

[2] Warming, R. F. and Beam, R. M., "On the Construction and Application of Implicit Factored Schemes for Conservation Law", SIAM-AMS Proceedings, Vol. 11, 1978, pp 85-129.

[3] Merkle, C. L. and Choi, Y.-H., "Computation of Low-Speed Compressible Flows with Time-Marching Procedures", International Journal for Numerical Methods in Engineering, Vol. 25, 1988, pp 293-311.

[4] Yee, K. S., "Numerical Solution of Initial Boundary Value Problems Involving Maxwell Equations in Isotropic Media", IEEE Transactions on Antennas and Propagation, AP-14, May 1966, pp 302-307.

ACKNOWLEDGEMENTS

This work was sponsored by the Air Force Office of Scientific Research under Contract No. 89-0312.

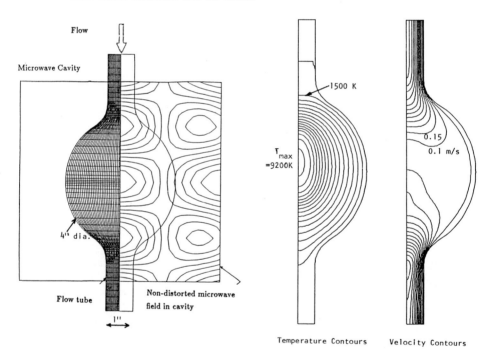

Fig. 1 Representative geometry for microwave-gasdynamic interaction showing generalized nonorthogonal grid in flow domain and superimposed microwave cavity with electric field lines.

Fig. 2a Representative solution of sphere-cylinder configuration. P = atm, T = 1000 K, u = 0.336 m/s, m = 1 x 10^{-5} kg/s, Re = 10, T_{inc} = 3 KW, P_{ref} = 0.8 KW.

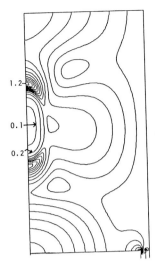
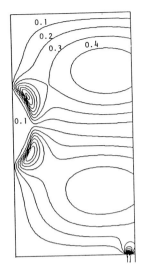

Axial Electric Field   Radial Electric Field

Fig. 2b Distorted electric field solutions corresponding to the case on Fig. 2a. The field lines are normalized with respect to $E_{ref} = 25000$.

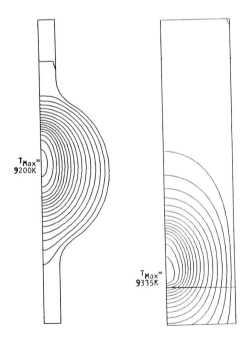
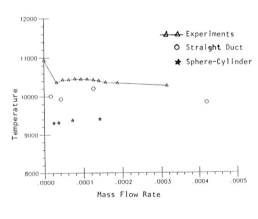

Fig. 3 Temperature contours for 3 different flow geometries -- the experimental configuration, an intermediate sphere-cylinder and a straight pipe. $m = 1 \times 10^{-5}$ kg/s.

Fig. 4 Comparison of computed peak temperatures with experiments. Solid line is from experiments. The circles are straight duct calculations and the stars are the sphere-cylinder calculations.

COMPARISON OF SEVERAL NUMERICAL METHODS FOR COMPUTATION OF TRANSONIC FLOWS THROUGH A 2D CASCADE

Jaroslav Fořt[1], Miloš Huněk[2], Karel Kozel[3], Miroslava Vavřincová[3]

[1] SVÚSS Běchovice, Prague 9, 19000, Czechoslovakia
[2] ČKD Praha, Compressors, Klečákova 1947, 19000 Prague 9, Czechoslovakia
[3] Departement of Computational Techniques and Informatics, TU Prague Technická 4, 16607 Prague 6, Czechoslovakia

The paper deals with a comparison of numerical results of several computational methods of steady transonic flows through a 2D cascade. The first group of the methods is based on numerical solution of the full potential equation and other group is based on numerical solution of the system of Euler equations.

In this paper we compare the numerical results of simulation of the flows through DCA8% cascade and several compressor cascades as well as turbine cascades using Mach number distribution or isomach lines. The mentioned numerical results are also compared to the experimental results.

I. <u>Numerical Methods</u>

During last years the following computational methods based on numerical solution of the full potential equation have been developed:

a) Nonconservative method using computation in two periods between the profiles (FP1), see [1]

b) Nonconservative method using a local disturbance form, multigrid technique (FAS and CS algorithm) and computation in one period (FP2), see [2]

c) Nonconservative method using multigrid technique (FAS) and computation in one period, see [3]

d) Conservative method using AF1 and AF2 iteration process, see [4].

Other three methods of computation of inviscid transonic flows were developed. The explicit conservative finite volume methods are based on numerical solution of the system of Euler equations by

a) Mac Cormack cell centered scheme and multigrid technique (E3) see [5]

b) Ron-Ho-Ni cell vertex scheme, see [6]

c) multistep Runge-Kutta cell centered scheme and multigrid technique, see [4].

Remark:

When one computes transonic flows described by full potential equation through a 2D cascade, then downstream homogeneous conditions are in unique relation to upstream conditions and cascade geometry. In the case of 2D cascade flows described by the Euler equations one can change downstream pressure p_2 (for $M_2 < 1$ as well as $M_2 > 1$) and then change the solution through a considered cascade.

II. Comparison of computed results

We consider three types of cascade flows. Flows through DCA8% cascade of Institute of Thermomechanics of Czechoslovak Academy of Sciences, ČKD1 and ČKD4 compressor cascades and SE-1050 turbine cascade of Škoda Pilsen, Turbines.

A) Transonic flows through DCA8% cascade

Fig. 1 and Fig. 2 shows results of numerical solution with subsonic M_∞ (Fig.1) as well as with supersonic M_∞ (Fig.2) compared to interferometric experimental results of Institute of Thermomechanics. One can compare shape and location of sonic line in experimental and numerical

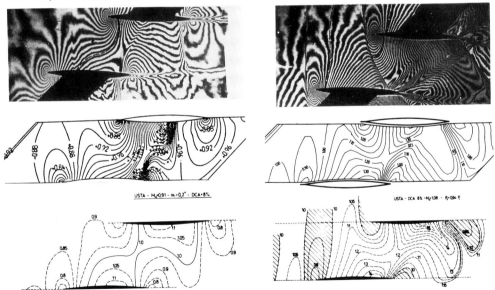

Fig.1 Comparison of experimental and computed results (FP1, E1) for DCA8% cascade with subsonic upstream flows

Fig.2 Comparison of experimental and computed results (AF2, E1) for DCA8% cascade with supersonic upstream flows

results or also shape and location of isomach lines and black or white strips in interferometric measurements.

B) Transonic flows through a compressor cascade

In the case of compressor cascade ČKD1 one can observe back pressure effect ($p_2 = p_v \cdot p_1$) for different p_v computed by E2 method and comparison of numerical results using E2 and FP3 method (Fig.3). Fig.4 shows comparison of numerical results for DCA-2S cascade using FP1 and FP2 (FAS) method. Fig.6 shows comparison of numerical results achieved by nonconservative method (FP2) and conservative method (FP4) for DCA-2S cascade and transonic flows near chocked regime. Fig.5 shows comparison of numerical and experimental results (of DFVLR Köln - Schreiber, Starken). Numerical result are denoted by full line.

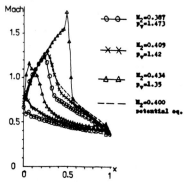

Fig.3 Back pressure effect for ČKD1 cascade computed by E2 method. Comparison of potential and Euler solution $M_\infty = 0.87$

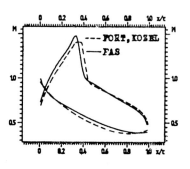

Fig.4 Comparison of two potential methods for DCA-2S cascade $M_\infty = 0.87$, $\alpha = 26.86°$, $\beta = 61.86°$

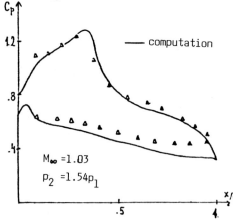

Fig.5 Comparison of experimental results (DFVLR Köln) and numerical results (E1)

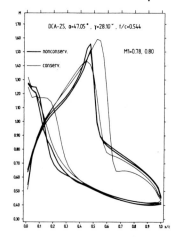

Fig.6 Comparison of nonconservative (FP2) and conservative (FP4) potential results for DCA-2S and $M_\infty = 0.78$ or $M_\infty = 0.8$

C) Transonic flows through turbine cascade SE-1050

We consider transonic flows through SE-1050 turbine cascade with subsonic upstream flows and subsonic or supersonic downstream flows. Fig.7 shows development of the flows for increasing M_2(E1) compared to experimental results of Institute of Thermomechanics. Fig.8 shows the case similar Fig.7 computed by E2 and mapped by isomach lines. The third case is compared to interferometric measurements of Institute of Thermomechanics (Fig.9). Fig.10 shows numerical results achieved by FP3 for M_2=1.2; but downstream (supersonic) part is incorrectly computed. It is caused by unsuitable approximation of boundary conditions near the trailing edge.

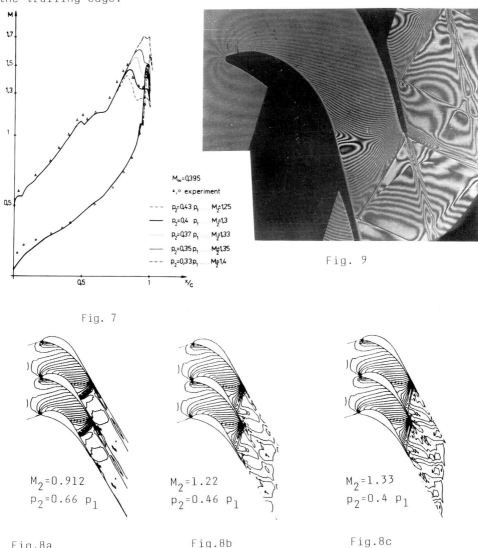

Fig. 9

Fig. 7

Fig.8a   Fig.8b   Fig.8c

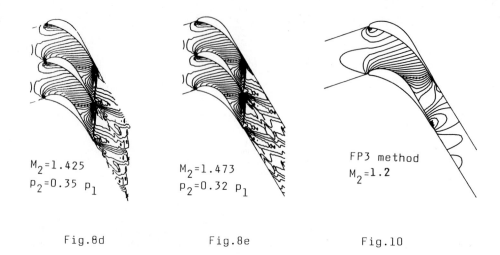

$M_2 = 1.425$ $M_2 = 1.473$ FP3 method
$p_2 = 0.35\ p_1$ $p_2 = 0.32\ p_1$ $M_2 = 1.2$

Fig.8d   Fig.8e   Fig.10

References

[1] Fořt J., Kozel K.: Numerical Solution of Inviscid Two-Dimensional Transonic Flows Through a Cascade, ASME Paper No 86-GT-19
[2] Huněk M., Kozel K., Vavřincová M.: Numerical Solution of Transonic Potential Flow in 2D Compressor Cascades using Multi-Grid Techniques, in Notes in Numerical Fluid Mechanics, Vol.23, pp.145-153
[3] Fořt J.: Private communication
[4] Huněk M., Kozel K., Vavřincová M.: Report ČKD Praha, Compressors; No KKS-TK 2-7-301, 1989
[5] Kozel K., Vavřincová M., Nhac N.: Numerical Solution of the Euler Equations Used for Simulation of 2D and 3D Steady Transonic Flows, in Notes in Numerical Fluid Mechanics, Vol.24, pp. 329-341
[6] Fořt J.: Numerical Solution of 2D System of Euler Equations by Explicit Scheme, Strojnický časopis, 1990

Numerical Simulation of Unsteady Turbulent Flow through Transonic and Supersonic Cascades

Satoru Yamamoto and Hisaaki Daiguji

Department of Mechanical Engineering, Tohoku University, Sendai 980, Japan

1. Introduction

A numerical method for solving the unsteady multi-dimensional compressible Navier-Stokes equations with the two-equation $k - \varepsilon$ turbulence model in general curvilinear coordinates is proposed, and its applications to the unsteady transonic and supersonic cascade flow problems are shown. The fundamental equations used in the present paper were derived in the previous paper[1] for analyzing the steady compressible viscous flow in turbomachinery. The equations take the momentums of contravariant velocities as the unknown variables and have an advantage of simple and accurate treatments of the periodic boundary condition. The numerical method based on the implicit time-marching finite-difference method is extended to the method for unsteady flow problems[2] and has the second-order accuracy in time. And in this paper, a new high resolution TVD upwind-difference scheme is also proposed for capturing oblique and normal shock more sharply. With regard to estimate the turbulent quantities, the $k - \varepsilon$ model considering the low-Reynolds number effect is employed and designed to generate the transition from a laminar boundary layer to the turbulent due to the shock. The computed results of a supersonic compressor cascade flow and a transonic turbine stator-rotor interaction are shown and also some animations of them will be shown.

2. Fundamental Equations

The 3-D ensemble-averaged compressible Navier-Stokes equations of contravariant velocity components in general curvilinear coordinates together with the two-equation $k - \varepsilon$ turbulence model can be written in the vector form

$$\partial \tilde{q}/\partial t + \tilde{L}(\tilde{q}) \equiv \frac{\partial \tilde{q}}{\partial t} + \frac{\partial \tilde{E}_i}{\partial \xi_i} + \tilde{R} + \frac{1}{Re}\tilde{S} + \tilde{H} = 0 \tag{1}$$

and can be easily derived by taking the linear combinations of the existing fundamental equations defined as $\partial \hat{q}/\partial t + \hat{L}(\hat{q}) = \hat{q}/\partial t + \partial \hat{E}_i/\partial \xi_i + \hat{S}/Re + \hat{H} = 0$, that is,

$$\partial \tilde{q}/\partial t + \tilde{L}(\tilde{q}) = B\,[\partial \hat{q}/\partial t + \hat{L}(\hat{q})] = 0 \tag{2}$$

where B is a matrix to transform the existing Navier-Stokes equations to the momentum equations of contravariant velocities. The vector of unknown variables $\tilde{q}$ is composed of

$$\tilde{q}^t = J[\rho \ \ \rho W_1 \ \ \rho W_2 \ \ \rho W_3 \ \ e \ \ \rho k \ \ \rho \varepsilon] \tag{3}$$

where J, ρ, W_i ($i = 1, 2, 3$) and e are the Jacobian, the density, the contravariant velocity components of relative flow and the modified stagnation internal energy per unit volume, respectively. k is the turbulent kinetic energy and ε is the dissipation rate of k. $\tilde{E}_i$ ($i = 1, 2, 3$) are the flux vectors. $\tilde{R}$ is an additional term introduced in order to express the principal part of eq.(1) in conservation form. $\tilde{H}$ is a source term which includes the centrifugal and coriolis forces for the relative flow, and also the production, the dissipation and the low Reynolds number effect for the $k - \varepsilon$ model. The value of pressure p is evaluated from the equation of state. The $k - \varepsilon$ turbulence model employed here is a typical model widely used for solving incompressible flows, and modified by Chien[5]. The reason why we employ the two-equation turbulence model is that turbulent flows having unsteady wakes like a internal flow in turbomachinery must be simulated by using the turblence model taking account of the transport phenomena. In recent, we could develop a numerical technique, so-called 'know-how', to simulate a transition to turbulent flow due to the shock/laminar boundary layer interactions by improving the numerical stabilization.

3. Numerical Algorithm for Unsteady Calculation

The above-mentioned fundamental equations are solved by an implicit time-marching finite-difference method for unsteady flow problems. The method is an extended version of the previous implicit method for steady flow problems developed by the authors, in which the delta-form approximate-factorization scheme, the diagonalization and the modified Chakravarthy-Osher TVD scheme[3] are employed. The present method has the second-order accuracy in time by means of the Crank-Nicholson scheme and the Newton iteration[4] at each time step. Finally, eq.(1) can be rewritten as

$$\{\tilde{S}_1^{-1}[I + \Delta t\theta(\Lambda_1^+ \nabla_1 + \Lambda_1^- \Delta_1)]$$
$$\cdot \tilde{M}_1^{-1}[I + \Delta t\theta(\Lambda_2^+ \nabla_2 + \Lambda_2^- \Delta_2)]$$
$$\cdot \tilde{M}_2^{-1}[I + \Delta t\theta(\Lambda_3^+ \nabla_3 + \Lambda_3^- \Delta_3)]\tilde{S}_3\}^m \Delta \tilde{q}^m = RHS^m \tag{4}$$

where

$$\Delta \tilde{q}^m = \tilde{q}^{m+1} - \tilde{q}^m \tag{5}$$
$$RHS^m = -(\tilde{q}^m - \tilde{q}^n) - \Delta t B \hat{L}^* \{ (\hat{q}^m + \hat{q}^n)/2 \} \tag{6}$$
$$= -(\tilde{q}^m - \tilde{q}^n) - \Delta t \tilde{L}^* \{ (\tilde{q}^m + \tilde{q}^n)/2 \} \tag{7}$$
$$\tilde{M}_1 = \tilde{S}_2 \tilde{S}_1^{-1}, \quad \tilde{M}_2 = \tilde{S}_3 \tilde{S}_2^{-1} \tag{8}$$

$\theta = 1/2$ or 1, ∇_i and Δ_i are the backward- and forward-difference operators, respectively. $\hat{L}^*$ and $\tilde{L}^*$ mean the difference operators of $\hat{L}$ and $\tilde{L}$. $\tilde{S}_i$ ($i = 1, 2, 3$) are the matrices composed of the eigenvectors and $\Lambda_i^\pm$ ($i = 1, 2, 3$) are the diagonal matrices composed of positive or negative eigenvalues $\lambda_{ij}^\pm$ ($i = 1, 2, 3; \ j = 1, 4, 5$). n is the number of time steps and m is the number of the Newton iterations. If $m = 0$, then $\tilde{q}^m = \tilde{q}^n$. And if $m \Rightarrow \infty$, then $\Delta \tilde{q}^m \Rightarrow 0$, that is, $\tilde{q}^m \Rightarrow \tilde{q}^{n+1}$ and we can obtain the second-order accuracy in time.

4. Numerical Algorithm for Shock Capturing

We have already proposed a high resolution shock capturing scheme based on the Chakravarthy-Osher TVD scheme and normal shocks occurring at the channels of compressor and turbine cascades could be obtained clearly. In this paper, we propose a new efficient scheme for capturing oblique and normal shocks more sharply. Now let us define the flux-vectors for the existing Navier-Stokes equations in general curvilinear coordinates at mid-point $\ell + 1/2$ as $\hat{E}_{i(\ell+1/2)}$. The derivatives of $\hat{E}_i$ with respect to ξ_i at point ℓ can be written as

$$(\partial \hat{E}_i / \partial \xi_i)_{(\ell)} = [\hat{E}_{i(\ell+1/2)} - \hat{E}_{i(\ell-1/2)}]/\Delta \xi_i , \quad (i = 1, 2, 3) \tag{9}$$

and employing the flux-vector splitting, the Roe's Riemann approximation and the finite-volume approach for the interpolation of metrics, the flux-vectors $\hat{E}_{i(\ell+1/2)}$ are linearized as

$$\hat{E}_{i(\ell+1/2)} = A^-_{i(\ell+1/2)} Q_R + A^+_{i(\ell+1/2)} Q_L \tag{10}$$

$A_i^\pm$ ($i = 1, 2, 3$) are the Jacobian matrices split according to positive or negative eigenvalues. Q is the vector of unknown variables defined as

$$Q^T = [q_0 \; q_1 \; q_2 \; q_3 \; q_4 \; q_5 \; q_6] = [\rho \; \rho w_1 \; \rho w_2 \; \rho w_3 \; e \; \rho k \; \rho \varepsilon] \tag{11}$$

where w_i ($i = 1, 2, 3$) are the physical velocities of relative flow. The subscripts R and L mean the right and the left, respectively. Finally, the linearized flux $A^\pm_{i(\ell+1/2)} Q$ can be rewritten as the following form composed of the sub-vectors.

$$A_i^\pm Q = J N^{-1} S_i^{-1} \Lambda_i S_i N Q$$

$$= J \begin{bmatrix} q_0 \\ q_1 \\ q_2 \\ q_3 \\ q_4 \\ q_5 \\ q_6 \end{bmatrix} \lambda_{i1}^\pm + \frac{J}{c\sqrt{g_{ii}}} \begin{bmatrix} 0 & + & \Delta \bar{W}_i \\ \xi_{i,1}\bar{p} & + & \bar{q}_1/\bar{q}_0 \cdot \Delta \bar{W}_i \\ \xi_{i,2}\bar{p} & + & \bar{q}_2/\bar{q}_0 \cdot \Delta \bar{W}_i \\ \xi_{i,3}\bar{p} & + & \bar{q}_3/\bar{q}_0 \cdot \Delta \bar{W}_i \\ \bar{W}_i \bar{p} & + & (\bar{\phi}^2 + c^2)/\tilde{\gamma} \cdot \Delta \bar{W}_i \\ 0 & + & \bar{q}_5/\bar{q}_0 \cdot \Delta \bar{W}_i \\ 0 & + & \bar{q}_6/\bar{q}_0 \cdot \Delta \bar{W}_i \end{bmatrix} \lambda_{ia}^\pm$$

$$+ \frac{J}{c^2} \begin{bmatrix} \bar{p} & + & 0 \\ \bar{q}_1/\bar{q}_0 \cdot \bar{p} & + & \xi_{i,1} c^2/g_{ii} \cdot \Delta \bar{W}_i \\ \bar{q}_2/\bar{q}_0 \cdot \bar{p} & + & \xi_{i,2} c^2/g_{ii} \cdot \Delta \bar{W}_i \\ \bar{q}_3/\bar{q}_0 \cdot \bar{p} & + & \xi_{i,3} c^2/g_{ii} \cdot \Delta \bar{W}_i \\ (\bar{\phi}^2 + c^2)/\tilde{\gamma} \cdot \bar{p} & + & \bar{W}_i c^2/g_{ii} \cdot \Delta \bar{W}_i \\ \bar{q}_5/\bar{q}_0 \cdot \bar{p} & + & 0 \\ \bar{q}_6/\bar{q}_0 \cdot \bar{p} & + & 0 \end{bmatrix} \lambda_{ib}^\pm \tag{12}$$

$$\lambda_{i1} = W_i, \quad \lambda_{i4} = W_i + \sqrt{g_{ii}} c, \quad \lambda_{i5} = W_i - \sqrt{g_{ii}} c \tag{13}$$

$$\lambda_{ia}^\pm = (\lambda_{i4}^\pm - \lambda_{i5}^\pm)/2, \quad \lambda_{ib}^\pm = (\lambda_{i4}^\pm + \lambda_{i5}^\pm - 2\lambda_{i1}^\pm)/2 \tag{14}$$

$$\bar{p} = q_0 \bar{\phi}^2 - \tilde{\gamma}(\bar{q}_1 q_1 + \bar{q}_2 q_2 + \bar{q}_3 q_3 - \bar{q}_0 q_4)/\bar{q}_0 \tag{15}$$

$$\bar{\phi}^2 = \tilde{\gamma}(\bar{q}_1^2 + \bar{q}_2^2 + \bar{q}_3^2 - r\omega v_u)/2\bar{q}_0^2 \tag{16}$$

$$\Delta \bar{W}_i = q_0(W_i - \bar{W}_i) \tag{17}$$

$$\bar{W}_i = (\xi_{i,1}\bar{q}_1 + \xi_{i,2}\bar{q}_2 + \xi_{i,3}\bar{q}_3)/\bar{q}_0 \tag{18}$$

$$W_i = (\xi_{i,1} q_1 + \xi_{i,2} q_2 + \xi_{i,3} q_3)/q_0 \tag{19}$$

where $i = 1, 2, 3$ and the matrix N transforms the fundamental equations in conservation form into the nonconservative form and S_i further transforms them into the ordinary differential equations. Now we define a vector of the overlined variables $\bar{Q}^T = [\bar{q}_0 \; \bar{q}_1 \; \bar{q}_2 \; \bar{q}_3 \; \bar{q}_4 \; \bar{q}_5 \; \bar{q}_6]$, which is capable to control flexibly, and some averaging techniques such as the Roe's averaging can be applied to it.

The above equations can be easily interpolated by using the MUSCL approach. In this paper, Q_L and Q_R are interpolated in the second-order accuracy, that is,

$$Q_L = Q_{(\ell)} + \frac{1}{4}[(1-\phi)\Delta^+ \bar{Q}_{(\ell-1/2)} + (1+\phi)\Delta^- \bar{Q}_{(\ell+1/2)}] \tag{20}$$

$$Q_R = Q_{(\ell+1)} - \frac{1}{4}[(1-\phi)\Delta^- \bar{Q}_{(\ell+3/2)} + (1+\phi)\Delta^+ \bar{Q}_{(\ell+1/2)}] \tag{21}$$

and also the streamwise TVD approach[6] can be extended to the present scheme as

$$\Delta^-\bar{Q}_{(l+3/2)} = (1-\alpha)\Delta Q_{(l+3/2)} + \alpha \text{ minmod}[\Delta Q_{(l+3/2)}, b\Delta Q_{(l+1/2)}] \qquad (22)$$

$$\Delta^+\bar{Q}_{(l+1/2)} = (1-\alpha)\Delta Q_{(l+1/2)} + \alpha \text{ minmod}[\Delta Q_{(l+1/2)}, b\Delta Q_{(l+3/2)}] \qquad (23)$$

$$\Delta^-\bar{Q}_{(l+1/2)} = (1-\alpha)\Delta Q_{(l+1/2)} + \alpha \text{ minmod}[\Delta Q_{(l+1/2)}, b\Delta Q_{(l-1/2)}] \qquad (24)$$

$$\Delta^+\bar{Q}_{(l-1/2)} = (1-\alpha)\Delta Q_{(l-1/2)} + \alpha \text{ minmod}[\Delta Q_{(l-1/2)}, b\Delta Q_{(l+1/2)}] \qquad (25)$$

$$\Delta Q_{(l+1/2)} = Q_{(l+1)} - Q_{(l)} \qquad (26)$$

where ϕ and b are the parameters of the TVD scheme[3], and α is the parameter for controlling from the TVD scheme to a non-TVD scheme according to flow angle[6].

5. Numerical Examples

We calculated only two types of 2-D cases because of the computer limitation. The calculated results of a 2-D unsteady turbulent flow through a supersonic axial-flow compressor cascade refered from Ref.[7] are first shown. The computational grid has 141 × 61 grid points and is H shaped grid. The inlet Mach number and the static pressure ratio are 1.59 and 2.12, respectively. Figures 1(a)(b) show the Mach number contours at the different time levels. Figure 2 shows the eddy viscosity contours at the same time as Fig.1(a). These figures show that the present method can capture the unsteady oblique and normal shocks clearly, and simulate the transition from laminar boundary layer to the turbulent due to the shocks.

Next, the calculated results of a 2-D unsteady turbulent flow through a transonic axial-flow turbine stator-rotor are shown. The computational grid is shown in Fig.3 and it has 71 × 41 and 91 × 41 grid points at stator and rotor, respectively. In this case, since the stator/rotor blade ratio is 1:1, it is sufficient to calculate each single passage of stator and rotor. Of course the case of a different blade ratio can also be calculated. Figures 4(a)(b) show the Mach number contours at 0 and 0.5 cycles, respectively. These results show that the present method can simulate the effect of the stator wakes on the rotor passage flow clearly and capture the unsteady shocks occurring at the trailing edge of the rotor.

References

[1] Daiguji, H. and Yamamoto, S., "An Implicit Time-Marching Method for Solving the 3-D Compressible Navier-Stokes Equations," *Proc. of 11th Int. Conf. on Numerical Methods in Fluid Dynamics, Williamsburg, Lecture Notes in Physics* , (1988), Springer-Verlag, pp.210-214.
[2] Yamamoto, S. and Daiguji, H., "A Numerical Method for Solving the Unsteady Compressible Navier-Stokes Equations," *Preprints of Int. Symp. on Comput. Fluid Dynamics-Nagoya* , (1989), pp.779-784.
[3] Chakravarthy, S.R. and Osher, S., "A New Class of High Accuracy TVD Schemes for Hyperbolic Conservation Laws," AIAA Paper 85-0363 (1985).
[4] Rai, M.M., "Unsteady Three-Dimensional Navier-Stokes Simulations of Turbine Rotor-Stator Interaction," AIAA Paper 87-2058 (1987).
[5] Chien, K-Y., "Prediction of Channel and Boundary-Layer Flows with a Low-Reynolds-Number Turbulence Model," *AIAA J.*, Vol.20, (1987), pp.33-38.
[6] Yamamoto, S. and Daiguji, H., "A Streamwise TVD Scheme for Solving the Compressible Euler Equations," *Proc. Third Int. Conf. on Hyperbolic Problems-Uppsala*, (1990), to appear.
[7] Tweedt, D.L., Schreiber, H.A. and Starken, H, "Experimental Investigation of the Performance of a Supersonic Compressor Cascade," *Trans. ASME, J. Turbomachinery*, Vol.110, (1988), pp.456-466.

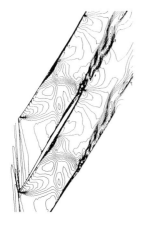

Fig.1(a) Mach number contours

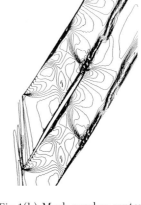

Fig.1(b) Mach number contours

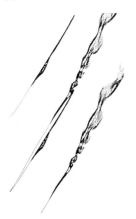

Fig.2 Eddy viscosity contours

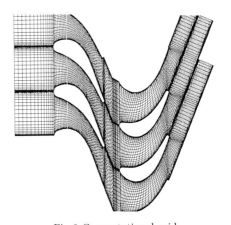

Fig.3 Computational grid

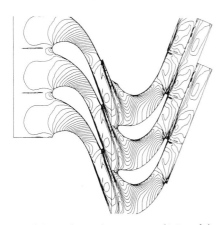

Fig.4(a) Mach number contours(0.0 cycle)

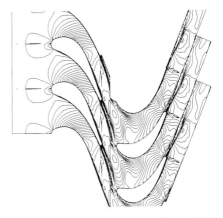

Fig.4(b) Mach number contours(0.5 cycle)

A COMPARATIVE STUDY OF THE LOW MACH NUMBER AND NAVIER-STOKES EQUATIONS WITH TIME-DEPENDENT CHEMICAL REACTIONS

G. Fernandez[◊], K. Z. Tang[▷], H. A. Dwyer[▷]
◊ INRIA Sophia-Antipolis, 06565 Valbonne, France
▷ University of California Davis, CA 95616, USA

INTRODUCTION

Although there has been considerable progress in the simulation of low Mach number reactive flows, there still exist fundamental and numerical difficulties which must be addressed if further progress is to be made.

For such flows the low Mach number limit [1,2] is very attractive because the large sound speed does not restrict the numerical calculation. Nevertheless, this approximated model reaches its physical validity limit when rapid chemistry and strong density changes occur and only the full Navier-Stokes equations can properly describe such a flow [7].

An efficient numerical approach based on M.U.S.C.L. and semi-implicit time integration has been developed for the full equations [3,5]. In spite of very promissing results, the combination of large Courant numbers and the steep temporal and spatial gradients associated with the flame can still cause significant over and undershoots near the flame.

In the present paper, the full Navier-Stokes equations (N.S.) are compared to their low Mach number limit (L.M.) for a one-dimensional, unsteady flame propagation problem. With the use of numerical examples we clarify and cure some of the difficulties associated with such calculations.

FORMULATION/METHODS OF RESOLUTION

With the assumption of a one-step chemical kinetics $R \longrightarrow P$ the flame propagation problem is governed by the following equations:

$$\rho_t + (\rho u)_x = 0 \tag{1}$$

$$(\rho u)_t + (\rho u^2 + p)_x = [(4/3)\mu u_x]_x \tag{2}$$

$$E_t + [(E+p)u]_x = [(4/3)u\mu u_x + kT_x]_x + Q\omega \tag{3}$$

$$(\rho Y)_t + (\rho Y u)_x = (\rho D Y_x)_x - m\omega \tag{4}$$

$$p = (\gamma - 1)\rho C_v T \ , \ E = \rho C_v T + \frac{1}{2}\rho u^2 \ , \ \gamma = \frac{C_p}{C_v} \ , \ \omega = \mathcal{A}\frac{\rho Y}{m}\exp\left(\frac{-\mathcal{E}}{\mathcal{R}T}\right) \tag{5}$$

where ρ stands for the density, ρu the momentum, E the total energy, Y and m the mass fraction of the reactant and its molar weight, T the temperature, $\mathcal{R}$ the ideal gas constant, C_p (resp. C_v) the specific heat at constant pressure (resp.

volume), Q the heat release, ω the reaction rate, $\mathcal{A}$ the Arrhenius prefactor and $\mathcal{E}$ the activation energy. The viscosity coefficient μ and the heat conductivity k are related through the Prandtl number: $P_r = \mu C_p/k$. The molecular diffusion coefficient D and the heat conductivity define the Lewis number: $\mathcal{L}e = k/\rho D C_p$.

The Low Mach number limit: can be formely obtained by expanding the N.S. equations in powers of the Mach number. The resulting equations have almost the same form as the full N.S. equations:

$$(1) \ \& \ (4)$$
$$(\rho u)_t + (\rho u^2 + p_d)_x = [(4/3)\mu u_x]_x \tag{6}$$
$$\rho C_p T_t + \rho u C_p T_x = (k T_x)_x + Q\omega + (P_T)_t \tag{7}$$
$$P_T = \rho R T \quad C^{te} \ in \ space \tag{8}$$

where the pressure is splitted as $P(x,t) = P_T(t) + p_d(x,t)$ with $p_d/P_T \sim \text{Ma}^2 \ll 1$. The thermodynamic pressure P_T is assumed to be constant in space everywhere and the dynamic part of the pressure field p_d, which appears in the limiting form of the momentum equation, must be obtained from the continuity equation.

A fundamental difference between the low Mach number equations and the equations for incompressible flow is the density derivative in the continuity equation, and this difference is the major factor in violating the assumptions of the model.

Numerical methods/outlines: the numerical methods utilized were mostly developed in previous works and only some features of the N.S. numerical method are discussed. Introducing $W = (\rho \, , \, \rho u \, , \, E \, , \rho Y)$ the equations (1)-(4) are rewritten as $G(W) = 0$ and solved by applying the iterative algorithm:

$$G'(W^\alpha)(W^{\alpha+1} - W^\alpha) = -G(W^\alpha) \quad , \quad \alpha = 0, 1, \ldots \tag{9}$$

where G' is a linearization of G and $W^{\alpha=0} = W^n$. Stopping this procedure when $\alpha=1$ defines an efficient linearized scheme and partly second-order accuracy can be achieved by using semi-implicit time integration as in [5]. Here, we consider a totally non-linear algorithm where the process (9) applies to:

$$G(W^\alpha)_i = \kappa W_i^\alpha + (1 - 2\kappa)W_i^n + (\kappa - 1)W_i^{n-1} + \delta t \left[\theta \mathcal{N}(W^\alpha)_i + (1 - \theta)\mathcal{N}(W^n)_i\right]$$

until the residual $|W^{\alpha+1} - W^\alpha|$ is small enough. Gear (resp. Crank-Nicolson) second-order scheme is recognized when $\kappa=1.5$, $\theta=1$ (resp. $\kappa=1$, $\theta=0.5$). The discrete operator $\mathcal{N}$ stands for the remaining spatial terms.

For multi-dimensional applications an iterative algorithm is generally used to solve (9) at each α. For instance, Gauss-Seidel algorithm fails when a centered scheme is used to discretize the hyperbolic terms (left hand side of eqations (1)-(4)). Actually, an upwind scheme improves the efficiency of the algorithm in addition to provide non oscillatory results. Therefore, after a careful examination of several upwind techniques [4], Roe's scheme [6] has been adopted with M.U.S.C.L. techniques for higher-order spatial accuracy.

NUMERICAL RESULTS

Three different flame propagation problems in an opened tube (a),(b) and (c) have been considered and range from a slow flame speed of .023 m/s to a fast flame speed of 2.5 m/s. The quantities C_p, k, μ, Q, ρD, m, $\mathcal{A}$ and $\mathcal{E}$ were assumed to be constant. The initial conditions for the unburned products were: $p_u=1$ atm, $\rho_u=1.3$ kg/m^3 and $T_u=300°$K. Moreover, we set $\mathcal{L}e=1$, $P_r=0.7$, $C_p=10^3$ and $\gamma = 1.4$. A closed tube problem with moderate flame speed (b) has also been considered. All calculations used a 10 mm long tube with 400 points.

Some of the results are summarized in Table I where T_b denotes the adiabatic flame temperature, V_f The burning velocities and Δp the pressure jumps across the flame. There is a very good qualitative and quantitative agreement between the two models. Both numerical methods used the same direct solver and for all speed cases, the N.S. model was slightly more efficient than the L.M. model.

Table I:

case	$T_b(°K)$	$k(J/m.K.s)$	$\dfrac{\mathcal{E}}{\mathcal{R}T_b}$	$\mathcal{A}(1/s)$	V_f (m/s) L.M	V_f (m/s) N.S	Δp (Pa) L.M	Δp (Pa) N.S.
(a)	1800	0.01	4.167	1.60 10^5	0.024	0.022	3.19 10^{-3}	3.23 10^{-3}
(b)	1800	0.10	4.167	8.00 10^6	0.494	0.496	1.59	1.60
(c)	2100	0.30	9.600	1.39 10^{10}	2.428	2.404	50.0	45.3

• Using case (a), we adress the problem of pressure over and undershoots obtained with N.S. model. In figure **1** are plotted the results for the non linear scheme with different spatial accuracies and CFL$\sim 10^5$. The maximum value of the profile obtained using the first-order scheme is one order of magnitude above the physical value. TVD techniques do not help the second-order scheme for this smooth numerical diffusion process. The overshoot can only be eliminated either by using mesh refinement or third-order interpolated arguments in the MUSCL scheme.

The plots in figure **2** clearly show that the linear algorithm causes both over and undershoots while the non-linear algorithm gives perfectly monotonic profiles. The extrema of the profile obtained with Crank-Nicolson scheme are several order of magnitude far from the exact values, but no differences would be observed in the temperature or the density profiles.

The temporal accuracy of the non linear scheme does not affect the results for all the opened tube problems but it strongly affects the results for the closed tube problem. Plotted in figure **3**, the mean pressures versus time show that used with a large CFL which corresponds to a "convective" CFL about 10, the second-order non linear scheme produced accurate results.

- The breakdown of the L.M. number model can begin to be seen for the fast flame speed case (c), where the problem of large acceleration appears during the flame formation process. Due to the assumption of constant pressure, changes in temperature are channeled directly into the gas density changes without interaction with the pressure field and result in excessive fluid particle accelerations. In a full compressible model and in the limit of very rapid changes in temperature, the density remains constant and the pressure and temperature follow each other. Shown in figure (4) is the relative change in the end wall pressure as a function of time, and it can be seen that the L.M. number model generates large pressure disturbances during startup, and these quickly decay. The N.S. results show a wave reflection process, which is very weak and this is missing in the L.M. number simulation due to the assumption of infinite sound speed.

- The final results presented are for a closed tube where the background thermodynamic pressure must be a function of time. The results of the simulations are shown in figures (5) and (6) and they again show very good qualitative and quantitative agreement. The big change is that the low Mach number model is much less efficient because of the constraint to calculate the background thermodynamic pressure, and the N.S. simulation is not influenced by this constraint.

CONCLUDING REMARKS

The full Navier-Stokes equations can be competitive in terms of efficiency and accuracy for problems involving rapid chemestry. This is particularly true for closed combustion chambers and is very relevant to internal flow calculations such as internal combustion engine dynamics and it suggests strongly that future simulations should consider the use of the full N.S. equations. However, the low Mach number model does give very good results, except for some short term transient processes and should be considered for external flows.

Aknowledgements: The first author has been supported by INRIA. The third author was partially supported by Sandia National Laboratories at Livermore.

References

[1] Chorin, A.J., Math. Comput. Vol 22, pp.745-762, 1968.
[2] Dandy, D. and Dwyer, H. A., Accepted J. Fluid Mech. , 1989.
[3] Fernandez, G. , Larrouturou, B., Notes on Numerical Fluid Mechanics, 24,pp. 128-137, Vieweg.
[4] Fernandez, G. , Inria report # 873, 1988.
[5] Fernandez, G. , Guillard, H., Troisième Conférence Internationale sur la simulation Numérique de la Combustion, Antibes, July 1989.
[6] Roe, P.L., J.C.P., Vol 43, p: 357, 1981.
[7] Tang, K., Dwyer, H.A., WSS/CI # 72, 1989.

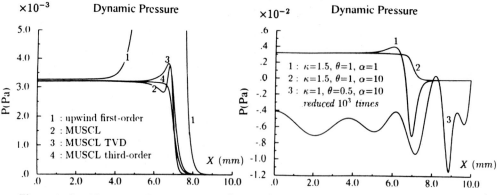

Figure 1: Spatial accuracy, flame (a), CFL=10^5
$\kappa = 1.5$ or 1, $\theta = 1$, $\alpha = 10$

Figure 2: Linear/non-linear schemes, flame (a)
CFL=10^4

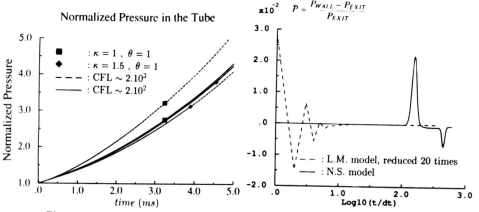

Figure 3: Temporal accuracy, closed tube

Figure 4: Wall pressure histories, case (c), CFL=10

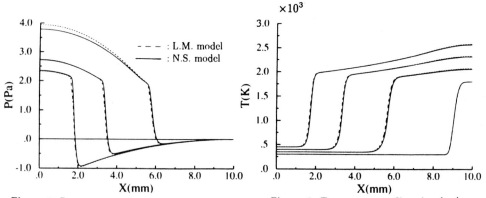

Figure 5: Dynamic pressure, closed tube, CFL=200

Figure 6: Temperature profiles, closed tube

AN ACCURATE AND FAST 3-D EULER-SOLVER FOR TURBOMACHINERY FLOW CALCULATION

S. Lecheler, H.-H. Fruehauf
Institut für Raumfahrtsysteme, Universität Stuttgart
Pfaffenwaldring 31, D-7000 Stuttgart 80

An implicit 3-D Euler finite-difference code is developed to solve the unsteady flow equations in turbomachinery blade rows. Special attention was paid to high accuracy, fast convergence and broad applicability.
The governing equations are transformed to a rotating general curvilinear coordinate system. The transformation maintains the absolute cartesian flow velocities u, v, w in the relative system:

$$\frac{\partial}{\partial \tau}\hat{q} + \frac{\partial}{\partial \xi}\hat{E}(\hat{q}) + \frac{\partial}{\partial \eta}\hat{F}(\hat{q}) + \frac{\partial}{\partial \zeta}\hat{G}(\hat{q}) + \hat{H}(\hat{q}) = 0$$

$$\hat{q} = \frac{1}{J}\begin{bmatrix} \rho \\ \rho u \\ \rho v \\ \rho w \\ e \end{bmatrix} \quad \hat{E}(\hat{q}) = \frac{1}{J}\begin{bmatrix} \rho U \\ \rho u U + \xi_x p \\ \rho v U + \xi_y p \\ \rho w U + \xi_z p \\ (e+p)U - \Omega(\xi_y z - \xi_z y)p \end{bmatrix} \quad \hat{H}(\hat{q}) = \frac{1}{J}\begin{bmatrix} 0 \\ 0 \\ -\Omega\rho w \\ \Omega\rho v \\ 0 \end{bmatrix}$$

$U = \xi_t + \xi_x u + \xi_y(v + \Omega z) + \xi_z(w - \Omega y)$ is the contravariant relative velocitiy. $\hat{F}(\hat{q})$ and $\hat{G}(\hat{q})$ are similar to $\hat{E}(\hat{q})$ with the corresponding contravariant relative velocities V and W and the metric derivations of η and ζ. Ω is the rotational speed, $p = (\kappa - 1)\left[e - \frac{1}{2}\rho(u^2 + v^2 + w^2)\right]$ is the static pressure and J is the Jacobian of the transformation.
The Euler equations are solved on H-grids or C-grids using the noniterative approximate factorization implicit finite difference method of Beam and Warming.
High accuracy is ensured by use of compatibility relations at all boundary surfaces, edges and corners. This guarantees a low entropy error and an accurate convection of total pressure /1/. Furthermore these characteristic boundary conditions increase the robustness of the code, especially if skewed grids or short inlet and outlet regions are used.
An accurate and nearly oscillation-free shock resolution is obtained by a combined second- and fourth-difference artificial dissipation model with nonlinear coefficients. These modell is similar to /2/ with an improved treatment at the boundaries.
The good convergence rate of the implicit algorithm is further improved by an implicit implementation of all boundary conditions. Especially the implicit treatment of the periodicity condition is very important for fast convergence.
Additionally, for time asymptotic calculations a simple reduction of the implicit time step can be used. This reduces the factorization error and thereby the convergence rate can be doubled again.
Typicall Courant-numbers are around 30 to 40. Converged solutions are obtained within 100 to 300 iterations for a 20.000 point grid without using mesh sequencing or multigrid techniques.
The code has been assessed for axial, mixed-type and radial blade rows.
Numerical results are shown for the DLR-L030-4 transonic compressor cascade and for a highly loaded transonic radial compressor rotor.

For the cascade a fine boundary orthogonal 201x25-C-grid is used with 141 points on the airfoil and 12 points at the round leading edge with a radius of about 0.4% of the length. The radial impeller has 28 blades with sharp leading and trailing edge. The used H-grid has 49x13x7 points.

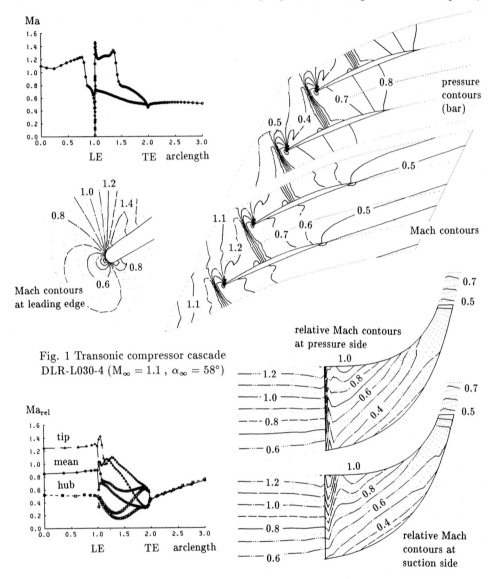

Fig. 1 Transonic compressor cascade DLR-L030-4 ($M_\infty = 1.1$, $\alpha_\infty = 58°$)

Fig. 2 Transonic radial compressor rotor (n = 50 000 min^{-1}, $\dot{m}$ = 2.8 kg s^{-1})

/1/ Küster U., Boundary Procedures for the Euler-Equations, Notes on Numerical Fluid Mechanics (Vieweg), Vol.13, Sep. 1985

/2/ Pulliam T.H., Steger J.L., Recent Improvements in Efficiency, Accuracy and Convergence for Implicit Approximate Factorization Algorithms, AIAA-85-0360, Jan. 1985

Computation of the Flow in Elliptical Ducts
by
Z. Bar-Deroma and M.Wolfshtein
Faculty of Aerospace Eng.
Technion, Haifa, Israel

Abstract

Three-dimensional, incompressible laminar flow in ducts of uniform elliptic cross sections is considered. This work is related to the flow in engine inlets of contemporary air-breathing propulsion systems. Very often the entrance to such inlets is elongated in the horizontal direction. Transformation to a cylindrical-polar coordinate system is very inefficient as it concentrates many mesh points near the center where the flow is fairly uniform. Transformation to a rectangle does not lend itself easily to the prescription of small mesh distances near the solid wall. Coordinate transformation to an ellipse appears to offer certain advantages for such cases. Therefore we chose to study the flow in ducts with elliptical cross sections having various ratios between the major and minor axes. The equations were transformed to an orthogonal coordinate system. The solution domain was covered by an orthogonal finite difference mesh, staggered in all three coordinate directions.

The reduced Navier-Stokes equations are solved, as proposed by Briley. This formulation is suitable for internal problems where a predominant direction of flow exists. We did not assume any symmetry in the solution procedure, and therefore the procedure is suitable for various kinds of initial conditions, representing angles of attack and/or side slip. ADI operator splitting scheme of Douglas-Gunn with the so called "delta" form was used to march the momentum equations along the duct. A Neumann problem for the velocity correction potential was solved by the SOR method. The program was successfully checked by comparisons with an analytical solution for developed flow in elliptical ducts.

Solutions were obtained for a combination of geometries, Reynolds numbers and initial conditions. Three types of initial conditions were used for the axial velocity: uniform (plug) flow, linear profile and parabolic profile. For the lateral velocities we used either zero or uniform values as initial conditions.

At least four longitudinal vortices are observed in the developing duct flow, one in each quarter of the pipe, as can be seen in figures 1-2. They can be observed even at very large distances from the entrance to the duct. The magnitude of the lateral velocities associated with them is very small compared with the axial velocity. The strength of the vortices increases with the aspect ratio.

The developed axial velocity distributions over the ellipse axes are similar in the developed part of the flow.

The distributions of the lateral and axial skin friction coefficients at various cross secitons, and for various Reynolds numbers and axes ratios were calculated. Typical results are shown in figure 3. The lateral profile shows some skewness, as can be expected. The difference between the skin friciton data at the various aspect ratios results from the fact that the Reynolds number based on the hydraulic diameter is identical for the two axes ratios shown, but the cross section and the flow rates are not.

The axial developments along the duct of the maximum axial velocity, the pressure gradient and of the mean axial skin friction coefficient for various Reynolds numbers were calculated as well.

In general we found the elliptic coordinate system to be very suitable for such problems. The mesh staggering used, required special attention and complicated bookeeping of the indices, particularly near and between the focii of the system. This applies also to other coordinate systems with two focii, that may be required for the geometries in question. Finally, we wish to comment that the mesh used is not suitable for ellipses with nearly identical major and minor axes, i.e. almost circular geometries. This is not surprising as in these cases the two focii degenerate into one.

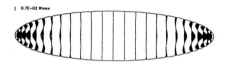

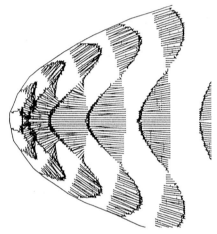

Figure 1
Distribution of the cross-flow velocity vector in a lateral cross section

Figure 1a
Zoom-in on the left part of figure 1

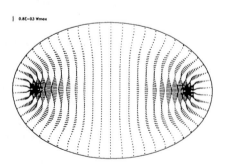

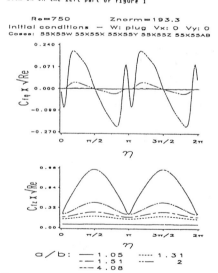

Figure 2
Distribution of the cross-flow velocity vector in a lateral cross section

Figure 3
Distribution of the skin friction coefficient in a lateral cross section as function of the shape of the cross section

A FULLY IMPLICIT METHOD FOR SOLVING THE 3-DIMENSIONAL QUASI-EQUILIBRIUM EQUATIONS ON THE SPHERE

M.H.Mawson and M.J.P.Cullen
U.K. Meteorological Office, London Road,
Bracknell. Berkshire. RG12 2SZ.

1. Introduction.

The Quasi-Equilibrium (henceforth QE) equations have been shown to describe many motions of interest in the atmosphere (see Cullen et al 1987). To fully evaluate their potential for global weather prediction it is necessary to extend the 2-dimensional solution procedure of Cullen (1989a) to 3-dimensions on a sphere. The solutions obtained by this method represent at each timestep a minimum energy state for the system. This ensures a unique solution except in areas where air parcels can be interchanged without affecting the total energy of the system. Cullen and Purser (1989) showed that this was the case if parcels were exchanged across the equator. To overcome this problem a frictional term is added to the equations in the way described by Cullen (1989b). It is hoped that this approach should allow solutions to be obtained numerically without altering their qualitative nature. An outstanding theoretical problem is to prove that as the frictional term tends to zero then the solution of the equations with friction tends to that of the equations without friction.

2. QE Equations and solution procedure.

We consider the hydrostatic form of the equations and use $\sigma=p/p_*$ as the vertical co-ordinate where subscript * denotes the value at the surface. The equations are written in terms of spherical polar co-ordinates (λ,μ,σ). The notation is standard except where explicitly stated.

The QE equations are as follows ;

Momentum,

$$p_*\frac{\partial u_0}{\partial t} + \frac{U}{a\cos\mu}\frac{\partial u_0}{\partial \lambda} + \frac{V}{a}\frac{\partial u_0}{\partial \mu} + W\frac{\partial u_0}{\partial \sigma} - f(V-p_*v_0) - \frac{Uv_0 \tan\mu}{a} = C_D(U-p_*u_0) \quad (1)$$

$$p_*\frac{\partial v_0}{\partial t} + \frac{U}{a\cos\mu}\frac{\partial v_0}{\partial \lambda} + \frac{V}{a}\frac{\partial v_0}{\partial \mu} + W\frac{\partial v_0}{\partial \sigma} + f(U-p_*u_0) + \frac{Uu_0 \tan\mu}{a} = C_D(V-p_*v_0) \quad (2)$$

Thermodynamic,

$$p_*\frac{\partial \theta}{\partial t} + \frac{U}{a\cos\mu}\frac{\partial \theta}{\partial \lambda} + \frac{1}{a}\frac{\partial \theta}{\partial \mu} + W\frac{\partial \theta}{\partial \sigma} = H \quad (3)$$

Continuity,

$$\frac{\partial}{\partial \sigma}\left(\frac{\partial p}{\partial t}\right) + \frac{1}{a\cos\mu}\left[\frac{\partial}{\partial \lambda}\left(\frac{\partial(pu)}{\partial \sigma}\right) + \frac{\partial}{\partial \mu}\left(\frac{\partial(pv\cos\mu)}{\partial \sigma}\right)\right] + \frac{\partial}{\partial \sigma}\left(\frac{\partial(pw)}{\partial \sigma}\right) = 0$$

which reduces to

$$\frac{\partial p_*}{\partial t} + \frac{1}{a\cos\mu}\left[\frac{\partial U}{\partial \lambda} + \frac{\partial(V\cos\mu)}{\partial \mu}\right] + \frac{\partial W}{\partial \sigma} = 0 \quad (4)$$

Equilibrium relationship,

$$-fu_0 = \frac{1}{a}\frac{\partial \phi_*}{\partial \mu} + \frac{1}{a}C_p\sigma^\kappa \theta \frac{\partial \pi}{\partial \mu} + C_D v \quad (5)$$

$$fv_0 = \frac{1}{a\cos\mu}\left[\frac{\partial \phi_*}{\partial \lambda} - \frac{1}{a}C_p\sigma^\kappa \theta \frac{\partial \pi}{\partial \lambda}\right] + C_D u \quad (6)$$

Hydrostatic,

$$-R\sigma^{\kappa-1}\theta\pi = \frac{\partial \phi}{\partial \sigma} \quad (7)$$

where $(U,V,W) = p_*(u,v,D\sigma/Dt)$, C_D is a frictional drag coefficient, H is a source term for heating, ϕ is the geopotential, $\pi=(p_*/1000)^\kappa$ is the Exner function where $\kappa=R/C_p$, θ is the buoyancy defined as $\theta= T^*(1000/p)^\kappa$ where T is the temperature, a is the radius of the earth. u_0 and v_0 are approximations to the full wind u and v and are called the equilibrium wind components. The Equilibrium relationship represents a balance between the largest terms in the equations of motion whilst the momentum equations contain the next order terms which are small in comparison. We can eliminate ϕ from equations (5) and (6) by using (7) giving

$$f\frac{\partial u_0}{\partial \sigma} = \frac{\pi R \sigma^{\kappa-1}}{a}\frac{\partial \theta}{\partial \mu} - \frac{C_p \sigma^\kappa}{a}\frac{\partial \theta}{\partial \sigma}\frac{\partial \pi}{\partial \mu} - \frac{\partial(C_D v)}{\partial \sigma} \qquad (8)$$

$$f\frac{\partial v_0}{\partial \sigma} = -\frac{\pi R \sigma^{\kappa-1}}{a \cos\mu}\frac{\partial \theta}{\partial \lambda} + \frac{C_p \sigma^\kappa}{a}\frac{\partial \theta}{\partial \sigma}\frac{\partial \pi}{\partial \lambda} + \frac{\partial(C_D u)}{\partial \sigma} \qquad (9)$$

The solution procedure is a predictor/corrector method based on that suggested by Meek and Norbury(1984) and is described in more detail by Cullen(1989a). The first four equations allow us to step the variables u_0, v_0, p_* and θ forward in time giving provisional values at the new time level. This still leaves us needing to find u and v at the new time level. This is done by noting that equations (5-6) applied at the surface and (8-9) at the levels above the surface will not be exactly satisfied by the provisional values at the new time level. We therefore solve these equations to obtain the corrections to u and v needed to satisfy them. This procedure is analogous to the pressure-correction method for the Navier-Stokes equations except that here it is the total wind field which is solved for implicitly rather than the pressure field. The equations to be solved are therefore ...

$$f\Delta u_0 + 1/a\ C_p\ \sigma^\kappa\ \Delta\theta\ \partial\Delta\pi/\partial\mu + C_D\Delta v = D_* \qquad (10)$$

$$f\Delta v_0 - 1/(a\cos\mu)\ C_p\ \sigma^\kappa\ \Delta\theta\ \partial\Delta\pi/\partial\lambda - C_D\Delta u = E_* \qquad (11)$$

$$f\frac{\partial \Delta u_0}{\partial \sigma} - \frac{\Delta\pi R \sigma^{\kappa-1}}{a}\frac{\partial \Delta\theta}{\partial \mu} + \frac{C_p \sigma^\kappa}{a}\frac{\partial \Delta\theta}{\partial \sigma}\frac{\partial \Delta\pi}{\partial \mu} + \frac{\partial(C_D \Delta v)}{\partial \sigma} = D_k \quad (2\le k\le N) \qquad (12)$$

$$f\frac{\partial \Delta v_0}{\partial \sigma} + \frac{\Delta\pi R \sigma^{\kappa-1}}{a \cos\mu}\frac{\partial \Delta\theta}{\partial \lambda} - \frac{C_p \sigma^\kappa}{a}\frac{\partial \Delta\theta}{\partial \sigma}\frac{\partial \Delta\pi}{\partial \lambda} - \frac{\partial(C_D \Delta u)}{\partial \sigma} = E_k \quad (2\le k\le N) \qquad (13)$$

D_*, E_*, D_k, and E_k are the residuals obtained from equations (5),(6),(8) and (9) respectively and N is the number of model levels. Equations (10) and (12) are solved in (μ,σ) slices with the corrections Δu_0, $\Delta\theta$ and $\Delta\pi$ written in terms of Δv by using the 2-dimensional (μ,σ) version of equations (1),(3) and (4). The resulting equations are elliptic provided that the potential vorticity is non-negative. The finite-difference scheme is not valid in the presence of discontinuities so exact solution of these equations leads to instability. Cullen(1989a) showed that a correct solution could be obtained by using a reduced system which is elliptic provided that

$$\partial \theta / \partial \sigma < 0 \quad \text{and} \quad -\partial u_0/\partial \mu + af > 0 \qquad (14)$$

which are the conditions that the data is statically and inertially stable. These conditions are enforced by modifying the data before solving equations (10) and (12). We obtain our provisional value of v at the new time level by adding on Δv calculated from this reduced system. The solution obtained by this method now may need to be iterated but if iterated to convergence will again produce instability in the vicinity of discontinuities. The solution may still diverge if the potential vorticity, defined as the determinant of a function P where $P = \phi + \frac{1}{2}f^2(x^2 + y^2)$ and $x = a\lambda\cos\mu$, $y = a\mu$ are displacements from the co-ordinate axes, is negative. This occurs infrequently at a few points and may be avoided by increasing the frictional force in the vicinity of these points. Periodic boundary conditions are applied in the μ direction and we set W=0 at $\sigma=0$ and 1. A similar approach is applied in the (λ,μ) plane using equations (11) and (13) to find the correction Δu with ellipticity conditions

$$\partial \theta / \partial \sigma < 0 \quad \text{and} \quad \partial v_0/\partial \lambda + af\cos \mu > 0. \qquad (15)$$

The three dimensional problem is solved by an Alternating Direction Implicit (ADI) method in which the (μ,σ) and (λ,σ) problems are solved alternately.

The theory behind the QE model (Cullen and Purser(1989)), in the absence of friction, shows that there is a unique u field which maintains the atmosphere in a minimum energy state. This computational procedure is an iterative method of calculating u. In practice it is found that some smoothing of the iteration is required to improve the convergence of the ADI on small horizontal and large vertical scales. This is done by averaging on a σ-surface Δu and Δv at point i as follows;

$$\Delta u_i = .5 * (\Delta u_i + .2 \sum_{j=i-2}^{i+2} \Delta u_j) \qquad (16)$$

with Δv being treated similarly.

The variables are held on a staggered grid which is the one which appears naturally when the equations are written out using centred finite-differencing, u and v on the MAC grid with v_0 and u_0 held at the same points as u and v respectively.

3. Future Improvements.

Early results from applying this procedure suggest the following improvements.
1. Replacement of the current direct solver with a multi-grid method.
2. Removing the periodic boundary conditions in the (μ,σ) direction to alleviate problems at the poles.
3. Improving the methods used in the prediction step to increase both accuracy and computational efficiency.
4. An algorithm to deal with points where the model is statically and inertial stable but still has negative potential vorticity. The present solution method will produce a result but not a solution of the true equations.

4. References.

Cullen,M.J.P., Norbury,J., Purser,R.J., and Shutts,G.J.	1987	Modelling the Quasi-Equilibrium dynamics of the atmosphere. Q. J. Roy. Meteorol. Soc. 113, 735-757
Cullen,M.J.P. and Purser,R.J.	1989	Properties of the Lagrangian semi-geostrophic equations. J. Atmos. Sci. 46, 2684-2697.
Cullen,M.J.P.	1989a	Implicit finite-difference methods for modelling discontinuous atmospheric flows J. Comput. Phys. 81, 319-348
Cullen,M.J.P.	1989b	On the incorporation of atmospheric boundary layer effects into a balanced model. Q. J. Roy. Meteorol. Soc. 115, 1109-1131.
Meek,P.C. and Norbury,J.	1984	Non-linear moving boundary problems and a Keller box scheme. Siam. J. Num. Anal. 21, 883-893.

The existence aspects of Dupuit and Boussinesq filtration models
using Finite Element Method

Migórski Stanisław, Schaefer Robert
Jagiellonian University; Computer Science Department
ul Kopernika 27, PL 31-501 KRAKÓW, Poland

1. Introduction

In this paper we study the free - surface filtration flow in well permeable stratum (see fig.1).

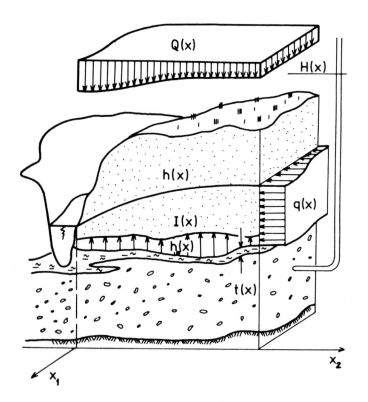

Fig 1. Free - surface ground water flow geometry

The stratum under consideration can be fed by
- rainfall of intensity displacement $Q(t,x)$,
- inflow displacement $q(t,x)$ from the geological structures laying behind the analysed region,
- inflow displacement $I(t,x)$ from the artesian stratum located below.

In case the artesian stratum is rich in water (i.e. it is well permeable and profusely fed) and the interbedded stratum is relatively thin and less permeable then

(1) $\quad I(t,x) = \dfrac{k_v(x)}{d(x)} (h(t,x) - H(t,x))$

where $H(t,x)$ is the artesian piesometric height, $h(t,x)$ is the free surface height in the main stratum and the quotient $k_v(x)/d(x)$ denotes the vertical permeability of interbedded stratum.

The main problem to solve is the prognosis of the evolution of free ground water surface (described by the function $h = h(t,x)$) under the influence of the environment factors mentioned above.

We can mathematically describe the filtration problem in two cases (stationary and nonstationary) using the modified Dupuit and Boussinesq partial differential equations, respectively (see [1]).

2. Statement of the theoretical results

Let Ω be a bounded domain of $\mathbb{R}^N$, let $0 < T < +\infty$. Given a partition $\partial\Omega = \partial\Omega_1 \cup \partial\Omega_2$ of the sufficiently regular boundary $\partial\Omega$, with
$\int_{\partial\Omega_1} d\sigma > 0$, $\int_{\partial\Omega_1 \cap \partial\Omega_2} d\sigma = 0$, let

$V = \{ u ; u \in H^1(\Omega), u = 0 \text{ on } \partial\Omega_1 \}$,

$W = \{ u; u \in L^2(0,T;V), u' \in L^2(0,T;V') \}$

Let $a_{ij}(x,s)$, $i,j = 1,\ldots,N$ be given functions such that

(i) $\exists M > 0 ; a_{ij}(x,s) \leq M, \forall x \in \bar\Omega, \forall s \in \mathbb{R}$

(ii) $\exists \alpha_0 > 0 ; \forall \xi \in \mathbb{R}^N, \forall s \in \mathbb{R} \quad \sum\limits_{i,j=1}^{N} a_{ij}(x,s) \xi_i \xi_j \geq \alpha_0 \sum\limits_{i=1}^{N} \xi_i^2$

(iii) $\mathbb{R} \ni s \mapsto a_{ij}(x,s) \in \mathbb{R}$ is continuous, $\forall x \in \bar\Omega$.

Finally, let us take

$$a(u;w,v) = \sum_{i,j=1}^{N} (a_{ij}(x,u) \frac{\partial w}{\partial x_i} , \frac{\partial v}{\partial x_j}) \quad \forall\, u, v, w \in V$$

where $(\ ,\ )$ denotes the $L^2(\Omega)$ scalar product.

Consider the following nonlinear evolution problem

$$(2) \quad \begin{cases} \langle \frac{d}{dt}\, h(t), v \rangle + a(h;h,v) = (f, v) + \int_{\partial\Omega_2} q\, v\, d\sigma \quad \forall\, v \in V, \\ h(0) = h_o \end{cases} \quad \text{a.e. } t \in (0,T)$$

We have following existence result:

Theorem 1 If hypotheses (i), (ii) and (iii) hold, then for each $f \in L^2((0,T) \times \Omega)$, $q \in L^2((0,T) \times \partial\Omega_2)$, $h_o \in L^2(\Omega)$ problem (2) admits a solution $h \in W$.

The main idea of proof consists in the application of the Galerkin method (see e.g. [4]). It allows to define the approximate solution which satisfy the system of linear differential equations. From the classical theorem of Cauchy – Peano, we know that such system admits a solution under assumption that its right hand side is continuous. But the latter it is easy to check using the Lesbeque dominated convergence theorem and our hypotheses. Next, owing to a priori estimates, we can pass to the limit in the sequence of Galerkin solution and observe that the limit function is the solution of problem (2).

We give now the formulation of filtration problem on the basis of the Boussinesq equation. To this end let us consider the problem (2) with the form $a(u;w,v)$ replaced by

$$(3) \quad \tilde{a}(u;w,v) = a(u;w,v) + (\frac{k_v(x)}{d(x)} w, v)$$

and

$$(4) \quad (f, v) = (Q, v) + (\frac{k_v(x)}{d(x)} H, v)$$

$$(5) \quad a_{ij}(x,h) = \begin{cases} (h_v - h_b)\, k_{ij}(x), & h < h_v \\ (h - h_b)\, k_{ij}(x), & h_v \le h \le h_{max} \\ (h_{max} - h_b)\, k_{ij}(x), & h_{max} < h \end{cases}$$

where $\Omega \subset \mathbb{R}^2$ is the filtration region (i.e. N = 2), the coefficients

$a_{ij}(x,h)$ and $k_{ij}(x)$ are the modified Dupuit permeability and the Darcy's tensor average all over the stratum thickness, respectiviely, h_v is the lower and h_{max} is the upper bound for the water surface location, h_b is the height of the main stratum bottom and the meanings of $H(x)$, $h(x)$, $k_v(x)/d(x)$, $q(x)$ are given in introduction.

Moreover, from the practical point of view the functions which appear in the problem are regular; in particular we can assume that:

(a) for $i,j = 1,2$ $k_{ij} \in C_I^0(\bar{\Omega})$ (the class of piecewise continuous functions on $\bar{\Omega}$, k_{ij} are symmetric (i.e. $k_{ij}(x) = k_{ji}(x)$) and uniformly elliptic (i.e. satisfy condition (ii)),

(b) k_v, H, d, $Q \in C_I^0(\bar{\Omega})$, $k_v(x) \geq 0$, $d(x) > 0$ $\forall x \in \bar{\Omega}$,

h_v, h_b, $h_{max} \in C_I^1(\bar{\Omega})$, $q \in C_I^1(\partial\Omega)$.

Theorem 1 shows that the filtration problem (problem (2) with the above modifications) has a solution $h \in W$.

3. Finite element formulation

In order to obtain the effective solution of the filtration problem we formulate the corresponding finite element approximated one as follows:

(6) $\begin{cases} < \frac{d}{dt} h_n(t), v > + \tilde{a}(h_n;h_n,v) = (f, v) + \int_{\partial\Omega_2} q v d\sigma \\ h(0) = (h_o)_n \end{cases}$

$\forall v \in L^2(0,T;Y_n)$

where (f, v) is given by (4), Y_n is the adequate finite element space which contains the functions from class $C_I^1(\bar{\Omega})$ and $(h_o)_n$ is the ortogonal projection of $h_o \in L^2(\Omega)$ onto Y_n.

We have the following

Theorem 2 Under the above hypotheses, problem (6) has at least one solution $h_n \in L^2(0,T;Y_n) \cap W$.

We consider also the corresponding free-surface stationary filtration problem of the form:

(7) $\tilde{a}(h_n;h_n,v) = (f, v) + \int_{\partial\Omega_2} q v d\sigma$ $\forall v \in Y_n$

Theorem 3 Given the conditions (a) and (b) there exists at least one solution $h_n \in Y_n$ of the problem (7). Such a solution is uniformly bounded by the constant depending only upon the domain Ω, and functions h_b, h_v, h_{max}, k_{ij}, H, k_v, t, Q defining the problem.

This result follows from the theorem of Browder (see [2]). In order to use the Browder's theorem mentioned above, we can prove that the nonlinear form (3) satisfy the following conditions
- for fixed u, $v \in Y_n$ $\tilde{a}(u;\cdot,v)$ is radially continuous,
- for fixed w, $v \in Y_n$ $\tilde{a}(\cdot;w,v)$ is continuous on bounded sets in Y_n,
- for each w, $v \in Y_n$ $\tilde{a}(v;w,w-v) - \tilde{a}(v;v,w-v) \geq 0$,
- there exists a real funtion γ on $\mathbb{R}$ with $\gamma(r) \to +\infty$ as $r \to +\infty$ such that for each $u \in Y_n$ $\tilde{a}(u;u,u) \geq \gamma(\|u\|) \|u\|$.

Under the above conditions the operator which correspods to $\tilde{a}(u;w,v)$ is surjective.

The theoretical results give a verification tool for the numerical models of filtration in
- valley fills of river and streams,
- glacial moraines.
- quaternarys covering of the valley sides.

Evolution filtration problem (6) can be solved using the space - time finite element technics, see for example Tzimpolos [5]. On the basis of the technics described above the computer package MUBS (multipurpose underground basin simulator) has been prepared and tested on some engineering examples (see [3]).

References:

[1] Bear J ; Dynamics of fluid in porous media. Elsevier, New York 1972.
[2] Browder F.E ; Existence and uniquennes theorems for solutions of nonlinear boundary value problems. Proc. Amer. Math. Soc. Symposium in Applied Mathematics. Vol 17, 1965 pp 24 - 29.
[3] Doniec M, Schaefer R.F ; The numerical approach to the little mountainous underground basin dynamics, taking into account the local parameters distribution. to appear in Technical Papers in Hydrology UNESCO / IAHS.
[4] Lions J.L ; Controle optimal de system gouvernes par des equations aux derivees partielles. Dunod 1968.
[5] Tzimpolos Ch ; Solution de l'equations Boussinesq par une methode des element finis. J. of Hydrology 30, 1976, pp. 1 - 18.

AN ACCURATE HYPERBOLIC SYSTEM FOR
APPROXIMATELY HYDROSTATIC AND
INCOMPRESSIBLE OCEANOGRAPHIC FLOWS

G. L. Browning, W. R. Holland, and S. J. Worley
National Center for Atmospheric Research*
P.O. Box 3000, Boulder, Colorado 80307 U.S.A.

H.-O. Kreiss
U.C.L.A., Dept. of Mathematics
Los Angeles, California 90024 U.S.A.

Abstract

It is well known that severe restrictions on accuracy and stability arise when using a numerical model based on the Eulerian equations to compute the low frequency motions of physical oceanography. We show that the loss of accuracy is due to the extreme skewness of the system. Although the system can be transformed to symmetric hyperbolic form, the diagonal transformation matrix contains factors that produce mathematical estimates indicating loss of accuracy and this occurs in practice. The stability restriction is just a result of the multiple time scales present in the system, i.e. the presence of sound waves. We discuss three alternatives to overcome these restrictions.

The first two alternatives are the well known primitive and quasi-geostrophic equations. Although models based on these systems alleviate some of the restrictions of the Eulerian model, they operate under a new set of limitations. We prove that a primitive equation model requires more resolution than a quasi-geostrophic model to obtain the same degree of numerical accuracy and is ill-posed for the initial-boundary value problem. And though the quasi-geostrophic equations are accurate to the order of the Rossby number, we show that there are cases when this error is on the order of 10 %. To reduce the error in these cases requires a considerable increase in the computing complexity of the system. The quasi-geostrophic equations also can not be used in the equatorial region and any improvements which would allow them to be used there would result in a computationally inefficient system.

The third alternative is to slow down the gravity and sound waves. Although the resulting approximate system alleviates the severe accuracy requirement, the stability requirement may still be unacceptable. By using the reduced system derived from the approximate system, we obtain a system which has all of the desired properties. The resolution requirements for a model based on the reduced system are the same as those of the quasi-geostrophic model, i.e. less than for a model based on the primitive equations. Since the reduced system is the proper limit of a hyperbolic system, a wide range of boundary conditions can be chosen so that the resulting initial-boundary value problem is well posed. We prove that the reduced system analytically describes the low frequency

---

* sponsored by the National Science Foundation.

solutions to two digits of accuracy (even when the Rossby number is .1), yet only requires the solution of a linear, constant coefficient, three dimensional elliptic equation. The reduced system can also be used in the equatorial region.

1. Nondimensional System

The dimensionless equations describing mid-latitude oceanographic motions which are geostrophic, hydrostatic, and fully three dimensional are

$$\frac{ds}{dt} - \tilde{s}\, w = 0 , \tag{1.1a}$$

$$\frac{du}{dt} + \epsilon^{-1}(p_x - fv) = 0 , \tag{1.1b}$$

$$\frac{dv}{dt} + \epsilon^{-1}(p_y + fu) = 0 , \tag{1.1c}$$

$$\frac{dw}{dt} + 10^{4-n}\epsilon^{-2}(p_z + g\, s) = 0 , \tag{1.1d}$$

$$\frac{dp}{dt} + 10^{4+n}\epsilon^{-1}(u_x + v_y + \epsilon\, w_z) = 0 , \tag{1.1e}$$

where ϵ is the Rossby number, t is time, $s = \rho - 10^{-2}p$ is essentially potential density, (u, v, w) is velocity, p is pressure, $f = f(y)$ is the Coriolis parameter, g is the gravity acceleration,

$$\tilde{s} = -10^n(\rho_{0z} + g) ,$$

$$\frac{d}{dt} = \frac{\partial}{\partial t} + u\frac{\partial}{\partial x} + v\frac{\partial}{\partial y} + \epsilon w \frac{\partial}{\partial z} ,$$

and ρ_{0z} is a known function of z. The variable n is used to ensure that $\tilde{s}$ is of order unity and determines the local length scale with $n = 0$ ($n = 2$) corresponding to 10^5 m (10^4 m). Reasonable values of ϵ for $n = 0$ and $n = 2$ are $\epsilon = 10^{-2}$ and $\epsilon = 10^{-1}$, respectively. Although (1.1) is a hyperbolic system, it is not symmetric. In fact it can be proved that perturbations can be amplified by max $(10^{4-n}\epsilon^{-1}, 10^{4+n})$. This means that the direct application of a numerical approximation to (1.1) is very delicate, and one has to calculate with very high accuracy. For example, when $n = 2$ at least six digits of accuracy must be carried. Methods to avoid this stringent accuracy requirement will be discussed.

2. Primitive Equations

If the smooth solution of (1.1) is determined from the first terms of its asymptotic expansion, then the hydrostatic and incompressible assumptions yield the primitive equations

$$\frac{ds}{dt} - \tilde{s}\, w = 0, \tag{2.1a}$$

$$\frac{du}{dt} + \epsilon^{-1}(p_x - fv) = 0, \tag{2.1b}$$

$$\frac{dv}{dt} + \epsilon^{-1}(p_y + fu) = 0, \tag{2.1c}$$

$$p_z + g\, s = 0, \tag{2.1d}$$

$$u_x + v_y + \epsilon w_z = 0. \tag{2.1e}$$

There are a number of problems with the primitive equations. First note that equation (2.2e) is used to compute the vertical component of the velocity w, but it should really be interpreted as an equation for the horizontal divergence $u_x + v_y$ which, in the limit as $\epsilon \rightarrow 0$, is zero. In the case that ϵ is not zero, a model which uses (2.2e) to determine w requires higher resolution than that required by models based on the equations in the next two sections.

A more serious problem is that by making assumptions (2.1a) and (2.1b) the hyperbolicity of the original system is lost so the primitive equations must be analyzed to determine their mathematical properties. The open boundary value problem for the primitive equations (2.2) is proved to be ill-posed.

3. Intermediate Systems

The second alternative to alleviate the severe accuracy requirement which oceanographers call the quasi-geostrophic system. This system is derived by requiring that the first order time derivatives of the vorticity and divergence formulation of (1.1) are bounded for all time. The constraints that arise from this requirement are satisfied at every time step which eliminates the fast time scale, but they require the solution of a three dimensional elliptic equation.

The quasi-geostrophic system

$$\frac{d_h s}{dt} - \tilde{s}\, w = 0, \tag{3.1a}$$

$$\frac{d_h \varsigma}{dt} - f_0\, w_z = 0, \tag{3.1b}$$

$$\nabla^2 p - f_0\, \varsigma = 0, \tag{3.1c}$$

$$p_z + g\, s = 0, \tag{3.1d}$$

where $\varsigma = -u_y + v_x$, $\delta = u_x + v_y$, $J = u_x v_y - u_y v_x$, $d_h/dt = u\, \partial/\partial x + v\, \partial/\partial y$, and for simplicity $\beta = 0$ has been assumed. Notice that the error in using (3.1) is of order ϵ which can be on the order of 10 % for the case where the representative horizontal length scale is $L = 10^4$ m.

The quasi-geostrophic formulation has not been applied in the tropics since (3.1c) is not accurate in that region. To overcome both the accuracy problem discussed earlier and this problem, the Jacobian term can be retained in the balance equation, i.e (3.1b) can be replaced by

$$\nabla^2 p - f_o \zeta - 2\epsilon J = 0. \qquad (3.2)$$

In the mid-latitudes (3.1a), (3.1b), (3.1d), and (3.2) can be combined to form a three dimensional, variable coefficient, elliptic equation for the vertical velocity. This system also requires the solution of a number of two dimensional elliptic systems for each value of z. In the equatorial region the equation for w becomes highly nonlinear because the Jacobian term in (3.2) is the same size as the pressure term. In the following sections a system of equations which can be used globally and only requires the solution of a linear, constant coefficient, three dimensional elliptic equation will be introduced.

4. Approximate System

The third alternative to alleviate the severe accuracy requirement for system (1.1) is to slow down the gravity and sound waves. Here system (1.1) is assumed to have a smooth solution, i.e. a solution where the function and its derivatives are of order unity. The approximate system is given by

$$\frac{d\bar{s}}{dt} - \tilde{s}\,\bar{w} = 0, \qquad (4.1a)$$

$$\frac{d\bar{u}}{dt} + \epsilon^{-1}(\bar{p}_x - f\bar{v}) = 0, \qquad (4.1b)$$

$$\frac{d\bar{v}}{dt} + \epsilon^{-1}(\bar{p}_y + f\bar{u}) = 0, \qquad (4.1c)$$

$$\frac{d\bar{w}}{dt} + \alpha_1(\bar{p}_z + g\,\bar{s}) = 0, \qquad (4.1d)$$

$$\frac{d\bar{p}}{dt} + \alpha_2(\bar{u}_x + \bar{v}_y + \epsilon\,\bar{w}_z) = 0, \qquad (4.1e)$$

where $\alpha_1 \ll 10^{4-n}\epsilon^{-2}$ and $\alpha_2 \ll 10^{4+n}\epsilon^{-1}$ are chosen to slow down the gravity and sound waves. The overbar notation is used to distinguish solutions of the approximate system (4.1) from the smooth solutions of the Eulerian system (1.1).

The error when using (4.1) to describe the smooth solutions of (1.1) is proved to be on the order of $E = \max[O(1/\alpha_1), O(1/\epsilon\alpha_2)]$. If $\alpha_1 = \epsilon^{-2}$ and $\alpha_2 = 10^2\epsilon^{-1}$, then there would be an error no larger than $\max[O(\epsilon^2), O(10^{-2})]$. Assuming $\epsilon \leq .1$, the approximate system with these values determines the smooth solutions of (1.1) over the entire mid-latitudes with an error term no larger than $O(10^{-2})$.

System (4.1) is a hyperbolic system so is automatically well posed for the initial value problem and a wide range of boundary conditions can be chosen so that the initial-boundary value problem is also well posed. Although the speed of the gravity and sound waves has been considerably reduced, the time step required by (4.1) may still be unacceptable. The next section proposes a method that relaxes this restriction.

5. Reduced System

In the oceanic case with $\alpha_1 = \epsilon^{-2}$, $\alpha_2 = 10^2\epsilon^{-1}$, disturbances of a smooth solution of the approximate system are amplified by max $(10, \epsilon^{-1})$ which is tolerable for ϵ in the range $10^{-2} \le \epsilon \le 10^{-1}$. The time step required by the CFL condition must be $10\epsilon^{-1}$ times smaller than that required by advection in the mid-ocean. For $\epsilon = 10^{-2}$, this is a factor of 1000. Thus, the approximate system may still not be an adequate basis for a model. Either a semi-implicit method for (4.1) or the reduced system of (4.1) introduced below is recommended. Note that a semi-implicit method applied to (4.1) is not the same as the semi-implicit method applied to (4.1), i.e. the coefficients of the elliptic equation for the pressure derived from the semi-implicit method for (4.1) are not all of the same size as they are in the corresponding equation for (4.1). The reduced system of (4.1) is given by

$$\frac{ds}{dt} - \tilde{s}w = 0, \tag{5.1a}$$

$$\frac{du}{dt} + \epsilon^{-1}(p_x - fv) = 0, \tag{5.1b}$$

$$\frac{dv}{dt} + \epsilon^{-1}(p_y + fu) = 0, \tag{5.1c}$$

$$\frac{dw}{dt} + \epsilon^{-2}(p_z + g\,s) = 0, \tag{5.1d}$$

$$u_x + v_y + \epsilon\,w_z = 0. \tag{5.1e}$$

The error in using (5.1) for the smooth solutions of (4.1) is on the order of 10^{-2}. Thus it can also be used to compute the smooth solutions of (1.1) with two digits of accuracy. Also note that the loss of accuracy factor due to skewing is only ϵ^{-1} since (5.1) can be symmetrized by introducing the new variable $\tilde{w} = \epsilon\,w$. Under the assumptions above this is less than or equal to 100 which is manageable.

As discussed earlier, equation (5.1e) should not be interpreted as an equation for w. Here the correct equation is obtained by applying the operator d/dt to (5.1e) (equivalent to bounding the second order time derivative of (4.1)) to obtain an elliptic equation for p of the form

$$p_{xx} + p_{yy} + p_{zz} = r,$$

where r is only a function of u, v, s, and their first derivatives. Note the similarity of (5.1) and the incompressible Navier-Stokes equations. Similar to those equations, the sound waves have been entirely eliminated. The reduced system only has high frequencies on the order of ϵ^{-1} which may be acceptable. If not, then the Coriolis terms need to be treated implicitly.

Vortex Simulation of Particulate Plume Dispersal and Settling

Ahmed F. Ghoniem
Massachusetts Institute of Technology
Cambridge, MA 02139

Howard R. Baum and Ronald G. Rehm
National Institute of Standards and Technology
Gaithersburg, MD 20899

In recent years considerable interest has developed in cleaning up large oil spills in open waters by burning off the resulting pools before they spread [1]. Although several researchers have shown that this can be a very efficient removal process (50% to 90% removal), the environmental nuisance caused by the dispersal of the resultant soot plume must be assessed. Since up to 10% of the fuel in a crude oil spill is converted to particulate matter by combustion processes, a mathematical model capable of predicting the dispersal and settling of the plume is needed. This paper concerns the development and application of such a model.
The principal assumptions underlying the model are:
1. The fire generating the smoke and hot gases is burning steadily.
2. The injection altitude and particle mass flux associated with the fire are known.
3. The gas temperature in the buoyant plume equilibrates with the atmosphere before any significant particle settling begins.
4. The ambient wind speed is much larger than the velocities induced by the negatively buoyant smoke particle plume.
5. Individual particle settling velocities are much smaller than the collective fluid velocities.
6. The stratification of the atmosphere plays no role in the settling process.
The last assumption limits the injection altitude for which the analysis is valid to at most one kilometer. The determination of the injection altitude itself is not part of the present calculation, although it does fall within the framework of the computational approach described here. The goal of the present study is the ability to predict the downwind deposition of the smoke particulate at ground level. The controlling parameters are the injection altitude and initial cross-section of the plume, the particulate mass flux and the ambient wind velocity. The resulting equations under appropriate non-dimensionalization become equivalent to the two-dimensional time-dependent Euler equations with gravitational forces. (The Boussinesq approximation is not made to allow future removal of many of the assumptions listed above.) The time-like coordinate is an appropriately scaled downwind distance.
The Transport Element Method [2],[3] is applied to the computation of the density and vorticity fields as the flow evolves downstream of the source. This is a Lagrangian, grid-free, field method in which the vorticity and density gradients, initially discretized among elements of finite area, are transported along fluid particle trajectories. The size of the elements change as the strain field evolves, while their strength is adjusted according to the source terms in the corresponding transport equations. In the scalar transport equation, the density gradient changes with the stretching and tilting of material lines. In the vorticity transport equation, vorticity is generated by gravity and baroclinic effects. The number of transport elements increases as vorticity generation intensifies and the plume shape becomes increasingly convoluted.

References

[1] Evans, D., Baum, H.R., Mulholland, G. and Forney, G., "Smoke Plumes from Crude Oil Burns", Proceedings of the Twelfth Arctic and Marine Oil Spill Technical Seminar, Environment Canada, p.1 (1989).

[2] Ghoniem, A.F. and Krishnan, A., "Origin and Manifestation of Flow-Combustion Interactions in a Premixed Shear Layer", Twenty Second Symposium (International) on Combustion, The Combustion Institute, Pittsburgh, p. 665 (1989).

[3] Krishnan, A. and Ghoniem, A.F., "Simulation of the Rollup and Mixing in Rayleigh-Taylor Flow Using the Transport Element Method", J. Comp. Physics, in press (1989).

A 3D Numerical Diagnostic Simulation for the General Circulation in the Gulf of Lion (France)

Kim Dan NGUYEN
Laboratoire de Mécanique de Lille

Introduction

This paper presents a 3-D numerical model to simulate the general circulation in the Gulf of Lion (Western Mediterranean sea), driven by winds, density differences in sea water, by continental shelf topography and mainly by the ligurian cyclonic circulation as a boundary condition.

Governing Equations

Continuity : $\partial u/\partial x + \partial v/\partial y + \partial w/\partial z = 0$

Momentum :
$\partial u/\partial t + \partial u^2/\partial x + \partial uv/\partial y + \partial uv/\partial t - fv$
$= -(1/\rho_0)\, \partial p/\partial x + \partial/\partial z(K_M \partial u/\partial z) + F_x$
$\partial v/\partial t + \partial uv/\partial x + \partial v^2/\partial y + \partial vw/\partial t + f u$
$= -(1/\rho_0)\partial p \partial y + \partial/\partial z(K_M \partial v/\partial t) + F_y$
$-\rho g = \partial p/\partial z$

Scalar Transport : $\partial \theta/\partial t + \partial u\theta/\partial x + \partial v\theta/\partial y + \partial w\theta/\partial z$
$= \partial/\partial t(K_M \partial \theta/\partial t) + F_\theta$

With
$F_x = A_M(2\partial^2 u/\partial x^2 + \partial/\partial y\,(\partial u/\partial y + \partial v/\partial x))$
$F_y = A_M(2\partial^2 v/\partial y^2 + \partial/\partial x\,(\partial u/\partial y + \partial v/\partial x))$
$F_\theta = A_M \nabla^2 \theta$

Where : θ may represent the temperature T or the salinity S ; K_M, A_M - vertical and horizontal turbulent viscosities ; K_H, A_H vertical and horizontal diffusivities.

Numerical Method

- Introduction of the dimensionless coordinates $\sigma = (h + z)/D$
- Two successive mode technique : the water levels are determined in the external mode and then, the velocity and scalar variables are evaluated in the internal mode.
- Implicit finite difference schemes for both modes, this permits to avoid the staggering between the two modes.
- Convection terms are handled by a characteristic method, combined with a 4th order interpolation to prevent numerical diffusion.

Validation : The proposed model is validated by the following test- cases :
- barotropic flow in a tidal bight. The values RMS of the mass transport and the water levels are computed (Fig. 1).
- 3D stratified flow due to density differences between two fluids in a tank. The initial conditions and the results of the problem are shown in Fig. 2.

Applications : 3D diagnostic simulation for the General Circulation in the Gulf of Lion. This Gulf is resolved on a 55 x 28 x 15 grid. The grid spacing is 5000 x 5000 m. The vertical distribution of grid points is irregulary. The observed wind field and the ligurian cyclonic circulations are used as the open boundary conditions. The circulation obtained from the model is in good agreement with the observations.

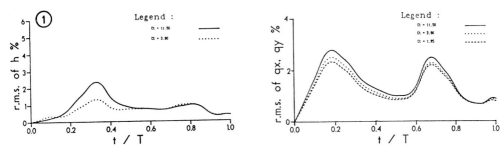

Fig.1 RMS of water levels and mass transport in the tidal bight test (Analytical solution obtained from Van de Krecke & Chiu, 1980) computed with courant numbers, $C_t=3.50$ and $C_t=11.50$

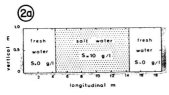

Fig.2 3-D stratified flow test : a) initial condition ; b) velocity fields on a vertical plan after 500 sec. and 1000 sec. ; c) horizontal velocity fields on surface and on bottom at t=1500 sec.

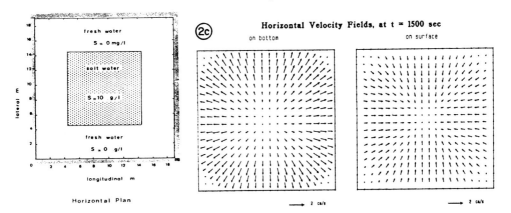

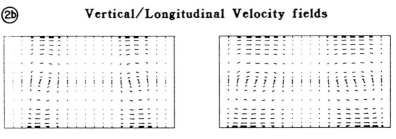

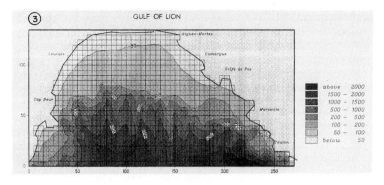

Fig.3 Topography of the Gulf of Lion.

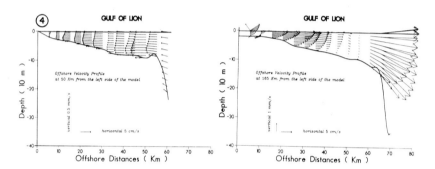

Fig.4 Computed off-shore velocity profiles at 50 and 165 kms from the western side of the model.

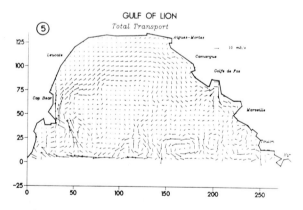

Fig.5 Computed mass transport in the Gulf of Lion.

Unstructured Grids, Adaptive Remeshing and Mesh Generation for Navier-Stokes

Ralf Tilch
CERFACS, 42 Avenue Gustave Coriolis, 31057 Toulouse cedex, France.

Abstract
The generation and automatic space adaptation of unstructured grids in two dimensions using the Advancing Front Technique(AFT) is presented. The AFT not only shows good results for the mesh generation and adaptation, it proves to be very efficient. The need for such a procedure is demonstrated for Euler- and Navier-Stokes computations. Furthermore some details of the implemented AFT are given.

Introduction
Over the past decades, interest in realistic and sophisticated computer simulation of processes involving fluid dynamics has increased drastically. The rapid development of flow simulation programs, as well as great improvements in computer speed and memory size, has made Computational Fluid Dynamics(CFD) a practical tool in many projects. As consequence, also the problems of interest to CFD researchers have changed substantially from those studied initially. Examples are, super- and hypersonic aircraft or re-entry spacecrafts, where for the latter extreme flow situations are encountered demanding for inclusion of air chemistry in the numerical simulation. Other type of problems which arises are complex geometries, like a complete aircraft with an abundance of sharp corners. Since every corner has an influence on the boundary layer, separation, etc., it can not be neglected under the high accuracy standard for the flow prediction of today. Often a numerical simulation includes both, physical and geometrical problems.
The development of numerical simulations was and is mainly limited by computer speed and memory size but, nevertheless the research has reached a level where stable high-resolution schemes are available. But with the progress in CFD a more serious problem appears, which is the mesh design. The efficiency, economy and quality for many studies in CFD depends on the used method. The relation between the quality of a simulation and the mesh arises from the fact, that the equations which are used in CFD are normally dominated by their advective terms. An error, which is introduced locally by a poor discretization can be transported out over the rest of the computational domain. Efficiency means, that the mesh generation should be fast and from this follows that the man/machine interaction should be reduced to a minimum.
Unstructured meshes offer the high flexibility needed for the mesh generation, which can reduce the costly man/machine interaction drastically. The mesh adaptation is simple and can be done automatically, only depending on the solution. Local high gridpoint densities are easily achieved, included its directional application. The mesh adaptation features are specially of importance for super-and hypersonic flow problems. They are characterized by small regions of rapid change embedded in large regions where the solution is smooth. The presented AFT include most of the discussed features. Further features of the AFT are listed below: 1) Stretched elements are easy to realize. 2) The mesh generation and the remeshing are using the same algorithm strategy. 3) It is not restricted by the complexity of the flow geometry. 4) The man/machine interaction is restricted and does not increase with the complexity of the geometrical flow configuration.

The Numerical Simulation
The flow is described by the unsteady compressible Navier-Stokes equations, assuming an ideal gas. The equations are used in their conservative form. For the discretization of the conservative equations a two-step Taylor-Galerkin scheme for Finite Elements is implemented. It can be classified as a Lax-Wendroff-type with a second order accuracy in

time and space. Due to the spatial coupling of the time derivatives it achieves an excellent phase accuracy. Strong gradients are captured with a simplified artificial viscosity model of Baldwin-Lomax, see [1]-[3].

Some of the mesh dependent problems are presented below, which can arises from a numerical simulation. They are taken from Tilch [5], where also an extended discussion is given.

Euler : Region with large derivatives in the conservative expressions, e.g. stagnation points, produce a large discretization error and increase the numerical diffusion, whereby the whole computational domain may be effected. It originate in the very short-wavelength harmonics which can't be presented by the mesh.

Navier-Stokes : The boundary layer thickness decreases with the square root of the Reynolds number. This yields for the values of technical interest $> 10^5$, to very thin boundary layers and strong gradients, which must be resolved to be able to represent the effect of the **natural** viscosity. In general it require the use of very stretched and small discretization elements, introducing new numerical difficulties.

The Advancing Front Technique

The two main features of the AFT are: 1) A point and the connected element are generated at the same time. This is a strongly local process which gives a nearly optimal control over the generated element and hence it can be easily adapted to the changing flow features. 2) The local parameter for a new element are defined by a background grid. From the background grid the needed parameter element size, stretching and stretching direction are interpolated linear from the respective node values. The background grid has the same structure as the actual generated mesh, with the only assumption, that it must cover the whole computational domain. These allows a very fast definition of a first background grid. If a computational result already exist, *no man/machine interaction is necessary* and the size parameter for the new mesh are computed with help of an error condition which fullfils a predefined threshold [2], using this mesh as a background grid. The following section shows the structure of the implemented AFT.

I   Pre-processing for the new mesh

   P.1 Definition of the boundaries of the computational domain

   P.2a Set up a background grid, including the definition of the element size, it's stretching and stretching direction for the new mesh (defined at the nodes).

   P.2b Within adaptive remeshing, a computational result already exist. The grid is used as background grid, to achieve a better adapted distribution of the elements. The needed parameter are defined by help of an error condition, which essentially uses the interpolation error to compute the size parameter.

II Generation of the mesh

   G.1 Discretize the boundaries using the information from the background grid. The discretized boundaries yields to a set of faces, which defines the initial front.

   G.2 Find the face from the actual front, which generates the smallest element. This avoids that large elements crossing areas with small elements.

   G.3 Define the "ideal point"(i.p.) for the new element and collect the surrounding points and their faces(using a point-face linked list) from the actual front.

   G.4 Determine if a point of the previous point list can taken in lieu of the "i.p." and if necessary, redefine the "i.p.".

   G.5 Validate the element formed by the selected point. If their is any given face crossing, go back to G.4 .

   G.6 Delete the used - and add the new face(s), connected to the new element and update the element-, point-, face- and their link-lists.

   G.7 Find the element-size parameter for the new face(s), using the background grid.

G.8 If there is any given face left, go back to G.2

III Post-processing for the new mesh

P.1 Smooth the mesh, to increase it's regularity.

P.2 If needed, interpolate the solution from the background grid.

The AFT is mainly characterized by finding, sorting and searching inside of an irregular pattern which inhibits the use of standard optimization techniques. Instead a data management was developed which gives the theoretical order in parts of $n \cdot \log(n)$. Practical tests shows that the needed time increases only linear with the number of generated points n, which is a order of $n \cdot$ (constant). The data management is constructed by using quadtree-, heap-list-, binary tree-structures and some special link lists. More information about the data management is given in Löhner [4] and Tilch [5].
The mesh generation for Navier-Stokes extended the previous described structure of the AFT by one point, which is introduced between G1 and G2.

G.N Within a Navier-Stokes simulation, a structured boundary layer mesh is generated. The discretization uses the generated boundary points by creating in every point a normal vector. Their length or the boundary layer thickness is taken from the background grid, which is included there as a fourth parameter. The difference angle between the normal vectors is smoothed, which avoid that they cross each other. Structured grid are used due to their much higher allowable stretching ratio. Afterwards the rectangular elements are triangulated, followed by the update of the initial front.

Results

The presented computational results show the hypersonic flow around a double ellipse, which geometry was proposed in Antibes [6]. Both cases, for an inviscid- and viscous flow, are computed under the assumption of equilibrium air and a free stream Mach number of 8.15.
The numerical solution of the inviscid flow was obtained for a double ellipse with a round afterbody and 0 degree angle of attack. Picture 1 and 2 show the starting - and the fourth remeshed mesh, respectively. The maximum allowed stretching ratio is 6 and the number of points after the fourth remeshing is 5514(10696 elements). Picture 3 presents the Mach isolines for the final mesh. With the final mesh the numerical solution of the flow behind the body start to be unsteady. This is due to the optimal adapted mesh, which reduces the numerical diffusion to the point where unsteady flow phenomena appears.
The next step was to include the natural viscosity terms and to compute the Navier-Stokes solution of the hypersonic flow. The same geometry is used without afterbody, at 30 degrees angle of attack and a unit Reynolds number of $1.67 \cdot 10^7/m$. The solution presented in picture 4 and 5 is obtained by applying three remeshings for the inviscid flow and a fourth one using the viscous flow solution. The number of points is 12351(24352 elements) and 349 points are on the profile. The boundary layer mesh is generated with 24 points in normal direction and a constant value for its thickness. The Mach number contours shown in picture 5 indicate clearly the separation at the canopy. Pictures 6 presents a magnification of the canopy, showing the velocity vectors. It shows two small recirculations under the large one. The continuation of the calculation shows that the flow started to be unsteady in the region of separation. Other tests with a decreasing thickness of the boundary layer mesh shows, that the structure of the separation changes. If the thickness reaches one third of its initial value, the flow separation disappears. A more detailed description is given in Tilch [5]. All the tests indicate, that an automatic adaptation would be rather helpful to detect local phenomena not known in advance.
It can be seen that about 70 percent of the overall number of points are placed in the structured boundary layer mesh. Hereby is the number of points, describing the physical boundary layer much smaller, which is due to the strong change in its thickness by a constant thickness for the boundary layer mesh. The physical boundary layer is approximately described by ten points, but in some regions it can happen, that the mesh is still

to thin, e.g. the separation at the canopy, or to thick, e.g. the stagnation point. An approximation gives that only 20 percent of the points from the boundary layer mesh are used to describe the physical boundary layer. An adaptation of the mesh for the boundary layer (which is under development) will surely reduce this percentage, which means also that the number of points can be reduced, by increasing the quality of the solution.

Conclusions

The presented AFT has shown its efficiency and simplicity for mesh generation. The results presented for the remeshing shows good results for the Euler- and Navier-Stokes solution, by reducing the number of needed points efficiently. Furthermore it can be seen that for a flow structure not known in advance an automatic mesh adaptation procedure will help to detect local phenomena and to represent them correctly and at the same time to optimize the number of needed points. This will be even more true for three dimensional flows where the technical and physical problems are much more complex.

References

[1] J. Peraire, K. Morgan, J. Peiro; *Unstructured Finite Element Mesh Generation and Adaptive Procedures for CFD*, AGARD/FDP 64th Meet. "Applications of Mesh Generation to Complex 3D Configurations", 24-25 May 89, Loen, Norway

[2] R. Löhner; *Adaptive Remeshing for Transient Problems with Moving Bodies*, AIAA-88-3737 (1988)

[3] P. Amestoy, R. Tilch; *Solving the Compressible Navier-Stokes Equations with Finite Elements Using a Multifrontal Method*, Impact Comp. Science Eng. 1, 93-107, (1989)

[4] R. Löhner; *Some Useful Data Structures for the Generation of Unstructured Grids*, Comm. Appl. Num. Meth. 4, 123-135, (1988)

[5] R. Tilch; *Unstructured Grids for the Compressible Navier-Stokes Equations*, Ph.D. Thesis, CERFACS, October, (1990)

[6] *Workshop on Hypersonic Re-entry Problems*; Organized by INRIA and GAMNI/SMAI, Antibes, France, January, (1990)

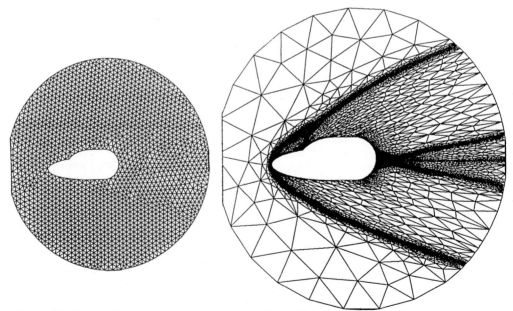

Pic. 1 : Starting mesh   Pic. 2 : Fourth remeshing

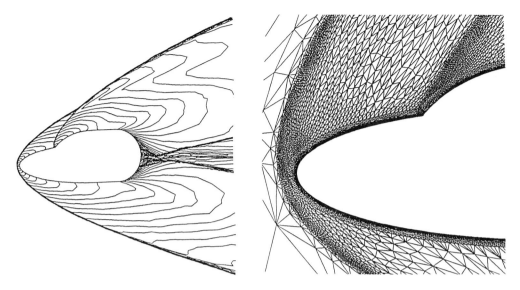

Pic. 3 : Mach contours (Euler, $\Delta M = 0.25$)   Pic. 4 : Fourth remeshing (Navier-Stokes)

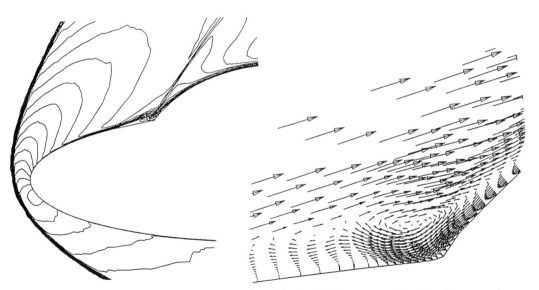

Pic. 5 : Mach contours (Navier-St., $\Delta M = 0.25$)   Pic. 6 : Velocity vectors (Magnification canopy)

A General Purpose Time Accurate Adaptive Grid Method

M.J. Bockelie[†*], P.R. Eiseman[@], and R.E. Smith[*]

[†] NRC Post Doctoral Research Fellow
[*] Numerical Geometry Laboratory (GEOLAB), NASA Langley, Hampton, VA 23665
[@] Columbia University, New York, NY 10027

Introduction

An adaptive grid method is described that has the ability to generate time accurate grids for a wide variety of problems. The solution method consists of three parts: (1) an algebraic grid movement scheme; (2) an explicit unsteady Euler equation solver; and (3) a temporal coupling routine that utilizes a grid prediction correction method to link the dynamic grid and the Euler solver. The ability of the method to perform time accurate adaptive computations has been previously established [1-3]. The purpose of this paper is to describe (a) a technique that improves the ability to adapt the grid to multiple solution features, and (b) the use of a coarse grid in the prediction stage of the temporal coupling that reduces the cpu cost for obtaining the new grid. The capabilities of the adaptive method are demonstrated on a shock vortex interaction problem.

Adaptive Grid

The grid movement scheme employs a "monitor surface" formed from the solution data, to identify regions requiring further grid resolution [1-5]. The grid points are repositioned on a curve by curve basis by requiring that a weight function be equally distributed over each curve. Including the monitor surface gradient and normal curvature in the weight function ensures that both the gradients and the transition regions of the solution are resolved. The weight function also includes a grid control that does not allow the grid cells to become too small. The grid movement is interwoven with a smoothing operation to enhance grid smoothness and to alleviate grid skewness.

To accurately adapt the grid to multiple solution features that merge when computing the solution requires that the monitor surface be defined as a vector function rather than a scalar function [2,3]. If the monitor surface is a scalar function of the multiple solution features (e.g., a linear combination) and the solution features being tracked should merge, the solution gradients will cancel each other and thereby destroy the grid in the region of merger. The vector function for the monitor surface is defined by the N dimensional vector $\Psi = (\psi_1(\mathbf{X}), \ldots, \psi_N(\mathbf{X}))$ where ψ_k is one of the N features of the solution that is to be tracked. The ψ_k are scalar functions and are typically formed from a single variable or property of the solution. With the vector monitor surface the gradients of the features being tracked cannot cancel and thus good grid resolution is maintained in the region of merger. The process for adapting a grid to a vector monitor surface is essentially the same as for a scalar monitor surface. Illustrated in Fig. 1 are the scalar and vector monitor surfaces for the case of two merging disturbances on a one dimensional domain. The resulting adaptive grid is indicated along the x axis; the improved resolution can be seen in the region of merger.

Temporal Coupling

The grid prediction correction method treats the time integration of the solution as a series of initial value problems over short time intervals in which the solution is first advanced to create a new grid and then recomputed on the new grid. Because the new grid is based on solution data forward in time, there is no time lag between the grid and the solution. The grid prediction correction method described here bears a philosophical similarity to a method presented in [6], but is not limited to using only PDE solvers that have a grid velocity capability. Here, the solver does not need grid speed terms because the solutions of the prediction and correction stages are computed on static - although different - grids. However, if the PDE solver had a grid velocity capability, the method would provide a means to accurately compute the grid velocity. To be cost effective, the prediction stage is computed on a coarse grid defined from the full grid used in the previous corrector stage. The resulting larger grid cell size allows using a larger time step based on the CFL constraint, making the cpu time of the prediction stage small compared to the time to the overall solution time. In the following description, the given full grid and solution at the initial time T are denoted by $\mathbf{X}$ and $\mathbf{q}(\mathbf{X},T)$, respectively.

step 1: choose a time interval τ. The time interval τ is determined by integrating forward p time steps (e.g., $p = 10$).

step 2: obtain a coarse grid from the full grid. The coarse grid, $\mathbf{XC}$, is obtained by deleting every second coordinate curve in each direction in $\mathbf{X}$.

step 3: transfer the initial condition to the coarse grid. The initial condition on the coarse grid, $\mathbf{q}(\mathbf{XC},T)$, is computed from $\mathbf{q}(\mathbf{X},T)$ using local bilinear interpolation.

step 4: predict the solution over τ on the coarse grid. The solution is predicted up to $T+\tau$ by integrating $\mathbf{q}(\mathbf{XC},T)$ forward in a time accurate manner. After each time step a monitor surface is computed and stored in a running sum. A monitor surface is also formed from $\mathbf{q}(\mathbf{XC},T)$ to enhance the temporal smoothness of the grid.

step 5: form a monitor surface over τ. The $p+1$ monitor surfaces computed in step 4 are averaged to yield a composite monitor surface on the coarse grid, $\psi^*(\mathbf{XC})$.

step 6: transfer the monitor surface to the full grid. The monitor surface on the full grid, $\psi^*(\mathbf{X})$, is computed from $\psi^*(\mathbf{XC})$ using local bilinear interpolation.

step 7: adapt the grid. The new grid, $\mathbf{X}^*$, is generated using two cycles of the curve by curve scheme. Before starting the grid movement a smoothing operation is applied to the $\psi^*(\mathbf{X})$ values (not the grid positions) to ensure the monitor surface is smooth.

step 8: transfer the solution data to the new grid. The old initial condition $\mathbf{q}(\mathbf{X},T)$ is transferred to $\mathbf{X}^*$ to yield a new initial condition $\mathbf{q}(\mathbf{X}^*,T)$. The data can be transferred with either local bilinear interpolation [2,3], or a method that maintains the conservation properties of the solution but assumes only piecewise constant solution variation [7,8].

step 9: recompute the solution over τ on the new grid. The initial condition $\mathbf{q}(\mathbf{X}^*,T)$ is integrated forward in time to yield $\mathbf{q}(\mathbf{X}^*,T+\tau')$ where $\tau' = 0.95\tau$ to ensure the solution does not outrun the resolution provided by $\mathbf{X}^*$.

step 10: go to step 1. The solution is marched forward through time to the desired final time level T_f by successively repeating the procedure.

Disscussion of Numerical Results

unsteady shock tube

A series of numerical test have been performed for the unsteady shock tube problem described in Sod [9]. Solutions have been computed on uniform stationary grids (100 × 2, 500 × 2 cells) and adaptive grids (100 × 2 cells) in which **step** 8 of the grid prediction correction method utilizes bilinear interpolation and the conservative data transfer method. The monitor surface for the adaptive solutions is formed from the internal energy per unit mass. The time interval of the prediction stage (τ) is prescribed to obtain a specified number of grid adaptations. The adaptive solutions generally have much closer agreement with the exact solution than the 100 cell stationary grid solution, whereas the 500 cell stationary grid solution is comparable to the adaptive solutions. Table 1 summarizes some solution properties for the shock tube studies. Notice that the stationary grid solution maintains the total mass, but only obtains the correct speed of propagation of the shock on the fine (500 cell) grid. The adaptive solutions computed using the bilinear interpolation method have a small loss of mass and an $O(1\%)$ error in the shock propagation speed. Furthermore, the errors increase as the grid is adapted more frequently. In contrast, the conservative data transfer method maintains the total mass and propagates the shock at the correct speed.

shock vortex studies

The flow field is assumed to lie within a solid walled channel and consists of an initially planar shock wave marching toward, and eventually over, a simple solid core vortex [1-3, 10-13]. The shock is initially located at $x = 0$ and the counter clockwise rotating vortex ($r_c = 0.1$) is located at $x = 0.65$. To the left of the shock wave is the uniform (subsonic) flow field that would follow behind a propagating shock wave with a relative Mach number (M_s) slightly greater than one. Ahead of the shock, the initial velocity field of the vortex is $\mathbf{V} = (0, u_\theta(r))$ where $u_\theta(r) = u_o\, r/r_c$ for $0 \leq r \leq r_c$ and $u_\theta(r) = u_o\, r_c/r$ for $r \geq r_c$ and r is measured relative to the vortex center. The density and pressure of the initial vortex flow field are determined by assuming the vortex is a constant enthalpy flow (see [12]). For boundary conditions, tangential flow is assumed along the channel walls and simple extrapolation is applied at the inlet and outlet.

Illustrated in Figs. 2-3 are the grid and pressure field for solutions computed on a grid adapted to both the shock wave and the vortex for the case of a strong interaction ($M_s = 1.3$, $u_0 = 0.30$). The adaptive grid (80 × 64 cells) is generated with a vector monitor surface formed from the density field (to track the shock) and the circulation about each grid point (to track the vortex) [1-3]. The solutions are computed using bilinear interpolation (Fig. 2) and the conservative data transfer method (Fig. 3) in **step** 8 above; in **step** 1 $p = 10$, resulting in 36 grid adaptations. There are two points to note from Figs. 2-3. First, the adaptive grid correctly captures the time dependent location and shape of the shock wave and the vortex. In the grid plots, the concentration of grid points at the shock wave occurs over a relatively broad band because the solution of the correction stage is computed on each grid for several time steps. The second point is that the two solutions are essentially the same despite the lower order solution variation in the conservative data transfer. A quantitative comparison of the two solutions indicates that the shock wave in the solution using bilinear interpolation is located about 1.1% further down the channel than in the other solution.

References

1. P.R. Eiseman and M.J. Bockelie, *11th International Conference on Numerical Methods in Fluid Dynmaics*, Lecture Notes in Physics, Vol. 322, p.240-244, Springer-Verlag, New York, 1989.
2. M.J. Bockelie, PhD. Thesis, Columbia University, 1988.
3. M.J. Bockelie and P.R. Eiseman, NASA Langley TP-2998 (1990).
4. P.R. Eiseman, *AIAA J.*, 23, 551-560 (1985).
5. P.R. Eiseman, *Comput. Methods Appl. Mech. Eng.*, 64, 321-376 (1987).
6. J.G. Blom, J.M. Sanz-Serna, and J.G. Verwer, *J. Comp. Phys.*, 74, 191-213 (1988).
7. J.D. Ramshaw, *J. Comput. Phys.*, 59, 193-199 (1985).
8. M. Kathong, PhD. Thesis, Old Dominion University, 1988.
9. G.A. Sod, *J. Comput. Phys.*, 27, 1-31 (1978).
10. S.P. Pao and M.D. Salas, AIAA 81-1205, 1981.
11. M.Y. Hussaini, D.A. Kopriva, M.D. Salas, and T.A. Zang, *AIAA J.*, 23, 234-240 (1985).
12. G.R. Srinivasan, W.J. McCroskey, and J.D. Baeder, *AIAA J.*, 24, 1569-1576 (1986).
13. K. Meadows, A. Kumar, and M.Y, Hussaini, AIAA 89-1043, 1989.

Table 1. Comaparison of Shock Tube Solution Properties.

Solution	Grid Cells	No. of Grid Adaptations	Δ Mass * (%)	Shock Speed † (M_s)	M_s Error ◇ (%)
Exact	n/a	n/a	0.00	1.7522	0.00
Stationary Grid	100	n/a	0.00	1.7642	0.69
	500	n/a	0.00	1.7524	0.01
Adaptive Grid Bilinear Interpolation	100	20	-0.13	1.7714	1.10
	100	100	-0.28	1.7810	1.64
	100	200	-0.31	1.7904	2.18
Adaptive Grid Conservative Data Transfer	100	20	0.00	1.7529	0.04
	100	100	0.00	1.7520	-0.01
	100	200	0.00	1.7523	0.01

*: $\Delta\text{Mass} = (\text{Mass}_f - \text{Mass}_i) / \text{Mass}_i$ and ◇: $\text{Error} = (M_s - M_s|_{\text{exact}}) / M_s|_{\text{exact}}$
†: M_s computed from least squares fit of shock wave location as a function of time.

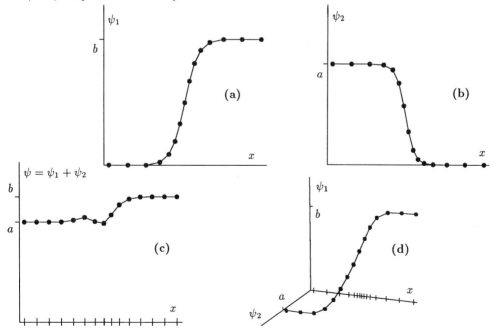

Fig. 1 Comparison of scalar and vector monitor surface for 1D problem. Illustrated are two solution features (a,b), a scalar (c) and vector (d) monitor surface. Resulting grid shown on x axis.

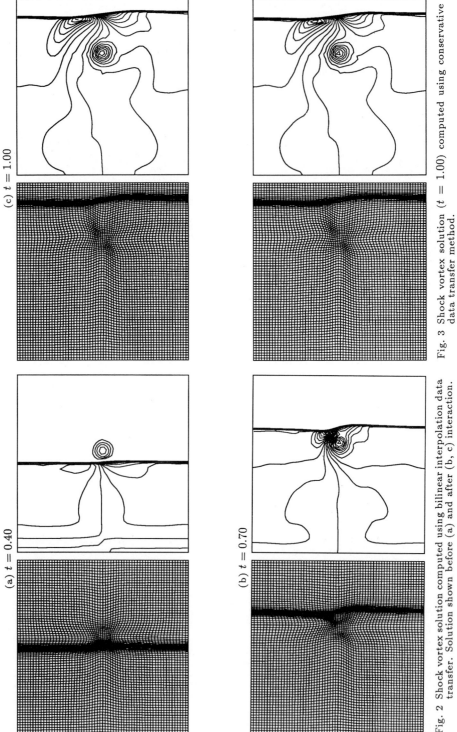

Fig. 2 Shock vortex solution computed using bilinear interpolation data transfer. Solution shown before (a) and after (b, c) interaction.

Fig. 3 Shock vortex solution ($t = 1.00$) computed using conservative data transfer method.

A FINITE DIFFERENCE FORMULATION FOR FREE SURFACE FLOW PROBLEMS USING A FREE SURFACE CONFORMING GRID SYSTEM

Satoshi CHIBA
Institute of Computational Fluid Dynamics, Tokyo, Japan
Kunio KUWAHARA
The Institute of Space and Astronautical Science, Kanagawa, Japan

1. INTRODUCTION

The majority of numerical algorithms of finite difference methods developed for incompressible fluids with free surfaces [1,2] are based on the Eulerian formula. Free surfaces are tracked on a fixed grid system by special devises, such as markers or a scalar variable which plays a role analogous to the density. An alternative way is a straightforward application of the transformation technique for general curvilinear coordinate system [3] to free surface problems [5,7]. A restriction of this method is that the one-to-one correspondence between a physical and a computational domain has to be maintained. Therefore, the free surface configuration is limited to a simple geometry (i.e., the fluid region has to be a simply-connected region). By utilizing this transformation method, formulations of general two-layer (stratified) flows will be described. The free surface flows will be shown to be a limiting case of this generalized formulation.

2. GOVERNING EQUATIONS

A schematic of the flow geometry under consideration is depicted in fig. 1. The two immiscible fluids are divided by a distinct interface surface, to which a grid surface (ξ^3 = const.) is conformed. In the case of free surface flows, layer 2 is only analyzed. For brevity, the summation convention is invoked unless otherwise noted. The symbol ' $\sim$ ' refers to the physical quantities of layer 1, whenever such a discrimination is required. The governing equations are the time-dependent incompressible Navier-Stokes equation and Pressure-Poisson equation. These equations are solved such that the continuity restriction and boundary conditions are satisfied [1]. The transformed forms for these equations are written as:

$$\rho\left[\frac{\partial \mathbf{v}}{\partial \tau}+\left\{\mathbf{a}^i \cdot (\mathbf{v} - \mathbf{v}_g)\right\}\frac{\partial \mathbf{v}}{\partial \xi^i}\right] = -\mathbf{a}^i \frac{\partial p}{\partial \xi^i}+\mu\left\{g^{ij}\frac{\partial^2 \mathbf{v}}{\partial \xi^i \partial \xi^j}+\left(\mathbf{a}^i \cdot \frac{\partial \mathbf{a}^j}{\partial \xi^i}\right)\frac{\partial \mathbf{v}}{\partial \xi^j}\right\}+\rho\,\mathbf{g} \quad (1)$$

$$\left\{g^{ij}\frac{\partial^2}{\partial \xi^i \partial \xi^j}+\left(\mathbf{a}^i \cdot \frac{\partial \mathbf{a}^j}{\partial \xi^i}\right)\frac{\partial}{\partial \xi^j}\right\}p = \rho\frac{\partial}{\partial \tau}\left(\mathbf{a}^i \cdot \frac{\partial \mathbf{v}}{\partial \xi^i}\right)-\rho\left(\mathbf{a}^i \cdot \frac{\partial \mathbf{v}}{\partial \xi^i}\right)\left(\mathbf{a}^j \cdot \frac{\partial \mathbf{v}}{\partial \xi^j}\right)$$

$$-\rho\left\{\mathbf{a}^i \cdot (\mathbf{v} - \mathbf{v}_g)\right\}\frac{\partial}{\partial \xi^i}\left(\mathbf{a}^j \cdot \frac{\partial \mathbf{v}}{\partial \xi^j}\right)+\mu\left\{g^{ij}\frac{\partial^2}{\partial \xi^i \partial \xi^j}+\left(\mathbf{a}^i \cdot \frac{\partial \mathbf{a}^j}{\partial \xi^i}\right)\frac{\partial}{\partial \xi^j}\right\}\left(\mathbf{a}^k \cdot \frac{\partial \mathbf{v}}{\partial \xi^k}\right) \quad (2)$$

where ρ, $\mathbf{v}$, $\mathbf{v}_g$, $\mathbf{g}$ and μ are the density, fluid velocity, grid velocity, gravitational acceleration and viscosity, respectively; ξ^i and τ represent the coordinate and the time of the curvilinear coordinate system; $\mathbf{a}_j$, $\mathbf{a}^i$ are the covariant and contravariant base vector; g_{ij}, g^{ij} are the covariant and contravariant metric tensor.

The kinematic condition at the interface surface can be expressed as:

$$\frac{\partial f}{\partial \tau}+\left\{\mathbf{a}^i \cdot (\mathbf{v} - \mathbf{v}_g)\right\}\left(\frac{\partial f}{\partial \xi^i}\right) = v^3 \quad (3)$$

In the above, f represents a single-valued function of x^1 and x^2; the interface surface is described by the equation, $f-x^3=0$. x^i denotes the coordinate of Cartesian system. Consequently, the present formulation will be restricted to the flow configurations which contain simply-connected regions.

The boundary conditions for the velocity fields in the NS equation at the interface surface are the zero divergence of layer 2 and the balance of tangential stresses in the ξ^1 and ξ^2 direction, i.e.,

$$\mathbf{a}^i \cdot \frac{\partial \mathbf{v}}{\partial \xi^i} = 0 \tag{4}$$

$$\frac{\mu}{\sqrt{g_{kk}g^{33}}}\left\{\left(\mathbf{a}^3 \cdot \frac{\partial \mathbf{v}}{\partial \xi^k}\right) + g^{3m}\left(\mathbf{a}_k \cdot \frac{\partial \mathbf{v}}{\partial \xi^m}\right)\right\} = \frac{\tilde{\mu}}{\sqrt{\tilde{g}_{kk}\tilde{g}^{33}}}\left\{\left(\tilde{\mathbf{a}}^3 \cdot \frac{\partial \tilde{\mathbf{v}}}{\partial \xi^k}\right) + \tilde{g}^{3m}\left(\tilde{\mathbf{a}}_k \cdot \frac{\partial \tilde{\mathbf{v}}}{\partial \xi^m}\right)\right\} \quad (k=1,2) \tag{5}$$

These can be summarized in the form of the Neumann condition as follows:

$$\frac{\partial \mathbf{v}}{\partial \xi^3} = A^{-1}(B+C) \tag{6}$$

$$A^{-1} = \frac{1}{\sqrt{g}g^{33}}\left(\sqrt{\frac{g^{11}}{g^{33}}}(\mathbf{a}_2 \times \mathbf{a}_3), -\sqrt{\frac{g^{22}}{g^{33}}}(\mathbf{a}_1 \times \mathbf{a}_2), -(\mathbf{a}_1 \times \mathbf{a}_2)\right),$$

$$B = -\begin{pmatrix} \frac{1}{\sqrt{g_{11}g^{33}}}\left\{\mathbf{a}^3 \cdot \frac{\partial \mathbf{v}}{\partial \xi^1} + g^{13}\left(\mathbf{a}_1 \cdot \frac{\partial \mathbf{v}}{\partial \xi^1}\right) + g^{23}\left(\mathbf{a}_1 \cdot \frac{\partial \mathbf{v}}{\partial \xi^2}\right)\right\} \\ \frac{1}{\sqrt{g_{22}g^{33}}}\left\{\mathbf{a}^3 \cdot \frac{\partial \mathbf{v}}{\partial \xi^2} + g^{13}\left(\mathbf{a}_2 \cdot \frac{\partial \mathbf{v}}{\partial \xi^1}\right) + g^{23}\left(\mathbf{a}_2 \cdot \frac{\partial \mathbf{v}}{\partial \xi^2}\right)\right\} \\ \mathbf{a}^1 \cdot \frac{\partial \mathbf{v}}{\partial \xi^1} + \mathbf{a}^2 \cdot \frac{\partial \mathbf{v}}{\partial \xi^2} \end{pmatrix}, \quad C = \frac{\tilde{\mu}}{\mu}\begin{pmatrix} \frac{1}{\sqrt{\tilde{g}_{11}\tilde{g}^{33}}}\left\{\tilde{\mathbf{a}}^3 \cdot \frac{\partial \tilde{\mathbf{v}}}{\partial \xi^1} + \tilde{g}^{13}\left(\tilde{\mathbf{a}}_1 \cdot \frac{\partial \tilde{\mathbf{v}}}{\partial \xi^i}\right)\right\} \\ \frac{1}{\sqrt{\tilde{g}_{22}\tilde{g}^{33}}}\left\{\tilde{\mathbf{a}}^3 \cdot \frac{\partial \tilde{\mathbf{v}}}{\partial \xi^2} + \tilde{g}^{13}\left(\tilde{\mathbf{a}}_2 \cdot \frac{\partial \tilde{\mathbf{v}}}{\partial \xi^i}\right)\right\} \\ 0 \end{pmatrix}.$$

As can be readily seen, the formulations for free surface flows can be secured by letting $\tilde{\mu} = 0$.

The boundary condition for the pressure field in the Pressure-Poisson equation at the interface surface is depicted by the balance of the pressures, normal viscous stresses and surface tension as:

$$p = \tilde{p} + \frac{2\tilde{\mu}}{\tilde{g}^{33}}\tilde{g}^{3i}\left(\tilde{\mathbf{a}}^3 \cdot \frac{\partial \tilde{\mathbf{v}}}{\partial \xi^i}\right) - \frac{2\mu}{g^{33}}g^{3i}\left(\mathbf{a}^3 \cdot \frac{\partial \mathbf{v}}{\partial \xi^i}\right) - \gamma\left(\frac{h_{11}g_{22} + h_{22}g_{11} - 2h_{12}g_{12}}{g_{11}g_{22} - g_{12}g_{12}}\right) \tag{7}$$

In the above, γ is the surface tension; h_{ij}, defined as $h_{ij} = (\mathbf{a}^3/\sqrt{g^{33}}) \cdot (\partial \mathbf{a}_j/\partial \xi^i)$, is the second fundamental quantity of a curved surface. In the case of free surface flows, $\tilde{\mu} = 0$ and $\tilde{p}$ is constant.

3. SOLUTION PROCEDURE

The NS equation(eq.(1)) and kinematic condition (eq.(3)) are discretized by the Euler-backward time differencing scheme. The second order central differencing scheme is adopted for space derivatives and metrics. The third order upwind scheme [4] is applied to the convective terms. The other equations and boundary conditions are discretized by the second order central differencing scheme, except for the space derivative '$\partial/\partial \xi^3$' at the interface surface. The first order forward and backward differencing schemes are utilized for this derivative.

The entire procedure of the computation is as follows.

1) The PP eq. for layer 1 is solved with a Neumann condition; e.g., $\partial p/\partial \xi^3 = 0$, at the interface surface.

2) The PP eq. for layer 2 is solved with the Dirichlet condition (eq.(7)).
3) The NS eq. for layer 2 is solved with the Neumann condition (eq.(6)).
4) The NS eq. for layer 1 is solved with the Dirichlet condition at the interface surface.
Steps 3) and 4) are iterated until converged velocity fields are obtained.
5) The position of the interface surface is updated by solving the kinematic condition (eq.(3)).
The complete solutions are marched in time by repeating the steps 1) through 5). Obviously, for free surface flows, steps 1) and 4) are not needed.

As to the grid system, the algebraic grid-generation method was utilized.

4. NUMERICAL RESULTS

1) Capillary-gravity progressive wave (2D problem)
Fig.2 a,b) display typical computed stream lines of the regular wave train. The waves were generated at the left side boundary and they advance in the right direction. Fig.3 a,b) show the effect of grid spacing on the dissipation of the surface waves. The number of grid points of fig.3,b) is one-fourth of that of fig.3,a), and the other computational conditions are identical. It is clear that at least 20-30 grid points are required to resolve one wave length. Fig.4 shows the dispersion relation of the computed waves. For comparison purposes, the dispersion relation of a linear wave is also included. The excellent agreement exhibited in fig.4 gives credence to the validity of the present method.

2) Sloshing of two-layer fluids in a rectangular tank (2D problem)
As an example of the two-layer fluids, the flow inside a rectangular tank was simulated. Fig.5 depicts the time evolution of velocity fields for density ratio 0.8. The initial interface shape was a straight line inclined 8.5 degrees to the horizontal axis. In the middle of the interface surface, Short waves due to K-H instability are observed. The time histories of the vertical elevation of the interface are plotted in fig.6. Clearly seen in the figure is the presence of the fundamental modes. Spectral analyses of the data were performed, and the frequencies of the numerical results were consistent with the eigen frequencies obtainable from the linear wave theory (table 1).

3) Free surface flows around a vertical circular cylinder (3D problem)
Free surface flows around a vertical circular cylinder standing on a flat plate were simulated. The Reynolds and Froude numbers, based on the diameter of the cylinder and the uniform upstream surface velocity, are $Re=3.4 \times 10^3, Fr=1.18$. The computed free surface configurations and the experimental data are compared in fig.7. The bow and stern waves in ship terminology are clearly captured. However, short waves were mostly damped by the numerical dissipation. In order to improve the resolution of the free surface waves, the number of grid points has to be increased by at least one order of magnitude.

We would like to thank Prof. S.Komori and R.Nagaosa of Kyusyu Univ., who provided the experimental data, and also thank to Prof. Hyun of KAIST for his useful advise.

5. REFERENCES

(1) Harlow,F.& Welch,E.,The Physics of Fluids,vol.8,No.72,1965
(2) Nichols,B.D.,Hirt,C.W.& Hotchkiss,R.S.,LA-8355 UC-32&UC-34,1980
(3) Thompson,J.F., et al.,"Numerical Grid Generation Foundation and Application",North Holland,1982
(4) Kawamura,T.& Kuwahara,K,AIAA paper 89-0294
(5) Miyata,H.,Sato,T.& Baba,N.,J.of Computational Physics,72,1987
(6) Hirt,C.W.,Amsden,A.A.,Cook,J.L.,J.of Computational Physics,14,1974
(7) Ryskin,G.& Leal,L.G.,J.of Fluid Mechanics,vol.148,pp.1-17,1984

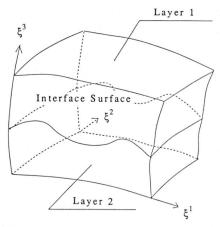

Fig.1 Schematic of Two-Layer Fluid

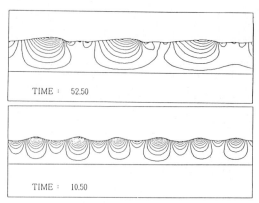

Fig.2 Stream Lines of 2D Progressive Wave;
$\rho = 1.0 \times 10^3 (kg/m^3), g=9.8(m/s^2), \nu=1.0 \times 10^{-6}(m^2/s)$
a) Freq. 62.8(rad/s), Depth 0.01(m) (upper)
b) Freq. 6.28(rad/s), Depth 1.0(m) (lower)

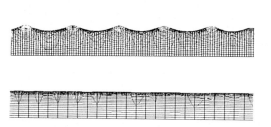

Fig.3 Effect of Number of Grid Points;
grid points a) 400×20 (upper), b) 100×20 (lower)

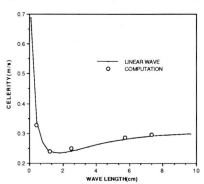

Fig.4 Dispersion Relation (Depth=0.01(m))

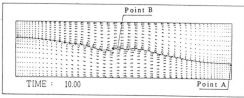

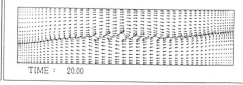

Fig.5 Velocity Fields of Two-Layer Sloshing
Size of the rectangular tank 1.0×0.25(m), grid points 200×50
$\rho_1 = 0.8 \times 10^3 (kg/m^3), \rho_2 = 1.0 \times 10^3 (kg/m^3), g=9.8(m/s^2), \nu_1=1.0 \times 10^{-5}(m^2/s), \nu_2=1.0 \times 10^{-6}(m^2/s)$

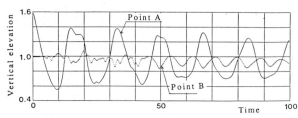

Fig.6 Time History of Vertical Elevation of The Interface Surface

Table 1 Comparison of Eigen Frequencies

	Mode 1	Mode 2
Linear theory	6.09×10^{-2}	1.14×10^{-1}
Computation	6.21×10^{-2}	1.20×10^{-1}

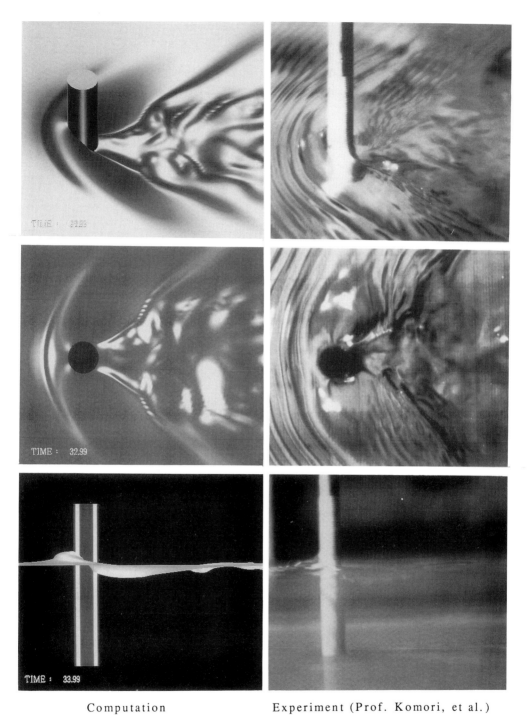

Computation　　　　　　　Experiment (Prof. Komori, et al.)

Fig. 7 Configurations of Free Surface around a Vertical Circular Cylinder
$\rho = 0.8 \times 10^3 (kg/m^3), g=9.8(m/s^2), \gamma=7.25 \times 10^{-2}(N/m), \nu=1.0 \times 10^{-6}(m^2/s)$
Diameter=0.01(m), Depth=0.037(m)

On the Use of Dynamical Finite-Element Mesh Adaptation for 2D Simulation of Unsteady Flame Propagation

Nathan MAMAN[*] Bernard LARROUTUROU[†]

July 27, 1990

1 Introduction

The benefits resulting from using mesh adaptation strategies for the numerical solution of stiff problems involving thin layers are now well known. We describe in this brief article a dynamical two-dimensional finite-element mesh adaptation procedure, and present its application to the solution of some stiff laminar unsteady combustion problems.

In the next sections, we first describe the adaptive algorithm itself, then address the question of choosing the scalar criterion which governs the refinement and unrefinement decisions, and briefly present the numerical method which operates on the adapted mesh, before showing some numerical results. The reader is referred to the bibliography for more details.

2 The Adaptive Procedure

The adaptive procedure is based on a multi-level conformal triangular finite-element mesh with a filiation hierarchy between two consecutive levels. "Multi-level" here means that the adapted mesh results from constructing embedded triangulations, using local element division. The basic refinement step is the division of a triangle into four subtriangles (called its sons) using the middle points of the edges. Beside this, some triangles have to be divided into two subtriangles to preserve the conformity (in the classical sense of this word in finite-element theory). Using the "father-son" hierarchy, the procedure can dynamically refine and unrefine the mesh.

The procedure is now written as a black box, which makes it particularly easy to use with any finite-element solver. To use this adaptation procedure, the user has to provide initially a (relatively regular) mesh, called the *macro-mesh*; this is the coarsest mesh the procedure will be able to manipulate. Then, during the computation, the procedure can be called at any time, and generates a new adapted mesh using refinement and unrefinement decisions made by the user. These decisions, which are based on the evaluation of a scalar criterion, are transmitted to the adaptation procedure as "−1" (which means *unrefine*), "0" (*keep that element*) or "+1" (*refine*) for each triangle of the current mesh. The procedure then returns the new adapted mesh, together with the values of all primitive variables at all nodes of the new mesh (thus, the procedure includes an interpolation algorithm, but the user can substitute its own interpolation routine), and the calculation can go on. The maximal number of divisions for the macro-elements is a priori prescribed by the user.

The basic features of the adaptation procedure have been described in [1], but many improvements have been implemented since that early description. We now present these new aspects, referring to [5] for the details.

An index associated with each triangle indicates the number of divisions that are required for this triangle (keep in mind that requiring n divisions for a triangle means that it is divided into four

---

[*]INRIA, Sophia-Antipolis, 06560 VALBONNE, FRANCE.
[†]CERMICS & INRIA, Sophia-Antipolis, 06560 VALBONNE, FRANCE.

sons and that $n-1$ divisions are required for each son). When the procedure is called, it changes this index for all triangles of the current mesh (whose index was previously 0), and, level by level, changes the index values for the fathers, grand-fathers and so on, taking at each time into account the new values of the sons' indexes.

Then, in order to maintain enough regularity in the generated mesh, the procedure checks on each level if an element requiring n divisions has a neighbour requiring more than $n+1$ divisions or has more than one neighbour requiring $n+1$ divisions in which case it increases the element index by 1.

Let us emphasise some effects of this construction:

1. All triangles can only be divided into 2 or 4 subtriangles; no other kind of element division exists.

2. All sons of a triangle have to be asked to be unrefined in order to be actually unrefined. For example, if only one, two or three sons are asked to be unrefined, then none will be deleted because they have at least one brother which the user wants to keep. But on the opposite, all four sons of a triangle need not have the same index.

3. If a triangle which was created only for the sake of preserving the mesh conformity is asked to be refined, then it and its (sole) brother are cancelled and their father is asked to be refined. But only the really new points are created, so that there is no useless interpolation.

4. Thus, all but the elements created from the conformity constraint are similar to their father (this is why the macro-mesh should be regular enough).

5. Moreover, as the refinement decisions are made on the refined mesh itself (and not on the macro-mesh, as in an earlier version of the procedure [1]), we are able to suppress and create triangles and points precisely.

6. Lastly, instead of reconstructing from scratch the new adapted mesh, as other methods do, the procedure only takes care of the required modifications of the mesh.

3 Adaptation Criterion

As said before, the refinement decisions are based on the evaluation of a scalar criterion.

We use two basic kinds of criteria. On one hand, the criterion may be related to a discretisation error, such as the spatial truncation error of the scheme; it will then point out the elements where the error is large. On the other hand, the criterion may be a quantity which is directly derived from the problem equations, such as a simple estimate of the time derivative of a given variable; then, it will detect the elements where the phenomenon is more active.

But this criterion has to be normalised using some reference value, unless it is really of no help for making the refinement decisions. For this reference value, we use either the global maximum, or the local maximum, or the mean value of the criterion, or the value of a primitive variable itself.

Let us briefly comment these four possible choices for the criterion normalisation. If we use the global maximum of the criterion as the reference value, and if the criterion is almost constant (no matter it is big or small), we will refine almost everywhere at the uppermost level. Similarly, if we use as local reference value for one triangle the maximal criterion in some neighbourhood of this triangle, then the finest refinement level will be required around each local maximum of the criterion, no matter what is the amplitude of this local maximum. Thus, these two normalisations of the criterion (which have the advantage that the normalised criterion is known to take its values in the interval $[0,1]$) make it particularly difficult to decide what thresholds to use and to master the number of elements and nodes which the adaptation procedure creates.

Choosing the mean value of the unnormalised criterion as the reference value amounts to trying to equally distribute the criterion among all elements. In some sense, this choice minimises the CPU as it leads to refine almost nowhere if the variations of the criterion are small. But this can also be a drawback if an accurate computation is expected.

Thus, after numerous numerical experiments (see [5]), we have chosen to use as a local reference value the primitive variable upon which the unnormalised criterion is based. In other words, the normalised criterion measures relative errors, or relative changes in some variable. The threshold which is used for the refinement decisions is then more meaningful: one can decide to refine an element if the relative change of the considered variable in this element is greater than, say, 5%. Moreover, we only weight the criterion with the element area. In this way, we do not refine, say, "any triangle where the value of the normalised criterion is greater than X", but "any triangle where the integral of the normalised criterion is greater than X".

More precisely, we evaluate the normalised criterion in our calculations as the ratio $\dfrac{\|\vec{\nabla}T\|}{T}$, where T is the gas temperature. This criterion revealed to detect appropriately the shocks, contact discontinuities and reaction zones in all considered applications (unsteady flame propagation, or transsonic reactive gaseous jets interaction; see [5]), for both unsteady and steady problems. But mastering the number of nodes created by the adaptation procedure remains a difficult question, and choosing the threshold which governs the refinement decisions (-1, 0 or $+1$) is an important issue.

4 Approximation Scheme

The numerical method we are using for the computation of reactive flows is a mixed finite-element / finite volume method, which has the ability to operate on any (possibly unstructured) finite-element triangulation. We refer to e.g. [2] where the method is described in complete detail.

Because our dynamical mesh adaptation always creates unstructured triangulations with obtuse angles, we wish to use a spatial approximation of the combustion equations which remains robust and accurate on such deformed meshes. In particular, some particular care of preserving the maximum principle for the mass fractions of all gaseous species has been taken, using the results in [4]. Our approximation scheme preserves the positivity of the mass fractions (on any finite-element triangulation) as far as the reactive and convective terms only are concerned, whereas some remaining difficulties related to the discretisation of the diffusive terms when obtuse angles exist are currently attacked (see [5]).

5 Numerical Results

The physical problem we are concerned by is the interaction of two gaseous transsonic reactive jets. The geometry is simple but involves rather disparate length scales, since the dimensions of the injectors are very small compared to those of the computational domain. In our experiment, the flow created by the chemically inert hot jet coming from the left-hand side with a 45 degrees angle is first stabilised, before the top injector of the cold reactant jet is opened. We refer to [3] for more details.

The figures first show a global view of the (very coarse) macro-mesh, with an enlarged view of the interesting part. Then we present the flow created by the first jet at two time levels. The last figures show the interaction of the two jets at later times. These computations were made on a nine-level adaptive mesh, which means that some elements of the macro-mesh are divided into 4^8 subtriangles.

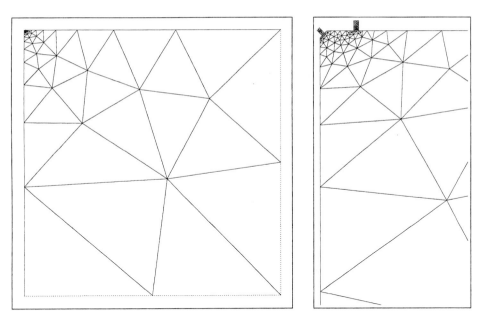

Figure 1: Macro-mesh, 162 nodes, 247 elements
On the right-hand side, enlargement of the top left part

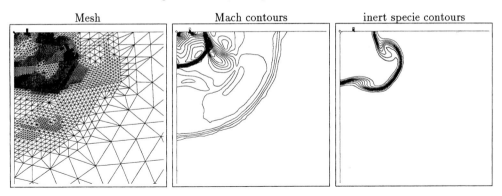

Figure 2: Unsteady flow generated by the hot jet only ($t = 2$)

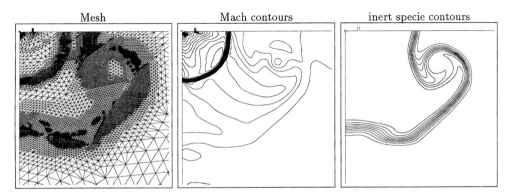

Figure 3: Unsteady flow generated by the hot jet only ($t = 8.5$)

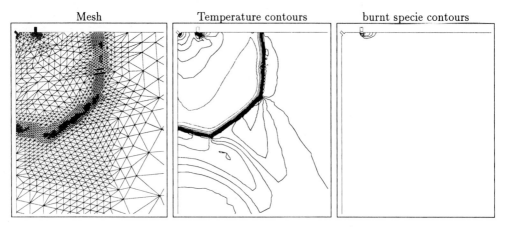

Figure 4: Unsteady flow after opening of the reactant injector($t = 0.2$)

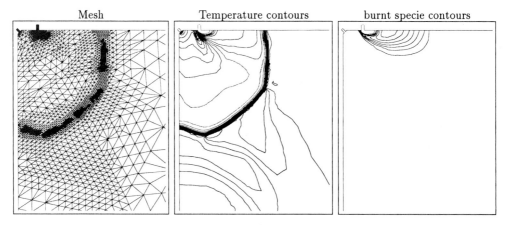

Figure 5: Unsteady flow after opening of the reactant injector($t = 0.6$)

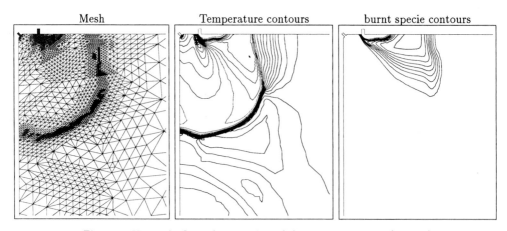

Figure 6: Unsteady flow after opening of the reactant injector($t = 1.2$)

6 Conclusion

The presented method allows us to obtain accurate solutions, with a satisfying capture of thin layers, with an adaptation cost which is very low with respect to the cost of the overall computation (between 5% and 10%).

Though our numerical approximation of the diffusive term does not yet insure the positivity on the adapted meshes (because of the presence of obtuse angles), our adaptation procedure proved its efficiency by drastically diminishing the CPU time required for the computations. Some aspects related to the adaptation criterion, and the question of the scheme positivity still need further investigation.

Let us finally add that, since the adaptation method uses no specifically two-dimensional ingredient, no conceptual problem should arise for extending it to three space dimensions.

Acknowledgements

We wish to warmly thank here F. Benkhaldoun, D. Chargy, A. Dervieux, T. Fernandez, M. Loriot and P. Leyland who all helped at some stages during the studies reported in this article.

Références

[1] Fayssal Benkhaldoun, Thierry Fernandez, Bernard Larrouturou, and Pénélope Leyland. A dynamical adaptive method based on local refinement and unrefinement for triangular finite-element meshes: Preliminary results. Research report 1271, INRIA, Sophia-Antipolis, 06560 VALBONNE, FRANCE, 1990.

[2] Didier Chargy, Alain Dervieux, and Bernard Larrouturou. Upwind adaptive finite-element investigations of two-dimensional transsonic reactive flows. *Int. J. Num. Meth. Fluids*, to appear in 1990.

[3] Didier Chargy, Alain Dervieux, Bernard Larrouturou, Mark Loriot, and Nathan Maman. Simulation numérique de l'interaction réactive de jets gazeux transsoniques. Research report, INRIA, Sophia-Antipolis, 06560 VALBONNE, FRANCE, to appear in 1990.

[4] Bernard Larrouturou. How to preserve the mass fractions positivity when computing compressible multi-component flows. *J. Comp. Phys.*, to appear in 1990.

[5] Nathan Maman. *Algorithmes d'adaptation dynamique de maillages en éléments finis. Application à des écoulements réactifs instationnaires.* Thesis, Université Paul Sabatier, to appear in 1990.

ON WEAK AND STRONG COUPLING BETWEEN MESH ADAPTORS AND FLOW SOLVERS

B. PALMERIO (*), A. DERVIEUX (**)

(*) University of Nice and INRIA-Sophia-Antipolis
(**) INRIA-Sophia-Antipolis, B.P.109, 2004 route des Lucioles, 06560 VALBONNE, FRANCE

The purpose of this paper is to study how to adapt by deformation the mesh to a given solution in order to allow finding a good mesh with only a very few calls to the flow solver.

The calculation of more and more complex compressible mean or high Reynolds flows makes arise a strong need of mesh adaptors that can be combined to existing flow solvers. With an efficient combination of a mesh adaptor and a flow solver, it should be possible to capture some rather thin layers; one prerequisite is that the flow solver is repetitively used in order to produce a more and more accurate location of the layer.

Note that another way would be to solve in a fully coupled formulation the system of both flow and mesh by a global approach as in the MFE method [1], but this option does not enough take into account existing software (existing flow solvers, existing mesh adaptors for new mesh solvers).

1 ALGORITHM FOR COUPLING FLOW AND MESH

Let us consider the coupling between flow and mesh. In mesh deformation algorithms, the new location of nodes, denoted by X^{n+1}, is calculated from X^n and the flow solution W^n. Conversely, in the flow algorithm, the flow W^{n+1} is computed from X^n and from an initial condition $\bar{W}^n$.

We can define several ways of coupling X and W:

Option 1: Weak non-consistent coupling

We put $\bar{W}^n = W^n$.

This means that, after moving the nodes, the new couple (W^n, X^{n+1}) is not a consistent representation of the solution (while (W^n, X^n) was) and is therefore a bad initial data for computing W^{n+1}.

Option 2: Flow-consistent error-frozen coupling

In this second option, $\bar{W}^n$ is an interpolation of W^n from X^n to X^{n+1}; this can be obtained by (at least) two ways:

(a) In the ALE formulation [2], the mesh velocity is taken into account by an advective equation for W:

$$\bar{W}^n - W^n + (X^{n+1} - X^n).\nabla W = 0.$$

The authors considered the application of this principle in a previous work [3]; one main disadvantage of ALE is a loss of accuracy for large mesh variations.

(b) In the standard static interpolation, we put

$\bar{W}^n = value\ of\ W^n\ at\ X_i^{n+1}\ for\ i\ node.$

This can be efficiently coded by drawing the straight path between X_i^n and X_i^{n+1} from one triangle to one neighboring one of mesh X^n, like in characteristic schemes for the advection equation.

With both (a) and (b), flow-consistent coupling allows time-consistent calculations with a moving mesh and are more efficient for steady calculations than the weakly coupling option; however, since $\bar{W}^n$ is only recomputed after X^{n+1}, the new mesh X^{n+1} has been iteratively adapted to $(W^n, X^{(\alpha)})$ better than to $(\bar{W}^n, X^{n+1})$; in other words, we have frozen the error criterion relatively to mesh X^n and the adaption is not at this time consistent with the physical flow; consistency will be obtained at convergence i.e. when $(W^{n+1}, X^{n+1}) = (W^n, X^n)$.

This again may result in loss of efficiency.

Option 3: A fully consistent coupling, with frozen flow

In the second algorithm that we consider, the flow is re-interpolated and the error criterion is recomputed in each inner iteration applied for solving the mesh system. This option satisfies the following properties:
(i) W^n is always a consistent discretisation of the flow over mesh X^n,
(ii) X^{n+1} is always adapted to the interpolation of W^n on mesh X^{n+1} (that is itself).

2 ONE DIMENSIONAL STUDY

The problem of advection diffusion is considered :

$$u_x - \epsilon u_{xx} = 0 \quad u(0) = 0 \quad u(1) = 0 \tag{1}$$

For the advective part, a finite volume (vertex centered) MUSCL scheme is applied ; for the diffusion part, the Galerkin method is used.

We concentrate on the strategies in which the equation (??) is solved up to complete iterative convergence (or by direct solution) so that options 1 and 2 are equivalent ; we present in Fig. 1 the mesh variation versus the adaption iteration number; we observe that these options do not allow convergence of the global mesh-flow scheme.

Option 3 has been tried with the usual linear (continuous) interpolation
It is shown in Fig.2 that the usual linear interpolation brings fast convergence. In Fig.1 and Fig.2 we put on the X-axis the number of remeshing, on the Y-axis the logarithm of mesh residual, that is the l_2 norm of difference between the initial mesh location and the final one at each remeshing. The resulting solution is depicted in Fig.3.

A good adaptation was already obtained in 5-10 global iterations .

3 TWO DIMENSIONAL EXPERIMENTS

We have constructed a 2-D code that applies Option 3 with linear interpolation. We restrict to concertina-like meshes, i.e. we only adapt y-coordinates. The flow is computed using a compressible Navier-Stokes solver relying on a MUSCL-FEM scheme defined on triangles as described in [4,5] ; the iterative solution is fully converged at every flow-solver call. The mesh adaptor uses a spring system extended to triangles described in [3]. The error criterion is computed between two nodes as a directional second-order derivative of the local Mach number. The mesh-system solution is performed by Jacobi iteration; after each Jacobi sweep, the flow variables are re-interpolated on the new locations of nodes and the mesh-adaption criterion is re-computed from these new values.

This program is applied to the calculation of a flow past a flat plate; the farfield Mach number is 3. The mesh is a 43×25 structured one; the initial mesh is stretched x-wise around the plate edge and uniform w.r.t. coordinate y, which will be adapted.

The mesh is fastly adapted, after 3-4 global cycles. The resulting solutions and meshes are presented for Reynolds numbers of 1000, 100,000, and 10,000,000.

From the examination of the meshes and flows obtained (Fig.4 to Fig.9), it appears that the adaption is really powerful, although not yet enough to capture the higher Reynolds boundary layer; indeed, the main part of the mesh is gathered into the boundary layer, but, due to numerical viscosity involved in the flow approximation, the resulting numerical boundary layer is much thicker than the physical one. Since a similarity argument can be applied to these calculations, it is conjectured that one main limitation of this adaption process in the above experiments is due to the frozen x-distribution.

The main part of the CPU effort is taken by the flow solver, since the adaption step is between one and two orders of magnitude less costly (on a vector computer), although vectorization was only applied to the flow solver.

4 CONCLUSION

A global strategy for mesh adaption is applied to an example of boundary layer. The main idea is to recompute, after every node updating, the flow solution by interpolation and the mesh adaption criterion; the interpolation is a linear one. We emphasize on the fact that convergence between mesh and flow is generally fastly obtained by iteration of this adaption algorithm with flow solution. The mesh is then concentrated on the boundary layer that is captured by the numerical scheme, with an accuracy determined by the approximation, for the given mesh topology and number of nodes

Since the coupling between flow and mesh is obtained performing only a few flow solver calls; we call this a static point of view: it allows an interactive adaption phase while flow solution is done on a batch mode; thus this kind of tool can be applied in the same conditions as any mesh adaptor relying on local enrichment or reconstruction.

It is planned to extend the present approach to fully-multidimensional adaption and to apply it to separated flows.

5 REFERENCES

[1] MILLER K. and MILLER R.N., Moving Finite Elements I, SIAM J. Numer. Anal., 18, p. 1019, (1981).
[2] HIRT C. W., AMSDEN A.A., COOK J.L., An arbitrary Lagrangian-Eulerian computing method for all flow speeds, J. Comp. Phys. 14 (1974), 227-253.
[3] PALMERIO B., A consistent ALE-rezoned mesh adaptation for compressible flow finite-element calculations, INRIA Research Report 829 (1988).
[4] FEZOUI L., Résolution des équations d'Euler par un schéma de van Leer en éléments finis, INRIA Research Report 358 (1985).
[5] FEZOUI L., LANTERI S., LARROUTUROU B., OLIVIER C., Résolution Numériques des équations de Navier-Stokes pour un fluide compressible en maillage triangulaire, INRIA Research Report 1058 (1989).

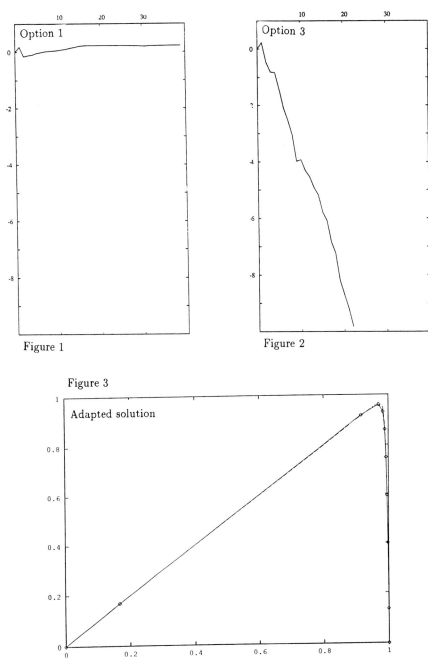

Figures 1 to 3: Steady 1-D diffusion-convection problem Mesh convergence to fully adapted solution

Figure 4

Re=1.E+3

Figure 5

Figure 6

Re=1.E+5

Figure 7

Figure 8

Re=1.E+7

Figure 9

Figures 4 to 9 2-D test-case: Supersonic flow over a flat plate

A FINITE-DIFFERENCE METHOD FOR INCOMPRESSIBLE FLOWS USING A MULTI-BLOCK TECHNIQUE

Hiroshi Takeda and ChingZou Hsu

Recruit Co., Ltd, 14F Inui Bldg.
1-13-1, Kachidoki, Chuo-ku, Tokyo 104, Japan

A finite-difference method using a multi-block technique is developed for 3-dimensional incompressible flows. In this method, continuity of location of grid points and metrix are not required in block interfaces, and therefore, computational grids can be generated in each block separately without any restriction. As a result, computational modelling of flows around/in complex bodies is simplified compared to previous finite-difference methods. Applying the present technique to some numerical examples shows that numerical solutions are connected smoothly near block interfaces.

Introduction

Boundary-fitted coordinates are frequently used when flows around/in bodies with complex geometries are solved using the finite-difference method. However, it is sometimes difficult to generate computational grids even if boundary-fitted coordinates are used so far as the coordinates consist of one block (or zone). On the contrary, mesh generation for complex geometries becomes much easier if one can divide the computational region into multi blocks and generate meshes separately in each block. There is also an advantage in using the multi-block technique that fine mesh can be generated locally only in the region where fine resolution is required.

Several multi-block approaches have been tried so far for compressible flows [1, 2]. On the other hand, Miki and Takagi [3] attempted to apply a multi-block technique for incompressible flows. However, there is a restriction in their multi-block method that grid points belonging to two adjacent blocks coincide on a block interface and also on a grid surface adjacent to the block interface; this can be a large restriction when generating meshes for complex geometries. In the present study, we develop a multi-block technique for incompressible flows in which computational meshes can be generated in each block separately without any restriction.

Outline of method

A newly developed implicit finite-difference method of $O(\Delta t^2)$ [4] based on the MAC method is used in the computation. Furthermore, we use physical components of contravariant velocities on a staggered mesh system defined by

$$U_i = (\partial \xi_i / \partial x_j) u_j / |\nabla \xi_i|,$$

where x_i represnets the Cartesian coordinates, u_i is the x_i component of the flow velocity, and U_i denotes the physical ξ_i components of contravariant velocities [5]. As a result of using physical components of contravariant velocities on a staggered mesh system, the continuity equation can be satisfied within the error produced by solving the Poisson equation for pressure.

In order to compute the velocities and the pressure on block interfaces and outside the block, we define virtual grid points on block interfaces and outside the block (Fig. 1), and interpolate the dependent variables at virtual points by using the variables at grid points in other blocks. We solve matrices for the velocities and the pressure using iterative methods such as the SOR method. We, therefore, perform the procedure of evaluating the variables at virtual points by interpolation at each step of iteration.

Computational examples

Figure 2 shows results for a 2-dimensional cavity flow with Reynolds number 100. The left figure represents the computational grids (number of the block is 2), the middle represents the velocity vector, and the right represents the pressure contour. It is found that both the velocity and the pressure are connected smoothly at the block interface.

Figure 3 represents a flow in the case where two plane Poiseuille flows join at 45 degrees (figures are computational grids, velocity vector, and pressure contour). This case also demonstrates that the velocity and the pressure are connected smoothly at the block interface regardless of the grids intersecting at 45 degree.

Figure 4 represents a fluid flowing in two pipes joining perpendicularly. The left figure shows the computational grids on the block surface, and the right figure shows the velocity vector in the center section. The computational regions are divided into 2 blocks; the cylindrical coordinates are used in a larger block (pipe) and coordinates obtained by modifying the cylindrical coordinates for fitting the interface are used in the other block (pipe). The results show that the computation is performed smoothly near the block interface even in 3-dimensional computations.

Conclusions

A finite-difference method using a multi-block technique was developed for 3-dimensional, time-dependent incompressible flows. In this method, continuity of location of grid points and metrix are not required in block interfaces, and therefore, computational grids can be generated freely in each block. As a result, computational modelling of flows around/in complex bodies is much simplified compared to previous finite-difference methods.

The present technique was next applied to three numerical examples. It was then confirmed that numerical solutions are connected smoothly near block interfaces in all the examples.

References

[1] J. F. Thompson: Grid generation. "Hand book of numerical heat transfer (ed.: W. J. Minkowycz et al.)", Wiley, 905-948, 1988.

[2] K. Fujii: A method to increase the accuracy of vortical flow simulations. AIAA-88-2562-CP, 1988.

[3] K. Miki and T. Takagi: Numerical solution of Poisson's equation with arbitrarily shaped boundaries using a domain decomposition and overlapping technique. J. Comput. Phys., **67**, 263-278, 1986.

[4] H. Takeda: A finite-difference method of $O(\Delta t^2)$ for incompressible Navier-Stokes equations. Proceedings of the 3rd International Symposium on Computational Fluid Dynamics, 1989 (in press).

[5] I. Demirdzic, A. D. Gosman, R. I. Issa and M. Peric: A calculation procedure for turbulent flow in complex geometries. Computers & Fluids, **15**, 251-273, 1987.

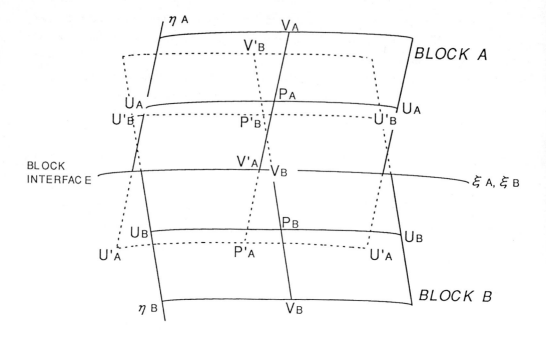

Fig. 1. Grid arrangement near a block interface. Subscripts A and B denote variables for block A and B, respectively. Prime indicates variables at virtual grid points outside the blocks and on the block interface.

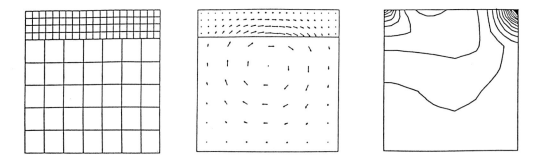

Fig. 2. A 2-dimensional cavity flow with Reynolds number 100. The left figure represents the computational grids (number of the block is 2), the middle figure represents the velocity vector, and the right figure represents the pressure contour.

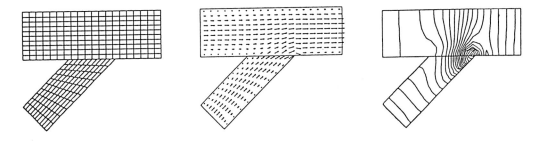

Fig. 3. A flow in the case where two plane Poiseuille flows join at 45 degrees. Figures are computational grids, velocity vector, and pressure contour.

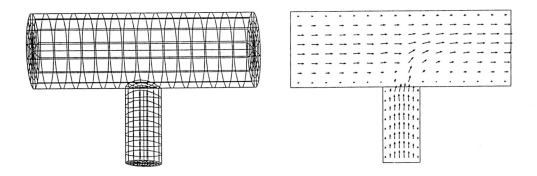

Fig. 4. A fluid flowing in two pipes joining perpendicularly. The left figure shows the computational grids on the block surface, and the right figure shows the velocity vector in the center section.

EFFECTIVE METHOD FOR CALCULATION OF GAS FLOW ABOUT BLUNT BODIES IN A NEW APPROACH TO CONSTRUCTING THREE-DIMENSIONAL GRIDS

S. L. Kabalkin
Keldysh Institute of Applied Mathematics, Moscow, USSR.

Let us consider an unsteady supersonic gas flow about blunt bodies. We introduce the Cartesian coordinates x, y, z. The point O at the nose of the body is an origin of the coordinates. The axis Ox is directed downstream. The axes Oy and Oz are orthogonal to Ox. The boundaries of the calculation domain Ω are the surfaces Π_B, Π_I and Π_E. Π_B is the body surface, Π_I separates the free stream and the calculation domain, Π_E closes the calculation domain (Fig. 1).

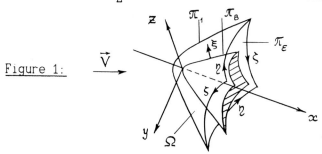

Figure 1:

Let us introduce the curvilinear coordinates η, ζ, ξ. The surface $\xi = const$ is located between the surfaces Π_B and Π_I. The variable ξ changes from $\xi = 0$ (the body surface) to $\xi = 1$ (the surface Π_I). We set one to one correspondence between the coordinates x, y, z and η, ζ, ξ on the body surface as:

$$x_B = x_o(\eta,\zeta) \qquad y_B = y_o(\eta,\zeta) \qquad z_B = z_o(\eta,\zeta) \qquad (1)$$

We set the angles $\alpha_x, \alpha_y, \alpha_z$ between the ray ($\eta = const$, $\zeta = const$, $\xi \geq 0$) and the axes Ox, Oy and Oz respectively, as the functions of variables η and ζ. We set function $\Delta(\eta,\zeta)$ that is equal to the distance along this ray from the surface Π_B to the surface Π_I. The full transformation may be written as:

$$\begin{array}{l} x = x_o(\eta,\zeta) + \Delta(\eta,\zeta)\cdot\Psi(\xi)\cdot\cos(\alpha_x(\eta,\zeta)) \\ y = y_o(\eta,\zeta) + \Delta(\eta,\zeta)\cdot\Psi(\xi)\cdot\cos(\alpha_y(\eta,\zeta)) \\ z = z_o(\eta,\zeta) + \Delta(\eta,\zeta)\cdot\Psi(\xi)\cdot\cos(\alpha_z(\eta,\zeta)) \\ \Psi(0) = 0 \qquad \Psi(1) = 1. \end{array} \qquad (2)$$

Let the intersection of the surfaces Π_B and Π_I be the ellipse:

$$\frac{y^2}{a^2} + \frac{z^2}{b^2} = 1 \qquad x = X(y,z) \qquad (3)$$

Then there is the convenient transformation:

$$y_B = a \cdot \sin(\eta \cdot \tfrac{\pi}{2}) \quad z_B = b \cdot \sin(\zeta \cdot \tfrac{\pi}{2}) \quad x_B = x_o(y_B, z_B) \quad -1 < \eta, \zeta < 1 \qquad (4)$$

In the new coordinates the ellipse is transformed into the square:

$$|\eta| + |\zeta| = 1 \qquad \xi = 0 \qquad (5)$$

The grid of $N \times N$ points (N is an even number) is built in the following way (we must satisfy the condition demand (4)):

$$h_1 = 2/N \qquad \begin{array}{l} \eta_i = h_1/2 - 1 + (i-1) \cdot h_1 \quad i = 1,2,\ldots,N \\ \zeta_j = h_1/2 - 1 + (j-1) \cdot h_1 \quad j = 1,2,\ldots,N \end{array} \qquad (6)$$

The disposition of the grid on the body surface by $N = 12$ is shown in Fig. 2. The $2 \cdot N$ marked points in Fig. 2 are on the ellipse (3). If it is a circle, the points divide the circle on $2 \cdot N$ equal arcs.

The important peculiarity of the transformation (4) is that the calculation domain for a gas flow about blunt nose is consistent to the one past the side surface of the body. The disposition of two jointed grids on the body surface is shown in Fig. 3. The new grid is used near the blunt nose. The second grid is built in cylindrical coordinates.

A series of calculations is fulfilled by the above method. The comparison with the analogous calculations shows that the method is effective and economical. The method simplifies the numerical algorithm and helps the best adaptation of the grid. The simple transformation from the Cartesian coordinates to generalized ones is realized analytically. The transformation (4) may be used independently for the 3D calculations of a gas flow in the elliptical channels.

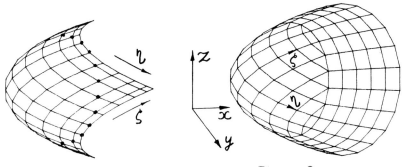

Figure 2.   Figure 3.

An Adaptive, Embedded Mesh Refinement Algorithm for the Euler Equations

James J Quirk[*]
College of Aeronautics, Cranfield Institute of Technology
Cranfield, BEDFORD MK43 0AL, ENGLAND

July 1990

During the development of computational methods that solve time dependent shock hydrodynamic problems two underlying strategies have emerged that enable flow features to be resolved adequately. One, employ a numerical scheme of inherently high resolution, usually a second order Godunov type method. Two, locally refine the computational mesh in regions of interest. It has been demonstrated by Berger & Collela [1] that a combination of both strategies is necessary if a solution of very high resolution is sought. The present study combines Roe's high resolution flux difference splitting scheme [2] with an Adaptive Mesh Refinement algorithm, AMR, which is developed from the ideas of Berger. Space does not allow for a full description of our scheme, full details are available elsewhere [3], here we simply attempt to impart some of the main features of the scheme and demonstrate that very detailed results may be obtained using relatively small computer resources.

The AMR algorithm discretizes the computational domain into a hierarchical set of nested, structured meshes. The scheme refines in time as well as space; more, smaller time steps are taken on fine meshes than on coarse meshes. The process of adapting the mesh structure as the flow evolves is fully automatic. In effect fine grids glide along coarse grids. Figure 1 shows the density contours and corresponding computational grid for the computation of a plane shock escaping from an open ended shock tube. Three grid levels were used for this calculation, a base level grid of 75 by 40 cells, a medium grid with 16 cells to each coarse cell, and a fine grid with 256 cells to each coarse cell. Note that only a very small part of the computational domain is covered by the finest mesh. The density contours show that the main features of the flow field are well resolved, in particular note that the contact discontinuity emanating from the triple point near the Mach disk displays a Kelvin-Helmholtz instability. The calculation took less than 12 hours on a Sun SPARCstation 1. The quality and accuracy of these results may be gauged by comparing them to those of Wang & Widhopf [4] which were computed using a fixed zonal grid.

The AMR algorithm is especially suited to highly transient flows where it is necessary to adapt the grid many times as the flow evolves. For our shock tube problem it was necessary to adapt the fine grid some 520 times yet all the work associated with the adaption accounted for only 0.6% of the total processing time.

---

[*]This work was funded by the Procurement Executive, Ministry of Defence.

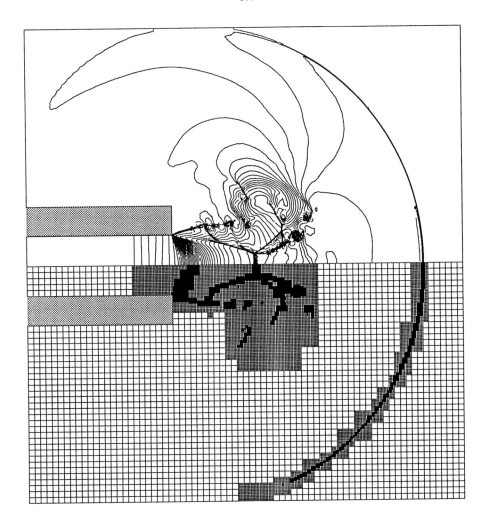

Figure 1: Spherical shock expansion from an open ended shock tube.

References

[1] M.J.BERGER AND P.COLLELA, *Local adaptive mesh refinement for shock hydrodynamics*, J. Comput. Phys., **82**(1989), pp. 67–84.

[2] P.L.ROE, *Approximate Riemann solvers, parameter vectors and difference schemes*, J. Comput. Phys., **43**(1981), pp. 357–372.

[3] J.J.QUIRK, *PhD Thesis*, To appear late 1990.

[4] J.C.T.WANG AND G.F.WIDHOPF, *Numerical simulation of blast flowfields using a high resolution TVD finite volume scheme*, Computers & Fluids, Vol. **18**(1990), No. 1, pp. 103–137.

Numerical Simulation of Flows over Complete Aircraft using Block Structured Grid Systems

Keisuke Sawada

Gifu Technical Institute, Aircraft Research Laboratory
Kawasaki Heavy Industries Ltd., Kakamigahara 504, JAPAN

In the aircraft industry, Computational Fluid Dynamics(CFD) is already recognized as an indespensable tool in aerospace research and development. However, there exist some problem areas which should be resolved before CFD can serve as a truely useful tool. The most immediate one may be the problem of grid generation around complex aircraft geometry. A less flexible method can not handle a variety of configurations which arise in various engineering applications.

The geometrical flexibility of the CFD method is strongly affected by the computational approach of the scheme: whether it adopts the conventional finite difference approximation with structured grid or the finite element approximation with unstructured grid[1]. The latter approach seems to be quite promising and many researchers are now making extensive efforts in this problem area. However, it is still true that it requires more memories and more computational resources than the former approach. It is then quite attractive if a finite difference approach can attain the comparable geometrical flexibility with FEM while retains its computational efficiency.

Of many approaches tried so far, a straightforward way to attain the required flexibilty within the framework of structured grid is the multi- block transformation technique[2]. Besides the geometrical flexibility, this approach is preferable since the resulting grid belongs to a class of single and structured grid system. In the present study, we adopt this multi-block transformation combined with an interactive grid generation scheme to treat various aircraft configurations ranging from a reentry vehicle to a four-engined short takeoff aircraft in an unified manner.

The numerical method for solving the Reynolds averaged Navier-Stokes equations is the upwind TVD finite volume scheme adopting the planar Gauss-Seidel relaxation technique[3] for a rapid convergence. We adopt the $q - \omega$ two-equation turbulence model[4] rather than the familier zero equation model because of the numerical compatibility with the multi-block grid system. In the multi-block grid system, a flow element can not recognize whether it is inside the boundary layer of one block or in the wake region of another one. Zero equation models change their form in the boundary layer and in the wake so that it is very difficult to apply them to the multi-block grid system having a general block arrangement.

A robustness of the code is particularly important in engineering applications since designers not necessarily familier with CFD may use it. A class of TVD scheme can provide this robustness without particular problem oriented tunning of the scheme. In the present study, we also apply the TVD formulation to the convective terms of the two equation turbulence model. A 7-components closely coupled formulation is adopted for the maximum numerical stability[5].

Together with multi-block grid systems, we show the results of numerical simulation of fluid flow over various complete aircraft configurations(fig. 1). The point we would like to emphasize is the numerical approach used in the present study, although not so sensational as FEM, can really provide a practical mean to construct a numerical simulation system for fluid flows over aerospace objectives.

This work is mainly supported by the joint research program on CFD conducted by Kawasaki Heavy Industries Ltd.(KHI) and National Aerospace Laboratory(NAL) and is partly supported by the basic technology research and development program of KHI.

1. Jameson,A. and Baker,T.J., "Improvements to the Aircraft Euler Method", AIAA paper 87-0452, 1987.
2. Rubbert,P.E. and Lee,K.D., "Patched Coordinate Systems", Numerical Grid Generation, North-Holland, 1982.
3. Chakravarthy,S.R. and Szema,K.Y., "An Euler Solver for Three-Dimensional Supersonic Flows with Subsonic Pockets", AIAA paper 85- 1703.
4. Coakley,T.J., "Turbulence Modeling Methods for the Compressible Navier-Stokes Equations", AIAA paper 83-1693.
5. Sawada,K. and Takanashi,S., "Numerical Simulation of Viscous Transonic Flows using Block Structured Grid System", Proc. of 19th Hydrodynamic Conference, held in Sendai, Japan, Oct., 1987.

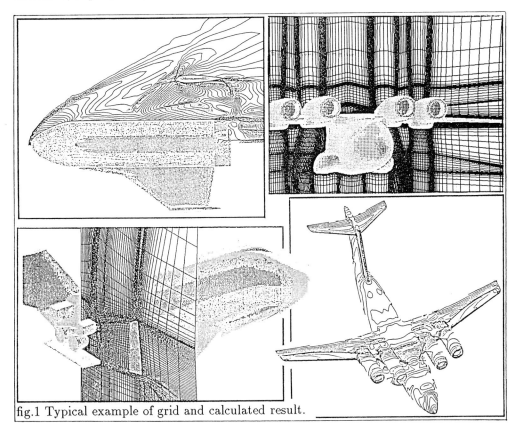

fig.1 Typical example of grid and calculated result.

FURTHER ALGORITHMIC IMPROVEMENTS OF ADAPTIVE H-REFINEMENT FOR 3-D TRANSIENT PROBLEMS

Rainald Löhner
CMEE, SEAS, The George Washington University, Washington, D.C. 20052
and
Joseph D. Baum
SAIC, McLean, Virginia 22102

Over the past year we have developed an adaptive finite element scheme for transient problems in 3-D [1]. The classic h-enrichment/coarsening is employed in conjunction with a tetrahedral finite element discretization. This initial capability has been improved further. For typical shock calculations, mesh adaption takes place every 5-10 timesteps. This implies that every stage of the adaptation process must be thouroughly optimized. The two areas that required the most intense optimization efforts were:
a) Error indicators for grids with large local variations of element size and shape, and
b) Faster construction of the new mesh.

Improved error indicators:
As an error indicator we use a modified interpolation theory error indicator:

$$E^I = \sqrt{\frac{\sum_{k,l}(\int_\Omega N^I_{,k} N^J_{,l} d\Omega \cdot U_J)^2}{\sum_{k,l}(\int_\Omega |N^I_{,k}|\left[|N^J_{,l}U_J| + \epsilon\left(|N^J_{,l}||U_J|\right)\right]d\Omega)^2}}, \quad (1)$$

where N^I denotes the shape-function of node I. By dividing the second derivatives by the absolute value of the first derivatives the error indicator becomes bounded, dimensionless, and the 'eating up' effect of strong shocks is avoided. This error indicator has performed very well in 2-D over the years [2]. However, when first used in 3-D, it proved unreliable. The source for this seemingly inconsistent behaviour was found to stem from the large *local* variations in element size, shape, as well as number of element surrounding a point encountered in typical 3-D unstructured grids. These will produce large variations of the second term in the denominator which are not based on physics, but on the mesh structure itself. The solution was to modify this error indicator as follows:

$$E^I = \sqrt{\frac{\sum_{k,l}(\int_\Omega N^I_{,k} N^J_{,l} d\Omega \cdot U_J)^2}{\sum_{k,l}(\int_\Omega |N^I_{,k}||N^J_{,l}U_J|d\Omega)^2 + \epsilon MMAT_I h_I^{-2}|U_I|}}, \quad (2)$$

where $MMAT_I$ is the lumped mass-matrix at point I, and h_I the average element length at point I. This error indicator proved to be remarkably insensitive to local variations in element size and shape, while still yielding the correct indicator values for physical phenomena of interest. We attribute this good performance to the smoothing effects of two averaging operations working simultaneously: the lumped mass-matrix and the point-lenghts.

Faster construction of the new mesh:
The main CPU-intensive operations performed during one mesh change are: finding the sides and the sides of each element, determining the refinement and coarsening patterns, correcting boundary points, and renumbering the elements. Although seemingly trivial, these operations account for a significant percentage of total CPU time. We developed new, optimal algorithms for them. Two examples are described here in more depth:
a) Determining the List of Sides for New Points
Given the side/element information, and a list of elements to be refined, a first set of sides on which new gridpoints need to be introduced is determined. In most cases, it will not lead to an admissible refinement pattern. Therefore, further sides are marked for the introduction of new points until an

admissible refinement pattern is reached. This is done by looping several times over the elements, checking on an element level whether the set of sides marked can lead to an admissible new set of sub-elements. Practical calculations revealed that sometimes up to 15 passes over the mesh where required to obtain an admissible set of sides. An 80%-90% reduction in CPU was achieved by presorting the elements as follows:
- Add up all the sides marked for refinement in an element;
- If 0,1 or 6 sides were marked: do not consider further;
- If 4 or 5 sides were marked: mark all sides of this element to be refined;
- If 2 or 3 sides were marked: analyze in depth.

b) Element Renumbering

In order to vectorize the element assembly as much as possible, the elements are renumbered, such that within each assembly pass a point is accessed only once by the elements. This renumbering has to take place after every mesh change. Before optimization, it took over 10% of the total CPU time for typical runs. The renumbering subroutine was optimized by b1) working only on the remaining elements, and b2) extensive scalar optimization, minimizing the number of operations and memory access. With the new renumbering algorithm, the CPU-time required for this operation dropped to less than 1%.

Numerical Example:

To illustrate the performance of the new techniques to a realistic problem, we consider the interaction of a weak shock with a tank. The shock impinges head-on, and is subsequently reflected from several surfaces. The surface grid and pressure distribution at T=8.0 are shown in Figure 1. The average grid size for this problem was of the order of 1.25 Mtetrahedra. Speed-ups observed for the adaptive refinement portion of the code employing the new techniques were in excess of 1:7.

References

[1] R. Löhner - Adaptive H-Refinement on 3-D Unstructured Grids for Transient Problems; AIAA-89-0365 (1989).
[2] J.D. Baum and R. Löhner - Numerical Simulation of Shock-Elevated Box Interaction Using an Adaptive Finite Element Shock Capturing Scheme; AIAA-89-0653 (1989).

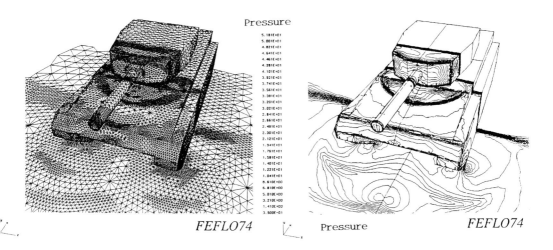

Figure 1: Shock-Tank Interaction

AUTHOR INDEX

Ahmad J : p 389
Aki T : p 90
Andersson H I : p 172
Arina R : p 162
Aso S : p 459
Bar-Deroma Z : p 497
Bardina J : p 361
Basdevant C : p 369
Bassi F : p 387
Baum H R : p 514
Baum J D : p 556
Belotserkovskii O M : p 177
Bergman C M : p 457
Billdal J T : p 172
Biringen S : p 190
Blinova L A : p 359
Blottner F G : p 421
Bockelie M J : p 524
Borelli S : p 416
Boris J P : p 197, 297, 318
Brandsma F J : p 152
Bristeau M O : p 373
Brown E F : p 297
Browning G L : p 509
Bruneau C-H : p 447
Cambier L : p 51
Cannizzaro F E : p 385
Canuto C : p 162
Caporaletti J : p 361
Carpenter M H : p 254
Casper J : p 325
Causon D M : p 329
Chang S-H : p 327
Chen H L : p 207
Chiba S : p 529
Cho N-H : p 167
Clarke N : p 329
Codding W H : p 361
Côté J : p 63
Coquel F : p 411

Crumpton P I : p 243
Cullen M J P : p 499
D'Asaro E A : p 195
Dadone A : p 379
Daiguji H : p 485
Danabasoglu G : p 190
Daudpota Q I : p 461
De Palma P : p 352
Débit N : p 375
Deconinck H : p 273, 352, 426
Dennis S C R : p 142
Dervieux A : p 540
Di Giacinto M : p 381
Dick E : p 333
Dimitriadis K P : p 111
Doorly D J : p 357
Dougherty F C : p 449
Dubois T : p 116
Duck P W : p 199
Dukowicz J K : p 291
Dulieu A : p 211
Dwyer H A : p 490
Egami K : p 342
Egorov Yu E : p 359
Ehrenmark U T : p 371
Eiseman P R : p 524
Eliasson P : p 172
Elmiligui A : p 385
Engquist B : p 335
Erlebacher G : p 121, 192
Falle S A E G : p 367
Favini B : p 381
Favorsky A P : p 219
Fernandez G : p 490
Ferry M : p 186
Fiebig M : p 147
Fletcher C A J : p 167
Forestier A J : p 339
Fořt J : p 480
Foutch D W : p 363

Fruehauf H-H : p 495
Fujii K : p 406
Gatski T B : p 157
Ghia K N : p 215
Ghia U : p 215
Ghoniem A F : p 514
Glowinski R : p 373
Goorjian P M : p 377
Grasso F : p 387
Griffiths D F : p 259
Grinstein F F : p 197, 297
Grossman B : p 379
Grubin S E : p 181
Guruswamy G P : p 377
Hafez M : p 389
Hall M G : p 287
Hall R C : p 357
Hänel D : p 268
Harten A : p 387
Hartwich P M : p 394
Hayashi M : p 459
Hirsch C : p 238
Hitchings D : p 365
Holland W R: p 509
Hollanders H : p 411
Holschneider M : p 369
Hong S K : p 361
Hsu C-A : p 248
Hsu C-Z : p 545
Hudson J D : p 142
Huněk M : p 480
Hussaini M Y : p 121
Ichinose K : p 183
Ingram D M : p 329
Jauberteau F : p 116
Jiang T L : p 464
Johnston L J : p 344
Jones D A : p 318
Jones K D : p 449
Kabalkin S L : p 550

Kajishima T : p 202
Karageorghis A : p 179
Kato Y : p 209
Kerschen E J : p 157
Khosla P K : p 400
Kiris C : p 132
Kolbe R : p 297
Kozel K : p 480
Krause E : p 35, 469
Kreiss H-O: p 509
Kroll N : p 188, 442
Ku H-C : p 223
Kuerten J G M : p 152
Kuruvila G : p 137
Kuwahara K : p 469, 529
Kuznetsov A E : p 359
Kwak D : p 132
Lacor C : p 238
Laminie J : p 278, 447
Lardy P : p 426
Larrouturou B : p 534
Launder B E : p 1
Lavery J : p 348
Le Quéré P : p 211
Lecheler S : p 495
Lê T H : p 213
Lee T H : p 248
Leighton R I : p 205
Leland R W : p 308
Leschziner M A : p 111
Li B : p 371
Li C : p 297
Liandrat J : p 369
Liang T E : p 400
Löhner R : p 556
Lombard C K : p 361
Macaraeg M G : p 461
Mackenzie J A : p 243
Maday Y : p 375
Maman N : p 534

Mane L : p 303
Marmignon C : p 411
Mavriplis D J : p 228, 233
Mawson M H : p 499
McCarthy D R : p 363
Meinke M : p 268
Melson N D : p 385
Meltz B J : p 291
Merkle C L : p 475
Migórski S H: p 504
Mitra N K: p 147
Miyake Y : p 202
Morinishi K : p 437
Morton K W : p 243
Müller B : p 451
Naitoh K : p 469
Nakahashi K : p 342
Nakamura Y : p 127
Ng L : p 192
Nguyen K D : p 516
Nishimoto T : p 202
O'Brien P : p 287
Obayashi S : p 377
Oliger J : p 361
Oran E S : p 297, 318
Oshima K : p 207, 283
Osswald G A : p 215
Palmerio B : p 540
Pandolfi M : p 416
Pascal F : p 278
Périaux J : p 373
Perrier V : p 369
Pfitzner M : p 432
Phillips T N : p 179
Phuoc Loc T : p 211, 213, 303
Piquet J : p 186
Poll D I A : p 199
Pruett D : p 192
Pulliam T H : p 106
Qin N : p 453

Quirk J J : p 552
Radespiel R : p 188
Ravachol M : p 373
Rehm R G : p 514
Richards B E : p 453
Riley J J : p 195
Rizzi A : p 172
Roe P L : p 273
Rogers S E : p 132
Rollett J S : p 308
Rosenberg A P : p 223
Rossow C-C : p 350, 442
Rotinijan M A : p 359
Rozhdestvensky B L : p 19
Rubin S G : p 400
Rudgyard M A : p 243
Sakaguchi Y : p 437
Salas M D : p 137
Saltz J H : p 233
Samarsky A A : p 219
Sanchez M : p 147
Sato T : p 209
Satofuka N : p 183, 437
Saunders R : p 329
Sawada K : p 554
Sawada T : p 209
Sawley M L : p 455
Schaefer R F : p 504
Schulkes R M S M : p 355
Shaw G J : p 243
Shi J : p 365
Shur M L : p 359
Simakin I N : p 181
Sjögreen B : p 335
Smith M G : p 371
Smith R E : p 524
Sobieczky H : p 449
Speziale C G : p 121
Srinivas K : p 167
Staniforth A N : p 63

Stolcis L : p 344
Stoynov M I : p 19
Streett C L : p 190
Strelets M Kh : p 359
Struijs R J : p 273, 352
Suhr S : p 361
Swanson R C : p 313
Swean Jr T F : p 205
Sweby P K : p 259
Takeda H : p 545
Tamura Y : p 406
Tanahashi T : p 209
Tang K Z : p 490
Taylor T D : p 80, 223
Tchamitchian Ph : p 369
Temam R : p 116, 278
Theofilis V : p 199
Tilch R : p 519
Tishkin V F : p 219
Tokunaga H : p 183
Troff B : p 213
Tromeur-Dervout D : p 303
Turchak L I : p 463
Van Ransbeeck P : p 238
Vasilevsky V F : p 219
Vastano J A : p 106
Vavřincová M : p 480
Venkatakrishnan V : p 233
Venkateswaran S : p 475
Veuillot J-P : p 51
Von Lavante E : p 385
Vos J B : p 457
Vyaznikov K V : p 219
Wang D : p 361
Wang J P : p 127
Weinerfelt P Å : p 337
Whaley R O : p 297
Wigton L B : p 313
Williams P S : p 371
Winters K B : p 195

Wolfshtein M : p 497
Worley S J: p 509
Wu H : p 283
Wüthrich S : p 455
Xiang Y : p 373
Yamamoto S : p 485
Yang J Y : p 248
Yasuhara M : p 127
Yee H C : p 259
Yoon S : p 132, 323
Zalesak S T : p 346
Zang T A : p 121